The Synchronic Fallacy

"Denne afhandling er den 23. maj 2000 af Syddansk Universitets humanistiske fakultetsråd antaget til forsvar for den filosofiske doktorgrad".

Flemming G. Andersen
dekan

Erik W. Hansen

The Synchronic Fallacy

HISTORICAL INVESTIGATIONS
WITH A THEORY OF HISTORY

ODENSE UNIVERSITY PRESS
2001

The Synchronic Fallacy
© The author and Odense University Press 2001
Printed in Denmark by Narayana Press
ISBN 87-7838-582-2
ISSN 1395-7236
Cover design by UniSats ApS.
Published with support from
Forskningsstyrelsen, Danish Research Agency

Odense University Studies in Linguistics vol. 12

Odense University Press
Campusvej 55
DK-5230 Odense M
Phone: +45 6615 7999
Fax: +45 6615 8126
E-mail: Press@forlag.sdu.dk
Internet: www.oup.dk

Distribution in the United States and Canada:
International Specialized Book Services
5804 NE Hassalo Street
Portland, OR 97213-3644
USA
Phone: +1-800-944-6190

CONTENTS

FOREWORD *11*

PROLOGUE
An importing discipline is to become immanent and perhaps an exporter. A reckoning with the 20th century linguistic tradition's synchronization of die geschichtliche Wirklichkeit *of the everyday language into* historische Zufälligkeiten *13-84*

0.0. A bird's eye view of the conception of history and change being contingent on synchrony and staticness: a critical look against a horizon of constructive synthesizing: preliminary exercises towards a theory of history *13*

0.1.1 The Moment-When problem *16*

0.1.2 Another kind of universal: synchronic frustrations over variability *18*

0.2.1 The beginning: the Jones-Hypothesis and the synchronic alternative *22*

0.2.2 Hermann Paul and the transcendence problem *26*

0.3.0 Three modern theories of sound change: are there *internal laws?* *32*

0.3.1 Abductive and deductive change; towards the transindividual and sign-functional *32*

0.3.2 Tenseness constitutes an internal law. The agentless transindividual *38*

0.3.3 The agentless transindividual *44*

0.4.1 *Vivo, ergo sum*: the existential function of our everyday languages and the historical nature of synchronic parameters *47*

0.4.2 Rehabilitation of the (spoken/written) substance. It is more than raw-material for change, it being a structural phenomenon *57*

0.4.3 The datum of historical analysis is the developmental series: an irreducible transsynchronic and sign-functional datum. Towards a definition of the developmental series: *loquor, ergo sum* and a practical critique of the slant-bracket notation *59*

0.4.4 The period and developmental series *69*

0.5. What really bothers the synchronic linguist: *contrast – distinctiveness – difference* *72*

CHAPTER I
The conceptual background of the issues involved in the construction of the theory of history. The first group of definitions *85-182*

1.1. Towards the sign-functional nature of historical phenomena. A bird's eye view of the history of the concepts of change and stability. On the role of the everyday language in our Western Civilization's dualist way of reasoning. On the Nominalist-Realist dichotomy in relation to history *85*

1.2. Transcendence and unidirectional dependency. The latter turns into an aition. The scope of the aition. No logical argument can be constructed to prove the priority of the concept of stability. The cause of change is part of data, not to be equated with the theory of data *100*

1.3. *The system at work*: synchronic processes. Ontological dualisms are not a thing of the past. The generativist proves Hermann Paul right. Preliminary remarks on idealisation in synchronic linguistics. The existential and cognitive function of *parole*. Definition of *Linguisticism*. The ideal historical narrative *115*

1.4. On genetic explanation. 'What must a language theory may able to do?' Empirical relevance is denotation. The Critico-Philological Method. An analysis of 'All A>B'. An aitional interpretation of 'to come from' and 'to develop into'. The Phenomenal Error. The meaning of **ex**istence *126*

1.5. Cause and the existence of the everyday language. Cause and datum merge: cause is not a theory. The basic definitions of the theory. Simultaneity, historical and the spoken chain's. The affinity between the developmental series and *parole*. The Neogrammarian Paradigm is reinterpreted in the light of the advanced definitions *139*

1.6. Vertical syntheses. Four ideologies. Two dualist theories compared. 'To be history' and 'to have history'. *Langue* and *parole* are equiempirical in the L-continuum. The everyday language is always *en ergeiai*. More criticism of the phonemic principle: assimilation, loss and addition. Katalysis *166*

CHAPTER 2
A constructive view of Geneticism and a critical view of the synchronic dualism's two-state model *183-226*

2.1. More definitions. The structure of the historical phenomenon: synchrony and history unified in the spoken substance. Aristotle's four causes. The dualism makes 'variation' static and synonymous with 'difference'. The two-state model is basically no theory, but expounds a certain view of (the mechanism of) change *183*

2.2. An interlude: further criticism of the slant-bracket convention. Contingency. The synchronization of the speech situation. A summary of the essay's argument. What makes a system static? An alternative *190*

2.3. Serial relationships vs. generic history. The two-state-model presupposes directionality and prediction is impossible. The meaning of history: historical phenomena are sign-functional. Change involves a three-state model. More remarks on the concept of cause. The future is a linguistically motivated concept. The everyday language is always *en ergeiai*: *ein unaufhörliches Schaffen* *195*

2.4. The two-state-model and the generic mechanical cause-effect parameter. Its logic implies the concepts of multiple conditioning and a phenomenon's *Eigennatur*. Historical causes are biuniformal, constitutive, effective and non-effective. Answers to Questions 1 and 2 above. The concept of cause is redundant: Aitken's Law *203*

Contents

2.5. On the validation question. The immanent state is self-validating, as a historical state it is validated transcendently. The state concept is analyzed in analogy to the phenomenon; it implies the concept of breaks. The state seen in the light of contingency and potentiality; contingency leads to absurdity: the necessity of the contingent fact 209

2.6 On the logical geography of the A>B relationship of the two-state model. Historical necessity and criticism of contingency. Synchrony defeats the two-state-model. Cause is an acquired property. The category, which is not stable, makes for synchronic stability. The Commutation Test and a definition of change, an epistemic concept 217

CHAPTER 3
Structure and History are compatible – neither antinomical nor two different views of the same phenomenon 227-282

3.1. What is structuralism? Sartre and the structuralist-existentialist controversy. Sartre's critique of the ahistorical nature of structuralism. Contingency is necessary and history is not paradoxy, but 'deconstruction' and construction of structures *qua* totalities 227

3.2. Linguistic definitions of *structure*. The idea of the static and supraindividual comes from the metalanguage. The idea of change comes from the everyday language, not the system(s concept); a dynamic conception of language. Structuralism, man's sense of reality and history are correlative concepts. Analysis of the synchronic systems concept will eventually end in history. A quick glimpse at the language-learning infant. The Glossematic argument backfires. The structural programme is a metaphysical statement 241

3.3. The sociological analogy: the 'disappearance' of *le sujet* and *la trace de l'homme* from the social fact. What is *un fait social*? An analysis of Durkheim's views, which end in history. Linguistic parallels. Extensional classification is not explanatorily adequate. The social fact is interpreted signfunctionally 253

3.4 A close-reading of Saussure's structural postulates about the immutability and mutability of the sign. William Whitney's concepts of arbitrariness and convention. Saussure's arguments do not transcend the level of postulates, but Saussure affirms the historical nature of the everyday language and the incompleteness of the synchronic point of view 267

CHAPTER 4
On the structural totality concept and the logician's view of the everyday language. Polysemy and synonymy are language-constituting properties 283-333

4.1. 7 tenets determining structuralism. Towards the anthropocentric basis of language. Historical and structural man. Analysis of the synchronic totality concept. Immanence and autonomy are acquired properties. The logic of the part-whole parameter does not justify synchronic linguistics' criticism of historical linguistics for being atomistic. The individual is a scientifically relevant concept: a class as many or a class as one 283

4.2. The totality concept is ambiguous and polysemous. A critique of the logician's view of language, language being necessarily polysemous (ambiguous) and synonymous (precise). These two language universals are not equilogical. The *Erkenntniswert* of the everyday language. The everyday language transcends into reality in two ways *294*

4.3. A comparison between the historian, the nominalist and the 'poet'. The historian is poised between the poet-maker and author-augmentor. An outline of a historical theory in the light of polysemy and synonymy; the historian is the hearer, historical phenomena the speaker (the sign): historiographical transcendence is sign-functional. The historical phenomenon's form is not captured by the formed-unformed parameter *310*

4.4. The universality of synonymy and polysemy and their *raison d'être*. Structure and polysemy are language-unifying phenomena. The language-learning child is creative and transcends into reality with her language. Analysis and synthesis are both empirically relevant. The everyday language has *Erkenntniswert*, is not an expendable cognitive means *323*

CHAPTER 5
*Towards the theory and its static application. The **break** concepts are introduced. The interidiolectal network and the theory. The developmental series* *334-391*

INTRODUCTION *334*

5.1. Last words on the synchronic systems concept: it is an *explanandum*. The concept is tentatively replaced by collective orientation. The constitution of the synchronic system; origin is common origin, structure is history. Synonymy and polysemy are further generalized *335*

5.2. All language change is equiempirical. A theory of change must not commit the Phenomenal Error. The line-point imagery of synchrony is conditioned. The concepts of the series and the element are not parallel; both are theoretically relevant. An outline of data. Definitions of the developmental series of isolative sound 'change': *inertial* and *ertial* developments; *breaks*. Ertial transitions reveal the inadequacy of the two-state model *348*

5.3. The definition system continued; the historian's focus; chronological series and nonchronological relationships: resumption of the trancendence problem. The notion of self-difference vis-à-vis self-identity. The concept of a cut is provisionally defined by the concepts *a break around* and *no break around*, i.e. around the past of the present's here-and-now intersection. State-transitions are polysemous *364*

5.4. The material definitions of the other developmental series. Further justification of the period *qua* an analysandum in its own right. Different types of 'breaks around'. Serial relationships are not symmetrical. Gradualness and multiple conditioning *qua* factors that 'bridge' gaps solve a problem *372*

Contents

APPENDIX: *A summary of the developmental series* *380*

5.5. The empirical relevance of the notion of *small* sign-functional *differences*. Grammaticalization theory and in general synchronic linguistics operate with the notion. The sign-functional nature of the historical theory is adumbrated: the basic meaning of change is change of meaning. Parmenides was both right and wrong. The *attitudinal system* is the prime mover in the existence of the everyday language *381*

5.6. The final definitions and the concluding analysis of the transitions of the two-state model, seen against the horizon of Archimedean omniscience and (only) the material aspect of the transitions. The concepts *a break before* and *a break after* are not the respective contradictories of *no break before* and *no break after*; the break-concept is mediated by *a potential break*. History is not episodic *387*

CHAPTER 6

The immanent theory of history is implemented as a sign-functional theory of sound change: language development is the processes that constitute developmental series through time. It is semantic change through katageneity and endogeneity, and formal variation (replacement) through endogeneity. Concluding arguments *392-440*

6.1. Isolative shifts, both A>B and continuation A>A, are explained by the sign-functional theory of history. The theory explains short- and long-term development that consists of various types of ertial and inertial processes. The attitudinal system *en ergeiai* defines the coherence relations of the developmental series *392*

6.2. The process of change. Predictability: Empirical impossibility and empirical necessity. The Functionalist parameter reinterpreted. Examples of impossible developments. Arbitrariness. Conformity. Isomorphism *409*

6.3. Summaries of the sign-functional dynamics of the theory with consequences. Polysemy, synonymy, transcendence and consequences. Definition of traditional historical terms. The synchronic linguist confuses seeing and judging. The definition of constant change in terms of devaluation: continuity and gradualness. The logic of *use* as the cause of constant change *422*

6.3.A A sign-functional summary of the theory's elements with consequences *422*
6.3.B The scope of the sign function: synonymy and Polysemy versus Metaphor. The speech situation and analogies *426*
6.3.C Redefinition of some basic terms *430*
6.3.D Change and cause appear from a judgement, a mentally discursive process ... *433*
6.3.E Constant change: continuity, gradualness and devaluation *435*
6.3.F The logic of the instrumentalist-view of change *438*
6.3.G Impossibility and arbitrariness *439*

APPENDIX 1
   An overview of the transitions of the various developmental series *441*
APPENDIX 2
   Explanation of abbreviations, symbols used, and some of the applied
   theoretical terms *443*
BIBLIOGRAPHY *451*
INDEX OF DEFINITIONS AND TECHNICAL TERMS *473*
SUMMARY   Sammenfattende redegørelse på dansk *477*

FOREWORD

The Synchronic Fallacy refers to modern linguistics' dominant conception of the relationship between state/stability and change/variation, which makes history dependent on synchrony: language development is a function of and controlled by the synchronic system and historical linguistics is only scientific in virtue of synchrony because the latter is a statement of the essence of language through the state concept's logical and empirical priority over change with change being regarded as an accidental property of language. Thus there is no such thing as an immanent historical linguistics.

The essay's **critical** thesis is that this is an empirical and cognitive fallacy with no logic to substantiate it. The essay's **constructive** thesis is the possibility of constructing an **immanent theory of history** in order to rehabilitate **man**'s **historical sense** as an essential process in his dynamic sense of reality so that an adequate grasp of reality hinges on **man's temporally structured mind**: the analytic mind of modern linguistics does not yield an empirically relevant grasp of the reality of our everyday languages; in the analytic approach there is an in-built conflict between its inability to build a world of absolute timeless reality and its attempt at populating it with permanent and unchanging 'entities'. The synchronico-analytic mind has not decided whether 1) synchrony is a metaphysical statement about the nature of language, 2) synchrony is an epistemological-cognitive statement, 3) the everyday language is *made* static in order to fulfil its communicative purpose, or 4) it is made static in order that the analytic mind can acquire knowledge of it. The very number of modern synchronic descriptions of the grammar of a given everyday language suggests that synchrony substitutes conventions for (the empirical) facts of language.

The essay does not substitute one *a priori* for another one, but sets out from the assumption that synchrony/systematicity/order and history/individuality/chaos are not antinomical and that they can be shown to constitute a harmonious unity at the level of theory once ideology is removed from the way we comprehend reality as well as from the way historical existence occurs.

Historical Investigations refers **critically** to the fact that modern synchronic linguistics leaves something to be desired in terms of a well-defined and generally accepted theoretical nomenclature so that also non-synchronic linguists may gain access to the various synchronic strongholds ('paradigms') erected during the last hundred years. *E.g.*, the synchronic literature is rich in theories, but poor in theoretical unanimity: the historian looks in vain for agreement on the meanings of these four pillars of modern linguistics, **synchrony**, **structure**, **state** and **system**. **Constructively**, the historian subjects these essential synchronic terms to an analysis with a view to being able to understand them so that they may assume their proper places in the advanced theory of history. This means that the essay's analyses of concepts and specific linguistic theories subserve the aim of illuminating and advancing an argument; a theory of X is not discussed in order to improve that theory or advance a better theory of X, but to investigate the historical views contained in it as well as its historical ramifications.

A Theory of History refers to the theoretical and realistic possibility of constructing an immanent theory of history. This theory is a theoretical language, devoid of the linguistic tradition's synchronized and static categories/terms, and it enables us to **comprehend** what cannot be seen, heard or felt, the sign-functional existence of a part of our everyday languages. As such the theory is not causal – causes belong to the level of data – cause is a redundant concept when the everyday language is understood from the perspective of its **existential** function. The essay assumes that *mind, thought, reasoning* and *thinking* are inseparable from – their synonym – *language*.

Man's everyday language is a sign-functional structure through and through; its existential function is one of the primary shapers of the individual person's existence. The theory generalizes the sign-function ('form/expression' and 'meaning/content') so as to subsume all of man's existence under its scope: all aspects of human existence are sign-functional. Being a legacy from both Hermeneutics and Glossematics, this idea may not be particularly novel; but the present writer arrived at this sign-functional generalization through the study of language change: no part of the everyday language is not sign-functional, so that human existence and reality are eminently amenable to being 'accessed' or comprehended by the everyday language – and *vice versa*. This generalization will be implemented sign-functionally.

Since the essay is also a rehabilitation of Hermann Paul's well-known views on linguistics and a rehabilitation of structuralism to the effect that language is a historical phenomenon and structure and history are compatible, it also regards **synonymy** and **polysemy** as two language universals, not in an idealizing or abstract sense that displaces them to a shadow existence in *parole*, but as two substantive (non-formal) features without which there would not be any language. This also entails forays into the Realism-nominalism discussion, a discussion which has been absent from the predominant Anglo-American conception of linguistics since the mid1950s, although it has played an important role in Continental European linguistics from Hermann Paul onwards by way of Louis Hjelmslev's *Glossematics* to Eugenio Coseriu. The essay is therefore also a critical reckoning with central parts in Louis Hjelmslev's *Glossematics*.

Despite its attempt at being ideology-free, the essay develops a moral: the present itself and knowledge of the present are products of man's ignorance and knowledge, which jointly create the present in an existential process that makes what was the present the past of the present. It is more than a truism that evaluations of the present, including the present period's *Selbstverstehen* are the joint, but contradictory product of what the 'judge' knows and does not know, of what has been forgotten and what has been remembered – by the present. The progress of the essay's argument is controlled by its own synthesizing dialectics: questions are asked, possible answers given which entail new questions – not by an analytic-oriented dialectics that moves from doubt to certainty, and then from certainty to certainty.

Ribe Landevej 24 *January 2001*
DK 6100 Haderslev *Erik W. Hansen*

PROLOGUE

An importing discipline is to become immanent and perhaps an exporter. A reckoning with the 20th century linguistic tradition's synchronization of *die geschichtliche Wirklichkeit* of the everyday language into *historische Zufälligkeiten*

> *(...) what from the one-sided (static) Modern English point of view is an isolated fact, is seen to be (dynamically) related to a great number of other facts in the older stages of the same language (...). (...) languages are always in a state of flux (...). This is an inevitable consequence of the very essence of language (...).* Jespersen 1963:30-31.

0.0. A bird's eye view of the conception of history and change being contingent on synchrony and staticness: a critical look against a horizon of constructive synthesizing: preliminary exercises towards a theory of history

THE OVER-ALL PURPOSE of the present essay is to make the fact and concept of change – the latter more so than the former – the immediate subject matter of theoretical investigation. I must add this proviso: the discussion will be conducted from the point of view of language development, *i.e.* it is a language historian who is going to deal with change, neither a logician nor a synchronic linguist, not a person trained in either biology, philosophy, sociology, psychology, phenomenological hermeneutics (see note 140 below) or for that matter in semiotics or other -isms such as Structuralism and deconstructivism, and what I shall have to say about scientific principles and methodology in general will come as *necessary consequences* of the essay's immediate target, language development. The essay wants to open the windows of the grand house of *Sprachgeschichte* to an outside world which, in the present century, has so strangled the independence of the study of language development that questions like – What can the immediate study of the language universal: *all languages develop*, contribute to other branches of scholarship? – cannot even be pondered. The essay's general line of thought will therefore be to change the direction of influence 180 degrees: since its advent 200 years ago historical linguistics has imported views from other sciences in order to explain the WHY and HOW of language development with the natural sciences, formal logic, sociology and psychology being influential exporters; has historical linguistics something to export to such fields as epistemology, logic, psychology, sociology, history and, in particular, to synchronic linguistics so that knowledge of language development will be seen to constitute an independent body of statements, indeed a necessary corrective to synchronic linguistics[1]? In the light of the fact that the puzzle about change (Jones 1970:121) lies at the root of many, if not all, the grand philosophical schemes that our Western civilization has produced, it is not unreasonable to assume that a field of study that has as its primary target the comprehension of one significant aspect of the complex field of change may have something to contribute to other dynamic research areas where the human factor (*l'individu*) is more than a *quantité négligeable*, or something to be *écarté* (Durkheim 1927:125).

This approach to language change entails a conceptual reckoning with at least some of the fields of study mentioned above. But before this reckoning can be made, both actants in such a comparative study must be placed on equilogical footings. In a word, not only must the possibility of an immanent science of language development be envisaged, but such a programme must also have been carried out *in concreto* to a certain degree. In attempting this, I shall be following Løgstrup and Paul's views of what it takes to constitute an object as a scientific phenomenon (Løgstrup 1984:228-246; Paul 1968:16-17): this constitution requires a selection of only a few of the many components that otherwise make up a would-be scientific phenomenon in its entirety, and as a corollary of such a radical reduction of the phenomenon in question, I shall draw up a too generalizing picture of modern linguistics, but whose tenor has dominated the greater part of the present century. To be explicit: it is true that the earlier antinominous distinction between synchronic linguistics and historical linguistics has been so softened (*pace* Harris 1993) in recent years that a happy union within a grammaticalization framework may be accomplished in the foreseeable future[2]. But

---

1 The target and line of reasoning of my essay is diametrically opposite to Lass (1997), who continues – in the Glossematic sense – the transcendent tradition's inclination to import foreign (biological) terms into historical linguistics. As early as in 1895 (1927:126-130) and in 1898 Durkheim (1974:13ff.) pointed out the fallacies of understanding sociology in biological or psychological terminologies and advocated – in Glossematic diction – a 'sociological sociology', *i.e.* an immanent conception of social facts: *c'est (...) dans la nature de la société elle-même qu'il faut aller chercher l'explication de la vie social* (1927:126). Meillet (1975:1) imported this view into historical linguistics in his 1906 *Leçon d'ouverture*, making linguistics, as Togeby (1965a:5-6) notices, transcendent (see end of chapter 3.3. below). I do not agree with Togeby in combining immanence and the form-substance dichotomy so that immanence translates Saussure and Hjelmslev's form-concept; in doing so Togeby follows the traditional view of history and the sensuous world as being contingent and removes any substance from being part of the essential nature of language (p. 194): sound, the medium of spoken language, is not the sole domain of certain branches of physics; thus the spoken substance is not mere raw material; history is bound to have an empirico-sensuous 'substrate'; with Hermann Paul I shall argue that a function in a structure called *langue, qui s'oppose à parole ou à style* (Togeby *loc.cit.*), cannot enter into *Kausalnexus* or into what I shall call the *developmental series* of existence (see 0.2.2.).
2 This possible synthesis is far from unambiguous, though. Heine *et al.* (1991) mention several versions: 1) grammaticalization is used as an explanatory parameter to account for synchronic structures (p. 13); 2) grammaticalization explains historical facts (pp. 14; 22) or is a general and constant developmental tendency that neither starts nor ends (p. 15; *an evolutional continuum* (p. 16)); 3) grammaticalization has both a synchronic and a diachronic aspect (pp. 18); 4) grammaticalization is a *continual* synchronic(?) process (*movement*) in the *interaction between discourse and grammar* (pp. 20;21); 5) grammaticalization, a historical process of long standing in the linguistic literature, has been known variously as devaluation, bleaching, loss of content, semantic weakening, depletion, etc. (pp. 21;6-9); 6) the concept's muddled position can be summed up as follows: is grammaticalization an *explanandum* (synchronic or historical or neither), *i.e.* something to be accounted for (p. 21) or an *explanans*, something that accounts for facts (synchronic and/or historical ones) or is grammaticalization a discipline in its own right (p. 23), perhaps over and above synchronic and historical linguistics?

0.0. A bird's eye view of history and change

neither the less rigid stand nor the ideal can conceal the fact that the whole issue consists of two general layers, that of synchrony (synchronic principles) and that of history (historical principles). So a proper understanding of these two layers in the current budding symbiosis still presupposes independent analyses of each; and since it does not take much reading to see how the synchronic point of view has been and is still dominating modern historical linguistics – it provides the terminological if not the theoretical apparatus of many such studies[3] – it is both appropriate and relevant to set up a stringent and consistent horizon against which my argument will be conducted: **the dependence of historical linguistics on synchronic linguistics with the corollary that synchronic linguistics is an autonomous discipline**. I shall illustrate the latter point with reference to a recent book dealing with central problems in linguistics (Wardhaugh 1994 (third printing)).

As such the essay is also a **constructive critique of synchronic reasoning** (the synchronic-analytic mind) and in accordance with its opening words I shall alter Nietzsche's words: *die Übersättigung einer Zeit* [in Synchronic Reasoning] [*ist*] *dem Leben feindlich und gefährlich* (1966:237), where 'in Synchronic Reasoning' substitutes for Nietzsche's *in Historie*; and five Ariadnian Threads (see note 184 below) will guide us through the labyrinth of language development to this essay's theory of historical development

Hermann Paul bequeathed two guidelines to his successors: firstly, the scientific relevance of the word *historical theory*; secondly, the warning that we must liberate ourselves from the tyranny of words[4]. Thirdly, Karl Verner's[5] legacy *eine regel für die unregelmässigkeit* (Verner 1877:101; cf. Andersen 1973:787-788) – with his reminder (p. 112) that the language historian should not stop at the accidental fact – to the science of language tallies perfectly with Paul's emphasis on the theoretical basis of historical linguistics. A fourth guideline worthy of mention is Karl Luick's statement that the changes in a language must be treated in their chronological order so that *ihre Zusammenhänge und Wechelwirkungen klar hervorträten* (Luick 1964:3; 4): historical linguistics constitutes *eine Geschichte, kein System, eine Erzählung, keine Regelsammlung* (pp. 5-6)[6]. Fifthly, Luick's criticism of the synchronic two-state model – should also be kept in mind: *Die Entwicklung der Forschung drängt (…) gebieterisch dazu, über diese Stufe hinauszukommen* (1964:7 Anm.).

---

3   Hopper and Traugott (1993) with some reluctance follow this path, and despite his apparently sympathetic attitude to historical linguistics in his 1997-work, Lass's approach to language change is not empathetic: historical linguists are bound by *general linguistic theory of one kind or another* and are both *historians and linguists* (p. xiv; cf. p. 12). Nowhere in his book does Lass tell us what this theory is: what do *contemporary linguistic theory* (p. 9) and *current theory* (p. 178) refer to? We know that the theory is synchronic (p. 9), contains universals (p. 208) and is slant-bracket dualistic (p. xxi). – For a sympathetic rehabilitation of historical linguistics, see Anttila (1989; 1992; 1993), whose '*theoretical linguistics is genetic linguistics*' (1989:411) of course reminds us of Hermann Paul (see 0.2.2.). In this connection I must emphasize that it is not my purpose either to perpetuate the antinomy of synchrony and historical linguistics or to make synchrony or synchronic structure *less* '*interesting*' (Lass 1997:9).

4   We must cure ourselves of *[die] Krankheit der Worte,* making the 'Word-Factory' produce empirically relevant concepts (Nietzsche 1966:280).

5   Note Lehmann's assessment of Verner (1992:30).

6   In their, in my opinion correct rejection of the Great Vowel Shift *qua* a *sui-generis* entity, Stockwell and Minkova (1988) infringe this elementary requirement *qua* a necessary part of our historical sense, as they only locate the relevant phenomena in accordance with a vague 'x-is-later-than-y' parameter; that the chronology is crucial will appear from the following in that it decides which theory to apply: Stockwell

0.1.1 The Moment-When problem

Wardhaugh's book is a good example of how the 20th's century's dominant view of language change was continued unchanged into the last decade of that century[7]; despite the irrefutable fact that our everyday languages change *over time* (p. 4), there is here no attempt at viewing language change as an independent study, and when it comes to the question of theory construction, the possibility that history and theory are a relevant couple is not mentioned even in passing (pp. 9-10;5). Hence it is not surprising that in discussing the notion of language competence (pp. 20 – 26), Wardhaugh does not envisage the possibility (i) that man's purportedly *innate capacitas linguæ* may include a historico-dynamic component which sees to it that under no circumstances can our everyday languages lose their **collective orientation** over time – despite changes, and (ii) that it is precisely because it changes that any

---

and Minkova (pp. 376-377) argue that ME /e:/ and /o:/ were so close that they were in-gliding diphthongs with a raised first element (*i-* and *u-*), an interpretation supported by the short *i-* and *u-*sounds as the respective 'long' sounds' shortened products, cf. *breeches, silly, nickname; stud, flood, blood, mother*, etc. Stockwell and Minkova overlook a joker namely the lowering of short vowels in ME and the subsequent lengthening of some of these new lax sounds. The chronology entails **two** options: **A**) Stockwell and Minkova assume that OE/ME /i:, u:/ had become diphthongal by the latter half of the 14th century, outgliding diphthongs, [ɪi, ʊu], so that C1 and C8 had been vacated, whereby the raising of /e:, o:/ could be seen as a *gap*-filling proces. Their point is that when shortened, /e:, o:/ merge with short /i, u/ – lowered or still high? If the sounds had preserved their original height, near C1 and C8 from OE, /e:, o:/ must certainly have been raised: Stockwell and Minkova say *by Middle English times*. Thus the lowering of short vowels will occur after this shortening. However, the subsequent lengthening of certain short vowels, [i > ɪ > e:], [u > ʊ > o:] yields a sound that quickly catches up with the original tense long vowels: OE *wicu* > ME ... /we:kə/ <week(e)> and OE *wudu* > ME ... /wo:də/ <wood(e)>, in order not to be swallowed up by the – ?also moving – open C3/C6 sounds in *e.g. lead, beat, home, boat*. This is unlikely. Their theory – because of the early raising/diphthongization of /e:, o:/ – does away with the lowering process, at any rate for the high vowels and entails this process [i > i > i:/iə], [u > u > u:/uə]. Another joker is the shortening of /e:/ in *meet – met*, and *keep – kept*. The alternative is that the lowering law had been enforced, so that the now low *i-* and *u-*sounds would be the sounds closest to the shortening of the tense vowels. – **B**) The spaces of C1 and C8 have not been vacated – before the lowering and lengthening processes. Stockwell and Minkova 's diphthongal theory will then imply a variant-scenario like this with the variants being 'free': /i:/ [ɪi – i:] : /u:/ [ʊu – u:] :: /e:/ [iə] and shortened [ɪ] : /u:/ [ʊu] and shortened [ʊ]. If /i, u/ had not been lowered, then /e:, o:/ must have been somewhat raised and also have begun to encroach upon the domains of /i:/ and /u:/ so that the further change of /i:, u:/ must be seen as a *push*-chain process (Luick's theory, which Stockwell and Minkova are arguing against). The results of ME /e:/ and /o:/ in the 15th century were /i:/ and /u:/, not diphthongs. Otherwise the same can be said as under **A**): with /e:, o:/ raised and /i, u/ unlowered, the explanation entails unlikely consequent processes; with /i, u/ lowered, their theory cannot be proved by the shortened material.

The chronology of data is certainly essential to a given theory and the synchronic search for purported *internal coherences* (internal structural laws) will be contingent on the coherences that can be established between *nacheinander* phenomena, and such coherences will be expounded by a *historical theory*.

7 Trask (1996:ix) is a clear example of the confused, if not muddled view of historical linguistics that is not untypical of modern linguistics: Trask states that his book is *atheoretical*; despite this he continues: *The only theories introduced here are theories of historical linguistics and of language change*. If this is not a blatant *contradictio in terminis*, there is only one way to make Trask's position meaningful: the fundamental basis of all language decription and explanation is static, be it universal or just synchronic: history is dependent on synchrony. – McMahon (1994) is also a summary of traditional positions.

everyday language will always preserve its self-identity and always be 'functionally' viable[8]. The mentalist approach to our everyday languages assumes tacitly that when the word *language* is used, it is always in its static conception, so that what the human brain[9] controls is a stable *language 'module' within the mind*; this module[10] so enables children learning, say, modern English to manipulate linguistic structures that a passive clause is formed by <u>moving</u> *structural units like subject and object noun phrases* (pp. 23; 255) in relation to the implied synonymous active clause in question. But as Wardhaugh admits (p. 94), all such workings of the human brain can only be set up *inferentially*. The main problem here is not this particular view of the category of voice, but its consequences: if it is permissible to infer from a point-in-time type of empirical data to the way our mind works, it must be equally permissible to include other data into our inferential basis. If a baby learning a variant of modern English today knows a priori how to produce clauses in the passive voice, a baby learning English in the days of King Ælfred must have had the same a priori knowledge of 'passive transformations', although she had no use of it. The English data then did not prompt her to acquire these 'passive' rules with a view to their active use when she spoke and heard English[11]. But when the passive construction was grammaticalized at some point in time in one specific state among the sum of states that constitutes the history of the English language, how – why – was this brought about? If a baby cannot rely on the deficient data (p. 233) that surrounds her during her language learning days, then something else must have prompted English babies to activate this latent, a priori knowledge of passivization at a particular point in the history of English, and this something must be the brain's above-men-

---

8 The modern dominant view of our everyday languages (Chomsky 1963:3) disregards the fact that its transcendence into an ideal haven of perfection is not cogent: if performance is faulty and imperfect, then it cannot be the ultimate basis of mutual intelligibility or effective communication and constitute the only input to the child's creation of her grammar: hence there is an innate ideal, a *capacitas linguæ*, that makes learnability, communication and intelligibility possible. The alternative is: as no child does not learn a native language, the language phenomena that she is exposed to are necessary and in order as they are – whatever their (degree of) imperfection: hence, if performance data were perfect, then the language would be unlearnable and no language. There are only two ways out of this dilemma: if a language is to be functionally viable, its manifestations must be imperfect and faulty; since all our everyday languages are functionally viable, mutually intelligible, communicatively effective and learnable – an everyday language never loses its collective orientation – it makes no sense to call 'performance' (*parole*, the spoken substance, *Rede*) imperfect and faulty. The merger of competence and performance is a transcendent ideal, occurring in a Plato-like world to which we humans – and Plato does not disagree – have no access. Trask (1996:267) approaches this from a non-idealizing angle: the game of chess cannot be played *successfully* if its rules are being *changed constantly during play*. The analogy is false through and through: language and chess are not similar sign systems, and the moment Trask utters the analogy, he has given the answer: the everyday language is not a game, games not everyday languages – these change constantly while being spoken. – Lass (1997:11) separates the everyday language from the speaker.
9 The Generativist School (*e.g.* Chomsky 1986:xxvi *et pass.*) does not seem to be able to make up its mind: is it the brain or the mind that controls – or are the two words, *mind* and *brain*, synonyms?
10 Whatever this language module (widely used in Chomsky 1986) may be, the conception reminds us of Paul's *Sprachorganismus*, the empirical object of the linguist: *etwas unbewusst in der Seele Ruhendes* a precondition for *die Sprachtätigkeit* (Paul 1968:29; 32). See Kurcharczik (1998) for an overview of the changing meaning of this organism-metaphor in the 19th century.
11 This spontaneous creation by the language learner of a new grammatical rule has been aptly criticized by Andersen (1973:766-767).

tioned language module itself. But this is a vacuous, if not a circular explanation: passivization becomes an active rule in the grammar of English the moment when the English language begins to turn periphrastic (*parole*) constructions into passive constructions *qua* grammatico-systemic entities: the moment when a language learner has internalized G2 (with a passive rule) and not G1 (cf. Andersen 1973:767). Secondly, is the passive rule that was internalized in the (?)Old English period the same as it is today (Strang 1970:98-99)[12]? To conclude: we may infer from the English data or for that matter data from any language that an essential 'component' in our human mind is man's **historical sense**, a faculty that enables us to handle change, development from T1 to T2, to so interpret the **inherited** (be it systems (Lass 1997:12), structures; G1; or Andersen's *Output 1*) that something **new** (systems; structures; G2) has been **produced**, while **the old** has been ousted. It is an empirical universal that no grammar of any language is stable, and ignoring this does not make either the fact of change or the problems with the transition from G1 to G2 go away[13]. That *la langue* is the product of *la parole*, as Saussure concluded, is not in doubt (see chapter 3.4.).

0.1.2   Another kind of universal: synchronic frustrations over variability

The statements linguists make about language, as Wardhaugh puts it (p. 131), *can never be quite complete,* there being a *gap here, an exception there.* Why is this so? Wardhaugh, who is not aware of the reflexivity of his own postulate, answers the question with reference to an empirically irrelevant situation: no language is *a perfectly wrought system complete unto itself* (p. 131)[14], to which the language historian may counter: why should it be such a system? Not to mention not only the question: what is a perfect system? But also the fact that the perfect system must be described by means of our 'imperfect' everyday languages and will consist of categories that belong to imperfect systems. No wonder that Plato took refuge in his ideal haven, leaving the everyday language in changing reality[15].

That variability *creates problems* (p. 131) is a value-loaded statement which stems from the (synchronic) linguist's displacement of his frustrations from their proper object to the

---

12 In the Anglo-Saxon Chronicle (Year 409) we read: *Her Gotan abræcon Rome burg* (Parker MS) and *Her wæs to brocen Romana burh fram Gotan ymb xi hund wintra ond xi wintra þæs þe heo getimbred wæs* (Laud MS): is the passive rule behind the latter construction identical with the rule beind the modern translation: 'Here Rome was destroyed by …', as Pilch's methodology seems to imply (1970:199-204)? The same applies to Traugott's treatment of the passive in English, it being subsumed under the general concept of subjectivalization (1972:36; 81-83), despite the fact that she acknowledges that this synchronic process of subjectivalization is different in different languages (p. 36;81). – Cf. Brunner/Sievers 1965:§350 Anm. 1.

13 Sociolinguistics encounters the same problem at the phonological level: the step from variant to change; it is questionable whether the same criterion, social generalization (Milroy 1992:86; cf. 0.4.1), can be transferred to the syntactic level. – Significantly, Labov (1994:75; 57;65:79 *et pass.*) can describe the tensing of (New York City) /æh/ into [i:ə] by way of [ɛ:ə] and [e:ə] with no reference to (intermediate or resulting) phonemicization stages. Labov (pp. 111-112) does accept the notion of not only *phonological systems*, but also the *stability* of the underlying categories of such systems.

14 The historian may ask: where is the gap in this statement? Does Wardhaugh really mean 'A language is a perfectly wrought system complete unto itself'?

15 Jones (1970:194): a glimpse of real reality is obtainable by means of *a kind of intellectual osmosis*, because *meanings can be communicated without words*.

0.1. A bird's eye view of history and change

empirical facts of inherent variability (p. 135) and constant change (Sapir 1949:144): that a grammatical system (theory) cannot describe all its data – exhaustively – is only due to the grammatical system in question: and there are only two inferences, a general and a particular one, to be drawn from this: the particular alternative is that the theory at hand is not good enough; the general alternative is that the very approach to data is fallacious, namely the conception of the empirical relevance of the idea of there being a static entity called *a synchronic grammatical system*; synchronic linguists seem to confuse or rather obscure the distinction between the concepts of systematicity and system[16].

Language change and variability are apparently something to be conquered through an enumeration of factors – most often social ones – that cause variability, and these factors are formalized by statements of variable rules or as a linguistic variable. Such a rule, however, is static; it does not consider the actual origin of the variants of a variable rule (Trask 1996:281)[17], but only describes the type of variant shift and the causes of the shift. In this way sociolinguistics has set up an impressive array of intersecting factors governing the systematic use of variants (pp. 142-144; cf. Labov 1994; Milroy's summary hereof and his remarks on the origin aspect (1992:81; 86-87)).

Wardhaugh's exposition of the HOW (p. 145) and WHY (p. 152) of language variability and change follows traditional lines. In so far as questions about theory-building are asked, answers make a theory merge with causal factors or make do with brief statements which I venture to boil down into **two** generalizations: **all everyday languages are imperfect**[18], this concept (fact?) being understood in such terms as **1**) absence of complete grammatical consistency (cf. Sapir's concept of Drift (1949:147) and Wardhaugh (1994:131,152)), **2**) generational transition (Samuels 1972:1972:112-113), **3**) contactual interference at various levels (Milroy 1992:74; Samuels 1972:92; Lass 1997:184-185), **4**) human imperfection in speaking (Strang 1970:10), in hitting some abstract target (a systemic entity such as a phoneme (Wardhaugh 1994:153; Strang, pp. 8-9; Samuels 1972:17), **5**) language not being overtly geared to keep pace with extralinguistic (social) change (Samuels, pp. 1-2), and **6**) *copying error* (Lass 1997:176)[19]. Even Milroy's social model operates with a contrast that flouts the constant-change assumption in that he works with stable and non-stable language states

---

16 A systematic investigation does not imply the (?hypostatized) system, an object in its own right, as its final (resultative) goal, whereas a systemic one more often than not implies such an entity (*où tout se tient*). A similar ambiguity inheres in the use of *state* and *static*. The adjective refers to a property with *stable* as its synonym, whereas the state concept also implies a (hypostatized) *sui-generis* entity, which, of course, is stable. Milroy (1992) is a systematic paper, although its *change*-concept seems to imply a system; and Hjelmslev (1963:10) equates the *systematizing* point of view with a systemic one.

17 Trask admits that sociolinguistics can only implement its rules, not explain the start of a change (the so-called actuation problem). However, it is correct that variation is the *vehicle of change* (p. 276), but all variants (allophones) are not socially motivated: history is not a subbranch of sociolinguistics, as Trask (1996:268) implies; interestingly, Labov (1994) calls his book *Principles of Linguistic Change*, thereby subsuming sociolinguistics under historical linguistics.

18 This attitude ties up neatly with the view of language development as being destructive (cf. Saussure (1972:124): *La langue est un mécanisme qui continue à fonctionner <u>malgré</u> les <u>détériorations</u> qu'on lui fait subir.* – Emphasis added. And Gregersen (1991) has aptly demonstrated how Glossematics is inexorably driven to an empirically irrelevant *cataclysmic* view of change.

19 Following Lass (1997:112-113) change is the child's failure to replicate the language she is learning.

(Milroy 1992:85), where the former type is maintained by (imperfect) *speakers and not by languages*. The **second** generalization, inherent in the first, refers to the widespread, if not universal, way of conceiving (the nature of) our everyday languages: **a language is not X, but (unfortunately) Y**: a language is not invariant, uniform, fixed, homogeneous, stable, but varying, non-uniform, non-fixed, heterogeneous, non-stable (cf. Wardhaugh 1994:140), and Milroy (*loc.cit.*) sums this type of attitude up: *the natural state of language is to be divergent and not convergent*. If this is so: why mention a property that is not to be predicated of our everyday language and which it is not natural for it to possess? The extreme case in point is the point of view that turns linguistics into the study of an entity that does not exist, the ideal speaker/hearer's language (Chomsky 1965:3-4). Add to this Wardhaugh's *'perfect state of affairs'* = *'A single language used by speakers throughout the world with no variation'* (1994:166), then the step both to Postal's view of language change, it being irrational and by implication unnecessary (quoted in Samuels 1972:vii; Saussure (1972:110) disagrees with Postal), and to Wardhaugh's having resigned himself to the inevitability of language change and variation: *we can do very little about it*, is very short indeed.

I shall quote Wardhaugh *in extenso* (p. 167)[20] because the attitude that emanates from his words is fundamentally wrong and detrimental to significant advances being made in linguistics in particular and in historical linguistics in general: it creates an inadequate and insufficient horizon for historical linguistics to be conducted against; it regards change as basically negative[21], destructive, as a nuisance we have to put up with; it implies a scenario that does not exist, thereby preventing historical linguistics from creating a historical theory in its own empirical right. So the present essay is in fact an answer to Wardhaugh's 'what-else-can-we-do' **question** in:

> There is good reason to believe that language change and variation are inevitable[22] and that we can do very little about it. From time to time we can try to apply a brake here or give a little push there, but change itself goes on. Similarly, with language variation: we may deplore this or that bit of variation, but at the same time we are not even aware of considerable variation elsewhere and even participate – generally unconsciously but sometimes quite consciously – in actually promoting variation. However, if, as many linguists believe, languages inevitably change and vary by their very nature, **what else can we do**? Dealing with language change and variation may be part of the price we must pay to remain viable as a species. That price may seem excessive on some occasions when misunderstanding and conflict arise as a result, but when we realize the advantages that the ability to use language confers on us as a species it is probably a very small price to pay for the benefits unknown to any other species on earth. – (Emphasis added.)

---

20 A more sympathetic, nevertheless fallacious, view of this Saussurean paradox is Trask's answer to the latter (Trask 1996:267-268). It is to be noted, however, that Trask's starting-point is the traditional one: not only is change *mysterious* (pp. 64; 268), *embarrassing, strange* (p. 87) and a *puzzle* (p. 267): *How on earth* (p. 268) can a language remain functionally viable while it undergoes constant change?, but change is also a process with a start and an end – arrived at when it runs *out of steam* (pp. 289; 295) – as well as with a disruptive effect (pp. 27,85). Milroy (1993:215) is of the same opinion: sound change is *mysterious: there is no obvious reason why it should happen at all*.

the **answer** being: we can construct a **historical** theory through which to comprehend the existence of our everyday languages, instead of following the *two strands of* the dominant trend in our Western civilization since the days of Thales (*pace* Heraclitos): either the doomed attempt at conquering empirical, sensuous, irrational change and variation through (ontological, methodological) dualisms that turn change into something which it is not: seeming reality (appearance), not real reality, or the commonplace notion of the inexorability of the defeatist whatever-will-be,-will-be attitude. While the argument of the present essay will be steps towards such a theory, the above remarks have already established two concepts and factors in the theory, namely **synonymy** and **polysemy**, both of which will be regarded as empirically relevant language universals. Needless to say: neither plays significant roles in synchronic theorizing or in the nature of the synchronic systems concept in that synonymy (implying redundancy) is to be reduced to **mononymy** by means of the *langue-parole* (*Sprache-Rede*) dualism, and so is polysemy (implying ambiguity) to **monosemy** by means of the notion of there being an entity such as a central or core meaning (cf. Hjelmslev's *signification fondamentale* (1972:84) and Sørensen's semantic definienses (1966:36 *et pass.*).). Synonymy is interdependent with the fact of variation: phonetic variants (allophones), whatever their motivation, will always be semantically correlated with synonymy: the front-back variation seen in RP /æ/ is correlated with an age-content, according to Strang (1970:11-12)[23], with the front variant being typical of older RP speakers and the back variant typical of young RP speakers (NB. written in 1970); similarly, the variants [-ɪn] and [ɪŋ] of the modern English ending /-ɪŋ/ are semantically connected in a complex synonymous relationship (Wardhaugh 1994:143). Here sociolinguistics misses a significant generalization; any socially motivated variable is a sign (cf. Trask 1996:268-285); hence sociolinguistics is to be subsumed under a sign-functional theory[24]. As regards polysemy, the very notion of a synonymous relationship presupposes it: two synonyms will also differ in virtue of their respective contents. Thus historical theorizing involves not only the search for **identity in differences** (variation/multiplicity), but also the search for **differences in identity**[25]: synonymy and polysemy cannot be relegated to the status of mere *parole* phenomena, intimately tied up with the substance of the spoken (written) chain (the material, non-formal aspect of language).

---

21 Although Labov (1994) levels serious and pertinent attacks on some of the pillars of synchronic linguistics, *e.g.* on pp. 351-352; 367-369; in ch. 14, he cannot help calling language change *erratic* (p. 231).

22 A commonplace that is perpetuated by Trask, change being *unavoidable*, *inevitable*, even *remorseless* (pp. 12,361).

23 Gimson (1994:103) places this allophone near Cardinal 4 (/a/) for *younger speakers*. To anticipate the essay's sign-functional basis, I shall here call attention to the sign-functional nature of the so-called smallest elements (allophones) in language. This will be symbolized by the sign function ErC, *in casu* /a/rC. It will be part of my theory to demonstrate how such 'smallest' entities can be marshalled systematically into 'attitudinal' systems, independently of the synchronic phonological system.

24 Variation is a sign-functional (semiotic) process, not always socially motivated; apparently, sociolinguistics has no sign theory, despite conclusions to the effect that the use of a particular variant signifies a content (Trask 1996:274). As regards polysemy, a socio-variable is polysemous in that it may be used differently by different groups of speakers (p. 276 (Item 2)). And no variant in Labov 1994 is without a content.

25 Near-mergers are a paramount example of this: articulatory differences are heard or perceived as identical sounds (Labov 1994:ch. 12 *et pass.*). Cf. below 0.4.1.

0.2.1 The beginning: the Jones-Hypothesis and the synchronic alternative

In 1786 the British diplomat, William Jones, formulated his irrefutable theory (Hansen 1983:38-42), which I shall call the **Jones-Hypothesis**: the cause of the great (morphological) *affinity* which Sanskrit bears to Latin and Greek is that the three languages *have sprung from some common source, which, perhaps, no longer exists*. Before he could jump from sensuous data to explanation, Jones had to interpolate the mediating step that the affinity was of such a kind that it could not *possibly have been produced by accident*[26].

A linguist of a static, synchronic persuasion would have jumped from the immediate, sensuous and concrete (*anschauliche*) phenomena – the immediate morphological similarities that the three languages exhibit – to the postulation of a set of non-material, abstract, *i.e.* formal features as that which accounts for the affinity in question – a finite set of formal features to boot. This static-synchronic leap from the material to the abstract *qua* an underlying (finite) system is simpler than Jones's in that it is unmediated (cf. Hjelmslev 1972b:11-45[27]). Jones implemented the identity in the material differences with another material and concrete feature, namely another language from which Sanskrit, Greek and Latin had evolved, whereas synchronically the identity in the material differences of the empirical world was implemented dualistically by an idea of there being an abstract system *qua* a constant behind the complex, varying and changing, fluctuating and fleeting, in a word, nonsystematic empirical world. This dualism can only establish indirect relations between the empirical phenomena in that Sanskrit is identical with Greek, this being identical with Latin (despite obvious differences) *in relation to* a common system – or a principle of systematicity in the sense that each language can become subject to a systematizing-classifying principle of analysis. As this static point of view will only need one datum (language) in order to be applicable, such a dualistic methodology is also an expression of how an empirical phenomenon becomes a scientific object (Løgstrup 1984:231; 1976:172-173; Hjelmslev 1963:8-10), by a more or less radical reduction of all the components that otherwise appear to make up a phenomenon in its empirical, material and *anschauliche* complexity. The analytico-synchronic reduction ends, as indicated, in the notion of a system of elements whose number is finite and relatively small and its consequence is that historical linguistics is reduced to becoming a comparison of synchronic language states (L1, L2) with a view to establishing differences between L1 and L2, differences which must be explained, and similarities which remain unexplained, while the concrete development of L1 into L2 (L1 > L2) will be explained through **internal**

---

26 In many, if not all expositions of the comparative method the step from formal (and semantic) similarity to common origin is made *im*mediately (*e.g.* Lehmann 1992:143; ch. 7; Arlotto 1972:ch. 3; Trask 1996:ch. 8; Lass 1997:104-109). There is no necessary connection between similarity and common origin. To illustrate (from Trask, p. 208, items 1 and 3): Trask presupposes what is to be established, namely genetic relationships and a common ancestor: how to jump from systematic correspondences to the just-mentioned concepts? Note also Labov's sweeping generalization: *A word class is a historical accident. It is composed of a very large number of brute facts that have no explanation or connection with any other linguistic facts* (1994:311; emphasis added), an attitude that so denies historical development the basic right of being rational that it also defeats Labov's own rational exposition of language change (in progress).

27 In this introduction Hjelmslev does not present one cogent argument for the logico-empirical priority of the synchronic point of view, but makes unmediated jumps from premiss to conclusion, and it is a curious fact that Hjelmslev-scholars like Gregersen (1991) and Rasmussen (1992) do not consider this important aspect of Glossematics: how synchronic priority is established.

laws, which control the combination possibilities of the system's finite elements of L1: in L2 new combinations, permitted by L1, have been materialized[28]. The abstract consequence of this point of view is that the object of linguistic analysis will be regarded as (existing at) a point (in time in a certain place: "L2 of T2 & P2") with no extension, necessarily able to be torn out of its structural, *i.e.* historico-empirical context: "L1 (of T1 & P2) > ≠ L2 (of T2 & P2) ≠ > L3 (of T3 & P2)"[29]. The immanent point of view does not consider the structural ties that any language has and enters into with so-called extralinguistic reality; it builds a bridge, in Lévi-Strauss's words, whose ends are connected with no shore. The synchronic point of view may therefore be summed up as follows: any particular language is an exemplum of the superclass called a language; any language is (in) a state (cf. note 1) with a system (Hjelmslev 1972b:24) of finite elements that somehow determines its own material realization; the state is atemporal in that its existence at a given point in time is immaterial to its analysis and nature. It is spatial in the concrete sense that a state exists in many places at the same time, as well as in an abstract sense in that the systemic elements exist – at the same time (synchronically) – in a nonmaterial space independently of the temporally extended space continuum that constitutes *parole* (the spoken or written substance). The assumption is that any one particular element of the static system could have occupied that point-space in the spoken/written substance that is actually being occupied by a given materialized systemic element (Hjelmslev 1963:36-37). The assumption implies a **contingency** principle (Hoenigswald 1965:1-2). Thus the empirical world is not only complex and asystematic, it is also contingent, a domain where freedom and arbitrariness rule (Steffens 1968:109), where chaos – if controllable – is controlled by the system. I shall return to William Jones.

From the static synchronic point of view the data that Jones set up as historical was not that of the point, but that of the **period: X (origin) > Sanskrit, Greek, Latin**, although his immediate starting point was synchronic, the simultaneous, albeit temporally skewed presence of the three languages. The period's data is represented by the traditional monolinear formula A > B. Jones advanced two hypotheses: the ancestor of Sanskrit, Greek and Latin exists either (synchronically) with its daughter languages or exists no longer. Without exaggerating Jones's historical sense, I shall argue that the former alternative expresses a historical understanding in embryo[30]. The ancestor or origin of something must exist simultaneously with that something[31]. This will be symbolized in 0.2.1.1

---

28 This view of historical linguistics was severely criticized by Luick (1964:7). – Cf. below 0.3.
29 It will be a thesis of the present essay's that the synchronic state L1 is a *historische Zufälligkeit*, which its synchronizing garb aims to conceal, because it has been lifted arbitrarily out of its *geschichtliche Wirklichkeit*, where we find empirically relevant cause-effect-like connections (Gadamer 1972:14) that cannot be analyzed into parts by the synchronic linguist's so-called *synchronic cuts*.
30 This in sharp contrast to the ahistorical interpretation by Friedrich Schlegel, writing about 20 years after Jones, who regards Greek and Latin as derived from (*abgeleitet von*) [der] *indische*[n] *Sprache*. (Schlegel 1808:3). – A proper understanding of Schlegel will presuppose the precise meaning of his dynamic verb: *zu ableiten von*, and his use of the organism-metaphor (see Kucharczik 1998).
31 I do not disagree with Arlotto's interpretation of Jones (1972:41-42), but add the simultaneity concept in order to emphasize the structural aspect of historical development. I am not thinking of the possibility that the ancestor could have survived in writing with the daughter languages as is the case with the modern Romance languages and not only Latin but also their respective medieval extant ancestors. Trask's version of the Jones Hypothesis does not address the decisive point: where do the notions of origin, development-into or a common ancestor come from?

0.2.1.1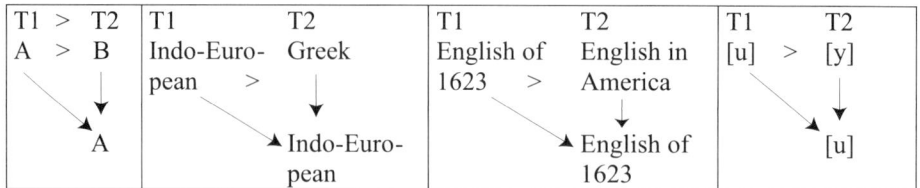

The historical development of a phenomenon A into B from T1 to T2 will therefore be in the form of a structural split process (its *Stammbaum*- and synonymy-aspects), in which B acquires part of its 'nature' from its origin, which so continues to exist in a simultaneous structural relationship with B that it influences the 'nature' of A continued[32]. The model captures historical processes at a macro-level, without implying that the structural processes involved are identical, such as the emigration of a group of people to another country: one could mention Horsa and Hengist's settlement in Britain in the 5th century, the Mayflower expedition in 1623 and the rise of Afrikaans in Africa's southern coastal areas (Reenan & Coetzee 1996:71), all can be viewed as split processes that do proceed along linear lines, A > B, (cf. Arlotto *loc.cit.*), but which also leave their respective origins to be continued into the same following moment, T2. At a micro-level, one could mention the origin of a new phonetic variant: PreOld English [y] was a novel sound which was created out of the old [u]-sound, but it did not oust this sound. On the contrary, they must have entered into a simultaneous structural relationship at T2[33]. From this type of example it follows both that the simultaneous structural relationship (at T2) is a synonymous one, which in turn implies polysemy, and that history and synchrony need not be conceived of in the antinomical terms of the Saussurean tradition (see end of chapter 3.4.).

The model will be so generalized as to also cover continuations proper, A > A (see 0.2.1.2), a type of language development that seems to have been accepted in certain sociolinguistic quarters today (Milroy 1992), but which has not influenced the fundamental way in which historical linguists and others conceptualize language development; the conceptual framework of historical linguistics is being dominated by the implications and consequences of the question: why does language change[34]?

The Jones-Hypothesis entails that the empirical data of historical theorizing is the period A > B, not the point, A or B. While the synchronic linguist moves from A or B to a formal level, the concrete peculiarities that the historian sets out from lead to another concrete, equally empirical phenomenon, and the two sets of equiempirical data will then be interpreted into a dynamic synthesis, A > B, characterized by an empirically relevant relation, which in turn

---

32 In the following '*nature*' will be replaced by *meaning* and *content*, whereby I indicate that the historical process is sign-functional.

33 The traditional version of the *Stammbaum* theory (as *e.g.* Trask 1996:183) misses the dynamics of the split process. The fact that a group of Englishmen emigrate to America alters neither their language nor the language of the rest of the population in England. To call this the mother language or ancestor of that of the emigrants is arbitrary. The languages of the two groups will of course be exactly the same as they were before the emigrants left England. The moment when the two languages changed (from T1 to T2), they become both descendants of the same ancestor (of T1).

34 Calling the two processes *persistence* in time, Lass (1997:117-118) merely provides a new etiquette.

is implemented by the idea of one empirical phenomenon developing into another phenomenon. It is this process, heterogeneous synthesis or A>B-'entity', that our linguistic tradition has denied scientific relevance *per se*; it could only be made a scientific object if the synthesis was split up into its purported autonomous parts *qua* homogeneous immanent totalities, because formal features, it was implied, can only be set up for static, hence point-in-time-existing entities, the class of which furthermore subsists in an abstract space. I shall demonstrate that it is possible to construct a formal theory of the period *qua* a synthesis of at least two elements of which one is the origin of the other, and the theory will be applicable to both types of historical development: A > B and A > A[35]. The notion of a historical theory will then be scientifically relevant, as well as for the reason that an element of necessity will be found in the empirical data, A > B: the development is not completely arbitrary or contingent; B will not be produced by accident[36]. William Jones's interpretation of his cross-language data's synchronic peculiarities will therefore be so generalized that the dynamic synthesis A > B is not wholly contingent, which means that Hoenigswald's above contingency view cannot be so generalized as to be applicable to his historical replacement concept:

0.2.1.2 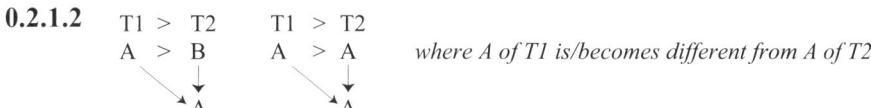 *where A of T1 is/becomes different from A of T2*

The model entails that there is some truth in the (traditional terminology of the) *Stammbaum* Theory in that A developing into B implies constant change, and the important point is that historical development is not a monolinear replacement process. There is nothing mysterious about replacement processes, as *e.g.* Trask (1996:268 *et pass.*) will have it: *a speaker must use one form or other, and that is the end of it*. Trask does not see that replacement processes are always structural ones in which a speaker does not discard the variant he does not select at a given point in time; the selection of a form occurs against the (synonymous) horizon of the form(s) selected in a given speech situation. The Neogrammarian formula, A > B, represents too simple a view of the structural process of variation and change.

---

35 Labov (1994) could be regarded as one long, not too long, criticism of this standard view of linguistic analysis – which he calls the *categorical view* (p. 348 *et pass.*). However, Labov does not set up a theory of the historical period in that his above-mentioned tensing of /æh/ is not captured by a comprehensive theory, which must include not only chain shifts, mergers and splits, but also simple vowel shifts (OE /a:/ > /ɔ:/) and continuations; the latter type seems to be subsumed under the notion of categorical stability (pp. 98;112). Labov's Expansion Convention (p. 264) appears to be both a non-generalized and perhaps non-generalizable concept.
36 In 1803 the Danish-German philosopher Henrich Steffens (1968:109-110) put it in a slogan-like manner: it is our task to 'recognize the Freedom of Reason in the Necessity of Nature' and 'to recognize the Necessity of Nature in History'. Thus the formula's B represents a successor that could neither be anything else but A nor not be.

0.2.2 Hermann Paul and the transcendence problem[37]

The tenor of the following statement (Paul 1968:1; cf. Lehmann 1966:23 note 4; Goldmann 1966:19): *Die Sprache ist wie jedes Erzeugnis menschlicher Kultur ein Gegenstand der geschichtlichen Betrachtung* is as well-known in linguistic quarters as the Jones-Hypothesis and as an appropriate continuation of the latter Paul draws some of the theoretical consequences of it[38]. He demonstrates how *die vergleichende Betrachtung* (*i.e.* Jones's synchronic comparison) naturally changes into (*verwandelt sich in*) *eine geschichtliche* on condition that the linguist wants to transcend the atomistic level of *Einzelheiten*. If we want to understand (*erfassen*; *begreifen*) how the single phenomenon exists in a coherent whole, we must adopt the historical point of view[39]. Even when we are confronted with a single language (*Entwickelungsstufe*), we shall be able to turn it into a scientific object in so far as we can demonstrate that the meaning(s) – polysemy – of a form point(s) to a *Grundbedeutung* or that related forms (synonymy) point to a *Grundform*; even the very attempt at explaining – in modern terms – synchronic allophonic/-morphic variation (*Lautwechsel*) entails the historical point of view (pp. 21-22). Thus a synchronic object is turned into a scientific phenomenon if it is placed in the horizontal synthesis of a period[40]. Paul cannot accept the synchronic linguist's unmediated jump from the material facts to the immaterial, abstract nature of the formal domain, because *eine Abstraktion*[41] cannot enter into *Kausalzusammenhänge*, *i.e.* into existential or *geschichtliche* connections. It is this aspect of Paul's view of not only language but also of what constitutes a scientific methodology that I shall focus on in the following.

For starters I shall sum up the problem in modern terms: is the synchronic systems concept, implemented by the above-mentioned construction of a system *qua* a finite class of formal elements, empirically relevant, or is the system a hypostatized entity *qua* the result of a process of abstraction[42]? Saussure (1972:25) is aware of the problem, saying that *la langue* is both *un tout en soi* and *un principe de classification*, but provides no particular discussion about it[43].

---

37 The word *transcendence* now refers to the relationship between theory and data.
38 Language is a cultural product, created by the interaction of man's psychical (mental) and physical forces (*Kräfte*) and all *Kulturwissenschaften* are *historische Wissenschaften* (p. 6). Paul, however, does not recognize the sign-functional nature of cultural products (see below), because his main concern is what may be called the Realist Fallacy, which vitiated the then new science of psychology. – Note also that *die Kulturwissenschaft ist immer Gesellschaftswissenschaft* (p. 7).
39 As far as I can see more than a hundred years had to pass for this position to be – if not universally recognized – reinvented: with linguistics being a historical science, *An accurate understanding of language, then, can only be achieved through mastery of historical methods and findings* (Lehmann 1992:ix;23;33; cf. Anttila 1989). Lehmann is being too modest here: knowledge of language (also) requires a historical theory. That such a notion is far from what Lehmann has in mind becomes apparent when he expects phonological theory to advance our understanding of sound change (p. 20; this position is repeated by Trask (1996:xi)). To repeat: a historical theory is a *sine non qua* for the understanding of sound change.
40 Paul's well-known equation of 'historical linguistics' and 'linguistics' is curiously reminiscent of Nietzsche's criticism of *Unsere moderne Bildung*, in which to be *gebildet* is identical with the tautological expression to be *historisch gebildet* (Nietzsche 1966:232).
41 The synchronic state – Lévi-Strauss's bridge – is like Nietzsche's *Blume ohne Wurzel und Stengel* (1966:280).
42 This doubt can be detected in certain sociolinguistic quarters (see below).
43 Hjelmslev, the nominalist, adopts the classification-principle and restricts it to the level of theory, while Saussure tries to make it belong to the level of data in accordance with his *langue-parole* view of *le langage*.

0.2. A bird's eye view of history and change

To construct a historical theory (*eine Theorie der Sprachentwickelung*) we must proceed from *das Wesen der historischen Entwickelung* (p. 5), including *die Bedingungen des Sprachlebens* (p. 6), a point of view *mutatis mutandis* not unlike the one proposed by Hjelmslev (1963:9) about 60 years later. History or to become historical is an acquired property that presupposes a social totality (*Gesellschaft*); and the basis of any cultural product is the principle of *Arbeitsteilung* and *Arbeitsvereinigung*, so that it will be created both through die *Wechselwirkung der Individuen auf einander* (cf. Goldmann 1966:152) and through the interaction between *der einzelne* and *die Gesamtheit*, and in this interaction the individual is both *empfangend und gebend, bestimmt und bestimmend* (Paul, pp. 6-7); finally the interaction is dynamic and periodic, not static and synchronic. Despite his extreme wariness of the transindividual *Gesamtheit*-concept, Paul does accept the structural dynamics that constitute the interaction of the individuals and the created cultural product *qua* a social entity (pp.16-18). However, he constantly emphasizes the irreducibility of the individual; there is neither such a transindividual entity as an ideal speaker/hearer nor a *Volksgeist* (pp. 10; 12-13), entities that do not have *konkrete Existenz*, and between and in which nothing can *vergehen* (p. 11). Paul's point was that the Medieval conflict between Nominalists (Occam) and Realists was not a thing of the past in that closet Realists (*die unbewussten Realisten* (p. 11)) were not an extinct race then, and it is hardly a coincidence that Hjelmslev also felt obliged albeit cursorily to advance a brief mention of the problem (1963:12; 13-15 *et pass.*). Paul's criticism of his contemporaries was indeed so pertinently sweeping that it is a curious fact that his successors did not continue this relevant metalinguistic discussion[44], and Paul labours his point in the following words: *(...) die Zeiten der Scholastik, ja sogar die der Mythologie liegen noch lange nicht soweit hinter uns, als man wohl meint, unser Sinn ist noch gar zu sehr in den Banden dieser beiden befangen, weil sie unsere Sprache beherrschen, die gar nicht von ihnen loskommen kann. Wer nicht die nötige Gedankenanstrengung anwendet, um sich von der Herrschaft des Wortes zu befreien, wird sich niemals zu einer unbefangenen Anschauung der Dinge aufschwingen*[45]. Such totality concepts as classes, species, families are only the products of the synthesizing and analyzing faculties of the human mind, and will always be subject to the arbitrariness of the adopted, dualistic point of view. Only the individual (speaker/hearer) has *reale Existenz* (p. 37). And in sharp contradistinction to the Glossematic hypothesis Paul concludes that the possibility of subsuming *die individuellen Sprachen unter ein Klassensystem* is no a priori evident presupposition (pp. 37-

---

44 Neither Hjelmslev 1963, Jankovsky 1972, Sommerfeldt 1962 nor Weinreich *et al.* 1968 resume the discussion, and it is also absent from such recent books on historical linguistics as Trask 1996; Hock & Joseph 1996; Labov 1994; Lehmann 1992; Hock 1991 as well as Lass 1980 and 1997; and despite his postulate that the Generativist School proceeds from the same predicament as Plato did, namely *Plato's Problem*, the transcendence problem, Plato's ontological dualism *versus* nominalism, is not even mentioned by Chomsky in his 1986-work; and the Generative School's historian Randy Harris (1993) is also silent about this subject-matter.

45 and he continues: *Die Psychologie ward zur Wissenschaft in dem Augenblicke, wo sie die Abstraktionen der Seelenvermögen nicht mehr als etwas Reelles anerkannte. So wird es vielleicht noch auf manchen Gebieten gelingen Bedeutendes zu gewinnen lediglich durch Beseitigung der zu Realitäten gestempelten Abstraktionen, die sich störend zwischen das Auge des Beobachters und die konkreten Erscheinungen stellen*. Modern (synchronic) linguistics has not subjected its systems-concept to such close scrutiny (Lévi-Strauss's characterization of superstructures as *actes manqués qui ont socialement "réussi"* (1962:336) is a pertinent starting-point).

38). Thus Paul is very much aware of the transcendence problem and not only is his criticism of the transcendent element in the then current grammatical terminology (pp. 31;33) relevant today, he also denies the possibility of setting up an empirical grammatical system that does not change constantly (pp. 36;38); therefore there is no unmediated correlation between grammatical categories and the empirically relevant psychological categories in our individual minds (pp. 36;31). The grammatical categories of the descriptive linguist are mere abstractions (pp. 23;26-27) while the *unbewusste Vorstellungsgruppen* in each individual's mind (*ein Organismus von Vorstellungsgruppen*) are psychologically real[46].

What makes the study of language a *Wissenschaft* and what makes it different from other *Kulturwissenschaften*? The necessary circumscription of the object of linguistics that Hjelmslev (1963:19; cf. Jankovsky (1972:150); Andersen's *homogeneous speech community* (1973:767); and §0.0 above) set so great store by in 1943 was anticipated by Paul by more than a half century (pp. 16-17). Furthermore he also anticipated the immanent stand (p. 18) and the dual nature of language: the object of linguistics is not language in its entirety (p. 17), but the particular relationship that a *Vorstellungseinhalt* enters into with certain *Lautgruppen*[47]. Linguistics rests on two *Gesetzeswissenschaften*, without becoming a branch of either natural science (cf. Jankovsky 1972:146-147) or psychology, a dual fact that makes language *das geeignete Werkzeug für den allgemeinen Verkehr* (p. 19). Linguistics is no branch of sociology, either, because *jede sprachliche Schöpfung* is a product of the individual, not his interlocutor(s). But in spite of this (consistently individual point of view) we can – also in linguistics – detect *eine Arbeitsteilung* and *Arbeitsvereinigung* – the prerequisite for any cultural process – in that the speech situation will involve processes in which a speaker's 'sprachliche Schöpfung' is imposed upon (*übertragen*) and then processed (*umgeschaffen*[48]) by the hearer. Paul's conceptual horizon is consistently the empirical period; linguistics deals with processes arising from *das Nacheinanderwirken* of individuals, while in the social disciplines the typical process is a conscious *Zusammenwirken* of individuals which produces the desired product against the horizon of time[49].

As a phenomenon in its own right language is neither physical nor mental, but a cultural

---

46 It is interesting to note how sociolinguistics today casts doubt on the empirical relevance of the systems concept. I am thinking of near-mergers of vowels (the so-called *Bill Peters effect* (Labov 1994:363 *et pass.*)), which Trask (1996:293-294) correctly so interprets that they *call into question* the empirical relevance of the systems concept (*in casu* phonemes): the purported (minimal) contrasts of two systemic units has no psychological reality to the speaker (cf. Labov 1994:20). – Following Paul (1968:31) the synchronic grammatical system has no psychological reality, either, it being manifestations of psychological categories (Paul, p. 263). Hjelmslev (1928:170) argues non-cogently against Paul: a psychological theory rests on grammatical premises. Lass (1997:365 note 31; 370-390) places this discussion in a hermeneutical context, thereby missing Paul's point that the system is always a thing of the past, not a statement or an expression of what is going on 'now'. Paul (p. 351) virtually rejects the empirical relevance of the two correlative concepts, destruction (*Verfall*) and construction (*Aufbau*) of 'entities', as, in modern diction, an evolutionary parameter.

47 Cf. Jakobson (1981 (1949):17): *That mysterious affinity that binds together the sound and the sense*, whose cosmologico(?)-historical match we find in Paul's *Urschöpfung, die uranfängliche Zusammenknüpfung von Laut und Bedeutung* (1968:35).

48 This emphasis on the individual, implying constant change, may be interpreted as an anticipation of the Glossematic connotation.

fact defined by on the one hand the constant interaction of man's body, soul and surrounding nature (p. 6) and on the other *die Wechselwirkung der Individuen auf einander* within a created social *Gesamtheit* (pp. 7;12 note 1). And Paul's view is again consistently periodic (p. 2): the object of linguistics is a synthesis, a complex entity, created by *das Ineinandergreifen der einzelnen Kräfte* and their *stetige Wechselwirkung* (pp. 2-3;7), not a system of elements (Kräfte) established through *ein Aneinanderreihen des Gleichartigen* (*i.e. Kräfte*), because if we emphasize the 'similar' we are bound to end up with abstractions (Hass 1979:120-121).

It has not been my purpose to demonstrate how Paul anticipates many important positions and points of view that we traditionally ascribe to 20th-century linguistics but one cannot help wondering how within the short space of 40 pages of his *Prinzipien* we find the following modern items established against the horizon of history:

1) As a cultural product language is an entity in its own right (whose scientific treatment presupposes a theory of history). – Cf. item 8.
2) The historical series viewed against the horizon of the period is empirically relevant, while the systems-concept is empirically irrelevant because it cannot enter into a causal nexus.
3) The structural principle was foreshadowed both in Paul's general treatment of the three actants in the historical process (any cultural activity), two individuals and their social and physical surroundings, and by his rejection of the atomism of synchronic state-description (Paul, p. 20).
4) The sign-functional nature of language is applied consistently (cf. Paul, p. 35).
5) The priority of the substance is implied by Paul's emphasis on Kausalnexus; therefore the meaning of a synchronic grammatical category is not a fact (*Tatsache*), only an abstraction from observed fact (p. 24; cf. item 2).
6) The individual is an irreducible actant in any cultural process(,) where language is involved, and the individual speaker is therefore the prime mover in the life of any language (p. 13;18); thus Paul is in agreement with modern sociolinguistic views (cf. Milroy 1992 and Labov's asymmetry of production (speaker) and perception (hearer) (1994:352 *et pass.*).
7) Language cannot be studied outside its extralinguistic surroundings (cf. item 3).
8) The methodological requirement of reducing the complex nature of language to a few components (cf. pp. 3;16), which involves the virtual use of the (Glossematic) immanence concept, is also present.
9) The state-concept (*ein Zustand*) is empirically relevant (p. 29 *et pass.*). With Paul a state-description is *einzelsprachlich* (p. 39) and correlated with the individual's *Sprachusus* (p. 29), not with a synchronic supraindividual systems-concept[50].

---

49 Paul does not succeed in working his focus on the individual and his otherwise pertinent observations of the dynamic social interaction of individuals and their 'totality' into a theoretical whole. It is as if he wants to see the simultaneity of the speech situation strung out in a monolinear temporal 'T1+T2' process.

50 This is not to say that Paul's norm concept is without arguable elements. On the contrary, Paul does not seem to escape the same problems as modern systems-oriented linguists are confronted with, namely the step from variant to change; to be precise: the latter operate with a process from individual innovation

10) It is hardly necessary to state that Paul also accepts the notion of constant change: the individual [*Sprachorganismus befindet sich*] *in stetiger Veränderung* (p.27).
11) The transcendence between theory and data.
12) It may be anachronistic to say that Paul also anticipates modern ideas of 'the semiotic process', but he certainly foreshadows the Glossematic *humanitas-et-universitas* goal (Hjelmslev 1963:127): [*die Sprache*] *erweist sich (...) als der Urgrund aller höheren geistigen Entwickelung im einzelnen Menschen wie im ganzen Geschlecht* (p. 20).

This is not to conceal the fact that Paul was caught up by his own strict empiricism: with reference to Goldmann (1966:152) Paul's speaker-hearer interaction (cf. his *Arbeitsvereinigung*) does involve a collective element, as a *Gesamtheit* will invariably be produced that cannot be reduced to either of a speech situation's two participants, and which therefore is transindividual; this, in Goldmann's words, *sujet transindividuel* (or *collectif*) must be (part of) the origin of what is received by and determines the individual; and the fact that no language has ever been seen to lose its collective orientation can be neither comprehended with reference to the individual *qua* an individual nor conceived of in a pure, horizontal causal nexus. I think that Paul would not have disagreed with Goldmann's definition of this transindividual entity in that it varies constantly because there are as many such entities as there are human activities, and the most important point is that it is not a hypostatized entity, the transindividual subject having no independent existence (*réalité propre*) in that it exists *dans les consciences individuelles engagées dans un ensemble de relations structurées*.

Paul did not see that his *bestimmt-bestimmend* interaction is a complex dynamic relation of participation, not only a causal-linear one along the lines of a stimulus-reaction model, in which the speaker 'influences' the hearer, who in turn 'influences' the speaker. The cultural product that is the inevitable outcome of the moment when people – begin to – 'interact' will accompany their continued – be it long or short – interaction and thereby determine their existence[51]. The problem with such a relationship is not conceptual, merely logical because our standard logic cannot cope with a subject that is both *bestimmt* (A) and *bestimmend* (nonA) *at the same time*. This individual-totality relationship may be symbolized by: $I \Rightarrow T(I)$, with T being created ('determined') by I (and other I's), and by: $T \Rightarrow I(T)$, with T both determining I and being a necessary condition for the continued (socio-historical) existence of the individual (see Kant 1923:20-23; Sätze 3-5; Hansen 1983:79-100):

(through spread) to successful supraindividual change, Paul's process is purely individual (p. 34). Thus Paul, albeit implicitly, operates with abrupt state-transitions *qua* changes in the individual's *Sprachusus* (p.34): how *verschoben* (p. 32) must a norm become for it to have changed? And he must also accept a cumulative change-view: (...) *die Summierung einer Reihe* [*von*] *Verschiebungen*, allowed by the speaker's free will, makes Usus 1 change into Usus 2, and on the whole his use of the imperfective-perfective verb *vollziehen* cannot but imply a mutative process and its result, if not also the beginning of the mutation. – Labov's 1994-work is a comprehensive documentation of the impossibility of creating a supraindividual system determining the language of a speech community (cf. Labov's item 3 on p. 359).

51 An answer makes no sense without its question, and vice versa; and in a conversation what was said at T1 provides the continued structural horizon for what is said at T2. Paul's empiricism could not cope with the simultaneous existence of two 'entities' in the same place (cf. 'T2' in figures 0.2.2.1 and 0.2.2.2).

0.2. A bird's eye view of history and change   31

0.2.2.1

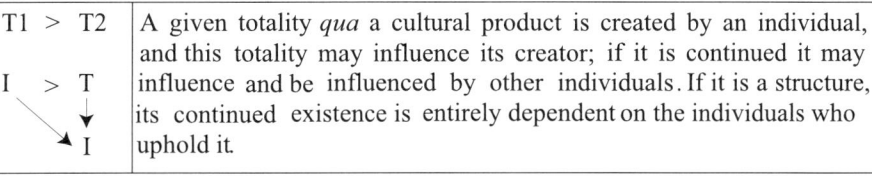

T1 > T2 I > T ↓ ▲ I	A given totality *qua* a cultural product is created by an individual, and this totality may influence its creator; if it is continued it may influence and be influenced by other individuals. If it is a structure, its continued existence is entirely dependent on the individuals who uphold it.

The diagram in 0.2.2.1 highlights the affinity between structure and history and will also capture the linguistic connotation:

0.2.2.2[52]

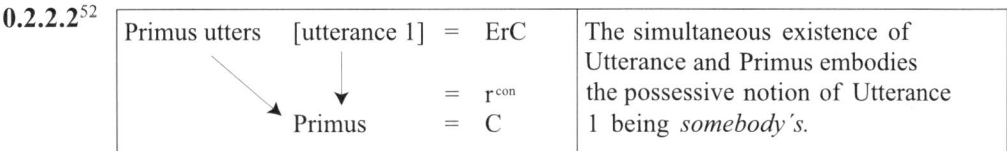

Primus utters [utterance 1] = ErC ↓ Primus = r^{con} = C	The simultaneous existence of Utterance and Primus embodies the possessive notion of Utterance 1 being *somebody's*.

Finally, these dynamic structural relationships also capture Henning Andersen's concept of phonological change, where Grammar 2/Output 1 is both determined (by the Output of Grammar 1/by Grammar 1) and determines (*i.e.* Output 2/Grammar 2)[53].

One cannot help wondering that if Paul had been oriented less towards the *Gesetzeswissenschaften* and had also sought inspiration from the theoretical discussions that then flourished in the historical disciplines proper, the theoretical orientation of historical linguistics might have taken a different turn from what it actually did at the turn of the 20th century, embracing a structure-concept that became static because of the eventual predominance, if not domination of the synchronic point of view in linguistics. Paul did not realize that *Kulturwissenschaft* was a science with its own principles and its cultural products *qua sui-generis* phenomena could be explained differently from their dual (psychic and physical) basis (Paul 1968:6-7). In short, he did not come to grips with the concept of structure which Dilthey had used to remove history from the domain of the natural sciences (Gadamer 1975:210). Structure and system, in the eyes of Paul (1968:3), must primarily have come from an *Aneinanderreihung* of arbitrarily selected data, and the science that created such entities ran the risk of committing the Realist Fallacy: a (static) system with no concrete existence could not control the dynamics (Kausalnexus) of a cultural product. He did not make the final step from his causal-dynamic understanding of, in Gadamer's words, *der Strukturzusammenhang des Lebens* to the empirically relevant idea of a totality-concept, let alone to its sign-functional nature (*ein Beziehungsganzes* (Gadamer *loc.cit.*); cf. Goldmann (1966:154) and his notion of the irresoluble unity of the science of *faits humains* (pp. 18-19)).

---

52 The diagram is strictly speaking only a model of the monologue (*Jede sprachliche Schöpfung ist stets nur das Werk eines Individuums* (Paul, p. 18), and as the speech situation is not a mere sum of two persons' language activity, it must contain a transindividual element (cf. Goldmann *loc.cit.*). – This diagram as well as those related to it will be further developed in 0.3.
53 Below I shall demonstrate how Andersen's model of change is too simple.

Paul's extreme empiricism entailed an awareness of, indeed, alertness to the intricate fallacies of the transcendence problem that prevented him from arriving at a consistently semiotic conception of the cultural process in that apparently he shied away from interpreting it sign-functionally (0.2.2.1 above); this view would have to place the words of our everyday language in a position where we may not seem to have liberated us from their *Herrschaft*. But whatever Paul's *Urgrund* may mean, I shall so generalize his position that his emphasis on the individual person points towards a semiotic[54] interpretation along the lines of the Glossematic connotation (0.2.2.2 above), where the individual becomes the (connotative) content of his own utterance *qua* the expression of that content: C^{con} rE(Er^C). *Urgrund* may therefore be comprehended relationally through the solidarity relation, r, contracted by the sign's *Laut* (expression) and *Bedeutung* (content): *Mensch* and *Sprache* are solidarily interrelated; from here it is but a short step to conclude that the historical process, not to be restricted to change, is sign-functional: language development is sign-functional. Paul has it that a change so completes itself in the individual's spoken chain through his *spontane Tätigkeit, Sprechen und Denken in den Formen der Sprache* (...), that *die Sprache hat sich ganz neu erzeugt und diese Neuschöpfung ist nicht völlig übereinstimmend mit dem Früheren, jetzt Untergegangenem ausgefallen* (p. 34)[55], a view that differs from modern aberrancy-notions.

0.3.0 Three modern theories of sound change: are there *internal laws*?

The five Ariadnian threads will all be broken by the following three synchronized versions of sound change: they exhibit no historical theory; they are heavily idealistic; the chronology does not transcend that of the two-state model; the changes have been lifted out of their *geschichtliche* reality, and the three theories of sound change highlight the (synchronic) tradition's arbitrary view of structuralism's *internal laws* as the prime shapers of a structure's development.

0.3.1 Abductive and deductive change; towards the transindividual[56] and sign-functional

With the synchronic turn that linguistics and language history took at the beginning of the 20th century the systems concept became the first and ultimate arbiter of hypotheses about language change. The synchronic system implies an interpretative alternative: phonetic inno-

---

54 As I am not concerned with semiotics either as a linguistic or as a non-linguistic discipline in its own right, I shall avoid using the word, *semiotic*, and replace it with the linguistic term, *sign-functional*.

55 This process will be called *the present's constant development into the present* and the notion of constant change will be inferred from it. It will be argued that the present's development into the present is always a sign-functional differentiation process based on synonymy and polysemy; Paul (p. 40) also regards his historical process as one of differentiation, but his view is restricted to change proper, not continuation: *Ohne fortwährende Differenzierung kann das Leben einer Sprache gar nicht **gedacht** werden*. (My emphasis; cf. chapter 6.3.); Anttila (1989:398) is reminiscent hereof. – Cf. The Functionalists' notion of *clearer communication*, *expressive needs of the present*, and the requirements of *communication* as in Samuels (1972:2; 136) and Martinet (1962:139) as well as Paul (p. 351).

0.3. Three moderne theories of sound change

vation (as well as the completed change) may be interpreted in terms of the predecessor state's system (S1). Henning Andersen (1973:791) interprets phonological change in virtue of the system that *gives rise to it*. Andersen's view implies a strict systems point of view, seeing that S2, the successor state, originates its own change through the innovations it motivates.

Andersen's theory (model of phonological change) follows Paul (1968:18; 34) in regarding the period when the child learns her native language as the primary *locus* of change, with a kind of *individual* restructuring taking place on the basis of acoustic data; but with its focus on the system it differs from Paul's as well as Luick's: the story, *eine Regelsammlung*, which a two state model narrates, hardly qualifies for a historical narrative[57].

Furthermore, Andersen agrees with the Neogrammarian conception of change in making ambiguity (homonymy[58]) an essential factor in his theory, but they disagree in their emphasis on the respective roles of the acoustic (the hearer) and the articulatory (the speaker) in the change process; finally Andersen's theory implies a degree of conscious alertness towards and awareness of her language on the part of the language-learning child that is hard to substantiate: that a child between the ages of 0 and 3, 5 and 6, should be able to exhibit any degree of comparative abilities vis-à-vis her own individual language and the individual languages of not only her parents, but in principle of all the persons with whom she will be in – regular and stable – contact[59] during these formative years, when she as an individual has to learn to become a social being with all the cultural conventions that such an existence entails[60], is yet to be proved.

Andersen's theory has the internalized grammar (Grammar 2) determine the speaker's spoken substance completely (Output 2), while this person's grammar (= G2) had been de-

---

56 The word *the transindividual*, borrowed from Lucien Goldmann (1966), will be used ambiguously in this chapter: it refers to something that transcends the individual entity (a person or a thing), say, a system in contrast to its elements; in the latter sense it corresponds to the logical term *a class as one* as against *a class as many*. To be precise: as a transindividual entity the system will be regarded as an entity in its own right, not only as a collection of phenomena or events.
57 Andersen agrees with Paul in subsuming language and language learning under the general notion of cultural activity (p. 776), a cultural product is acquired abductively, but against a static *déjà-constitué* horizon.
58 I shall not distinguish between homonymy and polysemy; polysemy will be the linguistic correlate of the nontechnical word *ambiguity*.
59 Andersen has this escape clause: his model only applies to a homogeneous speech community (p. 767); whether this means that his Czech speech communities from the 13th century onwards were completely homogeneous or that the language learning child at T2 has only the output of one grammar (G1) from which to infer her Grammar 2, is not clear. The former is perhaps more empirically relevant than the latter.
60 During the language-learning period the child is also creating and organizing her mental life into a rational existence, a period when she has to learn how to make the secondary processes check and govern the primary processes of her psyche (Jacobsen 1976:34-35). *E.g.*, it takes time for the child to acquire not only a rational conception of reality, but also an empically relevant sense of history (p. 27; cf. Nietzsche 1966:214-215). I am not denying the fact that during these years the child will be experimenting with her language in its metalingual function (Jakobson 1968:356) and will often check her spoken utterances against the languages of the grown-ups sharing her normal everyday life; but that such comparisons should result in the establishment of formal (phonetic/phonological) differences is unlikely. Finally, I fail to see the relevance of Andersen's §3-scenario: why should any responsible parent allow his child to say, *in casu*, [tivo] instead of her parents' [pivo], thereby allowing his child to commit an obvious speech error? Note that Labov (1994:47 note 4) rejects speech errors as a factor in language change. – See note 108 below.

termined by the spoken utterances (Output 1) that she had heard as a(n infant) child, and which was then determined by another speaker's grammar (G1). The state transition from G1 to G2 had then been mediated by the spoken utterances of G1, the relation between Output 1 and G2 being one of abduction (Andersen, p. 767). Innovations will therefore consist not in the speaker's production of a new sound [g2], but in the hearer's interpretation of the expected sound [g1] – articulatorily determined by G1 – as a sound to be determined by /G2/, not /G1/.

Thus Andersen adopts an imperfect-imitation model with the child of T2 being unable to correctly resolve the assumed ambiguity that [g1] constitutes (p. 769). With reference to his Old Czech material Andersen (p. 770) enumerates a number of points on which the language learning baby may make a *wrong decision* as to the number and nature of phonological oppositions underlying the acoustic manifestation of [g1] so that /G2/ will be internalized and not /G1/, this incorrect resolution *being tantamount to a structural innovation*[61]. Andersen does not tell why language learning babies of generation T2 cannot continue the phonological oppositions of G1, which other generations of babies before them had been able to imitate successfully[62]. The rise of the new G2-rule (p. 774) at T2 is temporally *spontaneous* (pp. 766-767)[63].

Andersen's model of phonological change has a logical basis in that the processual relationship between [g1] and /G2/ is one of abduction; Andersen compares the process of change with the following view of how we make correct conjectures about or perceive reality (p. 775)[64]: (a) a fact is observed, (b) a law is invoked, and (c) a noncogent inference is made (a guess). The diagram in 0.3.1.1 formalizes Andersen's examples (pp. 775-776).

The observed fact (a) that Socrates is dead is related to a law (b) saying that all men are mortal, from which we infer (guess) that (c) Socrates was a man. If (c) turns out to be true the law (b) is said to be validated. And it functions as an explanation of the observed fact (a). The syllogism on which abduction is based is a Second Figure syllogism; and the abductive uncertainty follows from the fact that only negative conclusions can be inferred logically from this type of argument (a second-figure syllogism, cf. Quine 1972:89-90):

0.3.1.1	P is M	All men are mortal (b)	M is P1, P2, P3, etc.[65]
	S is M	Socrates is dead (a)	S is P1, P2, P3, etc.
	S is P	Socrates is a man (c)	S is M *invalid inference*

---

61 The word *structure* will therefore apply to the domain of the system, not the spoken utterance. In chapter 4 I shall interpret the child's 'wrong' guesses **ex**istentially.

62 *I do not believe it is realistic or fruitful to declare that there are things human beings cannot learn because they are constitutionally incapable of doing so.* (Labov 1994:342).

63 In his reference to the Late Old English/Early Middle English mergers of Old English /y:, y/ and /i:, i/, Andersen does not consider the Kentish merger of the rounded front vowels with /e:, e/, nor for that matter the substantial complex of mergers that came to characterize this dialect in the 10th century (Campbell 1964:122): /æ:, æ/ (raising), /oe:, oe/ (unrounding), /y: y/ (unrounding and lowering) > /e:, e/.

64 In the light of Labov's near-merger concept, Andersen's perception-based theory may gain renewed relevance.

65 The predicates are *indistinctly recognized* (p. 775), while *M* is a *well-recognized kind of object*, and *S* is the *suggesting object*.

The explanandum is the minor premiss, and it will be explained by the major premiss if the (predictable) consequences (conclusions) of the argument can be verified through inductive and deductive processes. Andersen's abductive relationship between Grammar 1's output and Grammar 2 is based on the perceptual judgments made by the language learning child internalizing what is to become her grammar against the horizon of systemic change; the child acquires her grammar (G2) on the basis of certain observed facts, the *behaviour of the other members* of a given speech community. The inference she then makes follows from the assumption that the observed facts, (a finite, but open group of) [g1], have been determined by a system, and this system can only be 'seen' very indistinctly by her (cf. Samuels 1972:26). In fact this system (not the parents' grammar(?)) is an abstract *capacitas linguæ* that *completely determine[s] the nature of linguistic structure*, and all that is left for the child to do is to implement this *set* of innate '*laws*' (p. 776) or in logical tems, to construct a major premiss. Like Paul, Andersen reduces language acquisition to a *Gesetzeswissenschaft*, although we do not know the specific laws of this *capacitas linguæ*. However, the way Andersen describes the language learning child's efforts to construct an adequate Grammar 2 implies, as indicated above, a degree of metalingual rationality that is incompatible with the mental behaviour of a child between the age of 0 and 4/5[66]. My criticism may be a postulate, but less a postulate than Andersen's description of the child's abductive, inductive and deductive processes (pp. 776-778), and we are back to the problem with the spontaneous rise of G2: Andersen's description does not explain why one particular child all of a sudden begins to be unable to 'correctly' continue a set of phonological oppositions that a child or adult, say, 5 or for that matter 25 years her senior, did continue correctly. That systemic change is realized in the internalized grammar of the language learning child is unlikely unless the theory can explain why one specific child/generation of children spontaneously begins to restructure the grammar of their parents or – to imply an absurd consequence – for that matter older sisters or brothers. The argument backfires: that a child should so perceive and internalize a three-way opposition in her parents' grammar (G1): /p/ :: /p'/ :: /t/ as a binary opposition /p/ :: /t/[67] that while her parents go on distinguishing between (/pi:vo/) [pi:vo], (/p'i:vo/) [p'i:vo] and /ti:vo/) [ti:vo], her grammar (G2) would only produce (/pi:vo/) [pi:vo] and (/ti:vo/) [ti:vo] as its systemic output, while certain lexemes would be realized through the acquisition of an adaptive rule 'some [t] ⇒ [p']', which introduces a sound that the child was not able to perceive during her ordinary internalization process (of G2), is implausible. This adaptive rule would then be imposed by a *responsible* member or members of a speech community and the child, whose *innovations were censured* would of course *heed the corrections* (pp. 772-773)[68]. In short, the odds against a child's successful restructuring of the language (G1) spoken by all other speakers – except perhaps other language learning chil-

---

66 Cp. Andersen's analysis on p. 779 (in particular the 4th paragraph: *A learner* of what age?). Kiparsky (1995:655) follows suit, assuming that a learner will select the variant that *best* conforms to *the language*'s *system*. Below in chapter 4. I shall develop this functional view point into an existential theory of polysemy and synonymy.
67 With the sharped /p´/ merging with /t/.
68 *as a parrot would* (!) (p. 776). I fail to see why a learner, who constantly hears and listens to the output of G1 of her older family members, among which output she would be bound to perceive, in our case, the sharped-nonsharp opposition, should not internalize this opposition in a systemic manner.

dren, whom she will have no contact with – are extremely high, so high, indeed, that the opposite conclusion seems more likely: no systemic change (simplification) occurs while the child acquires her native language, but such changes will be introduced gradually into the languages of the individual speakers as they grow older through devaluation processes (see below). And the reason why languages do not change faster than they actually do is that the degree of variation or change is checked, if not stopped, when the older generation dies. This conclusion may not be an alternative, but a supplement to the standard view[69].

Andersen puts too great emphasis on the ambiguity-aspect; why should the level of ambiguity rise from T1 to T2? Any language-learning child is bound to be exposed to the grammar of their parents; and if their parents were able to perceive and internalize certain systemic oppositions, why should they not be able to do the same? If G3 of the youngest generation is a simplification of G2 of the previous generation, what kind of (complex) grammar (G1) must G2 be a simplified version of? In other words[70], of what grammar is Old Czech, *with phonemic sharpening in dental and labial consonants* (pp. 769-770), /t/ :: /t'/ :: /p/ :: /p'/, etc., a simplifcation and what are – were – the adaptive rules here[71]? Finally, let's look at a morphemic version of this hypothesis: in certain Middle English dialects, in the (east)Midlands, the nasal ending, *-en(n)*, appears as a new plural ending in the present tense of the indicative, instead of the original, inherited one with a thorn consonant, *-eth*. It is more than unlikely that a language learning baby/child should be allowed to say *we riden* while her parents say *we rideth*. Similarly, it is also unlikely that the LateOld English weakening of the same verbal ending: -aþ > -əþ, can be explained *entirely* in virtue of the *linguistic system that gave rise to it* (p. 778)[72].

Andersen's theory does not tell a story, being confined to describing two-state transitions characterized by change (p. 786); hence it cannot cope with continuations. Its basic conception is subsumptive[73]; historical development is negative-destructive – deviations from a norm (pp. 784-786) -, change being the (faulty) result of incomplete learning: G3 is the result of (a child's) incomplete learning of the predecessor system, G2 (the parents'), which is the result of (these parents') incomplete learning of the predecessor system, G1 (their parents'), etc. etc. Whether the change from G1 to G2 is abrupt or gradual, regular or irregular is pseudo problems (p. 787). I agree with Andersen here, but for different reasons: abruptness follows from the applied view of historical development, and so does irregularity, with irregularity being either defined basically in terms of what it is not, namely regularity, or regarded as intermediate steps in the process of change, and therefore subsumed under such concepts as *sound change in progress, a change progresses*, it *runs its course*, has a *disorderly progression* and *has an outcome*. This diction is an invariable consequence of the/ Andersen's synchronic basis of diachronic phonology: the synchronic system (a speaker's

---

69 that only the younger generation is innovative in respect of simplification and restructuring (cf. Samuels 1972:112; 11). Labov (1994) has demonstrated that to regard the young-er/-est generation as the sole innovators is too simplified a view of the change process.
70 Kiparsky (1995) has an argument of the same type (see below).
71 It would have been interesting if Andersen had told a story in Luick's sense.
72 Bynon (1977:5) uses similar diction: *language change operates as if its target had (...) been laid down in advance*.
73 Cf. the arbitrarily selected examples, from Old English to Latin and Russian, etc. (p. 781-784).

Grammar); from the minutest inception of change to its end result its progression will always be governed by the structure of its phonological system: *its end result is already given in the underlying representations* (p. 788; cf. Kiparsky 1995:659). This typically synchronic diction is also a step towards the reification of processes: the nominal construction *its end result* implies the prior existence of the change, the change that has not come into being until it is completed by 'acquiring an end': let 'A > B' represent the change, then *its result* is identical with both the whole processes's, A > B's result and a part of that process, B as the the end of A > B.

A historical theory deals with historical data – obviously. Is Andersen's theory historical and is its data historical? The synchronic tradition has ignored these questions. We may infer *e silentio* that continuations – existence in and through time without change (A > A) – are not historical data; hence no historical theory, nay, no theory at all is required to cope with these[74]. Thus Andersen's synchronic theory does not present us with a generalized view, and therefore no comprehensive theory of language development. Language data assumes the property of being historical when they can be subsumed under the general notion of **a change has taken place**. Andersen as well as the synchronic tradition does not see that this view of history, of the historical process, implies an entity that cannot be reduced to two immanent entities, the linguistic systems of G1 and G2 with their respective phonetic realizations (p. 790) – *parole* manifestations that are (selectively) ambiguous (Andersen) or – from the articulatory point of view – *synchronic variation in the speech of a community* (Bynon 1977:4; Trask 1996:276-285). – I shall clarify my point through reference to Lucien Goldmann's genetic structuralism (1966:151-165).

Synchronic vertical point-in-time relationships so imply a *manifestatio*-ontology that the individual relationship between G1 and its Output is autonomous and instantaneous[75], independent in relation to any other similar relationships. The synchronic datum will then turn out to be historical (cf. Paul) and the historical datum to be truly relational (cf. Luick above) in that the whole process is made up of the horizontal abductive relation between Output 1 and G2 as well as (following Andersen) of the horizontal-vertical adaptive relationship between Output 2 (of G2) and G1 with its output, seeing that the adaptive rule of G2 is created in relation to the structure of G1. What the synchronic point of view is bound to do is to break up this complex transindividual relationship, thereby presupposing that the historical event is a sum of synchronic phenomena, not an entity in its own right. Here I shall replace Goldmann's (socio-historical (p. 19)) *sujet transindividuel* with *transsynchronic phenomenon*. Synchronically, G2 *qua* the result of a change is a transsynchronic entity and will

---

74 Andersen (p. 790-791) confines the issue to a brief mention of what he calls the tradition's pseudo connections (between Output 1 and Output 2, or between G1 and G2 (p. 767)) as against his historical domain, the adaptive relation between Output 1 of G1 and (the internalization of) G2. – Note that his theory does not deal with split changes (*linguistic divergence* (p. 789)), either. – Bynon (1977:2) also misses this point in that she does not envisage the possibility (1) that the reason why a synchronic-based historical theory cannot yield *a true picture of the unbroken continuity of language in time* is that no such theory aims to cope with continuity in its general aspect; and (2) that this virtually universal assumption – that although change is basically an observable fact, the very process *elude*[s] *observation* – is wrong, change being not something to be observed, but comprehended. Finally, how is abduction to be applied to the child's correct guesses (A>A)?

75 The process of manifesting or realizing a systemic element in speech is not a temporal process.

therefore draw G1 into its sphere of transsynchronicity; thus two individual, synchronic facts *qua* Grammar-Output relationships, cannot be understood unless discursively in respect of the historical concept of the transsynchronic phenomenon. From here it is but a short step to this generalization: synchronic facts must be comprehended in terms of a historical theory: speech, communication, the use of the everyday language will always create transindividual phenomena because it involves in Paul's terms *eine Arbeitsteilung* and *Arbeitsvereinigung* (Goldmann's *la division du travail* (p. 152))[76]. I shall demonstrate that the fundamental elements in this transindividual entity are in fact themselves transindividual *qua* transsynchronic entities[77]. And since the transindividual entity, including the internalized grammar of any one language user, is a man-made phenomenon, it is also sign-functional (*significatif*; cf. Goldmann, p. 154). The relationship between this man-made system and its creator (*in casu*, the language-learning child) will be so interpreted sign-functionally that the sign function (ErC) is a connotation[78], in which the creator becomes the content of her language: $Cr^{con}E(ErC)$.

A historical theory constructed on a synchronic basis implies that the process of change is a two-state one with the *structure* of the (phonological) system of the synchronic state in complete control of the process of change. The historical process is reduced to being *diachronic correspondence*[s] *between two consecutive stages of a phonological system* (p. 791) or to *correspondences between consecutive states of a language* (p. 790). However, these three fundamental concepts in the synchronic conception: the state, the stage, and (the structure of) the system, are left more or less undefined. Apparently, *a stage* and *a state* are synonyms with the former being the historian's view of the synchronic linguist's state; therefore the stage concept is not synchronic, but transsynchronic, a state always implying more than a given point-in-time entity. The system is perhaps a more ambiguous concept: if it denotes the purportedly internalized grammar of the individual speaker, it need not be an idealized entity although it will be continued through time and has a particular *structure* that determines its different and various manifestations; in any case, it is a transsynchronic entity. But if the system is an entity that a group of speakers shares and somehow participates in, it becomes transindividual, and its relationship with the language-learning child's internalized grammar is to be established. In other words, the immediate task of the synchronic linguist is to prove that a state, a stage, or a system is not hypostatized entities.

0.3.2 Tenseness constitutes an internal law. The agentless transindividual

Kiparsky (1995:640;660) grounds the process of change on synchronic *articulatory* principles and his reading of the Neogrammarian Hypothesis is sympathetic, but since he changes

---

76 As far as I know no socio-linguistic theory considers these two important processes in the socialization of a variant; see *e.g.* Milroy 1992, 1994, Labov 1994, 1992.
77 In a word, historico-structural or dynamic entities. In this connection it should be emphasized that Goldmann is no closet Realist in Paul's terms, because his transindividual subject has no independent existence outside a given social situation born by at least two individual persons. Cf. Gadamer (1975:210) about Dilthey: *es gibt nicht ein allgemeines Subjekt, sondern nur geschichtliche Individuen*.
78 Cf. Andersen (p. 785): *centralized diphthongs stand for island allegiance on Martha's Vineyard;* Trask (1996:276) also refers to this situation, without understanding the sign-functional nature of sound change and sociolinguistic variables. Other examples can be found in Kiparsky (1995:659).

0.3. Three moderne theories of sound change 39

the Hypothesis into something it is not, a dualist systems-oriented hypothesis, his vindication of it is not empathetic.

Kiparsky's general view of change follows that of our tradition: historical development has no, indeed, cannot have a theory of its own; hence for the puzzling (p. 641) facts of change to be rational(ized), they must somehow comply with synchronic systems-principles; since the mechanical paradigm of the Neogrammarians is no theory, and since the regularity of sound change cannot be denied, sound change must be provided with a proper theory, and Kiparsky proceeds from these three questions:

a) How to reconcile the regularity of sound change, mechanical (articulatory) and autonomous, with the *systematicity of synchronic phonological structure*?
b) With sound changes being in origin *gradual articulatory shifts*, operating *blindly, i.e. without regard for the linguistic system*, why *don't their combined effects over millenia yield enormous phonological inventories which resist any coherent analysis*?
c) Why *does no sound change ever operate in such a way as to subvert phonological principles, such as implicational universals and constraints on phonological systems* (pp. 641; 654-655)?

Questions (a) and (b) are logically redundant statements of the same assumptions: since the assumed effect in (b) is never seen, sound change does not operate blindly; hence there is nothing to reconcile in (a), which has the following logical structure with the implicit and tacit assumption from (b): if sound change is mechanical (blind), then no synchronic phonological structure is possible; but since we do have orderly phonological systems[79], sound change is not mechanical – blind (p. 653). This type of systems-based argument is arbitrary:

Premisses: there is such an entity as a system (= p) [80];
there is such an entity as regular sound change (= q):

if p, then q is not blind, but structure-dependent, *whose alternative is*:
if q is blind, then non-p; but we do have p (non-p is false), hence q is not blind.

My point here is that Kiparsky has no nonarbitrary way of justifying his argument; the language historian may counter that sound change is blind, hence there is no such thing as phonological systems. Then it will be for the nonhistorian to prove the existence of a system[81]. This is not tantamount to denying any notion of a systems concept, neither in the sense of

---

79 Obviously, the system has been hypostatized as a class as one.
80 The purported existence of the system is the sufficient condition (= p) for the nonblindness of regular sound change (= -q), whose synonymous position is: the blindness of regular sound change (= q) implies the non-existence of the system (= -p). In either case, the system, whose existence is to be proved, is taken for granted, and the nature of sound change is also taken for granted in a curious way: as a mechanical process sound change 'should' be blind; but the existence of the system defeats this prejudice: sound change is not blind. The position is odd: sound change cannot be of a particular nature, S, because if it had been S, which it is not, then the empirical state of affairs would not have been Q, which they are, but R, which they are not. This position will be called the WHAT-IF FALLACY, and is closely related to the implicit contingency assumption of modern linguistics (cf. chapter 3.3.10 for a similar type of argument).
81 To be precise: what is the denotatum of the (complex) sign *a(n orderly phonological) system*?

some sort of systematicity nor of some kind of transindividual entity. The incontrovertible fact is the reality of regular sound change, and following Andersen's abduction argument, the abductive conclusion is not that (regular) sound change (S; the minor premiss) is structure-dependent (P) *qua* a consequence of this postulated major premiss: there is such an entity as orderly systems imposing regularity (M), but the conclusion should enunciate the existence of a system. Demonstrating the arbitrariness of the synchronic argument, the historian may construct two equally valid arguments (A and B): with language change being what it is, regular, indeed so regular that no language has ever been seen to lose its transindividual and individual[82] orientation (rationality),

A) we may wonder at the way a given system creates and imposes order and regularity (S is M) and then assume that historical processes create regularity and order (P is M), and thereafter conclude abductively that the system is a historical process (S is P), whose alternative is
B) regularity is created by historical processes (P is M); a system is created by historical processes (S is M), hence a system imposes or creates regularity and order (S is P).

And the question is: why is it so important for the synchronic linguist that the system *qua* a systematic creation must be a transindividual *sui-generis* entity i) controlled by (ahistorical) universals, and ii) controlling language change?

However, the real core of the issue is that the three questions above (a, b, c) rest on the ultimate hidden assumption (prejudice) that not only is historical development irrational, but in particular language change is destructive and contingent. The rejection of this common prejudice accompanied by an empirically relevant conception of the transindividual concept (man-made and *significatif* (sign-functional)) will pave the way for a rational conception of immanent language development, independent of synchronic linguistics. If it can be shown that the notion of structure-dependence is neither an idealized abstraction (Paul) nor inheres in an autonomous supraindividual entity (cf. Goldmann), then the compatibility of exceptionlessness or 'blindness' and structure-dependence (p. 655) has been established with no recourse to Realist entities.

This leaves question (c). Again the existence of implicational universals has been grounded on the above biassed view of historical development (p. 654) seeing that such universal constraints are the objective factors that prevent sound changes from violating *every phonological universal in the book*[83]. The question is, to repeat, biassed: why should sound change behave in an irrational manner? – We need no universal to bolster the systems concept: let the synchronic systems stance first substantiate the empirical relevance of the subversion hypothesis. We cannot proceed **from** the fact that something 'fails' to occur – *cascades of secondary splits* (p. 667) – **to** the empirical existence of the entity which *qua* a cause prevents this

---

82 *Pace* a person suffering from mental or physical language impairment.
83 Kiparsky overlooks the fact that if sound change was blind, then its blindness could not be of such a controlled type that it would steer any sound change away from all routes that seemed to be controlled by various structure-dependences and universal implications. A blind sound change would just as often as not happen to follow a 'rational' trajectory.

0.3. Three moderne theories of sound change

state of affairs from occurring. The argument is doubly fallacious as it wants to have the best of two worlds: first, it assumes that historical development is irrational, destructive, etc.; next it assumes that sound change is not irrational because it is controlled completely by the system (p. 659; cf. Andersen's Grammar) or universals (cf. Andersen's laws of language). After the logical analysis of Kiparsky's argument, let's proceed with a critical analysis of his rationalization of the set of isolative changes that goes under the reifying sign *The Great Vowel Shift*.

Following Labov's characterization of vowel trajectories in chain shifts – tense[84] (Labov: long) vowels tend to be raised; lax (Labov: short) vowels tend to fall; back vowels tend to be fronted (Labov 1994:116) – Kiparsky makes the triggering feature in the *various phases* of the Shift the tenseness feature of the tense-lax parameter.

In the first place, following the diction of our tradition Kiparsky has *the Great Vowel Shift* denote a transsynchronic entity (*a type of natural sound change* (p. 662)) of this familiar structure: it is a phenomenon that has a start and an end connected by internally motivated processes. Kiparsky's theoretical starting point is this so-called *canonical three-height vowel system* (p. 664; with unrounded back vowels omitted):

0.3.2.1

+High, -Low	+Tense	i, u	[ ]
	-Tense	ɪ, ʊ	
-High, -Low	+Tense	e, o	[-High]
	-Tense	ɛ, ɔ	
-High, +Low	+Tense	æ, ɒ	[+Low]
	-Tense	a, ɑ	

Being a so-called *privative* or *underspecified* system, with no redundant feature assignment, only marked features (the right-hand column), this type of feature-assignment is the most economical one to distinguish between three elements.

The Late Middle English vowel system implements it as a four-height system with this *arbitrary* distribution of the tenseness feature (p. 665):

0.3.2.2

(+High, -Low)	(-Tense)	iː: *bite*, ɪ *bit*	uː: *bout*, ʊ *but*	[ ]
-High, (-Low)	+Tense	eː: *beet*	oː: *boot*	[-High]
	(-Tense)	ɛː: *beat*, ɛ *bet*	ɔː: *boat*, ɔ *pot*	
+Low	(-Tense)	aː: *bate*, a *bat*		[+Low]

Kiparsky regards ME /aː/ and /a/ as nontense, and unrounded *because* there are no unrounded back vowels in his system (p. 664). Tenseness was distinctive in the midvowels, and Kiparsky then *assume[s] that all other vowels were nontense* (p. 664). The arbitrariness of

---

84 While Labov (1994) is very wary of tenseness *qua* a systemic feature and how to correlate it with in particular the length feature, Kiparsky works with an unordered definition of the two, it being made up of *duration, height, peripherality, and diphthongization* (Kiparsky, p. 652).

the theory is conspicuous. Why does Kiparsky not proceed from the historical four-height system – because he must adhere to the phonological canon in question – and why do ME /e:/ and /ɛ:/ enter into a paired tenseness relationship while /i:/ and /e:/, /ɛ:/ and /a:/ do not? And as an apparent illustration of the arbitrariness of this systems point of view, we may set up an alternative[85]:

0.3.2.3

+High, -Low	+Tense	i: *bite*, ɪ *bit*	u:*bout,* ʊ *but*	[ ]
	-Tense	e: *beet* (or ɪ *bit*)	o: *boot* (or ʊ *but*)	
-High, -Low	-Tense	ɛ: *beat*, ɛ *bet*	ɔ: *boat,* ɔ *pot*	[-High]
-High +Low	-Tense	a: *bate*, a *bat*		[+Low]

which will be supported by the lengthening products of (nontense) ME /ɪ/ (*wike > weeke* 'week') and /ʊ/ (*wude > woode* 'wood') into (nontense) /e:/ and /o:/, respectively, as well as the shortening product of this nontense /e:/ into nontense /ɛ/, as seen in the modern forms, *meet/met, keep/kept,* and *leap/lept dream/dreamt.* However, that the tenseness feature was neutralized in the short vowels does not explain why this neutralization yields nontense vowels, the historical explanation for this being the general lowering of short vowels in Middle English. Kiparsky then adds 7 diphthongs to the above system, the developments of which are not touched upon.

The process consists of three interconnected stages, a height shift – completed by 1500 – followed by a tensing process – completed by 1650 – followed by a height shift in the 18th century, so that first /i:, u:/ and /e:, o:/ are raised (sic!) to /ei, ou/ and /i:, u:/, then /ɛ:, ɔ:/ and /a:, a/ are tensed into /e:, o:/ and /æ:, æ/, respectively; finally, these new /e:, æ:/ are raised to /i:, e:/.

Obviously, this theory leaves a lot to be desired, explaining only the processual *disjecta membra* (boldfaced below) of the whole system with the continuations unexplained. The basic mechanism (p. 663) is the loss (*unmarking* or *suppression*) of a *marked* feature concomitant with the assignment of default values to the vacated features (Kiparsky's note 26). This involves a peculiar type of raising-process: ME /e:/ [-High],[+Tense] > /e:/ [Ø], [+tense] > /i:/, because the result of the unmarking yields the default value […], which because of the tense-feature yields a raising of /e:/. The question is: what actually effects the new sound /i:/: automatic default-value assignment: […] → [+High] or the feature [+Tense] alone? On p. 666 Kiparsky's formulation may suggest a reversal of the causal chain: ME /e:/ and /o:/ were raised, thereby losing their two ME features. Similarly, it is not clear whether /i:/ and /u:/ were diphthongized because of the *activation* of a default rule: [-Tense] → [-High], or this rule-activation caused the diphthongization.

---

85 If the triggering feature is still tenseness, the system would now imply a drag-chain version of the shift with diphthongization being the first change, and the short *i*- and *u*-sounds should be paired with nontense /e:/ and /o:/.

0.3. Three moderne theories of sound change

0.3.2.4

1400	feature	>	1500 raising	feature	>	1650 tensing	feature	>	18th c raising	feature
i:	-tense		ei	?-High		ei			ei	?
u:	-tense		ou	?-High		ou			ou	?
e:	+tense		i:	-tense		i:	?+tense		i:	+High
o:	+tense		u:	-tense		u:	?+tense		u:	+High
ε:	-tense		ε:	-tense		e:	+tense		i:	+High
ɔ:	-tense		ɔ:	-tense		o:	+tense		o:	-High
a:	-tense		a:	-tense		æ:	+tense		ε: / e:	-High
a	-tense		a	-tense		æ	+tense		æ	+Low
System 1			System 2			System 3			System 4	

In the next process tensing does not seem to accompany any feature-loss in the open *e*- and *o*- sounds, these preserving the [-High] feature, but follows from the arbitrary implementation of another default rule which makes all long vowels tense (p. 664), including short /a/ > /æ/, this in contravention of another default rule which makes all short vowels nontense (p. 664). In the third stage, only certain tense front vowels, /e:/ and /æ:/ are raised, losing their respective features, [-High] → [Ø], /e:/ thereby becoming /i:/, and [+Low] → [Ø] (= [-Low]), /æ:/ thereby becoming /e:/.[86]

In addition to what has already been said, other objections to Kiparsky's narrative may be advanced: (1) why ME short *a* is part of the theory is odd seeing that the lowering of nontense ME /ʊ/ ('but', 'cut') is not; (2) that the diphthongization of /i:/ and /u:/ is a raising process is odd[87], and that it must be explained in the same way as the raising of /e:/ and /o:/ is no less odd; (3) the alternative system in 0.3.2.3 could be characterized as a consistent tensing process; (4) that /i:/ and /ɪ/ of system 3 are tense and nontense, respectively, is odd, seeing that ME /i:/ and /ɪ/ were both nontense (all short vowels except /æ/ are nontense); (5) no explanation of why /o:/ is not raised by the final tensing process, and it is impossible to see the eventual merger of ME /e:/ and /ɛ:/ as being due to the *selection* of *those variants which best conform to the language's system* and which *respect structure* (p. 655).

Kiparsky's theory is not a part of a narrative of English phonology; the theory is an idealized universal phonology which is manifested by a given phonology (LateME); it does not distinguish between cause and theory proper because the arbitrary distribution of (tenseness) properties over the historical system entails the cause(s); it simplifies the available empirical data[88]; it creates arbitrary breaks between the undefined systems or states; the sign *the Great Vowel Shift* has been arbitrarily so limited as to exclude the diphthongizations of /e:/ and /o:/ as in *way, day, name, take*; *change*; *home, bone, stone*, as well the continued history of ME /i:/ and /u:/ into *ai*- and *au*-diphthongs, respectively.

---

86 Processual gradualness is thus determined by the system.
87 Kiparsky explains the process by the *activation* of a *universal default rule* (p. 666) of this type [-Tense] → [-High].
88 *E.g.*, the system of the 16th century includes an open *o*-sound as in *taught, fought, paunch, bought*, as well as a low, long and unrounded *a*-sound as in *demand*; cf. the alternative in *staunch*. The monophthongizations of the ME diphthongs seem to have no influence on the processes. – The problems with the *great, steak, break* group are explained as contact influence from other systems which obeyed different internal laws – which ones?

This systemic explanation demonstrates (i) the nonstaticness, the historical nature, of phonological systems, as well as (ii) the adhockery of this type of explanation[89], and (iii) the unhistorical nature of the processes: nowhere does Kiparsky attempt to implement his general view of sound change – which hardly anybody will disagree with – that sound change originates through *synchronic variation in the production, perception and acquisition of language* (p. 653-654).

0.3.3   The agentless transindividual

Systemic explanation finds the cause of change within the system itself, in one of its elements, arbitrarily selected by the theory (linguist)[90]. This change-triggering element stands out against all the other elements, creating some sort of dissonance, variously called asymmetry, imbalance, markedness, something that is not optimal so that the consequent change can redress this imperfection. Kiparsky set great store by (the ?nonsymmetrical application of) the feature of tenseness within a universal phonological system, while Anderson (1992) works within a universalist conception, too, now implemented by a notion of *system geometry*. This geometry governs the selection of a vowel, the *unspecified* one, which is to play the triggering role in a change. While Kiparsky's version involves the uneven distribution of the [+/-tense] feature, Anderson appeals to the concept of symmetry so that the odd vowel out in relation to an abstract symmetrical system is the unspecified one; *e.g.* in Modern English /ɔː/ *law, four* enters into no symmetrical relationship with other vowels and is therefore isolated, a result from *the asymmetrical application of the GVS* (p. 105) and caused by the unspecified vowel at the start of the Great Vowel Shift.

In order to bring out both the Realist element in our two versions of the Shift and the arbitrariness associated with systems explanations, I shall compare the two causal chains, including Kiparsky's and Anderson's views on the historical process itself.

In Kiparsky's system the causal chain of the Great Vowel Shift is *triggered by both distinctive and nondistinctive tenseness* (p. 666): (1) ME /eː, oː/ lose their [-High] feature, while keeping their redundant feature [-Low]; (2) because of the system's default rules, tense vowels will raise, being assigned the default value [+High], > /iː, uː/; (3) this activates another default rule making /iː, uː/ diphthongize by unmarking of their [-tense] feature so that they are assigned the feature [-High]; (4) the tensing stage (all vowels being nontense) will be activated by a default rule making all long vowels tense, (?)and all short vowels nontense (p. 664)[91]; (5) this tenseness *feeds* the third stage of the shift, and /eː/ and /æː/ are therefore raised to /iː/ and /eː/ through the unmarking of the respective features [-High] and [+Low],

---

89   Despite his conclusion (p. 666) Kiparsky cannot distribute the tenseness features at the phonological level nonarbitrarily or cogently.

90   In the preceding section Kiparsky's theory could not decide cogently whether the Great Vowel Shift had a top-start (Jespersen's view (1965:233; cf. Strang 1970) or a mid-start (*e.g.* Luick and Lass 1992).

91   It would not have been unnatural if this tensing process had been applied in connection with the Middle English lowering of short vowels, which made short and long vowels qualitatively distinct: OE *feld* /feld/ > /feːld/; ME *stelen* /stelən/ > /stɛːlən/; OE *climban* /klimban/ > /kliːmban/; ME *wike* /wikə/ > /weːkə/. – The nontensing of short vowels is redundant in respect of the short *i, u, e, o*-vowels, and seems not to apply to /æ/, which remains tense. Therefore it is not correct to say that <u>tenseness is entirely predictable</u> (Kiparsky's emphasis, p. 666); cf. next note.

0.3. Three moderne theories of sound change

with the resultant merger of ME /eː/ and /ɛː/ left unexplained and with /æ/ left dangling as a tense short vowel[92].

The idealist fallacy consists in Kiparsky's extensive use of agentless passives and nomina actionis[93]; there is no attempt at placing the processes within a coherent time- and space-continuum (cf. note 39 above). Anderson's causal chain is perhaps an example of the ultimate reification of a concept:

> (…) various asymmetries in the application of two major historical processes ("sound changes") to the phonological system of English, specifically the vowel system, can be attributed to the selection of unspecified vowel at the period concerned, and specifically to interaction of (the phonologisation of) these processes with that selection (p. 104; cf. p.111).

Where and who are the agents of *application of*[94] (the GSV); *attributed to*; *selection of*; *unspecified* (vowel), as well as of *interaction*. Let's rephrase Anderson's processual description into proper dynamic terms:

> "The fact that X SELECTS a vowel (*in casu* ME /ɛː/) which Z does not SPECIFY (within the given theory), and (the fact) that the GVS INTERACTS with X SELECTING this vowel at T1, somehow determines/governs the way a historical process (the GVS) is APPLIED ASYMMETRICALLY BY Y at T1 to the vowel system of English[95]."

Like Kiparsky's three-state process, Anderson's reified transsynchronic process – as well as sound system – has not been placed in an anthropocentric context:

1) The vowel system at T1, being the input to the change (A > B, the GVS), is an entity in its own right – existing independently of the sound change and speakers.
2) The vowel system at T1 is – must be[96] – in a state of nonsymmetry.
3) As Y applies the GVS to the system at T1, the sound change is also turned into an entity in its own right. – Note that *the GVS* refers to a set of categorical or conditional instructions to *each input vowel* (pp. 106, 109, 111)[97].

---

92 Why the unmarking of the tense [+Low] vowels does not hit the short tense /æ/ is not explained, unless a default rule makes it lax; cf. previous note. Kiparsky's selection-process at language transmission is not made to connect with the systems explanation: the selection process necessitates the presence of variants, variants that have no role in the explanation proper.
93 This is a kind of *syntactic exploitation* that serves many purposes, least of all explicitness (Bolinger 1973:543-544).
94 Hock (1991:165) also talks about an *irregular application of otherwise regular sound* changes, with no agents (see Keller 1985:216). And his idealism is so extreme as to ascribe historical reality to processes that appear to have the property of regularity.
95 This explains the asymmetrical nature of the modern English long vowel system with the vowel in *law*, *caught* etc. left dangling outside symmetrical constrastive relationships.
96 Or: the system *qua* an entity *has* this property, which follows arbitrarily from the theory.
97 In other words, Anderson makes the sign *The Great Vowel Shift* the expression (E) of a content (C) which is a complex sign (ErC), the instructions, *i.e.* a historical narrative.

4) The unspecified vowel assumes an empirical status different from all the other vowels of the system.

Anderson's other example is the 'failure' of West Saxon /æ:/ to be *I*-umlauted (p. 113). The reason for the continuation of PreOE */æ:/ into OE /æ:/, unaffected by a following *-*i/j*-sound, is that this vowel was the unspecified vowel in the PreOE vowel system; its component-marking was systemically nil, {...}, and therefore contained no components *I*-mutating factors could hit. But that an unspecified vowel takes no part in a given change is inconsistent with the triggering role of other unspecified vowels: the nil-specification of PreOE */æ:/ makes this vowel *invisible* to the umlauting process, and the nil-specification of ME /ɛ:/ does not make this vowel invisible to the Great Vowel Shift. Anderson is aware of this systemic discrepancy and explains it away by an ad hoc 'principle' (p. 113).

Anderson's application of this theory to the so-called input system relies on an after-the-fact distribution of features, this being particularly interesting both from a synchronic and a universalist point of view, as the synchronic/universalist interpretation of the ME synchronic input system will thus observe the material nature of the historical facts and be a transsynchronic process.

As regards the Great Vowel Shift, it is first to be noted that Anderson's system yields two versions, one with *an "unprincipled" selection of unspecified vowel* (p. 108) or one with an analysis of the symmetrical Middle English vowel system that arbitrarily turns it into an unsymmetrical one. The asymmetry of the otherwise symmetrical LateME vowel system is arbitrarily established by the open *e*-sound /ɛ:/, so that the back vowel in ME *bone* {a;u} is the congener of /a:/ ME *name*, now a pure front vowel, characterized by vowel-colour: {a;i} instead being neutral {a} as in the symmetrical system[98].

Secondly, that /ɛ:/ in ME *strem* 'stream' (< OE *stre:am*) is phonologically different from /ɛ:/ in ME *great* (< OE *gre:at); breke* (< OE *brecan*) is empirically irrelevant to its synchronic function[99]. Thirdly, deviating from Dobson's and Samuels's dialectal views of the

---

98 Anderson's arbitrariness flouts historical facts and scientific stringency: the Cardinal/Height-symmetry of ME vowels is broken: the height-symmetry of EarlyME /ɛ:/ and /ɔ:/ is corroborated by the merger of these originally long vowels and the lengthening products of the respective short ones; furthermore, both vowels follow parallel routes in their respective (*solidary*) developments, namely from /ɛ:/ and /ɔ:/ to /e:/ and /o:/. I agree with Anderson (p. 108) that ME /a:/ has a fronted and slightly raised quality (= {a;i}) rather than a pure low centralized one (= {a}). However, Anderson's analysis is inconsistent in relation to his analysis of both PreOE */a:/ {a} (p. 112) and of the modern low unrounded vowel /a:/, this being characterized by only *vocalicness* (= {a}), an analysis which yields the presupposed result, isolation of /ɔ:/ in the modern system (p. 104). Gimson (1994:106-107) classifies it as a back one – or as a front one (north of England). So Anderson's geometrical analysis of the modern back vowels /u:/, /o:/, /ɔ:/, /a:/ will yield symmetrical contrastive patterns. Note that his system with its three components acuteness {i}, gravity {u} and sonority {a} and combinations thereof seems to be bound to create an asymmetry when specifying four elements of the same category. On the parallel of his asymmetrical analysis of the ME front vowels /i:, e: ɛ: a:/ → {i, i;a, i:a, a;i}, we may analyse the modern long back vowels as follows: /u:, o:, ɔ:, a:/ → {u, u;a, u:a, a;u}. Like Kiparsky's arbitrary assignment of the [+/- tense] feature to ME /e:/ and /ɛ:/, Anderson has no objective basis for letting ME /ɛ:/, and by implication modern English /ɔ:/, differ from the other long vowels in their resepctive series. – Finally, note how /æ:/ of the West Saxon dialect is made the origin of the standard language's ME /ɛ:/ (p. 112).

99 Events posterior to a synchronic system are not relevant to this system.

whole process from LateME into the 18th century Anderson unempirically conflates dialectal developments into one system (p. 110). Fourthly, the single processes in Anderson's system (p. 110) do not follow their well-documented individual (gradual) courses in that his system merely considers the start (1400) and the presumed end (1700) of the GVS[100]: the raising of ME /e:/ and /ɛ:/ occurs simultaneously, yielding /i:/, while /ɛ:/ in the minority subset of long open *e*-sounds develops into /e:/[101]; similarly ME /i:/ and /u:/ diphthongize into /ai/ and /au/ rightaway, with ME /ai/ *day* (with ME /a:/ in *name*) and /au/ *law* simultaneously developing into /e:/ and /ɔ:/, respectively[102]. Thirdly, by his identification of the unspecified vowel in ME with the purportedly unspecified vowel in a PreOld English vowel system, /æ:/, (p. 110), which develops into West Saxon /æ:/ (= Non-WS /e:/) as in *mære* 'famous', *læce* 'physician', Anderson establishes an empirically relevant link between two historical phenomena separated by 700/800 years, phenomena whose lexical manifestations indicate their disparateness: OE *mære* (WS)/*mere* (Non-WS); OE *great, stream* (/e:a/); OE *brecan, stelan* > ME *breken, stelen* (e > ɛ > ɛ:), where only the first sound is unspecified in the PreOld English sound system. That these two *pre-GVS* sounds have been lumped together can only be justified *a posteriori*, while any synchronic analysis of the ME vowel system would have the same vowel, a long open *e*-sound, in words such as, *gret* 'great', *strem* 'stream', *speke* 'speak', *breke* 'break', and *stele* 'steal', *ded* 'deed' (*pace* dialectal differences), and synchronically Anderson has no means of singling out the words that had the unspecified vowel in PreOE from the class of ME words with /ɛ:/[103].

Anderson's analysis is a clear instance of how the systems point of view manipulates, if not abuses the historical data: there is no historical relationship between (Pre)OE /æ:/ and LateME /ɛ:/, let alone WS /æ:/ and the development of ME /ɛ:/ in the standard language; synchronically, structurally and dialectally, they are different; finally, Anderson displays no sense of history or historical directionality: theoretico-systemic needs preponderate over the periodicity of the historical facts. – After these negative analyses I shall put my critical analyses in a constructive setting in the following sections.

0.4.1 *Vivo, ergo sum*: the existential function of our everyday languages and the historical nature of synchronic parameters

The nomenclature of historical linguistics is, if not flowery, then loaded with various figures of speech (cf. 0.1.). Historical development and in particular sound change are **blind** (*aveugle*), **spontaneous**, **isolative**, **mechanical**, a sound change and its **incidence** are **inexorable** and **necessary**, and **exceptionless** or **conditioned**; a change is in **need** of **repair, destroys** and **redresses** some kind of **balance** or **symmetry**, is **therapeutic** and effects **ho-**

---

100 Anderson, however, must be credited for trying to incorporate the histories of the ME diphthongs into his system.
101 These 'two' synchronically identical phonemes are defined differently by the theory: the majority sound is specified by nil {...}, the minority one by {i;a}, while the specification {a;i} is reserved for ME /a:/, a specification that happens to apply to (Pre)OE /æ:/.
102 Anderson's two-state model is empirically irrelevant.
103 One needs hardly invoke the principle of the irreversibility of mergers.

meostatic regulation; in spite of this sound change is **regular**. Each of these terms, more negative than positive in tone, refers to particular aspects of the historical process, but they do have one thing in common: each of them implies that a theory of language development is **impossible**, a theory of history is a **contradictio in adjectis**: historical development and the historical phenomenon or event are **irrational**, man's very existence is a contradiction[104]. Furthermore, the historical domain is **contingent**, **accidental** and **haphazard**, in short: it need not be what it actually is, it could have been different, something else could have happened instead of, say, the French Revolution, the Great Vowel Shift, the Second World War, or OE /a:/ > ME /ɔ:/ or OE /-an/ > /-ən/. And the crowning commonplace of them all is the fact that the everyday language changes constantly, is *always* in a state of flux (*e.g.* Samuels 1972:2; Jespersen1963:31)) because time changes everything (cf. Saussure below in chapter 3.4.), often accompanied by an **It-can't-be-helped**. And why should it!

These characteristica agree with the imperfection-point of view (0.1.)[105] and the historian will try to reconcile the relevant ones with the constitutive factors in man's historical existence, an enterprise that will include a proper understanding of the interrelatedness of the fact of (constant) change and traditional synchronic functions of language (Jakobson 1968:353-357).

---

104 Unless (a kind of) rationality *qua* some kind of *order and connectivity* is imposed on them, historical phenomena and events are *irrational* (Lass 1992:152). Thus the inexorability of change – with its regularity – that inheres in the historical phenomena cannot be rational; this imposed rationality, which in the final analysis springs from the systems concept, will be called *modus-ponens* explanation: Lass (*loc.cit.*) is an obvious case in point: the Great Vowel Shift is so rationalised by the logical If-Then implication: NOT H(igh) V(owel) D(iphthongization) WITHOUT M(idvowel)R(aising), that what is regarded as the motivating factor (the first change) becomes a necessary condition for the following change: if ME *i:* and *u:* change (diphthongize), then ME *e:* and *o:* must have changed (risen) first. However, Lass's logic implies an equilogical, but oddly negative hypothesis: absence of MVR is the sufficient condition for absence of HVD: if there is no ME *o:* to change (rise), then ME *u:* does not change (diphthongize) first: as the raising of ME *e:/o:* is interpreted as the cause of the diphthongization of ME *i:/u:*, so the non-change of ME *u:* is interpreted as the cause of (?)the absence of ME *o:* (in the northern dialects). There is no cogency in either Lass's or Luick's defense of the operation of a push-chain mechanism, as against Jespersen's drag-chain one (Luick 1964:§482 Anm.1; Jespersen 1965:232-233), to which must be added Luick's own observation that in a border region – the north of Lincolnshire – we do find a merger of ME *u:* – the northern feature – and ME *o:* (> *u:*) – the southern feature. Furthermore, Luick (§517 Anm. 2) has an odd argument to the effect that in dialects with absence of ME *u:*, and hence with absence of ME *o:* (> 15th c. *u:*), the ME *ou*-diphthong rises and monophthongizes into an *o:*. Luick is thereby forced to operate with two causal mechanisms in the Great Vowel Shift: "o: > u:" is the ('pushing') cause of "u: > ɔʊ" and is the ('dragging') cause of "ou > o:", which merges with ME ɔ: in *home, stone*. Stockwell and Minkowa (1988:360-361) reject the above critical argument, orginally advanced by Western. But Luick's own counterargument is weakened by its very nature: the exception (borrowing from the north) being in a border- or transitory region. – At the existential level, man's *Dasein* [*ist*] *ein Ding, das davon lebt (...) sich selbst zu widersprechen*, and the reason for this is man's inability to exist oblivious of the past and the future (Nietzsche 1966:212); the necessary empirical relevance of the period concept has been asserted. – The opening statement, written in 1922, in Jespersen (1969:7) is virtually the linguistic translation of Nietzsche: *a language or word is (...)* [*the*] *result of previous development and at the same time the starting-point for subsequent development* (my emphasis); and Jespersen could even return to Paul: the science of language is historical.

105 Cf. Nietzsche's *ein nie zu vollendendes Imperfektum* (p. 212).

0.4.1 Vivo, ergo sum

The synchronic conception is logically static, and there is a static element in historical phenomena; the instrumental functions of language also lend support to the static or unhistorical part of language, but language is also socio-culturally dynamic and creative, a fact that lies behind Goldmann's view of the sign-functional nature of any *fait humain*. Thus a reconciliation of language's instrumental (communicative) functions and creative function, a synthesis of language's metalinguistic and object functions is to be attempted. The first step towards this goal will consist in the finding of positive traits in the fact that any language develops constantly, characteristica that will counterbalance or even 'disappear' the negative ones of our tradition.

When subserving man's life, and controlled by, not controlling man's existence, history is useful (Nietzsche 1966: 217-219). This positive and necessary aspect hinges on the separation of man's existence from man's knowledge (*Leben und Weisheit*) in the sense that the value of knowledge must be measured against its service to life, and the historical point of view does not hold the final and complete answer to historical phenomena, including man's existence (pp. 230-231). However, it would be anachronistic[106] and indeed a mistake to transfer Nietzsche's criticism of his contemporaries' belief in the historical point of view's cognitive-epistemological monopoly to the situation in modern linguistics: Nietzsche's opposition between the historical and the unhistorical cannot be mapped onto the modern history-synchrony dichotomy, because *the historical* is as necessary as *the unhistorical* (p. 214), but the interesting point is that much of Nietzsche's criticism of the historical point of view may *mutatis mutandis* be levelled at modern synchronic linguistics[107]: it is hard to see how the modern view both of language itself, in its object and metalinguistic functions, and of knowledge of language is a positive and dynamic contribution to man's existence, with its fundamentally defensive and limiting WHAT-IF approach to language: what if language and *la condition humaine* had been 'different', perfect – something that neither is -, and not a Nietzsche-like *imperfectum*. The opposition between what man's existence is and what it is not is a false one, expressing an empirically irrelevant alternative; with this alternative gone, the imperfection-perfection parameter or traditional horizon against which historical

---

106 This may not be quite true, in view of the way modern synchronic linguistics treats or rather selects its historical phenomena; this approach is certainly reminiscent of Nietzche's *monumentalische* function of history (p. 230), and Anderson's use of his data (PreOE /æ:/ being identical with LateME /æ:/, and Modern English /ɔ:/ being explained in terms of the GVS) is not a far cry from Nietzsche's *antiquarian* function of history; Lehmann (1992:184) is another case in point: because of their Indo-European origin as lenis (voiced) plosives */b, d, g/, the modern aspiration of unvoiced plosives [ph-, th-, kh-] must have occurred after Grimm's Law. Such a conclusion is only legitimate if Lehmann has a historical theory, which explains the continued existence of IE /b/, etc. in a developmental series spanning the period from 3,000/4,000 B.C. to *the present*: IE /b/ > ProtoGermanic /p/ > /p/ > … > OE /p/ or /ph/ > /p/ or /ph/ > … > PE [ph-] (cf. Nietzsche, chapters 2 and 3). The historical theory will be the historically and theoretically correct (counter)answer to Nietzsche's correct criticism of this use of history, which is not in the service of life.

107 Nietzsche's emphasis on the unhistorical aspect in our existence is due to the exclusive dominance of the historical point of view in his days, which naturally entails that it is the historical that is to be delimited and controlled; in our century, we may make a conceptual U-turn: the overdose of synchronicity destroys life; it must be delimited and controlled by a superior power!- Note that Nietzsche also distinguishes between *Leben*, *Historie* and *Wissenschaft*, i.e. scientific history (p. 231).

existence is seen, will be gone, too, with all its concomitant negative attitudes towards history or sensuous reality. The generativist view, called Plato's Problem (Chomsky 1986), will therefore also vanish into a subjective, value-loaded metaphysical statement: that man – the language learning child – is able (to come) to know so *much* as he or she actually does despite the poor in-put material he is exposed to is a postulate that is beyond scientific corroboration: how is *much* defined? How does the generativist know that we do in fact know 'much'? How do we know that sensuous reality is a vale of misery?

There is no opposition between man's existence and history, no opposition between language and history; if we take language into the service of man's existence (historical development), then we shall be able to arrive at an empirically relevant conception and understanding of language history. The everyday language is both **instrumental** and **creative**: it is an indispensable means and integrated component in our everyday lives and scientific lives. It is a means by which we communicate (Jakobson 1964: 353-357), it is also a means by which we exist, not unlike Nietzsche's *vivo, ergo sum*, and it is precisely this existential-creative[108] function that our linguistic tradition has overlooked, perhaps because it is a **participative** relationship. In its metalinguistic function, the everyday language **becomes** the **instrument** by which it is described: Meta-L(=O-L) describes O-L; as an object language it **becomes** a **means** to an end, namely knowledge of language. Thus knowledge of the object language is also knowledge of the metalanguage; in its metalinguistic function the everyday language **becomes** an **instrument** by which man creates something, and this something is also a component of the metalanguage: Meta-L creates knowledge of O-L $\Rightarrow$ Meta-L(= O-L) creates knowledge of O-L. This argument leads to the **cognitive** function of the everyday language, its *Erkenntniswert*[109]. Its consequences are (1), since the everyday language conveys knowledge of reality – not all our knowledge of reality – since the everyday language is a constituting element in man's historical existence – when we constantly exist out of the present into the present – and since constant change is an empirical universal, man's historical existence is a cognitive process; and (2), since man's development into the present (existence) in-

---

108 I shall restrict my use of the word *creative* because of its overt relationship with the poetic/literary functions of language; but when used, it is synonymous with *existential*. – Modern synchronic linguistics – the Generativist School (adopted by Bache and Davidsen-Nielsen 1998:22) – is impressed by the fact any child will acquire a native language. It would have been more 'impressive', if a child could avoid learning a language – and the grown-up to forget his native language. Secondly, the modern point of view is fallacious; it follows from the unsubstantiated postulate that the spoken language which the language learning child is exposed to is imperfect and deficient. Therefore it is a tremendous feat that the child performs. However, the spoken language is only imperfect seen against an entity that is nonexistent; if the spoken language were something else, it would not be what it is, the foundation on which any child builds her language. So the logical fallacy looks like this (to labour what was said above in 0.3.1.): if the 'imperfect' spoken language is the sufficient condition for the learnability of a native language, then what cannot be learnt is a spoken language that is not imperfect, the perfect language becoming a necessary condition for the acquisition of a native language! Thirdly, any child can learn any language; if this is so, there is no reason why she should not be able to learn the language of her parents: why should she change or restructure her parents' language? If she alters the language she is exposed to, there is only one sensible reason for that: it helps her to survive, less dramatically, to exist. – This will be elaborated in chapter 4, where an existential theory of synonymy and polysemy will be set up.

109 which logicians deny it, precisely because of its *distinctive feature* (Jespersen *loc.cit.*), historical nature, replacing it by logico-mathematical formal systems, whose cognitive value is not questioned.

0.4.1 Vivo, ergo sum

volves knowledge of this process – *vivo, ergo cogito* – and since man's development into the present is change – *vivo, ergo sum*, change is a cognitive process, involving knowledge. From this the notion of man having innate – or otherwise – knowledge of his own language will follow: when a person speaks, he will not only produce an utterance (or sign) at T2 that is different from the – perhaps same – utterance (or sign) he produced at T1, but the latter will participate structurally in the T2-utterance: modern English [phone] produced at T2 acquires part of its content from the fact that the speaker *knows* the older form [telephone] and establishes a structural relationship with it: he carries with him what is necessary from or of the past as a constituent part of the present through his historical sense: the interaction of man's ability to 'remember' - and to 'forget' – and to produce an utterance in the present will therefore involve a *transsynchronic* metalingual function (cf. Jakobson, p. 356)[110].

This view of change makes me disagree with Labov (1994:9) and agree with Milroy (1993:215). Maintaining that language change is *too obvious to be recorded or even listed among the assumptions of* linguistics, the fact of *language change being a given*, Labov confuses the perceptual immediacy of transsynchronic or transindividual variation and differentness – both on the synchronic axis: Primus's language [ε] (Refined RP/Cockney/Australian) differs from Secundus's language [æ] (RP), and on the diachronic axis: Secundus's [æ] (RP) is older than Tertius's language [a] (*younger speakers of RP*) – with what is to be established/answered, namely (the question of) historical links: is RP [æ] the origin of the other sounds [ε, a]? That a language has changed is knowledge mediated by a historical theory, a language change is not a perceptual or sensuous fact, but an event that is comprehended[111]. This is the reason why Milroy is not far from the truth when he concludes that *human beings* [must] *attach great importance to changes* (…), otherwise *they should not implement them at all*. Changes are necessary and unavoidable. Needless to say that I disagree with the way Milroy arrived at the above[112].

In its **existential** function the everyday language becomes an instrument by which man creates something of which the everyday language becomes a constitutive part. The existential function of the everyday language is simply man's constant development into or creation of the present: man's development into the present is a cognitive process because it is *significatif*, sign-functional, a semiotic process. This process is historical; therefore it is perfectly sensible why Nietzsche distinguishes between the historical and the unhistorical, while at the same time acknowledging their inseparability (cf. Nietzsche 1966: 214). The two aspects are structurally related and we may say with Nietzsche that the ability to feel, think or live unhistorically is the more important faculty as it is the basis of any cultural product (p. 214-215), but this is not tantamount to being able to separate the two faculties; it only indicates that the unhistorical possesses a perceptual immediacy that the historical does not, because

---

110 It is therefore not surprising that *Any process of language learning* (Jakobson, p. 356) is a metalingual operation, since language acquisition is also an existential process.
111 In A.S. Byatt's *Angels and Insects* (1995:18. London:Vintage) William Adamson also requires a theory in order to interpret *significant variations in the forms of* [certain butterflies], *which do suggest the species may be in the process of modification, of change*.
112 A change is neither *implemented* (Milroy) nor *applied* (Anderson), because a change does not exist before, well, it exists!- How is the French Revolution or the Destruction of *La Bastille* the implementation or application of itself?

the perception of the historical involves comprehension or interpretation (cf. below): it is *significatif*: history is not only seeing, but comprehending. History is therefore rational in so far as man's existence is sign-functional, and this existence is a priori not to be associated with any value-loaded or, in the final analysis, *logical* or *moral* properties: the present's development into the present is perfect or regular, as no alternative can be set up; it is a process as 'sovereign' as the primary processes of our mind, in which true-false- and right-wrong-parameters are suspended (Jacobsen 1976:30): the present is the *substance* or *subjectum* of man's existence. This concept of empirical necessity, an *absolutum*, is devoid of metaphysical overtones because it is neither perfect nor imperfect, neither regular nor irregular; no matter how this process has occurred, it is always comprehensible and unchangeable, it can be thought different, but not undone; historically, there is no point in attempting to justify the ways of history: the building of *La Bastille* (Paul's *Aufbau*) was a historical process equilogical with its destruction (Paul's *Verfall*): the former transindividual and transsynchronic process is just as much the destruction of something else as the latter is the construction of something else[113]. The destruction of *La Bastille* or the Viet Nam War may be praiseworthy or disgraceful events, nonetheless they remain historical processes, processes by which man has created the present (moment) out of the present (moment); the Great Vowel Shift and the grammaticalization processes in Middle English that from our modern perspective substantially changed the shape of the English language are perhaps awe-inspiring and momentous and crave special justification, but the insignificant change 'OE *o:* > ME *o:*' and split of OE *a:* into Southumbrian *ɔ:* and Northumbrian *a:* both possess a degree of necessity that is no less compelling. What synchronic theory can distinguish non-arbitrarily between *Verfall* or rule-breaking and *Aufbau* or rule-creation here?

The outlined view of history and the historical process also dispenses with the holy trinity of time: the past > the present > the future; the future does not exist equiempirically with the present. Man's innate knowledge of how to exist is his historical sense, a sense that needs no justification or *apologia*[114].

The present moment, be it ever so unhistorical, becomes historical the moment we comprehend it – against the horizon of existence, of 'forgetting' and 'remembering'. In this sense Nietzsche (1966:214-215; 281) is right in taking 'the unhistorical' as the foundation of history, and even as what is a cure against the historical disease (p. 281)[115]. However, the necessary unhistorical element will not justify the synchronic stance to the effect that it is the

---

113 In our predisposition for the analytical process we tend to separate the properties of a phenomenon and classify them into categories that from a logical point of view are incompatible; hence the synthesis of construction and deconstruction is not taken seriously. – Cf. the Jespersen quote in note 126 below.

114 The historical sense is basically nihilistic. – Nietzsche defines *den historischen Sinn* here in 1874 by: '[to have] *das Glück, sich nicht ganz willkürlich und zufällig zu wissen, sondern aus einer Vergangenheit als Erbe (…) herauszuwachsen* (p. 227)) so that history becomes an *apologia pro existentia* or *pro rebus gestis* (*Apologeten der Thatsächlichen* (p. 265)). The historical theory is nihilistic as it is neither *eine Entschuldigung* nor *Rechtfertigung* of historical development; it makes us comprehend the existence of our everyday languages. As long as it sticks to the dated distinction between change (A > B) and nonchange (A > A), modern linguistics is bound to come up with excuses (value-judgments) for the former aspect of a language's existence. And a historical theory (of language) that makes us understand (part of) our existence through time will certainly serve, not hamper man's existence, it being a true *Memento vivere*, not futile attempts at arresting time and changing *la condition humaine*.

0.4.1 Vivo, ergo sum

'system' that makes change rational or creates order out of chaos: historical data are neither chaotic nor irrational; they need no *apologia*, but a theory of history, a requirement whose alternative is a kind of theoretical nihilism that is incompatible with man's constant development into the present[116]. Nietzsche (p. 259) was again right in maintaining that the problem of history is history's, to be solved by history.

The existential function of language has largely been ignored in linguistics for two reasons (*pace* Rydén and Labov below, and perhaps Milroy above): **the static and simplifying conception man has of himself in relation to circumambient reality**, which is embodied in the semiotic triangle (language/symbol – language-user/thought – reality/referent (cf. Ogden and Richards 1969:11). The three components are regarded as givens, being there before any verbalization; thus the spoken utterance establishes a binary relationship between language and reality *qua* denotatum, in which relationship the speaker fades unobtrusively into the background, or between speaker and denotatum (including message and addressee), in which the spoken utterance seems to vanish. The dual function contracted by *discourse* and *universe of discourse* is static because denoting language only *verbalizes* what already exists (cf. Jakobson 1968:351)[117]. Man is thereby turned into a recipient, passive *patiens*, who can only assume a defensive attitude towards hostile reality; and man's *goal-directed verbal behavior* (Jakobson, p. 351) is merely a semi-active function that makes man *experience*, because verbal behavior does not create, only exploits what exists a priori, namely the system's systemic elements[118]. Speech – like the traditional metalinguistic function – establishes a *projected* world (cf. Heine *et al.* 1991:24;32; Hjelmslev 1963:8), – a Kant-like *für-sich* reality, leaving 'real' reality unaffected.

The second reason for the above oversight is that the existential function cannot be appreciated if the dualism of FORM (the exterior) and CONTENT (the interior) with its twin sister, SUBSTANCE and FORM, is taken as a dichotomy. (This will also apply to the implied separability of the below-mentioned binary functions). The synchronic epitome of this gap (separability) is expounded by phonology (phonemics), and to a lesser extent phonetics; here the so-called building stones of our everyday languages have been deprived of that which makes language language, the sign-functional property. In its existential function language becomes an integral part of man's existence, where history and synchrony (the present) cannot be separated, where knowledge and existence are inseparable, and where man and his

---

115 Cf. Paul's psychological basis of language; this was also ahistorical. – The modern cure against the Synchronic Disease will, correspondingly, be the historical.
116 I base this 'optimistic' position on the irrefutable fact that no instance of what I call the present's development into the present has ever been seen to lose its collective or structural, rational orientation or to lose its meaningfulness.
117 Although Rydén (1997:421) attributes a dynamic function to language – it *constructs* and *captures* reality as well as *concretizes* our mental nature, this function is static, not creative. The same goes for Gregersen's anti-immanent, but non-processual conclusion: a sociolinguistics must study both 'the life of language in its social setting and the life of society in language' (1991:2.255). Gregersen's brief comment on his view is couched in the static diction of language norm and standardization, with no references to similar parameters which we find in tensional fields of German Functionalism (Horn and Lehnert) as well as Samuels and Martinét. – See note 135 below.
118 Jakobson's *two modes of arrangement* are a case in point: *selection* occurs from elements existing prior to (both their actual selection and) their *combination* (1968:358).

language, thinking (mind) and language so merge that they fuse into one[119]. The everyday language is not the inadequate outward garb of thoughts, messages, knowledge, neither the inert material, be it ever so plastic and malleable, of its content, nor the means by which meaning is conveyed from a sender to a receiver; this will reduce the everyday language to mere convention and outward style, and imply a return to a Plato-like idealism[120]. Content without form, on the other hand, degenerates into pure abstraction[121]. In short, the existential function of language makes not only the semiotic triangle a dynamic entity, but it also epitomizes a structural principle that will be the foundation of the theory of history that I shall develop and is a generalization of the sign-function's two relates, content and expression.

It is a commonplace in the synchronic literature that synchronic linguistics is not to be *confused with statics*. If this is taken seriously, the synchronic point of view defeats itself, and this inconsistency cannot be concealed by notions such as the *temporal dynamics* (Jakobson 1968:352) of the synchronic state[122]. Jakobson's version of the inconsistency runs like this: *any stage discriminates between more conservative and more innovatory forms* (p. 352), where my point is that these two well-worn synchronic concepts can only be comprehended by our sense of history, a sense that enables us to grasp the structural interactivity of temporal sequence (process) and presentic staticness that produces a historical phenomenon. That a form is conservative, say, ModEng *telephone*, *omnibus* cannot possibly arise from the form's structural relationship with other newly coined forms, *phone*, *bus* **at a point in time**: both *to be conservative* and *an innovation* are historical concepts that presuppose the language user's mental faculty of comprehending temporal directionality. Another of Jakobson's examples (p. 354) is the *coded* subphonemic feature of short and long pronunciations of the same word as in [big] and [biːg] of ModEng *big*, where the prolonged variant connotes emphasis. How do we know that synchronically? We can of course observe the statistically different uses at a given point in time and register the former as normal style and the latter emphatic style. My point is: how can the synchronic linguist decide that [biːg] is the prolonged variant, and [bɪg] is not the shortened variant? Appeal to the purportedly systemic element /ɪ/ is circular, because such variant forms contribute to the construction of systemic elements. In a word, the synchronic use of both parameters presupposes historical knowledge: our [iː] has been prolonged in relation to its origin, short *i*, and both have been continued so as to contract a sign-functional synchronic relationship (cf. 0.2.1.1/2 and Labov below).

Another example is the 'marked-unmarked' parameter[123]. The marked element of a subphonemic function is the element that has *lost* a feature in relation to its unmarked counterpart. The question is: how does the synchronic phonologist know synchronically that [-p, -t,

---

119 Jacobsen (1976:28-29) does not accept this type of identification; I shall return to this below.
120 Cf. Hass (1979:98). – Nietzsche (pp. 232-237; 235) levels a pertinent attack on the consequences of the separation of form from content
121 Paul's *Weg mit allen Abstraktionen* is curiously reminicent of Nietzsche (pp. 235-236): *Wir Deutschen empfinden mit Abstraktion*; while Paul would hardly underwrite Nietzsche's continuation: *wir sind alle durch die Historie verdorben*.
122 Note also Jakobson's (vague) use of the notion of *stage* and *contemporary stage* (p. 352), and not the proper synchronic term *state*. *Stage* and *contemporary* as well as *state* remain undefined. *Static*, however, seems to imply a notion of duration or stability through time.
123 My discussion is based on Lehmann (1992:37-38; 143), and will be resumed in the following chapters.

0.4.1 Vivo, ergo sum

-k] in modern German *gab, wirt, Tag* have lost the feature of voice in relation to [-b-, -d-, -g-] in *geben, werden, Tage*? Similarly, Lehmann's grammatical examples fare no better: the forms *brethren, dove – diven* (the verb *to dive*), as against *brothers, dived – dived*, are marked because we know the history of the respective forms involved; this is corroborated by Lehmann's analysis of intervocalic voicing of /t/ in American-English (pp. 142-144) as in *atom, letter, better, little*, where he sets up criteria for unvoiced [-t-], the origin of [-d-]. Again, there are no synchronic non-arbitrary criteria for saying that the unvoiced variant is the systemic element, /t/, underlying both [-t-] (unmarked variant) and [-d-] (marked variant) in *atomic* and *atom*. Let the argument run its course: if in American-English we always find a voiced intervocalic sound in a certain group of words, why is it that this sound [-d-] is classified as a member of the class /t/, and not /d/ – in as much the pronunciation of this voiced sound follows the rules for the pronunciation of the phoneme /d/? Thus: synchronic functions reflect historical processes and are to be comprehended historically; the synchronic conception of /X/ [a], [b] is a historical process, which entails two questions: is the slant-entity or underlying form necessary or are they results of the **Phenomenal Error** (see Def. 1.4.10)? What is the diachronic link between [a] > [b]? The synchronic linguist still has to demonstrate that his or her rules are not hidden historical statements[124]. As regards the table below, it is interesting to note Daniel Jones's *concentrate*-examples: the variants may be ordered historically in accordance with fully predictable processes that have characterized the English language since the 9th and 10th centuries; the processes may be implemented by this

---

[124] A recent case in point is Optimality Theory, a theory based on a strength hierarchy of constraints, from inviolable to violable ones. Here a constraint, *output-output correspondence*, has it that there is a certain *reluctance* to reduce vowels to schwa as for instance in *cond<u>e</u>nsation* [e] as against a type-identical noun as *comp<u>e</u>nsation* [ə]. The synchronic reason for this is that *condense* with [-e-] is the underlying form of the noun, while the other noun is *derived from* the base *compensate* with a reduced vowel (Gilbers and de Hoop:1998:8). However, this theory has not demonstrated that the difference(s) is(are) not due to historical processes of devaluation or for that matter spelling-pronunciation or analogy. In other words, the reluctance rule of Optimality Theory has a historical basis. The below table illustrates not only the arbitrary nature of this synchronic rule but also its instability, and the constraint hierarchy may just as well be replaced by predictable mechanical processes of the just-mentioned type.

	Adv. 1948 12th impr	Adv. 1989 7th impr	Longman 1978	Longman 1995	Jones 1963	Jones 1997
condense	-e-	-e-	-e-	-e	-e-	-e-
condensation	-e-	-e-	-e-, -ə-	-e-, -ə-	-e-, -ə-	-e-, -ə-
condemn	-e-	-e-	-e-	-e-	-e-	-e-
condemnation	-e-	-e-	-ə-	-ə-, -e-	-e-, -ə-	-e-, -ə-
compensate	-i-	-e-	-ə-	-ə-	-e-, -ə-	-ə-, -e-
compensation	-i-	-e-	-ə-	-ə-	-e-, -ə-	-ə-, -e-
concentrate	-ə-	-n-	-ə-	-ə-	-e-, -i-, -ə-, -n-	-ə-, -i- *
concentration	-ə-	-n-	-ə-	-ə-	-e-, -i-, -ə-, -n-	-ə-, -i-, -e-*
dispense	-e-	-e-	-e-	-e-	-e-	-e-
dispensation	-e-	-e-	-ə-, -e-	-ə-, -e-	-e-, -ə-	-e-, -ə-

comp<u>e</u>nse 1393-1706 (1825)
compensate 1656 (1648)

* In *concentrate* and *concentration* the [-ə-] has become weakened in the 1997-edition.

developmental series (under weak stress), a series that is infringed by the theory's *maximal-input-output correspondence* rule, which prohibits deletion of segments in base-forms (Gilbers and de Hoop, 7-8;5): -ẹ- > – ï – > -ə- > Ø.

In Nietzsche's scathing criticism of the predominance of the historical point of view in his days we may recognize the seedbed of the eventual linguistic schism between synchrony and historical linguistics. Despite his care in striking a balance between the historical and the unhistorical, Nietzsche does separate the former from man's existence (*Leben*) by emphasizing the subservient and instrumental function of history (1966:219). This view can easily be recognized in modern linguistics, where it has become a sacrosanct dogma (Greenberg 1979; Hammarström 1981; Winters 1993); but while history is an integral part of both Nietzsche's *vivo, ergo sum* and *Memento vivere*, there is another parallel too obvious to leave unnoticed: unfortunately, following Nietzsche, history as a science has interfered with the *healthy* relationship between life and history (pp. 231; 234); history has become an end unto itself, so that what has been gained on the swings of scientific history has been lost on the roundabouts of life: the synchronic point of view has come between our everyday languages and their historical existence, it being *the* theoretical approach to language; this position is embodied in the linguistic literature from Saussure by way of Hjelmslev to recent versions of this early 20th century dichotomy: *Philology and theoretical linguistics* (Syrett 1994:16), *Theoretical linguistics* being 'uninterested in' *the history of language* (Beekes 1995:4-5), *historians* as against *formal linguists* (Labov 1994:10). Martinét (1964:63) also proceeds from an opposition between 'structural and philological linguistics' – to which one may ask: don't we all love language?

Although his 1994-work is set within the confines of our dichotomous tradition, Labov's analyses lead to so many empirical results and conclusions incompatible with the synchronic principles on which he works that it is surprising he does not reject the synchronic basis of historical linguistics. As he does not do this, he cannot work with the general and dynamic term, language existence, but only with a restricted and indeed value-loaded, hence limiting concept of language development, to be explained by basically static principles, this being indicated by his '[a] *theory of language change relate*[s] *language change to general properties of human beings or of human societies*' (p. 5), where the three elements of the semiotic triad appear to be mutually separable, albeit interdependent. However, like Martinét (1964:39)[125], Labov is right in emphasizing that the existence of a language cannot occur in abstract isolation from its anthropocentric moorings[126], and I shall so generalize this view dynamically that the existence of an everyday language becomes an integral part of and contributor to man's constant development into the present.

---

125 Martinét feels obliged to remind us that the *activité*-element in his dynamic definition of language *fonction* comes from the speaker and not from the immanent conception of the sign-function.
126 As Jespersen puts it: *(...) we should never lose sight of the speaking individuals (...)*. (1969:7; 8), relying on Hermann Paul's words in Henry Sweet's translation: *the actual language exists only in the individual, from whom it cannot be separated even in scientific investigation, if we will understand its nature and development* (Jespersen, 94-95; 95). Again we can detect a firm grasp of how not to analytically separate what must be understood synthetically.

0.4.2 Rehabilitation of the (spoken/written) substance. It is more than raw-material for change, it being a structural phenomenon

Animals are born with a genetic blueprint that predetermines which species-specific 'language' they can 'learn', while the infant baby is genetically endowed with the capacity to learn any one particular everyday language as his or her mother tongue; the difference between man and animal is greater when the former's ability to learn more than one foreign language is considered, and Nietzsche (1966:211; cf. pp. 281;260-261) noted an even curiouser fact, that man – as the only animal – is able to forget, but **incapable of learning to forget**; the moment which becomes a 'was' the moment it 'is' does not become *ein Nichts*. To this we may add that no child can learn not to learn a native language or to forget her native language. To remember is an innate human faculty and constitutes with the faculty to speak and the faculty to forget a curious mental nexus that I shall call man's **historical sense**. The historical sense chains present existence to *part* of the past. We can and, indeed, must forget the past, not all of it, in order to exist (Nietzsche, p. 213); and a proper sense of reality will necessarily include the **structural** fact that *jedes Lebendiges kann nur innerhalb einen Horizont* exist (p. 214; cf. p. 281), and this horizon will invariably include the past – because of the twin faculties of remembering and forgetting, and the everyday language. Nietzsche did not see how the remembered and the forgotten enter into a structural synthesis with the new present moment to produce a moment (an 'is') that is different from the previous moment (an 'is' that has become a 'was')[127]: as each new present moment will consist of something *Unhistorisches* and *Historisches*, so the spoken chain and its substance will consist of historical and synchronic components, and it will be the task of any historical theory to describe these two layers as a whole. So my objective is not to create a synthesis of synchrony and history – that is in fact impossible *per definitionem* – but to **redefine the notion of synchrony** in the light of the synchronic phenomenon's *geschichtliche Wirklichkeit*[128].

We could reverse Nietzsche's diagnosis of the 19th century, which suffered from a near-fatal bout of the 'historical disease' (pp. 231-232), saying that the modern dependency-view suffers from the Synchronic Disease: instead of interpreting everything in the light of history, the present century sees everything through the glasses of the present or the purportedly synchronic system, so that historiographical method becomes *dehistoricizing* (Lass 1980:54). According to the Uniformitarian Principle (Labov 1994:21-25; Lass 1997:24-32) the past records (real-time events) are to be corroborated by present assumptions and facts (apparent-time change) or the extreme version: what cannot occur in the present could not have happened in the past (Lass 1980:55)[129].

The role of the spoken chain is not in doubt: not only <u>must</u> a language be 'used' in order to change, a <u>new</u> variant must arise and be generalized by selection; this is the general mean-

---

[127] Thus Nietzsche's condemnation of the effect of the historical sense, it makes *seine Diener passiv und retrospektiv* (p. 260), is too sweeping.
[128] Jakobson's view of a *history of language* as a *superstructure* to be erected on *successive synchronic descriptions* is only a return to the priority of one extreme of the traditional alternative (Jakobson 1968:352).
[129] Historical considerations do not enter into Swiggers' discussion of theories of linguistic knowledge (1992).

ing of 'synchronic variation' and 'diachronic change'. But the variant is merely the raw-material for change, the change process being governed by the requirements of the predecessor system and the successor system: the variant either implements a phonological rule (Kiparsky 1988: 658;655;667) or is a *phonetic tendency* waiting to become 'phonologized' (Anderson 1992:106;108). Similarly, the spoken chain is also necessary in generational transmission of language. The ultimate *raison d'être* of the spoken chain will be its sign-functional role in the present's development into the present.

While both Labov (1994:43) and Samuels (1972) operate with a two-state version of the historical process and adhere to a Saussure-like *langue-parole* dualism – Samuels explicitly –, one may say with some justification that Labov does not draw the logical consequences of his philologically minute investigations[130] and Samuels abandons the dualism when he comes to grammaticalization processes (the cyclic swing from synthesis to analysis).

If Samuels had generalized the periodic basis that underlies his views of grammaticalization as well as circular shifting, and Labov had elevated the data of his study of changes in apparent time and real time to constitute the proper (periodic) datum of historical linguistics, the synchronic system or state with the two-state view as a corollary could no longer remain the fundamental datum. Trajectories of, routes to be taken by phonetic variants, or the process between the beginning and completion of a change are all notions that point to the same type of historical datum as the developmenal series. Examples from Labov (1994): New York City 1895 [ɛ:ə] > 1940 [e:ə] > 1960 [i:ə] (p. 75), these variants being tensing of /æh/; the /oh/ phoneme has undergone these changes recently: [ɒə] (= low) > [ɔə] (= lower mid) > [oə] (= upper mid) > [ʊə] (= high) (p. 48).

The necessity of the spoken chain in language change is a step towards what I shall call **rehabilitation of the substance**, the material and concrete aspect of the everyday language so that the variant becomes more than raw material for change; it has an **Eigennatur**, independent of its possible systemic affiliations. So instead of formal (abstract) features, which invariably imply a dualism, **material (substantive) features** will expound the nature of historical phenomena. Origin or common origin (the Jones Hypothesis) is such a material feature or trait, which does not take leave of material or *anschauliche* reality (cf. Løgstrup 1976:173).

Historical data is periodic, hence transsynchronic; I shall demonstate that a **developmental series of at least three elements** can be captured by a historical theory which respects the material Eigenart of our (spoken) everyday languages. This theory entails neither a rejection of the notion of synchronicity and staticness nor a return to the hegemony of the kind of historicism that Nietzsche criticized so aptly in his 1874-work. But it does entail (a) a redefinition of the word *synchronic* in such commonplace notions as *synchronic variation* (as *e.g.* in Kiparsky, p. 658), (b) the theoretical rehabilitation of the single 'isolated' isolative sound change (*e.g.* OE /a:/ > ME /ɔ:/)[131], (c) the isolative change: CONTINUATION (A > A), (d) a sign-theoretical understanding of the sounds of the everyday language; fi-

---

130 Labov assumes that a stable system underlies all variation (pp. 98-112). His adherence to the Uniformitarian Principle (p. 23) obliges him to such a stand, but what seems to constitute Labov's real stumbling-block is that the fact of change and (the notion of) system are not really compatible (pp. 10-11).

131 See Mitchell (1992:94) for a similar recognition of the single philological detail, here at the syntactic level: *They are there. They are the stuff of language*, *i.e.* the facts that *defy neat classification*.

0.4.3 Towards developmental series 59

nally, the notion of circular shift will also be redefined in accordance with the constructed theory[132].

To sum up: with Nietzsche's pertinent analysis in mind, the synchronization of history reduces historical existence to a subjectless *cogitale*: *Cogitatum, ergo est* (like the non-anthropocentric position of Lass (1997:xviii;12)), which, constituted analytically, cannot but yield *eine historische Zufälligkeit*; and with a slight change of the Austrian poet Franz Grillparzer's judgement on his contemporaries, history exists through *Abstraktion*[133], the formal features of synchronic theory.

0.4.3 The datum of historical analysis is the developmental series: an irreducible transsynchronic and sign-functional datum. Towards a definition of the developmental series: *loquor, ergo sum* **and a practical critique of the slant-bracket notation**

In practically all discussions of the Great Vowel Shift *qua* a Circular Shift the notion of the developmental series is implied in that the respective routes taken by ME /iː/ and /uː/ to develop into their modern diphthongal products /ai/ and /au/, as well as the route of ME /aː/ to become /eː/ c.1700, implement the notion, *e.g.* a > æ > ɛ > e > ɪ > i. Such a series, whose existence implies a period of time, will be the data of the theory to be advanced in chapter 6.

As all variants are not only physiological entities, to be determined acoustically and articulatorily or placeable in certain patterns and groups, but also sign-functional ones, the content of the developmental series will also be established[134]. This part of the process of change is virtually disregarded by the synchronic – and sociolinguistic – tradition; as far as I know only Samuels (1972) tries to explain the origin of a change (the actuation problem of sociolinguistics is only mentioned as a problem in Weinreich *et al.* 1968, and Labov 1992; 1994; Milroy 1992; 1994)[135]. *E.g.*, all Labov has to say about the origin of a change is *that a single phonetic feature of a phoneme* is 'transformed gradually' in regular sound change (Labov 1994:542; 568; 1992:43-44) and at this *initial* stage (in the process of isolative change) the choice of the intrasystemically motivated variant is not accompanied by *any degree of social awareness*, despite it being *change from below* (p. 542).

Samuels asks a question that is crucial to an empirically relevant understanding of language development. Unfortunately, he did not pursue its ramifications: (…) *are the variants* (originating in the spoken chain) *more than the raw material of change?* It is true that Samuels finds that *a tolerated margin of aberration* is characteristic of all *actual*, in contradistinction to *ideal*, speech production (1972:9-10), because the linguistic and nonlinguistic

---

132 This means that I shall be concerned with neither chain shifts as in Labov 1994 nor such 'monumental' processes as the Great Vowel Shift, whose reality is extremely doubtful (cf. Stockwell & Minkova 1988).
133 *Wir empfinden mit Abstraktion* quoted from Nietzsche (1966:235; cf. 235, 239; 242; 280 and below).
134 It may not be entirely anachronistic to say that had Nietzsche (pp. 232-234) been language-oriented, his emphasis on the impossibility of divorcing the form (*die Form*) from the content (*der Inhalt*) bei *allem Lebendigen* anticipates the interdependence of the sign function.
135 In 1922 Jespersen (1969:261) saw the actuation problem as a structural tug-of-war between the individual's inborn tendency to *laziness* (p. 263) and the fact that language is also the speech-community's business. – Cf. note 117 above.

context in question as well as systemic redundancy will ensure disambiguation or communicative efficiency. Samuels's answer is in the negative (p. 136) and is in perfect line with the dualistic tradition of our Western civilization: firstly, the empirical world is not the real world, as there is a real reality 'behind', 'above' or 'under(lying)' this sensuous (irregular) world, and that trans- or supra-empirical world controls and governs, determines and shapes the other world; secondly, for man to acquire knowledge, two or either of two conditions must be fulfilled: the object of cognition is unchanging and static, unvarying and unmoving, permanent and universal (regular; cf. Nietzsche's *ein Ewiges*; and chapter 1. below); knowledge can only be acquired by means of an unchanging and static, unvarying and unmoving, permanent and universal instrument which fixes and/or reduces its object to one (*in unum*). In any case, the everyday language in its immediate and material appearances and uses plays no significant role in either process, *die sichtbare Tat*, in Nietzsche's words (p. 235), being only raw material, which cannot as a forming factor do justice to what it manifests. Despite this, I shall interpret Samuels's position to the effect that he does try to rehabilitate the material aspect of our everyday languages in that the *tolerated margin of aberration* is the basis of *universals* in language change; but the value-loaded notion of aberration must be dispensed with as it reminds us too much about the equally value-loaded imperfection view of history[136]; this rehabilitation process contains five components and implies the existential function of language:

1) The variants are change itself – they are more than the raw material of change. They expound our sense of history.
2) The everyday language carries *Erkenntniswert* with it, and will as such constitute a non-transcendent cognitive system in its own right[137]. There is a middle road between nominalism and Realism.
3) The spoken substance is necessary vis-à-vis the system; *parole* and *langue* contract an aitional relationship in that **the former (exists) but not only if the latter**[138].
4) Synonymy and polysemy are universals rooted in the material aspect of language.
5) The possibility of a theory of the concrete/material, *i.e.* historical development, without having recourse to (phonological) systems[139].

---

136 *Aberration* implies deviation in relation to a fixed item *qua* an ideal so that it, too, carries moral – 'it-ought-not-to-be-what-it-is' – overtones. Samuels's *langue-parole* axiom lands him in a dilemma: his stylistic parameters: relaxed-forceful style, formal-informal style, etc. are universals and as such they are not aberrant (in relation to the nature of the everyday language), but the sounds that implement them are aberrant in relation to the manifested phoneme (Samuels, p. 29). Or does Samuels imply a distinction between 'expected variation' and 'aberrant variation'?
137 This, in sharp contrast to the generativist view of man's knowledge of his language as a cognitive system (Chomsky 1986), is in fact a reversal of Lass's *retreat from (...) excessive concreteness* (1976:83;51). My position tallies with the Glossematic conception that what has become the substance of a form may itself become the form of another substance. – With reference to the so-called irregularities in language (*historical junk*) Lass (1997:13-15) curiously accepts the cognitive value of data that cannot be formalized by the set of formal-functional features of a given synchronic theory.
138 There was no PreOE systemic element to determine the rise of *y*-sounds; and the question that proponents of synchronic systemic explanations should ask is: why does a system – from one (static) moment to the next (static moment) motivate the rise of a new sound? That one change may have some systemic connection with other changes (cf. Samuels 1972:20) is still to be proved.

0.4.3 Towards developmental series

The programme implies that there is *une science du concret* or *de la finitude* (Lévi-Strauss 1962:chapter 1; 353), so that the spoken utterance is understood, in Lévi-Strauss's words (pp. 294-295), the moment it is perceived, not by way of a system. Let the present moment be the concrete and finite, then I shall demonstrate the *possibility* of a theory of the present moment *qua* a dynamic entity. The existential function of language entails a reorientation of the traditional position: the everyday language will always keep its collective orientation and systematicity, not **in spite of the fact of change** but **because of change**: the system(s concept) is always a man-made entity, only a relative totality, a cultural product, hence a sign-functional, structural and historical entity: it can only spring from man's existence (Nietzsche, pp. 278, 281; cf. Lévi-Strauss's *superstructures* (1962:336)); and the modern opposition between 'being' and 'reason' as expressed variously by Descartes's *Cogito, ergo sum* and Nietzsche's *Vivo, ergo cogito* (p. 280) will be combined into an irresoluble merger against the horizon of man's everyday language: *Existo, ergo cogito* in virtue of *my everyday language*, and *Cogito, ergo existo* in virtue of *my everyday language*. Without venturing into a discussion with Gadamer's Dilthey analysis I would like to call attention to the latter's coupling of *Leben und Wissen* as *eine ursprüngliche Gegebenheit* (Gadamer 1975:223)[140], in order to emphasize that historical knowledge does not operate against the contradictory hor-

---

139 Lass (1976:51;84) is a firm believer in such *necessary primitives in phonological theory*. The problem with Lass's theory is that he must operate with the historical existence of 'a metarule', whose origin is unexplained (p. 85): it acts on a grammar 2 of T2, creating a rule that implements the metarule's *instructions*. The output of grammar 2 will then serve as the *primary linguistic data* for the new language learning generation, which will incorporate – into grammar 3 – the *results of the implementation of the metarule (…) – but not the metarule*. The absurdity of this view appears when following Luick we put the parts of the process in chronological order: at T1 there is a grammar, G1; at T2 a metarule arises and acts on – (is added to (cf. pp. 84, 85)) – G1, with the output of G1 now being different from the output of G1; at T3 there is a grammar, G3 with no metarule. As a three-state process, the whole event may be comprehensible, but if the process is to span a period of 300 years (the Great Vowel Shift), the programme will carry no empirical relevance. Lass (1997) does not mention metarules, and in his 1980-work, where he deals with the nature of *historical knowledge* and with what constitutes a *historical phenomenon*, there is no discussion of metarules, either. Furthermore, it is curiously absent from his 1988-rejoinder to Stockwell and Minkova (1988), as well as in his 1992-paper, where he continues to defend diphthongization (of ME *i:* and *u:*) and raising (of ME *e:* and *o:*) as a unitary phenomenon as '*something real*' (1976:51;53-54), as a *temporally localizeable "event" in the history of English* (1992:154). Hermann Paul's warning against the Realist fallacy is still pertinent.

140 I shall not be discussing recent hermeneutical approaches to language change as we see in Anttila (1989) and others, as such approaches are the importation into history of yet another transcendent conceptual apparatus, the primary objective of the language historian being the possibility of constructing an immanent theory of history. Coseriu (1975:94) rejected both an analytical and hermeneutical approach with reference to his *Norm* (correlated with necessity) and *System* (correlated with freedom) dualism, because this dualism makes *die Sprache* belong to history (p. 96; cf. Ottosson 1996:44). If hermeneutics is *a context of argument about understanding* (G.L. Bruns in Ottosson, p. 49), it is obvious that hermeneutical history becomes a '*science of the spirit*' and thereby lends itself to Karl Popper-criticism for being non-quantifiable and non-measurable, a non-scientific domain (Rozov 1997:343-344; against this see Gadamer 1975:317). On the other hand, I agree with Gadamer: *Jeder Begriff von Sprache, der sie ablöst von der unmittelbaren Situation derer, die sich im Rede und Antworten verstehen, verkürzt sie um eine wesentliche Dimension* (quoted from Ottossen, p. 46), to which I shall add Gadamer's dialectical-dynamic view of *das Verstehen* as *ein Geschehen* (Gadamer 1975:447), which agrees with my view of change as something to be comprehended, not seen, but which in my opinion does not attribute sufficient importance to

izon of either an empiricist or rational-idealistic *Standort*, which Lass evokes in his above-mentioned defense-article[141].

Part of Samuels's theory is a theory of the concrete established against the horizon of the principle of least effort (*in casu*, ease of articulation (1972:10-18)); conditioned change[142] is typically the selection of a synchronic variant that originated in colloquial or less careful ('careless'/relaxed style) register at the cost of the variant of formal register or forceful style, and in cases where formal similarity cannot be regarded as purely accidental (cf. the Jones Hypothesis), he establishes a diachronic link between the variants to the effect that the shorter 'comes from' the longer form (pp. 10-11). That the *precise diachronic link* between two such variants *escapes us* (p. 11) is due to the fact that Samuels makes do with the ease theory, not trying to construct a theory that will enable us to **comprehend** the connection between our two variants. Samuels is right in maintaining that there is **no alternative** to the dynamic interpretation: either **comes from** the other[143]. He does not exploit the fact that the forms of careless register (relaxed style; ModEng [phone]), explained by the principle of least effort acquire their meaning structurally through the simultaneous existence (presence) of their respective opposites, *i.e.* they enter into synchronic and synonymous relationships (ModEng [telephone]). Samuels, I assume, has the principle of least effort so remedy the synonymy (systemic redundancy) inherent in the synchronic style- or register distinctions that one of the variants is not continued, for *economy* or (absence of) *expressive needs* (pp. 137;65): Samuels's theory does not explain what happens to the continued variant apart from its generalisation into unmarked 'normalcy', a process that presupposes some kind of devaluation (p. 135).

The distinction between conditioned change and isolative change is one of quantification: Few A > B as against Many A > B. Borrowing the notions weak (*faible*) and strong (*fort*) from Lévi-Strauss (1962:346-347), we can characterize the explanation of conditioned change as a weak type of history, as its explanation is merely a question of precise description of what occurs: the conditioning factor in, say, LateOE *wimman* < OE *wifman* is overt, and once we say that ease of articulation subsumes the rise and subsequent selection of *wimman* (in ME), this assimilation process as well as the loss of OE *wifman* has been explained theoretically; cause and theory merge.

---

the rehabilitation of the spoken substance and the latter's role in the existential function of the everyday language. Finally, although I sympathize with Bubner's conclusion that hermeneutics does not stop at *resignation as the ultimate goal of theory* – as Wardhaugh did, my point is that the construction of an immanent theory of history must be the first alternative to theoretical resignation in the study of language (Bubner 1976:75).

141 (1988:405), where Lass virtually characterizes the Stockwell-Minkova rejection of the single-event conception of the Great Vowel Shift as *simple-minded empiricism (nihil in intellectu quod non prius in sensu fuerit)* as against his own – and Luick's – *more visionary or "mystical"* position: *nihil in sensu quod non prius in intellectu fuerit.*

142 This concrete basis agrees perfectly with Samuels's attempt at rehabilitating the Neogrammarians' mechanical view against Functionalism's vague notions of *Deutlichkeit, optimality, efficiency* as prime movers in grammaticalization processes.

143 Samuels does not ask the obvious question: where does the substance-based notion of **something coming from something else** come from? – The merger of theory and cause is due to the fact that the elements of the advanced theory (*in casu* Samuels's) are causal factors, and as such they belong to the same world (processes) as their affecta and effecta.

0.4.3 Dualism and change

With isolative change the over-all process is more complicated: the *parole*-based origin of isolative change, assimilation to suprasegmentals, makes it subsumeable under the principle of least effort; but the selection process that eventuates in a new systemic element needs a stronger history (Samuels, pp. 26-27 *et pass.*): the conditioning factors that determine synchronic variation, say EarlyME [ha:m] and [hɔ:m] < OE *ham* 'home' are hard to perceive or learn, hence only one variant is continued. This is only part of the story.

Samuels's theory contains a number of attractive points, but also weaknesses. First of all, it makes both types of change equiempirical and subsumeable under the Ease theory. Secondly, his theory is sign-functional, as the historical process consists of signs (to be developed below).

In order to demonstrate the absurdity of the *langue-parole* parameter in historical description, I shall interpret the two processes, in so far as it is possible, in accordance with it and visualized by the slant-bracket notation:

A. Conditioned change

As far as I know Stiles's 1988-paper is the only attempt in the literature at setting up a relative chronology of sound changes (in (Pre)Old English) based on the dualism, *i.e.* a *more accurate relative chronology of sound-changes*, because the *phonologization* of a sound change[144] *gives us a reliable dating for the **completion** of a change* (Stiles, p. 337). However, Stiles's use of the slant-bracket machinery is not consistent, brackets, *i.e.* allophones being virtually absent, and his reinterpretation of PreOE *I*-mutation omits possible stages in the process: Stiles's process consists of three steps (pp. 337-338):

T1 */fo:t/ (nom.-acc., sing.) :: /fo:ti/ (nom.-acc., plur.) > T2 /fo:t/ :: /fo:ti/ [fø:ti] > T3 (loss of short **i*) /fo:t/ :: /fø:t/.

Loss of short **i* (is it /-i/ or [-i] that is lost?) phonologizes *I*-mutation and phonemicizes [ø:] > /ø/, *i.e.* rounded front vowels. Clearly, phonologization refers to the purported grammaticization of vowel-alternation in athematic nouns, say, PreOE **boc* (sing.) :: **bec* (plur.), but Stiles overlooks the fact that the mutated form was both a singular and a plural form, so the form alone was not sufficient to convey number distinctions: **boc* (nom., sing.) :: **bec* (nom., plur.), but **bec* (gen., sing.) :: **boca* (gen., plur.). Secondly, Stiles's dualist-interpretation is too simple; I shall make do with the plural form:

T1 /fo:ti/ [fo:ti] (no mutated alternants) > T2 /fo:ti/ [fo:ti], [fø:ti] > T3 /fø:ti/ [fø:ti] (generalization of mutated variant; no truncated variant) > T4 /fø:ti/ [fø:ti], [fø:t] > T5 /fø:t/ [fø:t] (generalization of truncated variant) (… > OE /fe:t/).

---

144 not 'of an allophone': Stiles distinguishes between *a phonologized sound-change* such as *I*-mutation and *a phonemicized variant* (p. 351), thereby committing the Realist Fallacy with *I*-mutation being regarded as an entity in its own right; Stiles does not address the problem of how to cope with isolative change: what is the difference between the phonologization of the Old/Middle English sound change of *a:* > *ɔ:* and the phonemicization of the allophone [ɔ:] into /ɔ:/?

Stiles disregards the structural-sign-functional (synonymous) relationship that a new variant contracts with the old, and the longer process will have repercussions on the whole chronology issue; was *I*-mutation phonologized at T3 or T4/T5? If at T3, then the loss of final *-*i* is a functional process controlled by 'redundancy'; if later, then the process is a mechanical process. Finally, Stiles continues the tacit assumptions of the synchronic tradition, not addressing the crucial issue of how to establish the slant-form at T2/T4: why /fo:ti/ and not **/fø:ti/[145]?

Instead of applying structuralist principles to the so-called major language changes, I shall apply the method – with the necessary sign-functional additions – to a lexical detail such as the history of OE *wifman*, in which the motivating factors are overt and easy to perceive (Samuels 1972:25-26).

1. T1: /wifman/[146]   [wifman]   >
    T2: /wifman/   [wifman]   careful register/normal or formal style
                         [wimman]   careless register/relaxed style >
    T3 /wimman/   [wimman],   generalisation of the 'easy' variant due to *economy* with concomitant loss of the other (?redundant) synonym (Samuels, p. 136).

This scenario does not consider the possible intermediary variant [wivman], with the scope of voiced [-v-] expanded, which helps us to pose the questions: is any variant always the variant of a systemic element, or do we also have variants' variants? With reference to Stiles's example above, does *[fø:ti] develop into *[fø:t] immediately or by way of */fø:ti/? How many states, *i.e.* generalisations of a variant, are to be created? In other words, is [-mm-] a variant's variant: [-vm-]? A variant of /-vm-/? Or of /-fm-/? The dualist interpretation entails alternative scenarios:

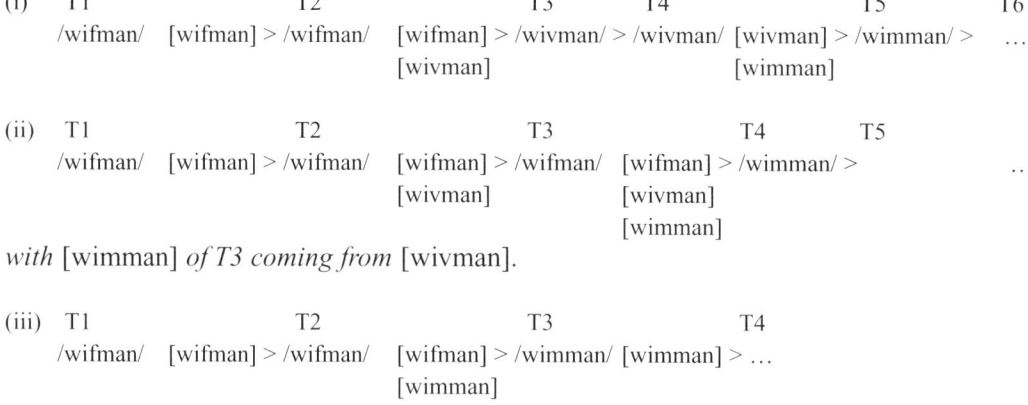

with [wimman] *of T3 coming from* [wivman].

(iii)   T1   T2   T3   T4
        /wifman/   [wifman] > /wifman/   [wifman] > /wimman/   [wimman] > ...
                                         [wimman]

---

145 Thus Stiles's exuberant praise of the dualist-synchronic approach to historical linguistics is an assessment to be taken with more than a grain of salt: *The application of structuralist principles to the study of historical linguistics must count as* **the** *major theoretical advance since the time of Luick.* (Stiles, p. 336).

146 The notation indicates that it is the singular, not plural form; shortening of *wi:f-* to *wif-* has been generalized. – The change is an exception to the isolative change: OE /f/ > /ME /f/.

0.4.3 Dualism and change

The Uniformitarian Principle is of no avail here; moreover, neither it nor the *langue-parole* point of view itself can explain whether we must postulate a cooling-off period in the creation of variants so that the generalisation of *e.g.* [wimman] of T1 into /wimman/ of T1+1 requires a period with only [wimman] as its manifestation before the variants [wumman] :: [wimman] arise at T1+1+1 out of /wimman/ [wimman] of T1+1? Or can the generalisation of [wimman] into /wimman/ be accompanied at once by the manifestations [wumman] :: [wimman]?

The two-state model makes such questions irrelevant, being concerned with the step from variant (at T1) to change (at T1+1), or the causal conditions for that process. This becomes even clearer when we consider other types of conditioned change.

It is a generally accepted phonological assumption that the conditioned variant in a split process assumes systemshood if it continues to exist without the presence of its orginal conditioning factors (cf. Samuels 1972:35; Labov 1994:331-333; cf. Hock 1991:27-28)[147].

2. *PreOE I-Umlaut* (cf. Campbell 1961: §§199; 288-289)
   OE *cyre* 'choice' : OHG *kuri* // OE *wyrm* 'worm' : Latin *verm-i-*s; Germ. *-iz*.
   OE *cyning* 'king' < PreOE *kun-ing-*

The rise of PreOE/OE /y/ is an exception to the isolative change of (All) PreOE /u/ > OE /u/, and its origin and generalisation look like this in schematic form:

T1 /u/ [u] > T2 /u/ [u] :: [y] (complementary variants) > T3 /u/ [u] :: /y/ [y].

Questions: did [y] make the step into changehood /y/ the moment when weakly stressed *-i* dropped (Campbell, §§345-354)? And/or the moment when (surviving) weakly stressed *-i* was lowered to *-e* <-e> (Campbell, § 369-370)? Or the moment when short*-i* dropped after a long syllable as in (Pre)OE **ly:si* (< **lu:si*). Does the *-y-* in OE *cyning* belong to OE /u/ or to /y/, whereas the short variant *cyng* manifests /y/? Is the systemic stage reached the moment when the original conditioning factors are not present everywhere? Finally, the Kentish development of these *y* sounds (into /e/) certainly indicates that they had some kind of independent status, and we may conclude that if a sound is part of a developmental series – Kentish: *u* > *y* > *e*, as against the later majority development: *u* > *y* > *i* – the empirical relevance of the sound cannot be questioned.

---

147 My aim here is different from Samuels's: Samuels argues against systemic-functional explanations of the phonemicization of splits in favour of mechanical factors, but he does not question the empirical relevance of the synchronic state- and systems-concepts. I question the relevance of the dualism (as embodied in *e.g. phonetics* versus *phonology* (cf. Fischer-Jørgensen 1975:327)), in particular that of the *typographical trappings* (Syrret 1994:17) of the phonemicists' slant-bracket notation. It is interesting to note that the majority of works on historical linguistics do not use the system-variant notation of the state-systems view consistently: we are never told how many states (stages), let alone the specific nature of these states or systems, that are needed to describe a period of any length. The same criticism applies to other notational conventions (see Hock 1991:28). A recent treatment of these voicing processes (Berg 1995) does not consider the theoretical issues, either.

3. *The rise of voiced fricative phonemes in Middle English as an example of the weaknesses of the dualism.*

The facts of these developments are not in doubt, although the individual sounds' trajectories from Proto-Germanic to historic times are complex (Campbell, §§444-448); I shall proceed from a synchronic cut in (Pre)OE, supposedly made at the moment when this complementary distribution was created: the voiced sound appears in voiced surroundings and the unvoiced one elsewhere (*pace* analogical influence (Campbell, §449)):

/ ?? /   [-V-]   = voiced sound
         [F-, -F]   = unvoiced sound[148]

Questions: what is to be put between the slants? This is a synchronic dilemma as old as the invention of, in Fischer-Jørgensen's words (1975:25-26), the *phoneme and variant* parameter. Parameters like basic-accidental, principal-accessory, or *variante fondamentale-variante accessoire*, suggesting that the former type of variant is the systemic one (Fischer-Jørgensen, pp. 326-327; 332), commit the Phenomenal Error. And there is no theory-consistent solution to the problem, and in his 1997 book Lass (p. 14) continues the tradition's arbitrary distribution of phonemic and phonetic status: OE *muþ* 'a mouth' is analysed as follows: /mu:þ /, *phonetically* [mu:þ] in the singular (uninflected forms), and in the plural (inflected forms) /mu:þas/, *phonetically* [mu:ðas]. Lass does not explain why the systemic forms contain the unvoiced fricative, and why variation is restricted to the *parole* level[149]? The problem is well-known and the following often-quoted examples are part and parcel of this synchronic conundrum: the complementary distribution of modern English [h-] as in *history* and [-ŋ] as in, *singing*, as well as of the two complementary Russian *i*-sounds, [i] and [ɪ] obliges the synchronic linguist to set up one common phoneme in both cases[150]; only practical solutions, based more or less on the phonetic substance in question can be offered, and not infrequently will they be based on historical considerations as in the Old English case, where we often read that it is the unvoiced variant that has systemic status. (Lass's use of the synchronic slant-bracket notation is due to historical knowledge). The synchronic problem is that the distribution (in syllables/words) of the respective sounds in modern English and Russian are predictable *by* [a] *general rule*. (Labov 1994:351. – Cf. Hock 1991:27-28).

As is to be expected, Campbell (§§ 449;444) recognizes the historical priority (of *s* and þ) of the unvoiced variants (*the basis of the alternations*), and the continued presence of /f/ and /v/, the systemic contrast having arisen when the accent shifted ('moved forward') in Germanic, certainly tells against making [v] a variant of /f/ in OE (a split plus merger).

---

148 Labov's version hereof is hardly correct (1994:333).
149 Furthermore the systemic-phonetic identity in the uninflected forms indicates the redundancy of the method.
150 Hjelmslev's solution to this problem – by the notion of *disinctive function* as against position – creates other problems: if (Hjelmslev's example) [h] and [ŋ] are two phonemes in English because both elements contrast commutatively with other phonemes as in *hat – cat* and *sing – sit*, then dark and clear *l* are also two phonemes: clear *l* in [laɪt] contrasts distinctively with *h* and *k* in [haɪt, kaɪt], and dark *l* in [kɪl] with *d* and *s* in [kɪd, kɪs].

0.4.3 Dualism and change

The next step in my argument deals with the (southern) voicing of initial fricatives from the 10th century onwards (Luick 1964:933-938), a process that disrupts the above complementary distribution[151]:

/ ?? /   [V-,-V-]   = voiced sound
         [-F]   = unvoiced sound

In these southern areas the dualist dilemma looks like this: the distribution of voiced/unvoiced variants is still predictable, but now the unvoiced variant is clearly the minority one.

The last steps in this phonemicization process is, partly, the voicing (lenition (Luick, §763 Anm. 10)) of weakly stressed fricatives [þə kíngəs] > [ðə kíngəz]; cf. *of* : *off*; *with*; *his*[152], partly the dropping of Final *-e* (Luick, § 782), which makes certain intervocalically voiced fricatives in OE appear in final position, cf. modern English *a life, alive, to live*; finally the early-15th-century loss of weakly stressed schwa (in inflections) disrupts the predictability of [-z] and [-s]:

/ ?? /   [V-,-V-, -V]   = voiced sound   → /V/
         [F-, -F]   = unvoiced sound → /F/

It is still to be proved that structuralist principles or a *langue-parole* description can tell us when the phonemicization of voiced and unvoiced fricatives in English was complete and whether completion and phonemicization are simultaneous. Finally, as to the relative chronology of changes the very logic of the changes is often sufficient to suggest a relative chronology. To illustrate with an example from Campbell (§444: OE *ræsde* [ræːzdə]: 1) intervocalic voicing of *s* is after PreOE rhotacism, which means that PreOE *ræsiða* had an unvoiced sibilant[153]; 2) OE *-de* indicates that *-ð- was voiced so as to develop into *-d-; 3) *-i- drops after *-ð- > -d; 4) voicing of *-s- is prior to loss of *-i-, otherwise OE would have developed *ræste.

---

151 The picture that Luick draws is extremely complicated, *e.g.* with French loan words retaining their unvoiced sound.
152 I contend that it is impossible for the *langue-parole* dualism to capture these mechanical processes:
   T1   T2   T3   T4
   /-əs/ > /-əz/ > / ?? / [-əz] (*box – boxes*) >   / ?? / [-əz] (*box-es*)
                           [-z] (*king – kings*; *duke – dukes*)   [-z] (*king-s*)
                                                                    [-s] (*duke-s*)
   What defines the transition from T2 to T3? Does the variant of T3 that was identical with the systemic element of T2 now become a variant of /-z/? Does the progressive assimilation that defines the transition from T3 to T4 require the complementary distribution of [-əz] and [-z], or the systemic element /-z/? Or in Samuels's diction: can variants which 'only' have existence in the spoken chain evolve there, *i.e.* independently of their eventual acceptance or otherwise into *langue*? – Does the language have 4 states with 4 systems, or one state with four different systems, or one state with one system? It should be noted that the systematicity of the modern English forms can be described exhaustively by means of predictable differential distribution in conjunction with the two universals, polysemy (the nominal and verbal contents of each form) and synonymy (the forms of each content).
153 Cf. also the OE compound *siþþan* 'since' as against PreOE gemination of **smiþ-jo-* > OE *smiþþe* 'smithy', and not OE **smidde*. Both words have unvoiced fricatives (Campbell, §§ 407; 444). – Note Modern English *smithy* [smɪðɪ, smɪþɪ]

To conclude: we have yet to see a narrative of this period in the English language's existence, a period flanked by PreOE *ræ:s-i-ð-, *cy:þ-i-ð-> OE ræsde, cyðde and the 15th-century development of *alive* [ə li:və] > [əli:v][154]: is this one voicing process? Is the 15th-century developments the completion of a process that started in PreOE, in the days of *Verner's Law*?

B. Isolative change or a simple vowel shift

It is noteworthy that Stiles's paper referred to above does not consider isolative change (*op.cit.*, p. 339); obviously, an isolative sound change cannot be phonologized, just as the conditioned voicing processes above are only a question of the status of certain (voiced-unvoiced) sounds: the phonologization of the sound changes is not different from the purported phonemicization of the respective allophones. The same applies to isolative change, and needless to say, the two processes make no sense in relation to continuations such as OE *a:* > ME *a:* in the Northumbrian dialects.

1. OE *a:* > ME *ɔ:* (South of the Humber).
   T1 /ha:m/ [ha:m] >
   T2 /ha:m/ [ha:m] careful (normal) register, [hɔ:m] less careful register/forceful style >
   T3 /hɔ:m/ [hɔ:m]. – Or: T1 /a:/ [a:] > T2 /a:/ [a:], [ɔ:] > T3 /ɔ:/ [ɔ:].

Following Samuels (1972:136) the systemic selection of [ɔ:] > /ɔ:/ is due to expressive needs or economy, its generalisation to the principle of least effort in the sense that the synchronic rule of the distribution of the above type of variants, being hard to perceive, gave rise to misunderstandings, all of which entails the loss of either variant in favour of the selection of the other (Samuels, p. 26).

The structural similarity of the two types of change is clear, each being a three-state process, with a start and an end. The only difference is that in the first type, the principle of least effort is allowed to run its immediate course, while in isolative change – and isolative continuation – the intermediate state is more complex. However, in either case the eventual selection into the successor state of T3 of the given variant entails the imperfection-notion in the shape of faulty imitation or transmission or repeated misunderstanding (Samuels, p. 10). The view contains an element of inconsistency: in conditioned change the synchronic (style- or register-)rule is easy to perceive; so why should this state of affairs, the successor state 2 of T2, not be continued, but eventuate in state 3? In isolative change, the synchronic (style- or register-)rule is not easy to perceive; hence the synchronic variation of the successor state of T2 is not continued, but eventuates in state 3[155].

The causal difference between the two types of sound change – some conditioned changes appear to have overt causes, while isolative change is covertly motivated, and in the voicing processes in Middle English we saw how the two types yielded the same phonological re-

---

154 A theoretical prerequisite would be agreement on how to define the phoneme (cf. Samuels 1972:38), a desideratum Fischer-Jørgensen's 1975-work is a monumental proof of. – In the above examples generalisation or phonemicization has been produced by 1) loss of a synonymous lexical item, 2) loss or change of conditioning factor, 3) extension of conditioning factors, 4) conditioned sound change.
155 The preceding discussion has not touched upon the question of phonemicization.

sult[156] – is obliterated if we make do with the three-step process of a structural or formal analysis: a (new) variant arises at T2 and this variant is phonemicized at T3. The main objection to this type of explanation is that it does not distinguish between theory and cause: the phonemicizing factors are empirical processes in their own right[157] and are as such in need of explanation (a theory) as well as the purportedly empirical process of [x] > /X/: the loss of PreOE *-i is the cause of – not the theory of – the phonemicization of *y, ø. And if phonemes (systemic elements) are empirical entities, then phonemicization of[158] *y, ø may just as well be the cause of – not the theory of – the loss of PreOE *-i. Thus we have ended up with the two so-called theories of language development, the Mechanical School and the Functional School, respectively: is it the mechanical processes that shape the system of the everyday language or does the system shape the spoken and written substance of it?

0.4.4 The period and developmental series

A theory based on notational conventions such as the slant-bracket convention of modern structuralist theories as well as the formal machinery of the theories I have dealt with runs the risk of ending in form without meaning: the conventions remain undefined. Furthermore they shape their data, if not beyond recognition, then to such a degree that the hallmark of history, the period and temporal dimension, is literally smashed to smithereens so that history becomes divorced from, in Lévi-Strauss's words (1962:342), its peculiar code, namely chronology. The reason for this is that development or history on the whole is envisaged as something with a start and with an end, and change may be handled adequately and sufficiently within the confines of two *per se* autonomous entities, called states. *E.g.*, the Generative School can only work with a two-state conception of sound development (cf. Hock 1991:27; Chomsky and Halle 1968; Labov 1994), the *langue-parole* dualism works with 3 states[159]. However, this period-concept must be taken with a grain of salt, as the three-state model only appears when we deal with change:

T1	T2	T3	T4	T5	T6	T7
OE 789	OE 790	OE 800	OE 900	OE 1000	11th c	12th c
/a:/	/a:/	/a:/	/a:/	/a:/	/a:/ [a:] [ɔ:]	/ɔ:/

So the dualist model reduces to the conventional two-state view of change in that the *parole* level is only invoked when we need to demonstrate where the change in question 'comes from', while continuation from, say, T2 to T3 needs no such superfluous level. However, the intermediary *parole* level is called upon for theoretical reasons in that for a language to

---

156 It is interesting to note that while intervocalic voicing of fricatives (OE) as well the voicing of weakly stressed fricatives in late Middle English is called conditioned change, the cause of the Kentish voicing is not initial position (Cf. Luick 1964:§703 Anm. 9).
157 Since their respective types are recurrent, they may qualify for substantive features.
158 Is the step from variant to change able to enter into a *Kausalnexus* with, *in casu*, the loss of a sound?
159 The state in the middle is transitory or intermediate, *i.e.* in relation to the beginning and completion of change.

change, it must be used (Samuels 1972:2); hence the developmental series above must be rewritten as follows:

T1	T2	T3	T4	T5	T6	T7
OE 789	OE 790	OE 800	OE 900	OE 1000	11th c	12th c
/a:/ [a:]	/a:/ [a:]	/a:/ [a:]	/a:/ [a:]	/a:/ [a:]	/a:/ [a:]	/ɔ:/ [ɔ:]
						[ɔ:]

a series that exhibits a degree of redundancy and some inconsistency, seeing that not all states exhibit variation. So the following model is both consistent and makes the structure of continuation and change equilogical, on condition of the universality of style-variants and if the universal of constant change is disregarded:

T1	T2	T3	T4	T5	T6	T7
OE 789	OE 790	OE 800	OE 900	OE 1000	11th c	12th c
/a:/ [a:]	/a:/ [a:]	/a:/ [a:]	/a:/ [a:]	/a:/ [a:]	/a:/ [a:]	/ɔ:/ [ɔ:]
[æ:]	[a]	[ɔ:]	[?]	[?]	[ɔ:]	[o:]

This example and the abovementioned developmental series 'a > æ > ɛ > e > ɪ > i' presuppose a period-concept, and if as it were we straighten out the dualist model, accepting the equiempirical nature of all historical data, then we shall have a developmental series like this

a > a > a > a > a > ɔ > ɔ > ɔ >

in addition to this type

a > æ > ɛ > e > ɪ > i.

With reference to Labov's New York City series, a third type is possible:

a > a > a > æ > æ > æ > ɛ > ɛ > ɛ.

This type of periodic data will be the basis of the historical theory that will be developed in the following chapters; the slant-bracket notation is superfluous.

In my exposition of isolative and conditioned change above I mentioned the selection process into *langue* of a particular variant only in passing, contending 1) (with Samuels) that the system plays only a minor role in this process, 2) (against Samuels) that the process can be viewed as purely phonetic controlled by the principle of least effort, and 3) that all change is sign-functional. Samuels' theory of the rise of conditioned and isolative changes was such a sign-functional theory *in embryo* seeing that the style parameter forceful style *versus* relaxed style constituted one of the conditions under which various types of variants arose; furthermore, Labov's – and Milroy's – sociolinguistic views of change also imply the assignment of a (social) content to a given form. Following Samuels our Middle English Southumbrian change originated in forceful style, and each of the two hypothetical series' elements above will be correlated with a forceful style content, too. Conversely, variants originating

0.4.4 The developmental series

in relaxed style will typically be lower(ed) variants; however, if we do not work with a simplified notion of language development, accepting continuation as an equiempirical process with change proper, then the simple FS-RS parameter is not adequate, seeing that series like 'OE (8th c) a > OE a > OE a > …. > OE (11th c) a > …' also deserve explanation; the elements of a developmental series characterized by continuation will be assigned a normal style content (NS). The normal-style content is in fact what Samuel's own theory presupposes, which will be shown in my criticism of Samuel's view of the Great Vowel Shift.

In the first place, the Great Vowel Shift is an empirical fact *qua* an example of *homeostatic regulation* in which functional and mechanical factors work together to preserve communicatively *adequate distinctions* (Samuels, p. 42); Samuels seems to be following Luick's push-chain view of the Great Vowel Shift (Luick 1932:89-90). Secondly, the unity of the Great Vowel Shift is suggested by the presence of mechanical factors for a certain period (c. 1400-1500) which are typical of forceful style; these factors create *stressed forms* which eventually become devalued (Samuels, p. 43); this devaluation creates instabilities (*imbalance* (p. 42)) that will be redressed by the functional factor of distance-maintenance.

On this view, the raising of vowels contains a dilemma: a vowel raised in *contexts calling for greater stress* (Samuels, p. 41) will – in respect of its over-all carrying power or acoustic prominence – lose on the roundabouts of sonority what it gains on the swings of greater muscular energy (*i.e.* pulmonary energy). As long as a vowel can continue to be raised (cf. the second developmental series above) this does not matter, but when the high vowels /iː/ and /uː/ cannot be raised further without losing their vowelhood, this forceful style variant will be devalued, and relaxed style variants, already existing in the spoken chain, may *in time be selected as more effective <u>stressed</u> forms than those nearer the original* (p. 42), owing to their greater degree of sonority. This accounts for the selection of diphthongal variants in the 15th c and their continued existence in the following centuries instead of monophthongs: [iː] > [i̯ː] > [ti] > [əi]; [uː] > [u̯ː] > [ʊu] > [əu].

While the two first variants in each series are FS, the diphthong(al sounds') development is clearly a continued relaxed style development and cannot as such assume the function of a stressed form: the conditions for their continued existence are incompatible with forceful style function (raising/fronting). Such originally RS variants must lose their stylistic differential by devaluation[160] for them to be reinterpreted – as normal forms – for reasons of expressiveness or economy (p. 137). Samuels's model (p. 137) does not address the following issues: can a variant exist and develop as only a variant (cf. above)? Is the step from variant to change involved in the selection process? Is this step bypassed by such imperfection factors as misunderstanding, faulty transmission (p. 10)?

---

160 In Samuels's theory the process of devaluation is 'imperfective' in that it creates pressure points or instabilities, something to be redressed. I shall demonstrate how it is possible to envisage a process in which the development of one content to another may occur gradually – without invoking the hearer's misunderstanding. – If Samuels had applied the slant-bracket notation to his selection-model on p. 137, the moot points in his theory would have been clarified; to labour the point: is the selection of a given *parole* variant into *langue*, or continued *parole* existence? And does the selection entail the loss of the other (redundant) variant? If this happens, selection for expressive reasons is also economical. Samuels's use of *restructuring* and *redistribution* (pp. 26; 137) explains nothing, the concepts being merely made to refer descriptively to the start and end of a process.

Samuels anticipates a seeming inconsistency (cf. Stockwell and Minkova 1988:357): how can opposite variants – RS and FS – be selected into a *single register* (p. 25)? The latter imprecise term must suggest the concept of normalcy, as it makes no sense to say that [i:] *qua* a forceful style variant of /e:/ and [əi] *qua* a relaxed style variant of /i:/ are selected into either a FS or a RS register. There is no inconsistency involved here in that Samuels's theory implies this acoustic equation: the carrying power or over-all auditory prominence of two differently produced variants may remain the same so that FS [i:] = RS [əi][161]. This equation dispels part of the mystery about sound change; from the point of view of **expression** (form) the continuation of either of such variants involves no break or imperfection in relation to rules (laws) and communication: they are perceived as 'likes'.

Samuels's theory observes the sign-functional expression-content distinction so strictly that he disregards the expression-content interdependency and therefore does not exploit the fact that his variants are not mere phonetic-material entities, but are fullfledged signs, existing in what I shall call *attitudinal* systems[162]. A clear indication of a sign-functional view of language development *in embryo* is seen in Luick (1932:89), who realised that push-chain mechanisms cannot be adequately grasped if the sounds (*Die Laute*) involved are taken as mere physical entities (*im naturwissenschaftlichen Sinne*); the historically active elements are *Die Phoneme*, because some kind of psychological *Faktor* is associated with the sounds of a language, and this content delimits the single phoneme vis-à-vis the other phonemes[163]. This brings us on to the question of contrast, distinctiveness and difference.

0.5. What really bothers the synchronic linguist: contrast – distinctiveness – difference

Structuralism's basic principle, *tout est différence*, is science's *concessive assumption*, according to Twaddell (1967): *No two things and no two events are exactly similar*. And as a corollary of this, all utterances are *phonetically different* (Twaddell, p. 69). Why this adverbial qualification? Apart from characterizing the scientific domain at hand statically, it implies that there is an element of similarity in the empirical differences (utterances) which permits us to say that certain utterances ('things' and 'events') are so similar that they are the same. This sameness is determined by *phonetically significantly alike* and different utterances will therefore be *phonetically significantly different* (p. 69). Why this extra qualification and how did Twaddell leap from the initial assumption to the twin corollaries? Like most members of our Western Civilization, and in particular of the modern synchronic circuit, he made this intellectual leap from the scientific convention that *if an overwhelming majority of observers agree in recognizing qualitative differentiations, the phenomena observed are different; if not, the phenomena observed are the same* (p. 68). Sameness is therefore absence of qual-

---

161 The application of greater/less stress entails less/greater sonority: [+stress → raising → -sonority] = [-stress →lowering → +sonority] in respect of acoustic prominence. Cf. Labov (1994:220) for a more detailed discussion of the inverse relationship between vowel height and sonority.
162 Borrowed from Lévi-Strauss, cf. below.
163 It is to be noted that this mental content is not sufficient to make the particular spacing of a system's phonemes continue.

itative *qua* significant differences. Unfortunately, nobody has been able to set up cogent criteria for (or to define) a(n in)significant difference – so as to enable scholars outside the consenting majority to concur with them in these (criteria). This means that modern English dark and clear *l* are similar, while /h/ and /ŋ/ exhibit significant/qualitative differences and insignificant/nonqualitative similarities: when in PreOE *[y] arose from *I*-mutated *[u], it exhibited so insignificant differences and significant similarities as to become a variant of */u/, and in historic times the two sounds became insignificantly similar – both were vowels – and significantly different – because something occurred to other elements of the Old English language: now certain qualitative differences appeared between (*)[u] and (*)[y].

What is the same and insignificant is established arbitrarily, by convention: Twaddell must have realised that the consequent entity, *in casu*, the phoneme, could not be an empirical phenomenon, as he called it an *abstractional fictitious unit* (p. 68)[164].

Twaddell's position is reminiscent of Bloomfield's. In his 1926-article *A Set of Postulates for the Science of Language* Bloomfield does not define our two crucial concepts, difference and similarity[165], but states that [s]*uch a thing as a 'small difference of sound' does not exist* (p. 157), it being understood that a small difference (in either sound or meaning) belongs to non-linguistic reality (cf. p. 155), and in addition to *significant* the concept of distinctiveness is introduced postulationally (a phoneme being a *distinctive sound* (p. 157, Def. 16)). Small differences are not distinctive, but if *tout est différence*, then we are bound to make this distinction (cf. Twaddell, p. 68): a significant difference[166] is (linguistically) distinctive, hence linguistically relevant; an insignificant ('small') difference is (linguistically) non-distinctive, hence linguistically irrelevant; a small difference refers to *non-linguistic shades of sound and meaning* (p. 155). The theoretical problem is that Bloomfield cannot tell us when a difference is small enough to be small, and that we must work with linguistically relevant *degrees* of differences: not all allophonic differences belong to non-linguistic reality. The linguistic problem is that there are such 'things' as small differences.

Trager and Smith (1975:19) come up with this common sense position: a non-significant difference between two sounds is linguistically relevant, if they are in complementary distribution (dark and clear *l* in modern English and PreOE *u* and *y*), and if they are phonetically similar (dark and clear 'l'); thus the phonetic difference between [h-] and [-ŋ] constitutes a significant difference, and since the two sounds also exhibit *pattern congruity* with other phonemes in modern English, they belong to the systemic inventory of modern English phonology, /h/ and /ŋ/. Twaddell has it that if sounds (utterances) are *phonetically significantly like (different) events*, they are *phonologically alike (different)*. **Differences** are therefore **1)** irrelevant, if too small; **2)** relevant, if small and non-significant (allophones), **3)** relevant, if non-small(!) and significant (phonemes).

The entities that we have been examing are devoid of meaning (cf. Bloomfield 1971:239-240). They are the meaningless elements (phonemes (Bloomfield 1926:157, Postulate 18)) of language that combined constitute signs – in Bloomfield's terminology *glossemes* (*loc.*

---

164 I shall not be concerned with Twaddell's micro-phonemes and macro-phonemes (pp. 69-70).
165 In his Fifth Postulate Bloomfield merely substitutes *same* for *alike*, and *different* for *not same*, that is the definiens and the definiendum are two synonymous words.
166 Significant phonetic differences correspond to articulatory differences (Twaddell, p. 68).

cit.; and 1926:161, Postulate. 50[167]). We saw above that Luick's view of the material aspect of language involved a content-like element, and that Samuels's view of sound change involved style contents. Somehow, these as well as the sociolinguist's content-elements are disappeared in the methodological process that moves from phonetics – the spoken utterance – to the systemic level of phonemes (phonology); the reason for this is that only – in Bloomfield's diction – forms (morphemes) can be analyzed into smaller, meaningless, entities, which contents (noemes) cannot (cf. Bloomfield 1926:Postulate 16).

However, Bloomfield (1926) was very close to a sign-functional conception of language, had he not made the arbitrary cut between forms *qua* morphemes (signs) and non-forms (non-signs), phonemes or *distinctive sounds* (p. 157, Postulates 16, 18; p. 155, Postulates 6, 9): this means that Bloomfield's synchrony does not make allophones appear from the analysis of phonemes – they are content-carrying elements of the spoken utterance, probably belonging to non-linguistic reality; but if we move these content-elements into the linguistic sphere, Bloomfield's Postulate 7 to the effect that [e]*very utterance is made up wholly of forms* can still be asserted on condition that the form-concept[168] be so defined as to include style/register-contents as well as 'social' contents. Such a redefinition entails a greater degree of parallelism than Bloomfield envisages between form and meaning: Bloomfield has it that a form is [a] *recurrent vocal feature which has meaning*; there is therefore a fixed and stable relationship between form (expression) and meaning (content). It goes without saying that the style contents of (some of) the 'forms' of the utterance are not to be correlated with a definite or recurrent set of *vocal features*; but if our perspective is that of the meaning, then I shall demonstrate how it is possible to state the structural conditions under which a style content, such as forceful style, relaxed or normal style, can be regarded as a recurrent, not vocal, but meaning feature. The concept of 'a small difference' will be reinstated as a linguistically relevant unit because it falls under the scope of what is sign-functional. This sign-functional conception of language leaves the concept of the phoneme in a precarious situation, because we must solve the paradox that if synchronic analysis analyzes the phoneme into smaller allophonic units, then this analysis goes from meaningless entities to meaningful entities; or in more realistic diction: how can a meaningless entity generate meaningfulness on its route from system to utterance, where it is manifested? All this boils down to the question: who needs the phoneme as an empirically relevant systems element[169]? And to the conclusion that the synchronic tradition has not come to terms with the empirical reality and relevance of (Bloomfield's) 'small differences' and the 'significant and insignificant differences' of the systems concept.

---

167 All meaningful elements are glossemes, and the meaning of a glosseme is called *a noeme*, with *sememe* the unalyzable meaning of a morpheme.

168 *The vocal features common to same or partly same utterances are forms* (Postulate 6).

169 The first prerequisite is that a generally accepted definition of it is current, which is not the case. – Jespersen (1970:213-215; cf. Twaddell 1975:77-78) even suggests another term *glottisch*. Secondly, the concept itself needs examining with a view to its over-all realistic, if not metaphysical status. I agree with Twaddell (pp. 55-56) that definitional *hair-splitting* is to be preferred to *methodological chaos*, as well as with Sapir (1925:43) that [i]*t is time to escape from a possible charge of phonetic metaphysics* (…); cf. Twaddell (p. 75), who warns against a mythological hypostatization of phonemes.

0.5 What really bothers the synchronic linguist

The historical investigations reported in Labov's 1994-work demonstrate that Bloomfield's stance cannot be upheld. Small differences, acoustically significant and linguistically relevant, do exist[170].

Categories are the *fundamentum divisionis* of synchronic linguistics, as Hjelmslev did not tire of stating in 1928. A category is a self-contained *sui-generis* entity so that the gap between two categories is absolute in the sense that there is (at least) a minimal[171] difference between them that establishes the two entities as discrete, totally independent of each other. As such there is no difference smaller than the minimal difference between two categories. Such a difference is significant, distinctive and contrastive.

Needless to say that through time no category's borders are sacrosanct; all languages have been subject to mergers and splits, and no category has not been exposed to expansion or otherwise of its scope ('elbow-room'), and synchronically discrete categorization has not been entirely successful; phonetic elements with restricted or predictable distribution (Labov, p. 348) defy attempts at phonemicization and the question of the step from variant to change is also left to the analyst's theoretical discretion, and the same applies to the definition of differences.

In order to establish empirically relevant contrasts Labov proceeds from the four mathematical possibilities that the pronunciation (the speaker's point of view) and the perception (the hearer's point of view) of two utterances yield: in a minimal-pair test and a commutation test two words may be

A   pronounced differently and perceived as different – cf. Twaddell above;
B   pronounced differently and perceived as the same
C   pronounced as the same and perceived as different
D   pronounced as the same and perceived as the same – cf. Twaddell above.

The production of and perception of a significant difference (A) and the production of and perception of a similarity (D) are the basics of the categorical view of synchronic linguistics and constitute the *raison d'être* of the symmetrical conception of the speaker and the hearer. (B) is regarded as impossible, and (C) barely possible under certain conditions, but irrelevant to the linguist (p. 357). However, (C) assumes special importance in generative phonology, and since the theoretical issues involved here throw light on my discussion in section 0.4.1 above, I shall discuss the C-scenario with reference to Hock (1991:247-251).

Hock (cf. also pp. 238-247) discusses devoicing of final -*d* in modern German, where the nouns *Rad* and *Rat* are pronounced the same [ra:t]. In spite of this native speakers claim that no merger has occurred, and the generative phonologist explains this through the native speaker's *subliminal knowledge* of his or her language: there is a systemic /d/ underlying all the declensional forms of the noun <Rad> which will be realised differently according to grammatical context. While the generative phonologist cannot prove his claim, the simplest

---

170 The following exposition (of the near-merger concept) is based on pages 348-418 in Labov 1994, and it must be noted that Labov's analyses only refer to acoustic-articulatory differences, not semantic ones, except indirectly.
171 Cf. the binary principle (Martinét 1964:73).

solution would be that some/all native speakers of modern German are influenced by the written language[172] – a type of solution that we shall be dealing with below.

Possibility (B) is impossible because it implies the empirical relevance of Bloomfield's small differences *qua* significant[173] in that the unperceived different sounds cannot always be regarded as belonging to the same phoneme. Examples: In *r*-less dialects in American English (New York City) /oh/ *sauce* and /ohr/ *source* merge in an in-gliding [-ə] diphthong; but Labov's acoustic measurements show that the two vowels are produced *with a marked phonetic difference* (often a formant-2 difference; p. 359). Labov calls this a near-merger as against the merger (D) and distinction (A); another example, the so-called *Bill Peters effect* (p. 363) may seem more convincing: an informant of Labov's Bill Peters (80 years old) maintained in his spontaneous speech the distinction between /o/ *cot* and /oh/ *caught* in a speech community (Pennsylvania) which otherwise was characterized by the merger of the two phonemes (the incoming norm); in a minimal-pair test situation Bill Peters showed a near-merger of the two phonemes, which Labov interprets as a case of style-shifting – due to the formality of the test-situation – between informal and formal speech, the informant having unconsciously become aware of and adopted the new norm, but he still produces markedly different sounds that are judged to be similar.

Labov interprets these facts, near-mergers, to the effect that the categorical, synchronic point of view cannot be upheld – *without some modification*, and since these modifications fatally touch the very conceptual basis of synchronic linguistics – he enumerates its components himself (p. 351): the concepts of (1) contrast, (2) distinctiveness, (3) discreteness, the notion that phonetics is irrelevant (4), the symmetry between speaker and hearer (5), and (6) the reliance on native speakers' intuition – synchronic linguistics needs more than modification of its fundamental assumptions, rather a general overhaul *de la raison synchronique*.

However, Labov does not – in my opinion – draw the necessary consequences of his own analyses, but uses the near-merger type of change, established from the living language (his uniformitarian axiom), to explain what he calls unmergings (pp. 371-390). I shall deal with few examples that are relevant to my purpose.

---

172 In OE the noun *muþ* 'mouth' exhibited similar morphological variation: the root consonant was unvoiced in final position, but voiced intervocalically (cf. the German genitive form [ra:dəs] of *Rad*). No synchronic phonologist can prove that the underlying fricative in OE was /þ/ and not /ð/; if we use the German example as a methodological analogy, the voiced sound is underlying in OE /mu:ð#/. Unfortunately there is no evidence for this postulate, which both contradicts the previous history of the word and is contradicted by the word's continued existence into modern English, where [þ] survives in the singular, and [ð] in the plural; cf. also *life, alive, lives*; *staff – staffs*; *stave – staves* from OE *stæf* (sing.) : *stafas* (plur.). The generativist (Hock) may counter: the example is not parallel, because at the relevant point in time OE did not have a systemic rule (like modern German) that made all final obstruents unvoiced: [+obstruent] → [-voice] / ___ #. Secondly, the English native speaker then may not have merged the two sounds and judged them to be the same. That is correct, but it does not dispel the arbitrariness underlying the systemic forms postulated by the generative method. So what the generativist really has to do is to prove these three statements false: (1) that the underlying form of German <Rad, Rad-es; Räd-er> is /ra:d#/ can only be substantiated through the synchronic linguist's historical knowledge of the word in question; (2) that the reason why some/all(?) native speakers judge [ra:t] not to be a merger of <Rad> and <Rat> is not due to their subliminal knowledge of the contrast between the two underlying forms /ra:d#/ :: /ra:t#/; (3) that the reason why some/all(?) Germans overrule what they actually perceive as different sounds is not due to their subliminal knowledge of underlying forms, but due to their knowledge of the written language and/or to analogical inference.

173 See David DeCamp's letter to Labov (in Labov 1994:368-370).

In generative phonology, we see a self-imposed problem; push-chain mechanisms are impossible, because the discreteness of its rules would imply a merger (Labov, pp. 199-200; cf. p. 224); Luick's *Verdrängungs* mechanism is impossible because if a sound /e:/ [- high, -low] is raised to become [+high, -low], it would merge with /i:/ [+high, -low]; the same type of argument as Dobson (1968:660) uses to reject this developmental series: ME /i:/ > /ei/ > /ɛi/ > /æi/ > /ai/, because at some stage the non-high sounds would merge with the raising of ME /ai/ > /æi/ > /ɛi/ > /ei > /e:/: *die = day*; *my = may*. Mergers are irreversible, so both viewpoints would imply impossible unmergings. Instead a so-called flip-flop rule (Chomsky-Halle 1968:255; Labov, pp. 147-148) is introduced that so explains away the merger problem that we are reminded of Nietzsche's criticism of the misuse of the historical method: Chomsky and Halle claim that ME /i:, u:/ and /e:, o:/, nonlow and tense vowels, underwent simultaneous lowering and raising, respectively, and changed positions, without merging: the high vowels were lowered (to nonhigh, nonlow) at the same time as the midvowels were raised (to high, nonlow) without meeting; and now the explanation: the environments of the two sets of vowels were different: the lowering rule does not hit the *i*- and *u*-sounds that <u>come from ME /e:, o:/</u>, and the raising rule does not hit the *e*- and *o*-sounds <u>that come from ME /i:, u:/, since only reflexes of ME /i:/ and /u:/ were diphthongized</u> (Chomsky and Halle, p. 255); the quote signals that the diphthongization rule could occur either before or after the lowering of the high vowels (p. 256). The whole explanation is an odd mix of a kind of historicism that has a successor sound remember its origin, an origin that determines something in a given present, and of unprincipled chronology.

The type of problem that confronts us here is solved by Labov's peripherality-concept, whereby the phonetic quadrilateral (the oral cavity), regarded as a formant space, is divided into three zones (pp. 170-177). The traditional threefold (articulatory) division of the oral cavity into a front region, a central region and a back region, which leaves us with a choice of three types of vowel, is further divided on the basis of acoustic measurements so that the front and back regions may each have two 'tracks' ([+/- peripheral]) in which sounds may move: sounds may move 'up' and 'down' without colliding, provided they move in different tracks. Thus the above mergers and subsequent unmergings present no breach of the principle of the irreversibility of mergers, if the respective sounds move in different tracks: /ɔ/ (near Cardinal 6) could develop into /u/ without merging with /o/ (near Cardinal 7), which was lowered to /ɔ/ in the Swiss Valais dialect (Labov, pp. 141; 232); the raising of the open *o*-sound occurred in the [+peripheral] track, while the lowering of the close sound occurred in the [-peripheral] track.

Labov uses the near-merger concept and the feature [+/- peripherality], whose acoustic definition is translated into articulatory terms, to make Bloomfield's small differences linguistically relevant; the small difference of near-mergers, in which speakers produce markedly distinct sounds which cannot be perceived as such, makes all the difference in the above and similar cases and in the unmerging of the 17th/18th c merger of ME /i:/ and ME /oi/ [ui] in [əi] (cf. Labov, pp. 371-376). The standard solution to this unmerging is that the reversal was motivated by the written language (p. 308):

ME /i:/ > ... > ...   > /ai/ <line>
             merge in [əi] which unmerges into
ME /oi/ > ... > ...   > /oi/ <loin>

According to Labov the two sounds never met (cf. Blake (1996:258) for the standard interpretation), because ME /oi/ followed the non-peripheral path of the back region, while the development of ME /i:/ was never retracted so much, remaining in the non-peripheral track of the front region; but at the crucial point in time they constituted a near-merger like the New York City example of /oh/ and /ohr/.

Labov's analysis is corroborated by the modern dialect situation in Essex (pp. 377-384), where we have three different types of locations: locations where the southern standard developments of ME /i:/ and /oi/ have been reached (/ai/, /oi/), locations where we have a near-merger of /ai/ and /oi/ in the inaudible difference between a very centralized and rounded *ai*-diphthong and a slightly backer *oi*-diphthong, and locations with a complete merger.

I have dealt extensively with Labov's 1994-work, not only because of his criticism of the synchronic point of view, but also because it has a notable affinity with Samuels's 22-year-older book.

Samuels's use of genetic relations between (two) synchronic variants and his generalization of the notion of synchronic variation ('styles') to be valid for all periods of a language (1972:10-11) are examples of how the present may explain the past (Uniformitarianism), but more importantly, the 'small' differences that Samuels applies are not the meaning-less entities of the American tradition (Bloomfield/Labov), but sign-functional in nature. I shall illustrate my point in the light of the so-called Great Vowel Shift.

The basic problem with explanations of chain shifts of this magnitude is that no single, uniform explanans can be found (cf. Stockwell-Minkova's criticism of Samuels above). In the Great Vowel Shift, the initial problem is how to reconcile diphthongization and raising. Samuels solved it by his prominence-equation[174] of style-contents and notion of devaluation. Generative phonologists (cf. Chomsky and Halle 1968) set up two distinct and unordered rules; Labov tries to reduce this number to one, making the diphthongization rule a kind of corollary of the Vowel Shift rule proper, which is a raising rule (pp. 247-252).

Labov's raising rule applies to all sounds (also /i:/ and /u:/) – so did Samuels's forceful style factor. While the rule can apply to /e:, ɛ:, a:, ɔ:, o:/ without problems, its output will defeat its own purpose when it is applied to /i:/ and /u:/, since the raising of these two sounds will produce semivowels /j/ and /w/. What the sounds gain on the swings of raising, they will lose on the roundabouts of sonority; no longer can they function as syllabic peaks or form a syllabic nucleus. The system(?) must therefore redress this output, and it(?) supplies[175] a syllabic nucleus: [i:] > [j] > [ij]. – In Samuels's theory the continued raising of the FS variant of /i:/ as well as /u:/ defeated its own purpose for reasons of sonority <u>and</u> devaluation; hence the more sonorous variant [ʊi], already existing in the spoken chain, is selected. The essential difference between the two points of view is that Labov creates an empirically irrelevant situation (cf. below), whereas Samuels's FS variant has grown less functionally viable than other variants, creating an imbalance at the level of the spoken substance, seeing that the FS-RS parameter is no longer implemented

---

174 Samuels (p. 22): the RP diphthong *[ɛɪ] does, at one point of its span, achieve the same acoustic effect as the monophthong [e:]* of modern Scots. Thus the small differences between the two sounds may be perceptually inaudible (cf. Labov's B-possibility).

175 Labov (p: 248): *In the actual development of these sound changes, a sonorous nucleus is supplied.* The diphthongization rule may have been eliminated, but it raises the question: who or what is the agent of the passive *is supplied*? Cf. below. – It is to be noted that Labov's semivowels are not syllabic.

0.5. What really bothers the synchronic linguist

The next step considers the communicative-phonological status of the variants [ɪi], [ʊu]. Apparently, Labov must here accept the irrelevance of a Bloomfieldian small difference, because these diphthongs are not sufficiently different from the raised products of ME /e:/ and /o:/ > /i:/ and /u:/ (pp. 238; 249), and thereby forfeited the possibility of a near-merger in the 15th c of ME /e:/ and /i:/ – all because he does not accept a consistent sign-functional view. A small difference established by [+peripheral] /i:/ (< ME /e:/) and [-peripheral] /ɪi/ (< ME /i:/). While Samuels does not consider this issue, we may apply his theory in the following way: the 15th-century situation in the region of the high vowels came to constitute a threatening merger (from the sounds' perspective) or a potential homonymic clash of some magnitude (from the content's perspective) (Samuels 1972:69-70), hence, in Labov's terms, the two high up- and outgliding diphthongs followed the 'downward' non-peripheral trajectory, thereby gaining in sonority what they *qua* relaxed style sounds will lose in terms of expiratory prominence (cf. Labov, p. 220).

The remainder of Labov's narrative of the Great Vowel Shift presupposes a bimoric conception of all long vowels, which therefore partly answers the question raised above: instead of a developmental series such as [i: > j > ɪi], the development looks like this [ii > ij > ɪj], and I shall quote:

> The immediate consequence of raising is a qualitative change in the relation of the two morae. As one mora is raised to ([j], [w]), it loses its [+vocalic] character; the other retains it. The raised mora is also more peripheral than the unraised one; like all phonological features, peripherality is contrastive, and the unraised mora automatically becomes [-peripheral]. (Pp. 248-249).

This is sheer *adhockery*[176]. The bimoric explanation does not apply to the other raised sounds, *e.g.* /e:/ = /ee/ > /ii/; here no dissimilation sets in, both morae are raised and keep their [+peripheral] feature; why should the height-contrast between [ij] – in Labov's system a [3 height] mora versus a [4 height] mora – be increased through an extra peripherality-contrast? And if peripherality is contrastive, then it must always be so contrastive that the two morae of ME /e:/ and for that matter /i:/ can never have parallel specification: [i+i] // [e+e] = ([+vocalic, +peripheral] + [+vocalic, +peripheral]) (cf. pp. 249-250).

In conclusion: like Samuels Labov envisages the process as simultaneous (p. 251), the raising factor being applied to all the sounds involved; but Samuels's diction is simpler and more realistic: *why do forceful-style variants preponderate* here in this period of the history of English (Samuels 1972:25)? It may be a causal answer to Labov's unasked question to the Vowel Shift's raising process. And two final comments on Labov's theory: (1) the theory is not simpler or less heterogeneous than a diphthongization+raising process, seeing that a dissimilation rule is necessary as a ?functional corollary of the raising rule; (2) this corollary is not consistent with Labov's claim that 'small' differences in peripherality are sufficient to keep phonemes distinct, *in casu*, [i:, u:] (< /e:, o:/) from [ij, uw] (< /i:, u:/); this near-merger could have been continued, and in order to make the change fit the facts, Labov (p. 250) must

---

[176] Labov follows Stockwell and Minkova's diphthongal conception of the long vowels in OE and ME, to which we must add this theoretical difference: Stockwell and Minkova (1988:377) apply a principle of diphthong optimization to explain the diphthongization process, from [ɪi] to [aɪ] and [ʊu] > [au], and in contrast to Labov they maintain that such small distinctions as [u:] :: [ʊu] are systemically *viable*.

appeal to a third causal-like factor: no phonological system has been registered to have contrastive distinctions between /ij/ and /i:/, /uw/ and /u:/.

Peripherality is not a feature of all sound systems, and there are limits to its applicability to high vowels (Labov, p. 238); so while maintaining Labov's near-merger concept, I shall argue that as a variant [ɪi] is sufficiently distinctive vis-à-vis the variant [i:] without having recourse to the peripherality feature. The curious thing is that Labov provides the mechanism himself in his analysis of the Belfast near-merger of *meat* and *mate* (Labov, pp. 384-390), a distributional explanation that Samuels (p. 69; cf. below) also applies.

In the (non-standard) vernacular of older speakers in Belfast (Labov, p. 385; Milroy and Harris 1980), the vowels in *meat* and *mate* are considered to be the same, *i.e.* a seeming merger of ME /a:/ and /ɛ:/ which took place by 1700 and continued into the present. Labov accepts the perceptual-judged identification and questions whether they are the same in production. He concludes that despite a 50% overlap in pronunciation, the other half of the examined instances maintain an articulatory difference between the two sounds. This distinction could not have been recreated after the alleged 17th-century merger, and the change then was a near-merger; thus it is possible for speakers to preserve a phonemic contrast that cannot be perceived or which is not judged as such. This maintenance is made solely on the basis of the existence of *parole*-variants. In *parole* the (phonemic) contrast was maintained, not because of physical differences, which certainly were small, but because of frequency of occurrence; either phoneme was defined by a distribution of its variants that was different from the other phoneme (differential distribution). I must labour the Belfast example, because of its significance to my argument and because it makes Labov's peripherality-definition of a near-merger redundant: the variants of the two phonemes are these, ordered according to nucleus height

ME [a:] /mate/	[ɪə]***	ME [ɛ:]: /meat/	Ø
	[eə]****, [e]**		[eə]***, [e]**
	[ęə]*, [ę]		[ęə]***, [ę]****
	Ø		[ɛ]**

It will appear from the variants that differential distribution may define the respective phonemes, and this purely material definition is underscored by frequential distribution (high/low frequency indicated by *'s). Such mechanical distinctions (cf. Luick's *naturwissenschaftlicher Sinn*) are of no avail to the listener; in the spoken chain [męət] is a homophone of *meat* and *mate*, and so is [męt]; but to the speaker these distinctions and differences are sufficient to continue the two phonemes in conjunction with the style-contents that follow their respective realization conditions.

Significantly, Labov does not venture into a phonemic interpretation of his data, and Samuels also avails himself of words rather than of his *langue-parole* theory. Apparently we do not need the dualism, let us therefore interpret the data in terms of developmental series starting in the 16th century (Labov, p. 389[177]), with raising and diphthongization

ME ɛ: ɛ: > ę > ęə > eə > eə >
ME a: æ: > ɛ: > ę > ęə > eə > ɪə >

0.5. What really bothers the synchronic linguist

Clearly, the two series exhibit much overlap; judging by the famous *enfants terribles*[178] in the standard dialect, where a merger did occur: cf. *great :: grate*, the over-all historical record points to a merger also in the Belfast material (cf. Labov, p. 385), unless a difference bigger than the small physical differences can be established. The 'small' differences that have apparently survived (Labov: *been maintained*) into the modern Belfast vernacular, must be correlated with a sign-functional interpretation: both the frequency-factor and the sounds' very realization conditions as well as register distribution will establish different contents[179].

In the literature there has been some debate about the route[180] that the diphthongization of ME /i:/ and /u:/ took in the light of the merger-question. The first step as outgliding diphthongs with non-peripheral slightly lowered nuclei, /ɪi/ and /ʊu/, exhibited too small a material difference from the rising products of ME /e:/ and /o:/, respectively. Instead of applying Labov's notion of differential distribution considered as only a material-mechanical relationship:

ME /i:/   early 15th c [i:] high-frequency variant : [ɪi] low-frequency variant
ME /e:/   early 15th c [i:] low-frequency variant : [e:] high-frequency variant

we can so assign psychological contents[181] to the respective variants that a distinctive pattern of differential distribution will be preserved:

ME /i:/   early 15th c [i:] old/conservative/formal : [ɪi] new/informal
ME /e:/   early 15th c [i:] new/informal : [e:] old/conservative/formal

The sign-functional interpretation of the sounds in the spoken chain – also a prerequisite for Samuels's devaluation- and misunderstanding-factors in his theory of conditioned and isolative change – has the meaning-differential of the various sounds preserve the original equi-empirical distance between 'phonemes' (Luick); and with reference to the realization conditions for the respective variants above (cf. Samuels 1972 pp.23;25-27;10), I shall set up the following types of developmental series:

---

177 Labov (p. 390) does envisage the possibility that ME /a:/ followed the peripheral track seeing that it caught up with ME /e:/ and even passed it: the latter sound may have developed in the non-peripheral track of the front region. – I am not implying that the series reflect the proper chronology, but the ingliding diphthongs could have arisen in styles with falling intonation (Samuels 1972:23).
178 *Great, break, steak, yea*; and perhaps *drain*; *Yeats* and *Reagan*.
179 Labov (p. 385) virtually concedes this much in that the (near-)merger of *meat* and *mate* is found *among older speakers*.
180 The routes of the diphthongizations of ME /i:/ and /u:/ were yet another issue that Jespersen and Luick could not agree on. - I shall not be considering the speculative scenario of generative phonology's flip-flop or exchange rules. (cf. Wolfe 1972), but it is worth noting that Dobson and Samuels seem to advocate a minority view, and that Labov's trajectory is yet another proposal. – Finally, the very conception of a sound's trajectory or route implies my concept of developmental series, the difference being that the latter is sign-functional and transsynchronic.
181 Such socio-temporal contents are well-known in the literature; Dobson (1968) uses them extensively, and virtually all Labov's developmental series in his 1994-work are characterized by such-like contents in addition to his acoustic analyses.

$$\text{i:} > \underset{\text{FS}}{\text{i̠:}} > \underset{\text{RS}}{\text{ɨi}} > \underset{\text{RS}}{\text{əɪ}} > \underset{\text{RS}}{\text{ə̞ɪ}} > \underset{\text{RS}}{\text{aɪ}}$$

$$\text{u:} > \underset{\text{FS}}{\text{u̠:}} > \underset{\text{RS}}{\text{ʊu}} > \underset{\text{RS}}{\text{əʊ}} > \underset{\text{RS}}{\text{ə̞ʊ}} > \underset{\text{RS}}{\text{aʊ}}$$

In suchlike series the devaluation factor does not play its role, and again with reference to Samuels's view of the process, the forceful-style variants were <u>devalued</u> because they could not rise any higher than Cardinals 1 and 8 without losing their respective vowel quality:

$$\text{i:} > \underset{\underline{\text{FS}}}{\text{i̠:}} > \underset{\underline{\text{FS}}}{\text{i̠:}} > \underset{\underline{\text{FS}}}{\text{i̠:}} > \underset{\text{RS}}{\text{ɨi}} > \underset{\text{RS}}{\text{əɪ}} > \underset{\text{RS}}{\text{ə̞ɪ}} > \underset{\text{RS}}{\text{aɪ}}$$

$$\text{u:} > \underset{\underline{\text{FS}}}{\text{u̠:}} > \underset{\underline{\text{FS}}}{\text{u̠:}} > \underset{\underline{\text{FS}}}{\text{u̠:}} > \underset{\text{RS}}{\text{ʊu}} > \underset{\text{RS}}{\text{əʊ}} > \underset{\text{RS}}{\text{ə̞ʊ}} > \underset{\text{RS}}{\text{aʊ}} > \text{RS}^{182}$$

Below I shall demonstrate that such series must be modified, not in the light of the purportedly systemic interplay between the origin and generalisation of a variant, but for reasons of compatibility between material form, content and realisation conditions. This means that I shall develop an empirically relevant **impossibility** concept.

The notion of the small difference (of the near-merger) has also been reinterpreted: the acoustic and articulatory difference between /u̠:/ and /ʊu/ may be too small to be audible so that they are perceptually identical [u̠:] = [ʊu]; but when we look at the realisation conditions for both sounds, common to both speaker and hearer, then these differences entail a common perceptual semantic differential. This accounts for how small material differences can be continued. Secondly, this semantic differential or style-content is so 'small' that it, too, will have a precarious existence, and it is easy to see how misunderstandings may arise in isolative change (Samuels, pp. 25-27).

As the developmental series above are not empirically relevant, I shall create a series with the ME Cardinal 3 vowel, where the **structural** aspect is obvious (cf. 0.2.1):

$$\varepsilon > \underset{\text{FS}}{\underaccent{\cdot}{\varepsilon}} > \underset{\text{FS}}{\underaccent{\cdot}{e}} > \underset{\text{FS}}{e} > \underset{\text{FS}}{\dot{e}} > \underset{\text{FS}}{\underaccent{\cdot}{i}} > \underset{\text{FS}}{i} > \underset{\text{FS}}{\dot{i}}$$

any one successor sound is forceful-style in relation to its predecessor; the content of the sound that has no predecessor cannot be determined: the empirically relevant historical point of view is always **from the present into the present**. The notion of the developmental series does not require a phonemic or systemic stage as an intermediary between two (synonymous forms) on condition that the transsynchronicity of the series can be subsumed under a historical theory. And we may ask: why do we need that '-emic' stage – why has it arisen? A

---

182 As indicated none of these series is empirically relevant, and a first indication of this is that the modern diphthongs are not typical of relaxed style, on the contrary they belong to neutral or normal style in Southern British English. Samuels's argument only considers the origin-aspect; his analyses are not detailed enough when it comes to the continued existence of sounds originating in a particular style.

0.5. What really bothers the synchronic linguist

preliminary answer will be given here; I shall take my point of departure from Lévi-Strauss (1962) and Christensen (1997:280-286) about Nietzsche.

Nietzsche (Christensen, p. 280) calls systems symptoms of weakness. Depriving the concept of this subjective characterization, I shall add rather uncontroversially that the system is a *Deus-ex-machina* totality that, *in casu*, the linguist introduces in order to cope with the chaos that the everyday languages purportedly constitute. As such it is an instrument which the linguist 'believes' (Christensen, p. 285) is useful – epistemologically, cognitively and pedagogically, but which also carries aesthetic overtones with it. It is superimposed on a given state so as to systematize (classify) part of that state's elements (cf. Lévi-Strauss's description of the system in note 45 above).

The phoneme *qua* a totality is a metaphysical entity, a hypostasized word (concept). Its roots are as old as our Western Civilization (cf. Hass 1979:116; Christiansen 1997:9). It satisfies a general human need for a reality that is in some sense better, more cope-able, controlable, less hostile and complicated, less mysterious and imperfect, more easily comprehensible than sensuous reality, the reality that we create ourselves (historical reality) and the reality we do not create (physical reality). In our search for this reality we only have the everyday language as our fishing net, and since our goal constantly seems to slip through our fingers and since we have no clear glimpse of what we are in search of, it is a natural conclusion that the everyday language is imperfect, is not capable of taking us to our goal.

We have seen two solutions to this dilemma – perhaps three. From the mid19th-century (*pace* Aristotle) mathematicians and logicians have been trying to develop a 'language' that does not contain the imperfections – changeability, redundancies and ambiguities – of the everyday language, namely formal or symbolic logic, and in scientific analysis the notion of 'pure form' and its less precise sibling 'formalization', of which a number of examples have been given in the preceding pages, began to gain ground as the end-all-and-be-all of any scientific description. This aspect will not be dealt with in the present essay.

The second solution was and still is the creation of a reality that is different from the reality that lies in the individual phenomena of sensuous experience: these individuals were regarded as entities **of** a larger entity so that the totality of individual phenomena now became the answer to our problems: the totality became the goal of our search, the real reality; the totality becoming hypostasized (in ontological dualisms) was equipped with the properties which unite all individuals, bring coherence to, are common to all individuals, are behind, underlie, (are ahead of; telos), are the cause of, are the reason for, are the basic substance (material), are the basic principle of all individual things, in short which make man control and understand the chaos of sensuous reality.

That totality – not unlike the everyday language (Ottosson 1996:75) – has had various locations: (1) in a transcendent world (Plato), (2) in or within the phenomena (immanence or inherence), or (3) in the human mind as something that is manifested or realized as we exist. The fourth possibility is that the totality is a man-made creation through-and-through, created in and by the everyday language (nominalism). Lévi-Strauss is therefore right when he says that classification *qua* a totality is an innate faculty in the human race, a means which all human beings, all societies avail themselves of – for various and varying purposes, reasons and causes. The decisive question is: what do linguists do with our totalities? How do they relate to them – while heeding Paul's Realism-warning? – Cf. Culler in chapter 3.1. below.

With the creation of a totality *qua* a reality behind the positive phenomena and events, a kind of distance between the two realities – types of entity – is created, a distance that is in accordance with the assumption that we cognize at a distance of what is cognized. The paradox is that ever since Thales explained, reduced everything to 'water', dualists have been trying to bridge what they themselves separate, to somehow create an 'attachment between' the two.

In synchronic linguistics, the state and the system are such two entities, the phoneme is also a totality: the phoneme unites individual sounds, brings coherence to sounds, and last but not least, it provides us with knowledge of the sounds, it tells us what makes a (language) sound a (language) sound, because it states the essential identity in chaotic differences (allophones) and with another phoneme it creates a significant contrast. The system of phonemes is then a totality (class) of totalities (classes) – with *mutatis mutandis* the same properties[183]. To rephrase the above questions:

1) what is the relationship between the system and its parts and the system and the phenomena of sensuous reality?
2) what is the nature of the totality that the system (with its parts) contracts with the phenomena of sensuous reality?

The problem with totality-entities is that they may be positive factors in our daily everyday lives or scientific lives, or they may become so petrified and sacrosanct that they become road-blocks; they may become shadows of the reality that they purport to tell us about. With a quick return to (a) Paul: we must liberate the object of our science from the veil of dogmatic totalities which themselves cannot enter into Kausalnexus or which prevent their individuals from entering into Kausalnexus; and to (b) Nietzsche (Hass 1979:116-124): the totality as a metaphysical interpretation of reality is turned into a scientific statement in modern linguistics, but it is also a statement of weakness or escapism: a refusal to realize that science's abstractions are in fact appearances – and that there is no 'whole' to be 'seen' **in, behind, above** or **under** sensuous reality[184].

---

183 Where is the state in this abstract network – what is its role? The linguistic literature is conspicuously silent about this.

184 *Der ganze Erkenntnis-Apparat ist ein Abstraktions- und Simplifikations-Apparat – nicht auf Erkenntnis gerichtet, sondern auf <u>Bemächtigung</u> der Dinge.* (Nietzsche 1966:442). – I am not advocating a Nietzsche-like nihilism; I only want to outline a position I believe it is wholesome for linguistics to be measured against, a position that dares to question some of the diehard cornerstones of modern linguistics. Lévi-Strauss's balanced account of the two ways of gaining access to reality is more in line with the tenor of the present essay: as historians we approach our data from the stand of sensuous reality, and from here we create a theory of history, a science of the concrete; thus Løgstrup's absolute division between science/ theory and philosophy does not mirror a dichotomy between synchrony-formalism and history (Lévi-Strauss 1962:356-357; Løgstrup 1976:173; cf. Hass 1979:120). All our knowledge begins with sensuous experience, but not all knowledge comes from it: this Kantian pronouncement is the historian's constant guideline together with the multilinear structural view of history to be voiced as follows: *we understand the predecessor A through its effects on its successor B* (cf. Gadamer 1972) **and** *B through its effects on A* (cf. the model in 0.2.1.1., which will appear all through the essay), *so that historical understanding is not the monolinear usurpation by the past of the present ('the historical disease') or the present's monolinear usurpation of the past (Uniformitarianism)*. - The issues raised in the *Prologue* will be discussed in chapters 1. and 2. from a conceptual point of view.

CHAPTER I

The conceptual background of the issues involved in the construction of the theory of history
The first group of definitions

> *ceram ab externis formis distinguo, et tamquam vestibus detractis nudam considero.* Descartes *Med*.ii.32

I.1. **Towards the sign-functional nature of historical phenomena. A bird's eye view of the history of the concepts of change and stability. On the role of the everyday language in our Western Civilization's dualist way of reasoning. On the Nominalist-Realist dichotomy in relation to history**

WHY HAVE PLATO'S and Aristotle's philosophies exerted such a persuasive power over and pervasive influence on the making of the intellectual climate of our Western Civilization? And why did neither Heraclitos's nor Parmenides's philosophies carry the day? After all, on reading the Presocratics we cannot help sensing an albeit willed primitive simplicity in their views that works two ways. The admiration we must feel for them and the intellectual recognition that is certainly their due cannot obscure the fact that their greatness does not rest on the substance – with apologies to Pythagoras – of their thought, but appears primarily when *we* interpret their philosophies in the light of their socio-historical context. So why take them seriously and study them at all (Barnes 1979:4-5)?

The Presocratics' thinking does not match the sophisticated depth and breadth of a Plato and Aristotle, but exudes logical stringency, consistency and coherence that is a paradigmatic model of scientific analysis, to which must be added the curious fact that much modern linguistic reasoning operates on and with the same type of views we find embedded in the ideas of the Presocratics[1]. The scientific meta-issues *they* tried to come to grips with are not unlike those that our 'linguistic' age has not succeeded in solving, namely those of change and stability seen against the horizon of reductionism and timelessness as an expression of the search for the immutable **identity** (of things) **in** (their) **differences**. The Saussurean *langue-parole* dualism is but an application to language phenomena of a Plato-like ontological dualism (Coseriu 1974:16-18; 25-26), and as such an answer to the questions raised by the above meta-issues.

---

1 In his *Metaphysics* (988b-23-28) Aristotle points out this inconsistency in the Material Monists' thinking: they wanted to explain everything, including the causes of generation, but they 'destroyed the cause of change': tò tês kinéseos aítion ànairoûsin. – For *kinesis* = '*change*', see Lear (1988:56 note 3).

Finally, the conceptual schemes that the Presocratics set up exhibit an exemplary intellectual progression and actual historical development in thinking and reasoning which perfectly exemplify the **Critico-Philological Method** (1.4.5): Anaximander was not slow in criticising Thales and creating a new synthesis out of the concepts he could use of the inherited ones and his own original contribution, and the same goes for Anaximines's critical and constructive thinking (Jones 1970:11; 12-13; 26-31; cf. Beck 1967:212). Platonism is also a synthesis of (part of) the old (Heraclitos and Parmenides), including a Socratean layer, and Plato's own orginal contribution (Aristotle *Met.* 987a29-987b8). The Critico-Philological method is a summary statement of man's existence: in Nietzsche-like diction, we 'forget' something (of the past) and create a new present out of what we 'remember' and what the present itself provides.

To return to the opening questions: Plato and, accordingly, Aristotle were the first fullfledged *ordinary language* philosophers; what I mean by this is nothing to do with this designation's modern referents (*language analysts* (Popper 1974:15; Sacksteder 1981)). Plato was not a speech-act philosopher, searching for consequential connections between speech-acts and behaviour (Sacksteder 1981:459; see chapter 1.2.3 onwards). What Plato did – must have done – was studying his own native language. In his Greek mother tongue he could detect the fundamental structure of what is describable as Platonism. Plato's dualism interpreted ontologically is a view of the world that has been built into perhaps all everyday languages[2].

Variants of the classical dualism – when compared with Material Monism – presuppose the idealising abstractness of the everyday language, its ambiguity or semantic opaqueness, its ability to transcend reality. – The Descartes vignette above and Descartes's famous wax example in his *Second Meditation* (30.3-32.6) unequivocally demonstrate these properties of the everyday language (Zeller 1963:230; 228-235).

This explains why the dualist parameter in general and perhaps Platonism in particular are indelibly embedded in our way of reasoning and comprehending reality. We are imbued with Platonism, dualist thinking and thought, from the moment we begin to learn our respective mother tongues. The more we study our everyday languages (as linguists), without remembering Paul's words, the more we become enmeshed and immersed in Plato's dualism; and who 'studies' what is to become her mother tongue more intensively and consciously than the infant baby, who produces utterances she has not heard before with words she has heard from her parents, and who produces utterances she has not heard before and which she will not like to be reminded of when she has acquired full competence in her native language?

---

2   Dualism as a general notion may be interpreted variously: ontologically & transcendently (Plato), ontologically & immanently (Aristotle), religiously (Christianity). As far as I know, only Glossematics (*pace* Togeby 1965a) takes the immanent-transcendent distinction seriously, while modern synchronic linguistics (Saussure onwards) do not consider the ontological implications of their respective versions of the dualism; Hermann Paul's warning has not been heeded (Paul 1966:11; 5). – The generativist's dualism of *Universal Grammar* and the *linguist's grammars* intended to match the so-called initial state of the mind's *language faculty* and the steady state of the individual's language *qua* knowledge of language is an ontological postulate (as *e.g.* in Chomsky 1988:46 *et pass.*).

1.1. Towards the sign-functional nature of historical phenomena

Escaping from dualist thinking or just recognising that we are constantly influenced by it requires great will-power and self-awareness, because the easiest thing to do is to let oneself be carried away by it, to subconsciously work and live within the conceptual confines that the dualist structures of our everyday languages provide.

Whether Plato pictured himself as a language philosopher is immaterial. My interpretation provides an immediate and logical reason why his philosophy has survived – and will survive – not only as an expression of a period in the history of philosophy, but as a forceful and influential factor in our lives – scientific and everyday ones. The enduring success of Platonism (and Aristotelianism) explain(s) the failure of perhaps all other philosophies in obtaining permanent as well as practical and theoretical relevance in our day-to-day lives[3].

The fact that a conceptual scheme is (has been) grounded on the everyday language means that it becomes all the more irresistible and immediately understandable as well as accessible to innumerable applications and interpretations. Understanding Platonism involves no comprehension; it has an inborn access to us and we to it. We may challenge, criticize and improve the logical geography of Plato's arguments concerning the nature of the concrete matters that he deals with, and we may apply Platonic views[4] to new material, but we shall be doing it on the ground of the fundamental premiss of Platonism itself, (part of) the nature of our everyday languages[5]. In a word, the study of Platonism like the study of language involves a participative relation. Hence we cannot escape Platonism: to escape it is to step outside our everyday language.

Platonism was built on the fundamental property of human existence, the everyday language. The Scholastic systems of the Middle Ages, Medieval Realism, presuppose an everyday language, otherwise we cannot grasp the importance of Occam's nominalism to the rise of empiricism (Jones 1969:320-322). Consequently, Plato-inspired interpretations of language should be criticized from a nominalist stance, and since nominalism is no immediate or viable historiographical basis, the historian must incorporate into his basic principles elements from the alternative to the effect that – in the diction of Glossematics – the everyday language carries existence postulates with it.

Even a superficial view of the everyday language reveals its dualist nature. The *langue-parole* dualism has been mentioned and the commonplace notion that we all (Danes) speak the same language (Danish) even though we do not speak the same language, points to a dualism, too, the ontological implication being that the word *Danish* refers to an entity that all Danes somehow participate in; in generative diction knowledge of language *qua* a cognitive system *is shared with others from the same speech community* (Chomsky 1986:55), but this 'sharing' process or relation remains undefined (cf. Chomsky, pp. 41-42). At another level one may mention the so-called nexus-substantives, which signify relationally complex notions: *embodiment, incarnation, manifestation, realisation,* and *formed substance; classification* and *subsumption* as well as *class* and not least Saussure's *form-substance* dichotomy

---

3   I am not saying that Platonism cannot, should not be improved. It is obvious that the historico-dynamic part of human existence has not received its correct place in Plato's ontology.
4   Scientific research is a Platonic idea, *unité abstraite et constant* (Hjelmslev 1972b:86).
5   Despite his belief in what Jones (1970:194; and above) calls 'intellectual osmosis' Plato followed Socrates in emphasising the importance of definitions, hence that of the everyday language (Aristotle *Met.* 987b).

are also dualist terms⁶. Platonism, being an aspect of what I shall call **Linguisticism**, leads invariably to the creation of hypostases, entities that cannot enter into *Kausalnexus*. Thus when criticizing Platonism, we shall be doing it on the implicit assumption that our venture presupposes the same categories as it sets out to investigate, and therefore we cannot construct a whole-sale refutation of Platonism, despite its sins of commission and omission (Jones 1970:203).

My position will contain an absolutist strand, because the role of the everyday language in our existence does not justify or support Nietzsche's claim that the Platonic-Christian metaphysics *qua the* interpretation of reality has been completely dissolved and abandoned (Hass 1979:98). This is impossible – nihilism cannot escape or do away with the meaningfulness that the everyday language attributes to our existence; so there is always an alternative to the one and only interpretation (theory). It is only under one interpretation of the logical universe A & nonA that the negation of A leads to 'nothingness' (*keinen Sinn im Dasein* (Nietzsche 1966:853; Hansen 1997:1594)) – if A is ALL, whatever this may mean.

The everyday language has the ability to transcend reality, but does so against the horizon of reality: escapism into a Platonic haven cannot 'forget' its original *resting-place*⁷. With the everyday language we transcend the present, creating the present, against the horizon of the present⁸. These transcendence-relations are participative⁹, and Platonism is participative, too:

1.1.1 The Greek everyday language was a necessary condition for and an integral part of Platonism. – Cf. 1.3.2.

The simplest precondition for a critique of Platonism is an investigation into the everyday language in order to find at least one property that Plato's dualism ignores or treats wrongly:

1.1.2 If a theory of Y and X has been constructed on the basis of principles said to inhere in X and these X-principles stem from a non-exhaustive analysis of X, then the theory is a priori inadequate and arbitrary. – Cf. 1.4.10.

---

6 Lévi-Strauss (1962: ch. 5) demonstrated how fundamental classification is to human existence, and how the triad – category : species : individual – can be given a sign-functional interpretation (pp. 159-160). The fundamentality of classification – man's urge to classify – is in perfect accordance with the language-based nature of it: it is one of two things that the infant baby first learns, that the 'same' sign can be used differently and in its different uses it creates the 'sameness' of a class: to classify equals to speak (to be elaborated in chapter 4.).

7 If this resting-place were the transcendent world, which Socrates mentions in *Phaedo* (101d-e), then Nietzsche would be right in regarding Platonism as pure escapism (Hass *loc. cit.*); but if we can unmask – with Hermann Paul – the escapist-transcendent uses of the everyday language and get a firm grasp of its existential function, then satisfactory and adequate knowledge of reality, including language development, is possible.

8 The potential denotator *a centaur* is meaningful in modern English because we know what reality is/is not (Climo and Howells 1976:10-11). – For the terms *a potential denotator* and *a denotator* see Sørensen 1958:ch.1.

9 The participative relation is not Plato's *méthexis* process as, *e.g.*, in *Phaedo* 99c-e (Aristotle *Met.* 987b7-15).

1.1. Towards the sign-functional nature of historical phenomena

These two principles, the principle of Participation and that of False Analogy[10], respectively, have grown out of the Glossematic concepts of immanence and transcendence (Hjelmslev 1963:3-6; 17-18; cf. Sørensen 1958:§14.1). Both Classical and modern Formal logic have partly evolved from non-exhaustive analyses of purportedly constitutive structures of the everyday language; hence none of the branches of logic can yield adequate analyses of the everyday language. – To illustrate the two principles:

1) Let Y represent reality and X the everyday language(,) on which a philosophy L has been constructed in accordance with 1.1.2, then the L-theory will yield a simplifying understanding of Y.
2) Let Y represent an everyday language and X the everyday language from which a linguistic theory (L) has been derived, which only considers either static-synchronic properties or dynamic-historical properties, then the L-theory will yield a simplifying understanding of Y.
3) Let Y represent the everyday language – or any other phenomenon – and X modern symbolic logic, (some of) whose principles have been derived from a non-exhaustive analysis of the everyday language, then the logical description of Y or the section of reality in question will yield a simplifying understanding of Y.

Obviously historical change and continuation, synonymous and polysemous variation are language properties, indeed, universals[11], whose non-linguistic counterparts have no positive constitutive place in Plato's view of knowledge; and in generativist linguistics (cf. Chomsky 1986) neither man's historical sense nor the above universals contribute to or play any role in the cognitive system that the ambiguous phenomenon, knowledge of language[12], is said to constitute. And Labov (1994:5;9) makes no bones about this mainstream view of change, stating that change is incompatible with the essence of language, rather surprisingly, as a matter of fact, in view of his damning criticism of the basic pillars of synchronic linguistics (pp. 349-353 and above).

Plato did not see that these empirical properties were fundamental to his native language, whereas he acknowledged the significant roles of their extralinguistic counterparts in sensuous reality, change and multiplicity, constant variation, redundancy (synonymy) and ambiguity (polysemy), relegating these to the lower world of the senses. He should have elevated neither the idea of seeing identity in differences nor the principle of definitions (*horismoí;* Aristotle *Met.* 987b1-7) containing only static concepts to their *ideal* status so as to exclude their logical alternatives: to see the differences in identity and the notion of the dynamic definition[13]. But being no professed ordinary language philosopher or linguist, Plato must be

---

10 The transference of abduction to language change as we saw in Henning Andersen (1973) above and Lass's biological analogy in his 1997-work are instances hereof.
11 While the structure of normal assertive, subject-verb-predicate/object clauses has been crucial to the rise of formal logic, the structure and function of change and polysemy and synonymy have not.
12 In his 1986-work Chomsky does not succeed in demonstrating that his 1986-version of the generativist's theory of language can be identified with the native speaker's innate and/or 'attained' knowledge of his or her mother tongue.
13 A static definition refers to things as single objects, only; it reifies reality. A dynamic definition subsumes phenomena (processes) which can only be grasped exhaustively through dialectical reasoning (cf. ch.6).

excused. Like any other native speaker, Plato was not particularly alert to the complex nature of his native Greek (Samuel's 1972:28-29), as Hjelmslev (1963:5) puts it: *(...) in our zealous haste towards the goal of our knowledge we may overlook the means of knowledge – language itself. (...) it is in the nature of language to be overlooked, to be a means and not an end, and it is only by artifice that the searchlight can be directed on the means of knowledge itself.*

To see and acknowledge the differences in identity is tantamount to being aware of the **identi*fying*** as well as unifying and dualising effect of the everyday language. The principle is, somewhat paradoxically, an essential aspect of analysis, whereas its opposite is basically a synthesising movement whose ideal is a sort of monism. The everyday language allures us into believing that differences are at second glance, *really*, (surface expressions of) identities (Kanngiesser 1972:72, 122, 180) or they will have to be explained away or simply ignored, as we have seen change and small differences be. Thus in deference to the dualism both Kanngiesser (1972) and Samuels (1972) introduce a tolerance concept to account for and to curb the degree of *parole* variation and heterogeneity.

The standard view of change and stability is that they are contradictory properties and/or concepts (*Phaedros* 245c-246a). Neither Heraclitos nor Parmenides disagreed about this, and Plato's ontological-dualism compromise reflects the same position; both Saussure and Hjelmslev (1928; 1972a; cf. Hansen 1982b) are in agreement with it. And modern linguistic theorising follows suit (cf. Labov *op.cit.*). Let's therefore take a look at a traditional argument that comes out in favour of the stability concept as the basis of mainstream theories of knowledge (cognitive systems).

1.1.3 Everything changes constantly, ergo no true statement of anything can be made.

1.1.4 True statements denote static, non-changing things.

Change and stability are contradictory concepts in relation to truth: truth can only be determined in relation to static objects (Jørgensen 1931:42-43). The two postulates in 1.1.3 and 1.1.4 constitute a contrapositive universe, either being the negation of the other ($p \supset q$ = non-$q \supset$ non-p). In fact the two are virtually logical synonyms. However, since a contrasting postulate can be constructed that finds in favour of change and against static objects, cognitive priority cannot be awarded to either view of reality. But the type of argument helps us to see how and why the appearance-reality parameter had to come into existence and assume cognitive prominence – as early as with the Presocratics. The argument may have run like this under the following condition that admits an intimate connection between knowledge and existence: it is possible to acquire (true) knowledge, hence static things exist:

1.1.5 All statements denoting changing facts are false.

1.1.6 The statement 'X changes' is false.

Hence:

1.1. Towards the sign-functional nature of historical phenomena

1.1.7 The statement 'X does not change' is true[14].

Parmenides or Zeno would say that X in 1.1.7 does not *really* change although we *see* it change; and this appearance-reality parameter also makes 1.1.3 and 1.1.4 empirically relevant but at the cost of reducing 1.1.3 to another cognitive level – and reality-level – than 1.1.4. Plato (*Rep.* 500c) would say that 1.1.6 may be true for all he cares; the statement denotes a fact of the world of senses, and epistemologically relevant knowledge does not spring from that world: being the sufficient and necessary condition for X, the idea in which X participates (metexei) does not change. The idea is the cause both of X's existence and of our knowledge of X (Jones 1970:128)[15].

Linguists following Plato assume in their search for a constancy (Hjelmslev 1963:8; Kanngiesser:1972:127; 122; Chomsky 1986:52-53 *et pass.*) that the fact that X changes is at best a dependent statement presupposing two object-based synchronico-static statements or is theoretically irrelevant. Hjelmslev (1972a11-45) does not present a cogent argument for the empirical priority of the static conception of the language system, his argument being based not on definitions, but on the common-sensical postulate that change must start with or from fixity. Hjelmslev does not realise that change and stability are equilogical concepts (1.1.8-1.1.9):

1.1.8 Change is the contradictory of stability = Stability is the contradictory of change.

Hence

1.1.9 "X is stable" is identical with "X does not change" = "X changes" is identical with "X is not stable".

But our tradition (1.1.3 – 1.1.9) has abandoned this equilogical position for the above common-sense stance, making the latter hinge on these three non-cogent assumptions:

1) the empirical relevance of the appearance-reality parameter;
2) the empirical relevance of dualist ontologies;
3) empirical difference and oppositions are defined binarily by logical contradiction: A or non-A.

Unless they are strictly nominalistic – as Glossematics is, synchronic theories operating on dualist assumptions presuppose 1) and 2) and imply 3). Bloomfield's view of 'small differences' reflects 1), and so does Daniel Jones's phoneme theory (1967:6-7). Chomsky's

---

14 To be precise: 'If the statement is true, then 'X changes' is false'.
15 Empiricism has now been sacrificed on the altar of speculation or metaphysics or to Socrates's principle of definition. This step in the history of Western Thought follows from the arbitrary making of the presupposition that makes knowledge and existence (cf. 1.2. below) merge: 'if we can make a true statement, then the object of knowledge does not change', epistemologically fundamental; this in turn makes 1.1.3 absurd. Obviously, Aristotle was not late in rejecting 1.1.3 and 1.1.4 as absurd (see Lear 1988:55-56; and Kierkegaard *Gjentagelsen*, p. 115).

version of Plato's Problem (1988:xxv *et pass.*) is but a variant on 1) and 2), and both Samuels (1972) and Kanngiesser (1972) presuppose 2). Item 3), a commonplace with Saussure and Hjelmslev, is not so obvious in the literature, but it is incompatible with any attempt at creating a 'unified linguistics' (cf. Kanngiesser 1972:4-6). I shall argue that this binary opposition is participative.

Let's ask these questions: where do the first two dualisms come from? Why is stability logically more fundamental than change, and why do two equilogical (and perhaps equiempirical) concepts not receive the same epistemological treatment? Why has epistemological/scientific and empirical priority been awarded to the concept of stability in our Western Civilization?

The dichotomy harks back to Socrates and Plato. The Material Monists could not explain physical reality consistently, and *when Socrates disregarding the physical universe (...) was the first to concentrate upon definition* (perì horismôn), *Plato followed him and assumed that the problem of definition is concerned not with any sensible thing but with entities of another kind; for the reason that there can be no general definition of sensible things which are always changing* (Aristotle *Met.* 987b1-7)[16]. Thus scientific analysis was not to be based on and true knowledge could not be obtained from physical nature – in our case from the spoken/written substance of the everyday language – only on and from 'mental' reality, which in turn implies: from the everyday language and only a section of the latter. Scientific analysis was restricted to being primarily definitorial in accordance with logical principles, in particular the law of non-contradiction with its *tertium-non-datur* corollary (Jørgensen 1931:41). Science became linguistically motivated (*i.e.* identical with the theory at hand) and the study of universals, *idéas*, by participation in which all sensuous things exist.

From then on physical reality was in need of rehabilitation, which occurred through William of Occam's Nominalist Revolt; physical nature eventually became the proper domain of scientific analysis. However, the logico-linguistic basis of the requirements for scientific analysis was not changed accordingly, and the Socratic-Platonic linguistic stance was refined by the logic of Aristotle and the Stoics to end with the deductive systems of the previous and present centuries (cf. Jørgensen 1931). Unless a strict nominalist position is upheld, *[Aristotle's] confusion of logic with metaphysics* will be continued (Jørgensen 1931:41). Furthermore the historical domain of human existence was by and large disregarded until the rise of historiography in the 18th century, but historical analysis was soon after its 19th century heyday discredited for being a nonscientific domain, because nomological-deductive methods could not be applied to the material phenomena of a domain characterized by novelty, uniqueness and changeability (Lass 1980). Thus the nominalist rehabilitation of physical reality did not include historical matter, because its entities were not amenable to the principle of *definitio per genus proximum et differentiam specificam* as a determination of the principle of the search for the identity in (sensuous) differences, and the introduction into the study of language of the *langue-parole* dualism implied that the physical reality (substance) of the everyday language was to be either controlled by its non-physical system (form) or totally excluded from linguistic studies. In either case, the linguistic universal that

---

16 This is the historical *raison d'être* of the immanent approach that we find in Glossematics and Togeby (1965a:194).

1.1. Towards the sign-functional nature of historical phenomena 93

no language does not change became irrelevant. For this universal to re-assume its scientific relevance, the physical reality of the everyday language must be rehabilitated and/or the inadequacy of the synchronic conception demonstrated. I regard Samuels's 1972-theory a step in this direction, as *parole* here is the prime mover in language evolution and a shaper of any language system. – I shall consider our problem from another angle.

Let's suppose that change and stability are not primitive concepts, but have a theorem-like status. Then no consistent logical system can be constructed so as to have our pair of contradictories derived from it; secondly, relegating changing phenomena to a shadowy existence in the world of appearance by means of axioms belonging to the domain of staticness is an instance of False Analogy formation; thirdly, the very starting point – there being both change and stability – presupposes the indubitable truth of at least one statement about change. If it is not the case that the statement "X changes" is a true interpretation of the two observations embodied in the complex statement: "This piece of wax is different from the piece of wax that was a moment ago", then Descartes could not have established the idea of *essentia*. We cannot doubt the sensuous fact that the piece of wax has changed its form from Time 1 to Time 2. Zeno must have taken the truth of: "Achilles caught up with the tortoise", for granted, otherwise he could not have inferred the deceptive nature of sense perception. Fourthly, the statement type: 'X changes' in "'X changes' belongs to the sensuous world, not to the world of forms", must be true; otherwise Plato's ontological dualism breaks down. Therefore Kant was right when he opened his *Kritik* with this essay's sixth Ariadnian Thread, a statement of **empirical necessity**[17] – *all our knowledge begins with sensuous experience*:

1.1.10 The transcendence from our deceiving senses – and everyday languages – to our non-deceiving intellect – and formal language of logic –, presupposes the indubitable truth of the facts that Achilles did win the race, Descartes's lump of wax was changeable, and all languages change.[18]

The necessity of empirical facts implies that so-called *contingent, apparent or accidental facts are necessary*; they are scientific phenomena in their own right. As knowledge *begins* with sensuous experience, it may be inferred that the structure of the human mind is (also) temporal so as to hold man's historical sense. Man's understanding of temporal processes embodied in the mental processes of forgetting and remembering is the faculty that uniquely distinguishes man from Nietzsche's animal, which sees each moment die the moment it is. Sensuous experience as well as forgetting and remembering is impossible without comprehension, and comprehension impossible without language[19]. This means that Parmenides

---

17 The correlative concept, **empirical impossibility**, will be determined sign-functionally in ch. 6.
18 Our deceiving languages lead us into seeing identities, where significant differences are in fact the case; and then *we* proceed from there to the promotion of the purported identities to the status of being the corner-stones in all cognitive processes and the 'causes' of reality.
19 Here I agree with Løgstrup (1978:113) that language does not modify sense-data (sensation), but disagree with him when he claims that there is something like language-less sensation; such pure sensation would be strictly synchronic and how animals sense reality. Any sensuous experience will involve language because our existence is sign-functional – now a postulate, but in chapter 4.3. it will be given a sign-functional basis.

and Zeno's arguments presupposed a set of dynamic mental (linguistic) categories whose epistemological and empirical relevance they denied, because they confused logic with metaphysics. And the result is well-known, with essentialism becoming the order of the day: our Western Civilization became saddled with a substance-accident metaphysics that could not transcend its object-hooked (static) basis, which, furthermore, tallied perfectly with the dominant status that analytical as against dialectical reasoning assumed.

Therefore, the study of change came to produce anomalous, even contradictory, statements; but the anomalies are not derivable from the nature of change alone, they stem from assumptions belonging to the staticness of the applied theory. The Aristotelian assumption has been tacitly accepted from the Peripatetic himself onwards, that deduction reflects an ontological order (Jørgensen 1931:44). This view is epitomised by the preference of logical form for the copula *to be, is* and the abovementioned essence-substance concept (Jørgensen 1931:43; Hammerich 1935:36): what does x in the formulae of the predicate calculus denote? Or: the logical identity of the generalisations of the propositional and predicate calculi presupposes an ontology that involves both an unknowable carrier, x, of properties and the independent existence of property classes (S and W below), a particularly gross example of language-based use of analytical reasoning:

1) All swans are white → All S is W.   (Categorical generalisation);
2) If there is a class of swans, then this class is white → $S \supset W$.
   (Conditional generalisation);
3) For all x it holds that if x is a swan, then x is also white → $\forall x\ (x \supset S) \supset (x \supset W)$.
   (Property-based generalisation).

The following non-cogent argument sums up conditions under which the concept of change creates inconsistency:

1.1.11 Stability and change are properties of the everyday language.

1.1.12 Properties are determined contradictorily.

Hence

1.1.13 To be stable (= q) is to be all that to be changing (= nonq) is not – and *vice versa*.

Hence

1.1.14 An everyday language cannot *really* possess both properties: $(p \supset q\ \&\ nonq)$ = nonp.

The dilemma and absurdity of 1.1.14 are created by 1.1.11 and must be dissolved as in 1.1.15

1.1.15 The nature of the everyday language is dualistic; the stable entities of language are the true, the real, the essential, the fundamental reality of language, while change is a non-necessary, an accidental and contingent property.

1.1.15 is ambiguous, implying either of the following positions:

1.1.16 Change and stability are two mutually exclusive properties which the everyday language so possesses at some time that it is either in a state of stability or in a state of change.

Change becomes a relation or a transitory state between two stable states. The alternative is:

1.1.17 Change and stability are two properties that the everyday language so possesses at any one point in time that either does not exclude the other.

From 1.1.15 and 1.1.17 it follows that change is always present (*pace* 1.1.14), but only as an evolutionary tendency controlled by the structure of the system in question. The notions of immanent tendencies, trends or propensities (Hjelmslev 1972a:20-22) imply that the everyday language is a calculus. Change is recombination of already (pre)existing entities, and the model underlying historical linguistics will therefore be a one-state one as against the commoner view of the two-state model L1 > L2 (Hjelmslev 1975:123-125), which turns into a three-state model, because of the transitory state, >L>, connecting stable L1 and stable L2 (Weinreich *et al.* 1968:101).

Assumptions 1.1.11 and 1.1.12 are the root of the epistemological anomalies, and the solution in 1.1.15 is a static assumption mirroring a Plato-like ontological dualism. Furthermore the argument does not tell us why change becomes dependent, and could be viewed as a process of hypostatisation in which the property *to be stable* is transformed into a non-property, an *essentia* in that the static element becomes the carrier of properties, including change.

Change being a fundamental property of the everyday language, a(n immanent) linguistic theory constructed on principles derived from the everyday language must also regard change as fundamental, in so far as the theory is to be applied to a changing phenomenon (1.1.2). This will be called the Amenability Principle[20]. Hjelmslev (1963:17-18) committed the sin of omission of not including the change universal among the *properties present in all those objects that people agree to call languages*. Hjelmslev neither generalised the property of change nor established it by definition because he was an idealist in a nominalist disguise: it is impossible to define sensible things (see above). Let's return to the Lyceum in Ancient Greece.

Aristotle adopted the alternative to Plato's transcendent idealism, namely immanent idealism (Jørgensen 1931:43). With Aristotle we see the first significant attempt at formalisation. The classical syllogism (Delong 1971:136)

---

20 This criterion is basic to Glossematics (cf. Rasmussen 1992:71-72) in this form: synchronic causes or reasons explain synchronic facts. Why, then, should synchronic reasons explain non-synchronic facts?

1.1.18   All M is P
　　　　<u>All S is M</u>   → ((M ⊃ P) **&** (S ⊃ M)) ⊃ (S ⊃ P)
　　　　All S is P

epitomises Aristotle's immanent dualism and view of change as well as the calculus view of change: the argument satisfies the scientific need for generalisation, the conclusion contains no more than is contained in the premisses so that it exists immanently in and simultaneously with the synthesis of the major and minor premisses, it being a recombination of preexisting entities. The conclusion implies neither novelty nor uniqueness; the temporal process needed to make the conclusion manifest is only a historical process at the methodological level, reflecting no empirically significant process (Saumjan 1971:24); in fact it testifies to human imperfection. However, the view is not flawless, when it comes to its transcendent application – to nonlogical matters: no logical rule of inference formalises an empirical process. I am thinking of Modus Ponens, Modus Tollens and the Law of Transitivity; the process of syllogistic reasoning is not an immediate step from premiss(es) to conclusion: the processes of **conjunction** and **inference** are implied in 1.1.18, and even a Zeno must admit that cognition does imply some change, from ignorance to knowledge (cf. Christiansen 1998:34): preestablished knowledge immanently present in our mind, before we know (of) it, is not knowledge.

　　With Plato (and Parmenides) our Western Civilization begins to see reality against an unchanging and stable time horizon. "Achilles cannot catch up with the tortoise" was perhaps too crude to justify the empirical relevance of the appearance-reality parameter. We need the sophisticated minds of a Plato and Socrates to construct a more empirically convincing synthesis of the inherited concepts and new concepts. But the argumentation of these founding fathers of the Civilization that Nietzsche thought was being destroyed in his days (Nietzsche 1966: III, p. 853; Hass 1979:98) demonstrates unambiguously how language-dependent their static conceptions of reality were. The everyday language was more than a means to the objective of obtaining knowledge. The material monism of Thales springs from his own everyday language: we see neither generalisations in the world of our senses nor that everything is water: the concept of generalisation is found in the everyday language and that water is the cause of, say, gold, could only be formulated by means of the reality-transcending function of the everyday language. This fact permits us to reformulate the above Kant quote to the effect that *not all we know springs from our senses*. The early monists did not realise the idealising function of the everyday language to the same extent as Plato did; they did not know the epistemological function of the concept of relation in the shape of *denotation* or *predication* (Socrates's *horismoi*), *participation* (Plato) and *inherence* and *predication* (Aristotle) (Jørgensen 1931:44).

　　Despite motion and the passing of time relations of things to each other may be viewed as stable; two syntheses (totalities) are (relationally) identical, although each consists of mutually different relates. The entity that the two relates are identical in respect of is a concept, a category (predicate), a class, a form, an idea, or a universal, in short a linguistic sign (Sørensen 1958:25; Løgstrup 1976:58). Like a class, relations transcend the level of the particular.

　　A category as well as a relation is a linguistic sign whose content is a man-made definiens. The category will therefore contract a participative relationship with its everyday lan-

1.1. Towards the sign-functional nature of historical phenomena

guage in that the latter is not only a precondition for the former, but also an integral part of it. Its definiens does not make the category transcend into an independent haven of categories: **a category(a language sign) :: a language sign**. A language sign exists irrespectively of its becoming (a part of) a metalingual sign.

Plato might have said that the particular somehow participates in the forms, Aristotle that the general inheres in the particular. Essentialism is, to repeat, in either case the centre of study (Popper 1961:26-34), but categories and relations spring from the everyday language, not from the sensuous world. If Thales and the other material monists had known the concept of relation or what amounts to the same had been more language-aware, they could immediately have seen the absurdity of this type of triangulation:

1.1.19
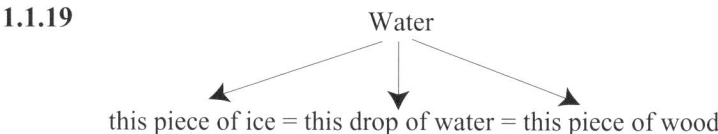

this piece of ice = this drop of water = this piece of wood

The triangulation in 1.1.19 exemplifies the participation relation: ice(water) :: water, as well as what I have called the Phenomenal Error (cf. 1.1.2): one phenomenon equiempirical with another phenomenon is made to explain the latter. Finally, Thales's word for water was *made to denote* through a specific ontology *qua* its definiens. The hypostatising triangulation of both immanent and transcendent dualisms (cf. Jones 1970:128) escapes the Phenomenal Error, replacing Water *qua* a cause with a form *qua* a language sign with a horismos that ousts the historically evolved content of the sign in question. That is the reason why Togeby (1965a:6) may characterize any theory that relies on a given substance as transcendent (cf. 1.1.2 above); the consequence is that history is no immanent science, only a transcendent one, that is: no science at all[21].

However, in all three cases above the triangulation process is the seeing of identity in differences through the suppression of what our linguistic tradition would call either 'small' or 'nonsignificant' empirical differences or contingent differences, and the sign-functional structure of a category (or a form, etc.) leaves us with two types of categories.

First there is ErC(ErC): the category is a complex sign whose expression E (*signifiant*) is the name of the content C, which is a sign *qua* a definiens. The sign-expression is a word of the everyday language having the man-made function of being a short-hand expression for the definiens: /a father/ r C(*a male parent*). The alternative is that a category may be the content, C, (*signifié*) of an expression (E), which in turn is a sign: CrE(ErC).

No matter whether a difference can be read into the two formulae, the alternative indicates that neither the expression nor the content can be regarded independently of a content and an expression, respectively. The interdependence between E and C makes a Platonic in-

---

21 There is an ambiguity here: history is no science <u>because</u> it relies transcendently on the principles of another science – *e.g.* sociology, or <u>because</u> it is impossible to establish a formal system of definitions that defines the formal features – now that sensible phenomena cannot be defined – 'underlying' historical phenomena, *e.g.* the beheading of Louis XVI, the destruction of *La Bastille*, Nazi Germany's invasion of Poland in 1939.

terpretation impossible which idealises meaning; if the category (= E or = C) was to exist independently of its definiens, it would reduce to a sign in the everyday language, or to no sign at all – either to an entity (C) that can only be 'seen' with the mind's eye or to a symbolic entity (E) that awaits interpretation.

Since it is a basic premiss of the present essay that the Glossematic parallelism between the sign's two relates cannot be upheld, a category will have this semiotic structure: ErC(ErC), while the structure CrE(ErC) will indeed be the epitome of the signs of the everyday language. The sign function will therefore not be a simple binary one, but participative, and the two complex signs do not reduce to ErC, as Hjelmslev has it. This entails that the metalingual relationship between theory and object (language) has the participative structure (cf. The *Prologue*): a scientific language(the everyday language) :: the everyday language, and not the alternative: a scientific language :: the everyday language(a scientific language), which implies the idealisation of the theory.

Implicit in the above exposition has been the subsumptive view of scientific analysis: generalisation is classification of instances under a word from the everyday language or a common name equipped with a definition, and if we remain within the extensional confines of language, upholding the distinction between language and nonlanguage, the theory is nominalistic, carrying no existence postulates with it. If, on the other hand, a category is an expression of *das Sein des Seiendes*, then the signs of the theory (language) are more than mere nomina.

The historian places himself between the arbitrariness of nominalism and the 'realism' of the idealist[22]. We know that the historian's material is gone, but we also know this material, – that there was a past. This knowledge springs from the everyday language, with memory and traditional historical sources regarded as having a sign-functional nature. And the historian is a mediator between the nominalist and idealist (Paul's Realist), accepting both the extensional (denotative) function of language and the existential function of it[23], without accepting either the potentiality and arbitrariness of the former or the hypostatising function of the Realist's language. To repeat what was phrased differently above, the historian can never construct a Glossematic theory, completely 'unmotivated' in relation to data, which, furthermore, is reduced to being a model of the theory.

The Realist element of the everyday language resides with the sign-functional interdependences it contracts with the extralinguistic world; it follows from the very nature of the everyday language. In the concrete speech-situation the everyday language contracts an interdependence with its speaker (a connotation), and an interdependence with its denotatum (a denotation). A concrete utterance becomes a sign in its own right: in relation to the speaker it becomes the expression of the sign whose content the speaker becomes, and in relation to the denotatum it becomes the expression of the sign whose content the denotatum is. This dynamic function of the everyday language may in some sense of the word be arbitrary, but a given utterance is always necessary (motivated), connotatively and denotatively, irreplaceable as it is. The arbitrariness lies in the fact that the speaker could have chosen another –

---

22 Between the extensional relationship of language and nonlanguage and the merger of language and reality.
23 When the historian talks about the destruction of la Bastille, he is talking about what really happened.

1.1. Towards the sign-functional nature of historical phenomena

not any other – way of 'expressing himself', but the utterance selected could not be replaced by any other utterance: *the everyday language denotes and connotes against the horizon of synonymy*. Thus Samuels's notion of there being a necessary degree of redundancy in any language system and his view of the relation between *langue* and *parole* (1972:139) has been generalised:

1.1.20 [C $r^{connotation}$ {E(ErC)} $r^{denotation}$ C}
Speaker N.N. is connoted by utterance E, which denotes denotatum C

This complex sign is N.N.'s view (interpretation) of a particular section of reality, and as such the use of the sign-function is predominantly static. A dynamic reading of it will run as follows: 'That's how N.N. develops, moves into or creates the present (moment).'

The preceding remarks, to be elaborated below in 1.6., are intended to demonstrate that history may be a *sui-generis* mode of signification different from the traditional dichotomisation of the sign-functional view of scientific analysis into either of the two nominalist modes (Løgstrup 1976:51-52; Clausen 1963:138-142) or a Realist mode. I shall sum up the three modes of signification in this way:

The scientific object of the **nominalist** is the language, categories or content of his theory, which is arbitrary, enunciating no existence postulates; the theory is a merger of language and theory-created reality. Reality has been dichotomised into the potentialities of the theory and contingent data, and the work of the theory is controlled by the three requirements in the Glossematic Empirical Principle. The theory has this structure: ErC(ErC), where the first E – if it is a word of the everyday language -, has been stripped of its content, to be replaced by a man-made definition. The cognitive system established in this way is conditional on the definitions and constitutes an analytico-deductive system of hypothetical statements. The nominalist's transcendence problem – the step from theory to reality – is solved by the theory being projected onto data – a process entirely undefined. – Cf. below 1.6. Def. 27.

The scientific object of the **Realist** is the language of his theory, which carries existence postulates with it; while the nominalist created scientific reality, the Realist creates reality in the sense that there is a merger of the cognitive system in question and reality: the denotative content (1.1.20) and, in modern terms, the definiens of the category, which Gellrich (1986:18) very appropriately calls the content of the basic text, merge. In the *manifestatio*-diction of Realism, also typical of modern linguistics, the essence (*manifestatum*) of a given section of reality is inferable from the things themselves (*manifestata*) <u>and</u> their names. Transcendence is immediate from language to reality. – Cf. below 1.6. def. 26.

The historian's object and **historical analysis** will then consist of elements from the other two modes of signification. While a distinction between language and reality is maintained, the historian's theory, a narrative written in the everyday language, does carry existence postulates with it. An illustration may be better than lengthy explanation: It is self-defeating and absurd to deny the empirical relevance of a text like the Anglo-Saxon chronicles, and a narrative of the French Revolution is also a text that narrates indubitable historical events; and in this sense historical reality is inferable from the everyday language, and the narrative *qua* a cognitive system merges with the reality it describes; this will be modified later. Unlike the nominalist the historian does not so define his categories that the definition ousts the meaning of the category which has been evolved over time, but the historian's definition strategy is dynamic, as it considers the sign's language-created meaning: a modern narrative of the French Revolution cannot merge with the realities of 1789, because the (French) language has evolved since then. This means that the modern historian's definition task is to retrieve the meaning of the French language of 1789. An understanding of the Anglo-Saxon chronicles based on a dialect of present-day English is not appropriate, because this dialect is not a constituent part of Anglo-Saxon reality. This difference is also the reason why we may have more (synonymous) narratives of the same past event[24].

The Realist element in history comes from the fact that historical matter is eminently amenable to being described by the everyday language, as the latter is both a precondition for and an integral part of all historical existence. The nominalist element in history comes from the fact that the expression of a narrative (= E) will be a sign in the historian's present-day dialect, whereas the content (= C(ErC)) will be the historian's attempt at retrieving the precise meaning of the historical events he is describing. – This will be elaborated below, where its consequences will be drawn.

I.2. Transcendence and unidirectional dependency. The latter turns into an aition. The scope of the aition. No logical argument can be constructed to prove the priority of the concept of stability. The cause of change is part of data, not to be equated with the theory of data

The notion of there being a gap (distance) between linguistic and nonlinguistic reality, accepted by the nominalist and, partly, by the historian, involves a transcendence problem. This problem appears in two fundamental guises: in its simplest version it involves a dichotomy consisting of the man-made theory and its non-man-made data, the level of logic or theory and the level of data, respectively. It implies two pretheoretical views of data, the latter being either 'mere' raw-material for the theory with no inherent meaning or raw-material with some degree of independent meaning. The former alternative entails that the theory conveys meaning to data (being the 'underdetermined' form, structure or system 'behind' data), the latter alternative's meaning concept says that data has some immediate influence on the theory in terms of verification- or falsification-processes, so that access to this data is by way of (*der Erkenntniswert* of) the everyday language.

With the (Saussurean) introduction of the dualism into linguistics, the study of language took a Realistic turn seeing that the level of data was dichotomised into a *langue* part and a *parole* part, into a controlling, governing, 'causing' entity as against a controlled, governed, caused entity. And synchronic linguistics was then back to square one: ontological dualisms. Synchronic linguistics would now have to explain how the two entities could form a unity, the everyday language, and how the transition from *langue* to *parole* occurs. Samuels saw the transition as a(n agentless) selection process; how synchronic grammarians see it is not clear.

---

24 Synonymy is therefore the product of the passing of time, which – in relation to a given fact or event – opens a space which may be filled by more than one text denoting the event. It makes no sense to appeal to the simplicity criterion of immanent linguistics (Togeby 1965a:15).

1.2. Transcendence and unidirectional dependency

A clear example of this complex situation in modern synchronic linguistics is the generativist's view of language (see *Prologue*). The empirical phenomenon called language consists of two (mental, biologically determined) entities *the initial state S_0* and *the realization of* this state *under particular conditions* (Chomsky 1986:37), the former being a so-called language *module qua* some kind of *genetic endowment*, the latter being a particular internalized language like present-day English – a dual entity in its own right (pp. 40; 41). With the initial state being a (mental/biological) *constant* any particular language will be *an instantiation of the initial state under different conditions* (p. 38). Thus the generativist faces the dual problem of explaining the 'creation' of the particular language *qua* both a cognitive system and *a state of mind* (p. 40), and criteria that bridge the gap between the bipartite man-made description of the universal grammar, which is to correspond to the initial state (of mind) and the particular grammar (*a finite system* (p. 40)) that the single speaker has internalized and which is *shared with others from the same speech community* (p. 55).

EMPIRICAL LEVEL OF FACTS
initial state of mind/language module → particular state of mind → substance, spoken/written
(a biological entity)   (a mental entity)   (a physical entity)
shared by all humans   shared by some humans   individual, imperfect
(a) universal grammar   a finite particular system   infinite manifestations

MAN-MADE LEVEL
Theories of the two mental states and of how imperfect substance realises a speaker's internalized language

Generative theory is not concerned with the transcendence problem, but it is obvious that the whole conception has not liberated itself from dualist ontologies, which stem from the everyday language. And it is hard to see how such an overelaborate scheme will handle the transition facts mentioned in the preceding, 'the moment when' one internalized language is no longer transmitted and a new one comes into being. The whole programme does not escape an element of absurdity, and with reference to Henning Andersen's above attempt at solving the generational problem I shall mention the factual details that generative synchronic theory must cope with: in all speech communities it will inevitably occur that one baby as the first internalises G2 and not G1, the language of her parents; new conditions on the mental/biological transition from the initial state S_0 to G create G2 and not G1. In other words, the language-learning child acquires a state of mind, a cognitive system, different from that of her immediate surroundings, older family members, and is it possible that the natural time gap between the births of two children within the same family may result in the older child's learning her parents' G1 and her younger sister's learning G2? And if this is the scenario that makes Labov claim that language change is incompatible with the nature of language, then the child learning G2 will be in communicative trouble because the majority of speakers in her speech community will share an internalized language that is different from hers – in fact the above-mentioned particular 'sharing'-relation is irrelevant – at first, *i.e.*' at the moment when'. Finally, the generativist theory should explain how the 'shared' language arises in a speech community?

The generative position faces two problems: the notion of there being a specific language 'module' is incompatible with the biological fact *there is no evidence for an "absolute" language area* (Lenneberg 1967:61), and that a particular behaviour cannot be related exclusively to some cerebral *morphological innovation* (p. 54) of the human brain. Secondly, it is hardly appropriate to call the internalised language a cognitive system unless *cognitive* is defined[25]. If a cognitive process is conscious and completely rational, then there is a large realm in human existence that is not governed exclusively by rational factors: the domain of feelings and attitudes (Lenneberg, pp. 356-357). The generativist does not recognise the conflict between man's two different ways of organising his mental life, ways that are reflected in the bipartite division of the brain into cerebellum and cerebrum (Jacobsen 1976:11-13). The way a child acquires her native language does not run along rational lines; the human baby's primary 'objective' is survival (see chapter 4. below), and it is entirely unlikely that the child reacts rationally to the so-called deficient language stimuli she is exposed to and uses her language rationally. On the contrary, a 'rational' case can be made for the universality of the three properties, synonymy, polysemy, and change, so that they can be said to function as mediators in the psychic-mental conflict in man caused by the evolutionary fact that cerebrum evolved so fast that the other inherited mental faculties (attached to cerebellum) did not have time to adjust and adapt to man's – in evolutionary terms – newly acquired mental faculties (Jacobsen, p. 12).

In 1.2.1 I have listed *synonymy*-like names that have been used to denote and characterise the two relates of the transcendence problem, when we exist in a rational and cognition-oriented mode:

1.2.1	THE LEVEL OF LOGIC		THE PHENOMENAL LEVEL
	Theory	is predicated of	Data
	Language (L)	subsumes, denotes	Extralinguistic reality (EL)
		generalises, classifies	
		explains, describes	
		'lies behind', 'underlies'	
		corresponds to, represents	
		is true of	
		is selected by, is projected on	
	(thinking, thought		experiencing, experience)
	able to denote		amenable to being denoted
	embodied, realised by		embodying, realising
	symbols, forms		data
	categories, predicates, classes		instances, particulars, members
	relations, *functions*		relates, *functives*
	concept – species		individuals
	a class as many or *as one*		*a class as many* or *as one*

---

25 Neither *cognitive* nor *rational* is defined by Chomsky (1986); cf. Jacobsen (1976:34-35) for a summary of the features that characterise rational behaviour (thinking) and behaviour controlled by the primary processes.

1.2. Transcendence and unidirectional dependency

hierarchy	*resolves*	*a syncretism*
analysans		analysandum
meta-		object-
formal logic		reality, all objects
laws, rules		instances
'ogres'		'observable reality' (Bernsen 1978:192)
constancy		*fluctuation*
form = system & process		*formed reality, substance*
schema		usage
manifestatum		*manifestans*
'what is behind'		data
implicans/implicatum		implicatum/implicans
Causa		Effectum
La langue		*le monde* (see note below)

Even without the three classical relations *participation, inherence* and *mimesis*, the very number of synonyms testify to the magnitude of the problem, man's transcendence into reality, and the addition of the modern dualisms merely complicates the initial issues: does the dualism belong to the phenomenal level or to the level of theory[26].

The absolute distinction between the logical and phenomenal levels hinges on the empirical relevance of the fundamental presupposition of **separation**, formalised by this nonparticipative principle: **what is a precondition for X cannot be a part of X**. If either level is a precondition for the other, it is not a part of this level: cause and effect, reason and consequence, antecedent and consequent do not merge. This view is to be a safeguard against any Realistic overtones, but implies the idealistic notion of the *presuppositionless beginning* of scientific analysis, which Cassirer (p. 296) argues rests on an unjustified premiss, namely the separation of *thought* and *perception*, in my diction, the everyday language and existence.

Modern, naturalistically based science (Bernsen 1978:15) takes it for granted that the logical basis of its methods is self-validating, and that the staticness or stability of the formal structure of a particular theory mirrors the fundamental structure of the analysed object (Pinborg 1976:254-255). This view epitomises essentialism and is not a far cry from Paul's Realism-warning; it is bound to yield a simplified[27] view of a given object, separating the essential from the accidental, significant from the nonsignificant as well as reducing multiplicity and variation to unity and invariance. Gellner (1963:30-31) may be right in claiming that

---

26 The Glossematic position is clear (Hjelmslev 1971:29; cf. below 3.2.1): the structural hypothesis does not put forward existential – *Ding-an-sich* – enunciations; the process-system dualism will therefore only refer to the level of logic. Note how Hjelmslev's position is curiously reminiscent of Cassirer (1953 (= 1923):295-297), who shows how the initial existential difference between subject and object can be translated into a *methodological* distinction, so that the existential postulates of the theory can be ignored in the early stages of the scientific process.

27 This is not merely a question of a theory being underdetermined (Gregersen 1991,vol. 1: 226;253;255) in relation to data; the significant-nonsignificant dichotomy transcends the methodological level in that it is a pretheoretical and as such a transtheoretical assumption that conceals the existential-metaphysical prejudices of a given theory.

dualism has been on the decline as an official dogma in recent centuries, but that is not universally true (Bernsen 1978:192). Linguistics is a domain where transcendentalism, dualist views have so flourished at least since Ferdinand de Saussure that Dualism must be called the official dogma of (synchronic) linguistics.

Classical dualist transcendentalism is, following Gellner, the transcendent reduplication of reality into a group of entities that are called the essence of a given object and into a group of entities that are called the accidents of that object. If the former entities lay any claim to having empirical relevance, being a part of so-called objective reality, the theory involves an ontological dualism. This dualism arises from the very method that our century has hailed as a theoretical or methodological panacea, modern symbolic logic. The chief exponent of this modern version of essentialism is Jan Lukasiewicz (1967:1), who claims that modern logic is applicable to all objects; a corollary is that the logical representation or logic-based explanation of things is the be-all and end-all of explanation. Thus the inborn validity and empirical relevance of the principles of formal logic are taken for granted despite the simplifying view of reality they yield. Lukasiewicz's claim is therefore only true in so far as and to the extent that the applied logical principles have been derived from the everyday language: only the everyday language can be 'applied' to all objects, all of reality. This fact will be so generalised that the everyday language is the ultimate form of reality, thereby carrying meaning to the latter, if not the meaning of the latter; thus the nominalist position is a simple, but simplifying corollary of this.

Through the *omnia-in-unum* movement Thales introduced with no little help from his native language the *All*-concept, generalisation, into the description of physical reality. I shall now take a closer look at generalisation, it being the raison d'être of subsumption, denotation, predication, and not least prediction[28], as well as one of the factors in the transtheoretical distinction between the significant and the nonsignificant sections of reality[29].

A-statements to the effect that *e.g.* 'All languages are static', couched in the words of the everyday language are regarded by our scientific tradition as basic, primitive, indivisible. They are scientific individuals, falsifiable in toto. An A-statement satisfies the formal requirements of various deductive systems (I shall not be concerned with the predicative calculus):

1.2.2 The categorical syllogism   The propositional calculus
All M is P $M \supset P$
All S is M $S \supset M$
All S is P $S \supset P$

The logician's problem is that these formalisations of an A-statement need not represent ex-

---

28 The exposition is a shortened version of a longer and more detailed analysis of the logical conditional and the four classical categorical statements and their interrelatedness: an A-statement has this structure: All A is B; an E-statement: No A is B; an I-statement: At least one A is B; and finally the structure of an O-statement is: At least one A is not B.

29 That generalisation became one of the critical issues in both the Neogrammarian controversy (All A become B) and the Naturalistic Reduction of the historical disciplines will only be mentioned in passing.

haustively, albeit logically adequately, the empirical facts or, what should amount to the same, the whole meaning of a corresponding statement *qua* a non-formal statement in the everyday language: the everyday language in its scientific (metalingual) function. My argument is that the traditional A-statement (generalisation) is a complex statement, ambiguous and equivocal and that dependences *qua* unilateral or unidirectional relationships (implications) are not empirically relevant: **they are interpretations of empirical connections**.

Givón's importation of the logical conditional of modern symbolic logic into synchronic linguistic description (1973) will be my point of departure for a brief excursion into the weaknesses of the transcendent use of logical methods[30]. In the discussion of Universal Implications we see a curious return to Luick's call for the inner coherence of historical processes, indeed, the causal connection between sound changes. Due to the fact that universal implications are regarded as empirical facts *qua* constraints on as well as shapers and prime movers in syntactic change, the theory, whose chief exponent is John A. Hawkins (1979; 1980; 1983; cf. Robinson 1993:164-167; Comrie 1989:86-103), has it that there are causal relations between the process (the cause) that changes the word order of a language from OV to VO and the process (the effect) that changes the word order AN to NA in the same language. Applied to Luick's push-chain view of the 'Great Vowel Shift', we have in Lass's formulation: No diphthongisation of ME /u:/ > /ʊu/ without the raising of ME /o:/ > /u:/.

Givón makes part of the truth table of the logical conditional the formal definiens for the two notions *presupposition* and *implication* (p. 891), defining *to presuppose* as in 1.2.3:

1.2.3 P presupposes Q just in case True(P) ⊃ True(Q) and False (P) ⊃ True(Q).

The definiens is the two first rows in the truth-functional definition of the logical conditional (1.2.4)

1.2.4a   P ⊃ Q   **1.2.4b**   The aition: Q * P
               t T t   t T t
               f T t   t T f
               t F f   f F t
               f T f   f F f

with the third row of the conditional saying that P cannot be the case (true) with Q not being the case (false). I shall demonstrate how Givón's analysis (of factitive verbs) presupposes not the logical conditional, but a new logical function, which I shall call *the Aition* (1.2.4b)[31]: Q but not only if P.

Following Givón a factitive verb presupposes the truth of its complement clause *qua* its presupposition. The table in 1.2.5 implements the truth table of a logical conditional (1.2.4a):

---

30 Whether my words are relevant to logic as an immanent science is immaterial; my criticism refers only to the trancendent use of logical methods, in particular, in historical disciplines.
31 I shall demonstrate the historical relevance of this relation in connection with the elements of the developmental series, A > B > C, as a relation that steers a middle course between historical inexorability (determinism) and historical contingency (arbitrariness).

1.2.5 'John regrets that Sheila was hurt' presupposes the truth of 'Sheila was hurt' =

<u>John regrets that Sheila was hurt ⊃ Sheila was hurt</u>
(1) John regrets that Sheila was hurt & Sheila was hurt   T
(2) John doesn't regret that Sheila was hurt & Sheila was hurt   T
(3) John regrets that Sheila was hurt & Sheila was not hurt   F
(4) John doesn't regret that Sheila was hurt & Sheila was not hurt   F

While rows 1 and 2 make sense – there being no necessary connection between John's feeling regrets for Sheila's accident and the accident itself – row 3 is obviously false. The same goes for row 4: it is meaningless that both Sheila wasn't hurt and John didn't regret that she was. In other words, Givón is not applying the truth-functional definiens for the logical conditional, but that of the Aition: it is the case that *Sheila was hurt* (Q) **but not only if** it is the case that *John has any regrets* (P) – John may also have no regrets for her mishap. To labour my point: Givón may believe that his presupposition rests on the material definition: Not P without Q (= -(P & -Q)); but the fact that he makes row 4 false (linguistically and empirically irrelevant) turns the implied logical conditional and the formal definition in 1.2.3 into an aitional function.

Another type of factitive verbs – neg-factitive ones, *e.g. to pretend* – presupposes the falsity of their complement clauses (p. 892):

1.2.6a 'John pretended that Sheila was hurt' presupposes the truth of 'Sheila was not hurt' =

<u>John pretended that Sheila was hurt ⊃ Sheila was not hurt</u>
(1) John pretended that Sheila was hurt & Sheila was not hurt
(2) John didn't pretend that Sheila was hurt & Sheila was not hurt
(3) John pretended that Sheila was hurt & Sheila was hurt
(4) John didn't pretend that Sheila was hurt & Sheila was hurt

Since rows 3 and 4 are impossible/false in modern English, the original implicational relationship is not a logical conditional, but an aition: *Sheila was not hurt* **but not only if** *John pretended that Sheila was hurt*. The antecedent clause is also true if John did not pretend that Sheila was hurt. Let's put the two sentences in their chronological order, then the fact that Sheila was not hurt is temporally prior, T1, to John's pretending at T2:

1.2.6b T1: Sheila is not hurt > T2: John pretends that Sheila is hurt

Sheila was not hurt (or: remains unhurt)

That at T2 John could just as well not have pretended that Sheila was hurt is a historically relevant possibility due to the lack of determinism between a successor and its successor; in structural terms: the fact that Sheila remains unhurt is the horizon against which the other utterance is understood.

1.2. Transcendence and unidirectional dependency

Of further interest to the argument of this essay is Givón's mention of the polysemy (ambiguity) of negated clauses such as *Socrates is not bald*[32]. The two senses are: the negation denies the assertion – Socrates is bald – without denying the clause's presupposition, namely *Socrates exists*.

1.2.7 The assertion 'Socrates is not bald' presupposes the truth of 'Socrates exists'.

This meaning implies that it cannot be the case that Socrates is not bald (= P) and Socrates does not exist (= -Q). Since there is no necessary connection between the existence of Socrates and him being either bald or not, rows 1 and 2 of the conditional's truth table make sense – as in the other abovementioned cases; likewise rows 3 and 4 are impossible or meaningless due to the negation of the presupposition. The second meaning of *Socrates is not bald* appears when the clause is paraphrased by *It is not true that Socrates is bald*, which *per se* is not necessarily different from the shorter negated version, but is made to differ by assuming the presupposition: *Socrates does not exist*. The assertion that Socrates is not bald *fails for lack of denotation* (p. 892). This is an artificial distinction[33], because the same argument applies to row 3 (the Not-P-without-Q interpretation): if Socrates does not exist, both clauses *Socrates is bald* and *Socrates is not bald* will carry no denotative power with them. This is summed up in the table in 1.2.8, the truth-functional implementation of 1.2.7

1.2.8 Socrates is not bald ⊃ Socrates exists
(1) Socrates is not bald & Socrates exists   true/meaningful
(2) Socrates is bald & Socrates exists   true/meaningful
(3) Socrates is not bald & Socrates doesn't exist false/meaningless
(4) Socrates is bald & Socrates doesn't exist   false/meaningless

Another example (1.2.9) will illustrate the problems that crop up when logical principles are correlated with existence postulates; that the former fail to provide us with a means of bridging the transcendence gap, of telling us what it means to exist, is clear from Givón's analysis. Let's assume that

1.2.9 Indo-European was spoken 5,000 years ago // *La Bastille* was demolished in 1789.

These statements presuppose the existence of our Indo-Euopean mother language and the former (undemolished) existence of this particular building in Paris then, respectively:

---

32 Givón's analysis of the negated propositions hinges on a special implementation of the logical universe A & nonA: are the classes of the conditionals' subjects nonempty classes and/or empty classes? (Hansen 1997:1594).

33 Furthermore Givón here adopts the strict nominalistic position: the everyday language carries no existence postulates with it, this being a property that can only be applied to the theoretical level; it cannot be generalized to the object level, seeing that the everyday language would only consist of potential denotators.

1.2.10[34]   Indo-European was spoken 5,000 years ago ⊃ Indo-European existed
(1) Indo-European was spoken 5,000 years ago & Indo-European existed
(2) Indo-European wasn't spoken 5,000 years ago & Indo-European existed
(3) Indo-European was spoken 5,000 years ago & Indo-European didn't exist
(4) Indo-European wasn't spoken 5,000 years ago & Indo-European didn't exist

The absurdity of row 3 (not P without Q) is evident; the truth of row 1 appears equally evident. With row 4 things are different; apparently the value truth must be assigned to it, just as truth must be assigned to row 4 in the well-known example in 1.2.11. Now, row 2 in 1.2.10 is the moot point and will illustrate the pitfalls of existence and presupposition. Logically row 2 is true; empirically, this value assignment makes no sense: how can a language exist without being spoken? However, if we follow Hjelmslev (1963:39-40), the system of language does not require the presence of a process or text; so if *Indo-European* is defined by the system of this potentially spoken language, then the truth of row 2 will follow[35]. This is no way out of the dilemma, because the subject of the four clauses will not be stable. The problem may be so identified: the existence presupposition is contained (analytically) in the predicate of the antecedent and so is the predicate in this proposition's subject. The moment we try to assert the possibility of the nonexistence of language, indeed, of spoken language, we shall be denying the proposition. The negated existence postulate *qua* a presupposition will automatically turn the presupposing clause into a non-denoting one. The alternative is this version of Lewis's startling theorems: if an existence postulate is the necessary condition for a conditional, **then this postulate (Q) is implied by anything (*i.e.* both P and nonP)**: if the existence of the everyday language is the necessary condition for any statement about the nature of the everyday language, there is no argument that can make synchrony and stability logically prior to history and change: logically the existence of language *qua* a necessary condition for any statement about the everyday language will entail logically contradictory statements about the everyday language, because the existence of the everyday language will be implied by both 'Language is a historical phenomenon' and 'Language is not a historical phenomenon'. Even this well-known type of the purportedly empirical relevance of logical formalisation in 1.2.11 carries no cogency with it and certainly demonstrates the arbitrariness of the method: the empirical relevance of the second row may be salvaged by a condition – a hydrant has been destroyed – that is no less far-fetched than a condition on the truth of the third row to the effect that the streets had been covered with large tarpaulins. At best 1.2.11 will reduce to an interdependent relation.

1.2.11   It rains ⊃ the streets are wet
(1) It rains & the streets are wet   true
(2) It doesn't rain & the streets are wet   true
(3) It rains & the streets are not wet   false (not P without Q)
(4) It doesn't rain & the streets are not wet   true

---

34 Only row 1 is relevant to the historian: the truth of the destruction of *la Bastille* implies analytically that the building existed, and no historian is interested in this piece of logical information: on condition that *La Bastille* exists, the building may or may not be demolished.
35 Whether this is the argument behind Hjelmlev's claim that the Indo-European parent language may never have been used (Hjelmslev 1970:84), I cannot decide.

1.2. Transcendence and unidirectional dependency

The lesson I want to draw from this exposition is that the empirical relevance of unilateral determinism formalized by the logical conditional ('if – then'; 'not – without' -; '⊃') has to be demonstrated; I have shown that Givón's use of it as the definiens for his presupposition and implication concepts cannot be upheld in that some conditional relationships are aitional ones[36]. Furthermore, I shall argue that the logical conditional is only a conditioned interpretation of empirical data, a view that is corroborated by the formal relationship between the logical conditional and the aition: the former is inferred from – is an **interpretation** of – the latter:

1.2.17 (Q * P) ⊃ (P ⊃ Q)

```
t T t   t   t T t
t T f   t   f T t
f F t   t   t F f
f F f   t   f T f
```

All dependences in the empirical world are interpretations of two situations: an entity (q) has been registered in conjunction with another entity (p) and without this entity. Then the logician so interprets these facts that the latter (p) cannot possibly appear alone, it presupposes the presence of the other entity (q). The interpretation implies the unmediated transition from what has been registered as existing or as possible within a definite space- and time-continuum to an unlimited continuum: what is always and everywhere the case, from which a relation of impossibility is inferred: entity p alone is impossible. Synchronic linguists have yet to demonstrate that their universal implications or constraints on in particular what is historically possible have been set up as unconditioned or uninterpretative rules.

The narrative that the language historian tells (Luick) must first of all observe the chronology of the narrative's events; the ideal situation is that Kausalnexus (Paul) are established. Luick saw such a nexus in the Middle English changes of /e:/ and /o:/ in relation to /i:/ and /u:/, respectively (1964:§477); the change of the non-high vowels caused the change of the high vowels in a push-chain (§482, Anm. 1): no diphthongisation of ME /u:/ (and by analogy /i:/) without the rising of /o:/ (and /e:/). This monumental process is part of what may be called the French Revolution of English language history. I shall now analyze the logical interpretation of this particular causal nexus in greater detail (cf. Lass 1988 and *Prologue*).

If the change of /o:/, the cause, is interpreted as the sufficient condition we have this formalisation: **ME /o:/ > /u:/ ⊃ /u:/ > /ʊu/**; the problem here is that diphthongisation may occur without the rising of /o:/, a situation that has not been registered. The alternative interpretation makes the cause a necessary condition for the effect: **ME /u:/ > /ʊu/ ⊃ /o:/ > /u:/**. The problem here is that vowel-rising is allowed without the concomitant diphthongisation. This brings Luick's dilemma into the open: his Kausalnexus is not without exceptions. Since Stockwell and Minkowa try to explain away this discrepancy, I had better quote *in extenso*[37]: *Eine Berührung mit dem Ergebnis des me. /o:/, nämlich [u:], zeigt sich im nördlichen Lin-*

---

36 In the logical sequel to this essay I shall analyze the other examples that Givón applies logical methods to.
37 Notational forms different from Luick's have been used for typographical reasons.

colnshire (...); hier ist in nordhumbrischer Weise me <ou>[38] *bewahrt und in südhumbrischer Weise [u:] aus me /o:/ entwickelt. Von diesem Übergangsgebiet und vielleicht einem oder anderem Punkte abgesehen, erfolgt also die Diphthongierung des me. [u:] nur dort, wo me. /o:/ zu [u:] vorrückte, sie unterblieb, wo me. /o:/ schon früher zu einem [ü]-Laut umgebildet worden war (....).*(§482, Anm. 1). In this transition area ME /o:/ and /u:/ merge in [u:]. Since we do not have a situation where ME /o:/ remains (cf. §479), our logical relationship is not one of implication, but an aitional one: **ME /o:/ > /u:/, but not only if ME /u:/ > /ʊu/**. Again we must conclude that symbolic logic is not immediately applicable to all objects.

Writing some forty years later than Paul (1880), Luick seems to be influenced by idealizing tendencies in his days; his first mistake was that he connected the cause-effect relation with generalisation instead of working with a multiconditional cause-effect concept: a process may occur, but not only if another specific process occurs. His second mistake was that he did not follow his own story-telling principle: if the rising of /o:/ is the cause of the diphthongisation of /u:/, where and what, then, is the cause of the raising of /o:/? There is no empirically necessary reason to create this break in the sound-history of English in the fifteenth century. His third mistake, which follows from his start-end point of view, was that he had to hypostatize the factors that limit the process, turning 'three impulses' into supraindividual entities: Luick delimits a period of about 300 years by the operations of *drei Impulse* (§§ 477, 479), which, incidentally, remain undetermined; this corresponds to the historiographical chapter-treatment of a period from, say, 1789 to 1815 in French history. The problem is that such relative concepts as beginning and end are taken as if they are absolute[39]. In English sound history we cannot attribute any empirically relevant meaning to the concept of beginning: any beginning is also an end, thus both the effect and the cause of something. The French Revolution and the Great Vowel Shift are the products of historical hindsight. The analysis of our two sample events operates within the limits of a foregone conclusion: perhaps the respective historians do not commit a circle, but they do not question their starting point, whether there is such an entity as the French Revolution and the Great Vowel Shift (*pace* Stockwell and Minkova) or to what extent their respective linguistic '*prejudices*' preform their analysis. To make my point clear: I am not saying that historians are not aware of the pitfalls of the idealising and Realistic nature of their everyday language, only that if a historian sets out to explain the French Revolution (1789-1815) or advance a theory of the Great Vowel Shift (1400-1700) he has not escaped the above-mentioned pitfalls – the latter less so than the former. Both Jespersen and Luick as well as Anderson and Kiparsky (above) created theories of X on the preconceived assumption that there **is** an X, which sends us back to the intricacies of Givón's existential presupposition.

The aition is a logical function that has been ignored in both logic and linguistics in spite of the fact that it is implied in generalisations or statements that are not equivalences, but unidirectional dependences as seen in the linguistic subject-predicate structure. Consider these statements (clauses); clauses 1, 2, and 3 were dealt with by Jespersen (1963:150-154)

---

38 The Middle English spelling of the long *u:*-sound.
39 Linguistics is saddled with such irrelevant issues as inception- or actuation-problems (Weinreich *et al.* 1968). The parameter has not only eschatological overtones, but it presupposes a historical space continuum that is discontinuous.

1.2. Transcendence and unidirectional dependency

and it is his analysis that will be further analyzed to show the ambiguity of generalisations and next to criticize the transcendent use of the contradictory universe of logic

1) All lions are mammals $P \supset Q$
2) Bill is a grocer
3) Love is beautiful
4) Truth is beauty, beauty truth. $P \supset Q \,\&\, Q \supset P$

We can understand sentences (1, 2, 3) unidirectionally because we infer – or know beforehand – that (1) not all mammals are lions, that (2) there are more grocers than Bill, that (3) there are more beautiful phenomena, feelings or states of mind than love. In (4) Keats established an equivalence between the subject and the predicate, with the subjects exhausting the predicates. Aitionally we shall say on the basis of an immediate inference that the standard system of logic cannot formalise truth-functionally: (1) Not all mammals are lions; (2) Not all grocers are Bill; (3) Not all beautiful states of mind are love. In a word, the class of mammals is not exhausted by the class of lions: mammals but not only if lions.

Before we go on with the aitional analysis, an interlude on the notion of differences and concepts being contradictory; our universe is still change and stability: static (q) = nonchanging (q), change (nonq) = nonstatic (nonq); and the class of languages (p) versus the class of nonlanguages (nonp). The generalisations in 1.2.18 exhaust two universes: p and nonp, q and nonq:

1.2.18 All languages are static :: All nonlanguages are nonstatic[40].
       $p \supset q$ $nonp \supset nonq$.

On the assumption that nonp is all that p is not and vice versa, and the same goes for q and nonq, we infer that a language has one and only one property, while a phenomenon that is not a language may have more properties. The alternative position is 1.2.19

1.2.19 All languages change :: All nonlanguages are non-changing.
       $p \supset nonq$ $nonp \supset q$.

Now the everyday language has only the property, to be able to change. However, since *nonstatic* is equivalent to *changing*, and *nonchanging* to *static*, the two generalisations are rephrased as follows:

1.2.20 All languages are non-changing // All languages are non-static.

1.2.20 carries with it the absurdity that if language is static/changing, it has only one property, but if it is nonstatic/nonchanging, it may have more properties except being static/changing. The logical equivalence in

---

[40] Togeby (1965a:195-196) operates with a similar universe, in which *langues quotidiennes* are defined by sign-functional properties, which do not characterize the class of *non-langues*. Togeby, however, does not consider the Glossematic consequence that turns non-semiotics into semiotics (cf. 1.2.24).

1.2.21   All languages are static = All languages are nonchanging

is no equivalence at all. The logical system entails this absurd alternative: in a universe of more than two properties implemented by static/nonchanging and nonstatic/changing, a language has either one and only one property or all properties but one. With reference to the aitional remarks above we can now demonstrate the polysemy of generalisations in that 1.2.18 has this complex meaning

1.2.22   All languages are static such that not all static phenomena are languages in that some static entities are languages and some are not languages (= non-languages)

with 1.2.22. being formalised by

1.2.23   (Q * P) and (Q * nonP), from either of which (P ⊃ Q) follows, as a dependence-interpretation.

This argument has demonstrated that the standard system formalizes an empirical statement into the absurdity of monism or into a statement that allows an entity to have more properties than one. The next part of the argument will demonstrate that the monism of extreme synchrony cannot be upheld: the generalisation: All languages are static/nonchanging, implies that 'to be static' must be predicated of more entities than languages, namely about the class of non-languages, in order to uphold the unidirectional dependence between subject and predicate (cf. (4) above). The same applies to the predicate of non-languages; and since the only alternative class to the class of nonlanguages is the class of languages, then members of the latter class may all be characterised by the property to be nonstatic, hence changing.

1.2.24[41]

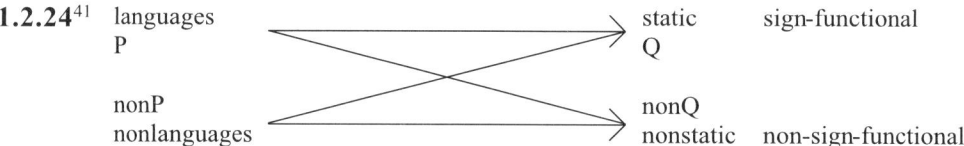

languages   static   sign-functional
P   Q

nonP   nonQ
nonlanguages nonstatic   non-sign-functional

If P is both Q and nonQ, then the standard system creates an absurdity, which I shall resolve by the aition: P but not only if Q/nonQ: the everyday language is both static and changes.

My argument does not prove that the everyday language is a changing or a static phenomenon, but it corroborates the analysis in chapter 1.1. – no cogent argument can be set up in favour of the priority of either stability or change; hence the monism of synchrony cannot be upheld. Finally, the meaning of a generalisation is not precise, but polysemous.

I shall now return to the transcendence problem. If the language of formal logic is no theoretico-methodological panacea, there is one alternative, and that is the everyday language; only the everyday language can interpret Lukasiewicz's generalisation, and if we accept this interpretation, it will also be possible to establish a relation between theory and data that

---

41   The addition of the (non-)sign-functional properties refers to my discussion with Togeby and Hjelmslev.

1.2. Transcendence and unidirectional dependency

transcends the level of postulate, and which will combine properties from Realism and Nominalism. The Glossematic notion of science being a Platonic idea must be so interpreted that the Platonic idea is the everyday language, and not *the* linguistic theory (Hjelmslev's) (1963:127), which Gregersen very appropriately calls the *state religion*. An empirically relevant act of transcendence has been effected in so far as the everyday language has been used 'communicatively' or denotatively successfully. Plato's methexis (to exist by participation) and the Pythagoreans's mimesis (to exist by imitation) must be replaced by successful denotation, to be defined by *to exist by the everyday language*. While, according to Aristotle (*Met*. 987b 7-15), Plato and the Pythagoreans did not determine their respective transcendence relations, *to exist by the everyday language* will be further determined by what is characteristic of historical phenomena or man's historical existence: the everyday language is both a contributing cause of and a part of any historical phenomenon, a position that is not a far cry from the views of Socrates and Plato concerning the use of language in ontologico-epistemologcial enquiries (*loc. cit.*; cf. 1.6.). Phenomena of which the everyday language is a part are therefore preeminently amenable to being described by the everyday language. This means, in turn, that the cause of X is different from a theory of X[42], a conclusion that is corroborated by the above discussion of existential presupposition: let the existence of the linguistic monument, the Great Vowel Shift above be a necessary condition for a theory X of GVS – X implies the GVS -, then the theory cannot be equated with a causal statement, because the existence of the Great Vowel shift is also implied by theory nonX: nonX implies the GVS. **A theory of history is not a causal statement**.

While I agree with the nominalist that the theory conveys meaning to its data, but disagree with him in his view of the contingent nature of data (cf. above), because they are signs before they are made so by the theory, I must disagree with the extreme Realist position and agree with it precisely because of the existential function of the everyday language, a function that guarantees the cognitive potential of the everyday language, a potential that no other theoretical language has. All our historical knowledge does not come from the the everyday language, but some does, and I shall argue that if we want to acquire knowledge of language development (change), it is not sufficient to proceed from and rely on the eyes of the senses. I shall therefore accept Hjelmslev's generalisation that there is no 'nonsemiotic' that is not a 'semiotic', too, but <u>not</u> his view of the all-embracing potential of the scientific language[43]. Togeby's position above must therefore be modified to the effect that nonlanguages may acquire sign-functional properties.

As regards the aitional interpretation of the process of change, I said above that aitional

---

42 A causal description of a historical process, *i.e.* an explanation of X in terms of causes, does not transcend what it sets out to explain, the causes being constituent parts of the historical event at hand.
43 In 1938 Hjelmslev (1971:173) was less restrictive, where he says that *la langue* is *la forme* by which we understand (*concevoir*) reality (*le monde* (see 1.2.1)); this language form is organized between the expression-substance and the content-substance (p. 161). And he continues (p. 173*): Il n'y a pas de théorie de la connaissance, objective et définitive, sans recours aux faits de langue*. Fair enough, if the last three words are to be replaced by *the everyday language*. As such no epistemologist will be surprised, and current generative theory will also concur since what *la langue* above refers to is a cognitive system. But when Hjelmslev concludes with this quantum leap, apparently from the level of data to that of theory: *Il n'y a pas de philosophie sans linguistique*, then the above *Prolegomena* position has been anticipated, and the leap is entirely arbitrary.

relationships are neither strictly deterministic nor strictly contingent; I shall illustrate my point with our sample change, OE *a:* > ME *ɔ:*, placed in a developmental series of three members: OE *a:* > ME *ɔ:* > N(ew) E(nglish) *o:*. Our point of view will be the Middle English sound as both a successor and a successor-turned-into a predecessor.

1.2.25 ME *ɔ:* * OE *a:*   Formalization of the relation 'comes from' (= '*')

Logically, the Middle English sound may come from any other sound than OE *a:*, historical development being contingent. This is empirically untenable, because we know that OE *a:* is in some sense of the word the cause of its successor; this causal synthesis cannot be broken up analytically, and it will be the task of the historical theory both to put constraints on and preserve the synthetic and transsynchronic nature of historical development. The formula says in general terms and in accordance with the principle of multiple conditioning: a successor, but not only if one particular predecessor[44].

1.2.26 ME *ɔ:* * NE *o:*   Formalization of the relation 'develops into' (= '*')

Here the aition: 'a predecessor, but not only if a particular successor', means that the open Middle English sound develops into a closer sound, but it could have developed into another, not any other sound; logically, the sound could have developed into any other sound, the logical system implying strict contingency.

Historical development – from either point of view – is as such polysemous as regards the origin of a successor and this successor's successor. In relation to its successor, a successor is also synonymous, presenting us with a choice. A historical theory will add constraints on the historical process without invoking strict determinism and remove strict arbitrariness. The synchronic alternative is ambiguous:

The two formulae describe events at the phenomenal level and must be related to the theoretical. Synchronically, the events are related to a fixed entity in order to be understood, and this fixed entity is the synchronic system at hand. Does this system belong to the theory or to the level of data, being the cause of the processes? We are now back to square one, the transcendence problem. If the system belongs to the level of theory, we shall be moving within the confines of nominalism; if it belongs to the level of data, an ontological dualism is the order of the day, of which one relate, the system, is the cause of the other, *parole*; thus theory and cause merge.

I shall combine the two (three) positions as follows: like all historical phenomena a language change is integrated in the everyday language so that the latter is a causal factor in the rise of the change, which becomes a causal factor in the continued existence of the everyday language. The perceptual basis of a change, the registration of a variant relationship between two forms, say, *ha:m* and *hɔ:m* 'home', will then be placed in a system of empirically relevant judgements, the historical theory, which, attributing meaning to data, permits us to **comprehend** the *geschichtliche* link between our data at hand[45].

---

44 As a historical *Zufälligkeit* the Middle English sound may have been the lowered product of a long tense *o*-sound.

45 This will be elaborated in chapters 5. and 6. For a static version of this, see Cassirer (1953:296-297).

My view can be generalised to history proper: the destruction of La Bastille in France in 1789 is an integral event in the history of France so that the latter (the continued existence of France) is a causal factor in the particular event, and the latter is a causal factor in the continued existence of what is called France, or just in the continuation of the historical existence of Paris. What makes us comprehend the particular event as both a product (effect) of a larger totality and a necessary contributing factor (cause) to that totality may be called a theory of the event.

In both cases, the (meaning-)contribution of the theory must not consist in the reification of the event so that it is equipped with the two unhistorical properties: a start and an end.

To conclude: I agree with Hjelmslev (1971 (= 1938)) (but not with his static version) that the everyday language has a forming function in relation to reality: the everyday language does contribute to the creation of reality, whereby a nonsemiotic is turned into a semiotic (to be elaborated below in 4.3.). I do not agree with Hjelmslev's panacea-like, cure-all conception of (his) linguistic theory. I fail to see how any non-semiotic turned sign-functional can be described by linguistic theory (= Glossematics), for the simple reason that Glossematics presupposes objects of a premised nature (point-in-time existing entities).

I.3 The system at work: synchronic processes. Ontological dualisms are not a thing of the past. The generativist proves Hermann Paul right. Preliminary remarks on idealisation in synchronic linguistics. The existential and cognitive function of *parole*. Definition of *Linguisticism*. The ideal historical narrative

In the previous section it was suggested that the everyday language is involved in bridging the gap between a theory and data. The reason for this is that the everyday language is implied by all phenomena, by historical as well as nonhistorical (natural) phenomena[46]. Since the everyday language is a contributing factor to the existence of historical phenomena, we shall formulate the above generalisation aitionally: the everyday language but not only if historical phenomena. It has also been argued that no logical argument can be set up for the priority of either change or stability, with the corollary that the systems concept in synchronic linguistics is an ambiguous concept.

The seeming contradiction in the truism, 'Danes speak the same language in spite of the apparent fact that they do not speak the same language' is resolved at the cost of introducing a dated (Gellner) dualist ontology which hypostatises the first occurrence of *the same language* by creating a supraindividual entity, denoted by *the Danish language* or less specifically, *a common language* shared by a group of people forming a so-called speech-community. That two or more people communicate – can understand each other – does not allow a cogent and unmediated inference to a shared, supraindividual, system. A pertinent question: how many – if a plurality, at all – particular systems go to make up the system of what is to be the denotatum of such particular signs as *the English language, the Danish language,* or *the French language?* – I have deliberately left out explicit historical designations: *modern English, 1968-English,* or *1999-English*.

---

46 Only *qua* denotata do natural objects imply the everyday language (cf. 1.2.24 above).

In a recent version of the generative approach to phonology Harris (1994:113) operates on the basis of a synchronic calculus view with a universal alphabet of phonological primes (features) which combine through **synchronic processes** to provide a *specification of the vocalic constrasts utilized in English*. The problem outlined above appears here at its simplest: the inventory of contrasts varies from dialect to dialect, leaving us with unexplained specifications of the synchronic processes from the universal alphabet to the dialect: is the inventory of a particular dialect processed by the inventory of what Harris calls *English*, a particular language in its own right – with an inventory of contrasts different from that of the particular dialect? Or does the dialect select directly from the universal alphabet? An answer was given in the previous section: the initial state of mind (So) controls the construction of the particular state of mind (S1), which controls the speaker's output.

Interestingly enough, this conceptual basis is not a far cry from what we find in a standard phonetics book such as Gimson (1994 (= 1962)). Here (p. 4) we read that a London- and a Manchester-pronunciation of the word *cat* will vary as to vowel quality, but in spite of this *we are likely to feel that we are dealing with a 'variant' of the same word*. Thus a two-kinds-of-reality dualism has been set up, namely the *concrete, measurable reality* of the sounds of speech, and a reality where the products of the identi*fying* mental *abstraction* by which we reduce the number of sounds to a '*manageable' number of categories* exist. These categories constitute a *system of significant sound units*, the phonemes, and with the other traditional linguistic levels (grammar, meaning, syntax), the phonological system combines to form a single system of ?English[47]. How does Gimson prove the existence of such a common system: *An utterance, an act of speech, is a single concrete manifestation of the system at work?* The existence of the system is proved by the abovementioned 'feeling' of sameness and by the fact that the 'same' sound cannot be produced twice (pp. 4, 44: *No two realizations of a phoneme are the same*), and the reason why this is so is human imperfection (p. 5).

Gimson (p. 43) makes an interesting concession: (…) *it is frequently possible to make several different statements of the phonemic structure of a language, all of which may be equally valid from a logical standpoint. The solution chosen will be the one which is most convenient as regards the use to which the phonemic analysis is to be put.* Such relativism is incompatible with the existence-by-mental-abstraction of the constructed system. Gimson thus goes back on his own ontological dualism and implies a traditional transcendence relationship in which the theoretical (logical) relate consists of more possibilities and the phenomenal relate is a monoplane entity.

Gimson cannot prove the existence of a single English phonological system, let alone its relations to the dialectal systems that his descriptions (pp. 97-196) set up. That there are *slight* differences between any two *realizations* of the *phoneme sequence* /k+æ+t/ proves neither the empirical existence of /k+æ+t/ nor that the variants can only be analysed with reference to an entity different from them. The same goes for the generativist's *alternants* (Harris, pp. 2-3; 118-126), which can be incorporated into the traditional phonetic-phonemic view:

English /n/ [n], [m]; /d/ [d], [Ø] (loss/truncation from the point of view of morphemes).

---

47 *This book describes the sound system of English* (…). (p. 5).

1.3. The system at work: synchronic processes

Harris claims that [m] is the assimilated product of /n/ preceding a labial: *te[m] birds*, with the conditions for the other two alternants being a following vowel: *send Anne*, and a following consonant *sen[d] two* (Harris, p. 2). The problem is that such statements – either with reference to phonemes or with reference to morphemes, *ten*, *send*, – have no predictive value: they are not generalizations about the sound structure of modern English (Kenstowicz 1994:89): the morpheme /send/ will not always be truncated to [sen], if the following word begins with a consonant. Gimson (p. 261) adds further non-obligatory conditions: 1) *rapid, colloquial speech*, 2) *especially at or in the vicinity of word boundaries*, which clearly belong to the level of the spoken utterance; and the negative element /-nt/ is realized by a set of systematic rules different from the above-mentioned one (Gimson, p. 262). Thus the task of the generativist when he leaves his systematizing stance to establish systematic rules for his everyday language becomes self-defeating or the theory becomes so underdetermined that its empirical relevance must be questioned, unless the theory rests on consistent and nonarbitrary determinations of identity and (non)significant contrasts.

With the risk of labouring my point I shall take a look at the way the phoneme is established in another authoritative work on generative phonology (Kenstowicz 1994:65-82), which opens a discussion about the phoneme as follows:

> One of the factors that initiated the development of phonology as a branch of linguistics distinct from phonetics was the discovery that native speakers often judge sounds to be identical that are clearly distinct phonetically – sometimes quite radically so (p. 65).

Firstly and fundamentally, the ability to detect identity in differences is an innate faculty as elemental as man's classifying faculty (Lévi-Strauss 1962:285n; ch. 2), this being a universal inherent in the everyday language (cf. Plato in chapter 1.1.); that native speakers judge different sounds to be identical is not surprising. The interesting thing is why and how as well as when and where they do so. Apart from the answer implied in the preceding lines, Kenstowicz himself provides no answer but makes do with this type of postulate: in many American dialects the coronal stop phoneme – /t/ – has *eight distinct pronunciations*, these being allophones of the same underlying sound (phoneme) (see 1.3.1 below), and the relationship between the phoneme and the realizations is a causal one: the variant is the *product of a systematic rule, and this rule modifies the segment (?the phoneme) depending on the context in which it finds itself* (p. 66; see 1.3.1 below). In other words, a Kausalnexus has been established with a rule affecting an affectum /t/ to produce a predictable effectum […]. This is absurd as long as we do not know the empirical reality of the affectum. Following the generative programme the native speaker has somehow learnt these rules, because, it is claimed, *No one has taught them to us*. (p. 66). In the first place, why should anybody teach the infant baby such rules; she can listen to what she hears! Secondly, and this is the important part of it all: why must we learn to *omit* [t] in the plural of *tent-s* [tens], if it has never been there? Why must we not learn to insert a [t] in the singular, *a tent*? The answer to the latter questions is that /tent-/ is the underlying representation of this word. This solves nothing: why is this the UR and not /ten-/? Why is /send/ and not /sen/ or /ten/ and not /tem/ the URs of the verb *to send* and the numeral *ten*? The answer is (p. 103): *Given that [x] alternates with [y], the decision whether [x] or [y] underlies the alternation is in general resolved by appeal to the criteria of predictability and the naturalness of the corresponding rules*

– *reflections of a more general criterion of simplicity.*[48] Whether /ten/ or /tem/ is the simpler, the more natural or predictable UR is hard to see. The problem is aggravated when we come to phonology proper: Kenstowicz provides no cogent criterion for the selection of /t/ – whatever its properties, and, significantly, he sets up no systemic entity / ? / underlying the allophones in 1.3.1, and since he emphasizes that it is not a question of a common core of features (p. 74), one must ask: what does *coronal stop* mean in: *in many dialects of American English the coronal stop [t] has as many as eight distinct pronunciations* (p. 65)? It cannot be what the eight variants have in common, and the absurdity is further highlighted by it being a variant or allophone [t] that is realized by eight variants or allophones[49]:

1.3.1[50] [t] "plain"   as in *stem*
       [th] aspirated   as in *ten*
       [t.] retroflexed   as in *strip*
       [D] flapped   as in *atom*
       [N] nasal flap   as in *panty*
       [tʔ] glottalized   as in *hit*
       [ʔ] glottal stop   as in *bottle*
       [Ø] zero   as in *pants*

As Hermann Paul put it, the synchronic grammarian is bound to turn historian the moment he tries to explain the empirical facts beyond what the synchronic rule does, a fact that generativists have been obliged to admit (Chomsky 1986:41-42; Harris 1994:120)[51]:

1) If *bottle* is never pronounced with a [-t-]-like sound in a given dialect, why is there a <t> in <bottle> or why is the glottal stop in [bɑ́ʔl] subsumed under the UR element /t/? Are all glottal stops to be subsumed under /t/? This is contradicted by Kenstowicz himself (Kenstowicz and Kisseberth 1979:211-219; 215): a phoneme /x/ that *is always realized phonetically as identical to the realization of some other phoneme /y/* can never be the systemic element of the word in question. Consequently, a postulated phoneme in an underlying representation that is not (never) realised in a dialect (language) is not a systemic element of that UR[52]. The next problem is: when a speaker from a *t*-less dialect meets

---

48 A point of criticism that will not be elaborated is the vague, indeed, a-generalizing diction that generativists use. To this we must add that Kenstowicz does not come up with anything like an exhaustive description of one single language or dialect in his voluminous book.
49 This type of argument carries heavy suggestions of Thales and the triangulation of material monism.
50 Harris (1994:119-121) provides a more detailed analysis of the same phenomena in Scotland and England, heavily relying on historical knowledge.
51 Both Harris and Chomsky set up developmental series (historical trajectories) for their synchronic alternants, so that their systematic rules reflect historical processes; in other words, the reason for, say, /send-/ being the UR of this morpheme is historical facts: the independence of the dynamics of the trajectory of synchronic processes vis-à-vis the facts of the trajectories, developmental series, has not been established.
52 Modern English presents a case which highlights the problems: *The Thames, Anthony, an author*: the respective URs are: /temz/ and not */þemz/, because <h> is never realized phonetically; /ænþəni/ or /æntəni/ are equally possible; /ɔ:þə/ because */ɔ:tə/ is not heard. The three cases clearly illustrate the historical nature of the generativist's systematic rules (SR), as these three <h>'s are late, early modern English additions

1.3. The system at work: synchronic processes

a speaker from a *t*-dialect, the generativist is bound to become historical: has the former dialect lost a *t* or has the latter dialect inserted a *t*? The problems are doubled when we move to the verb *to send*.

2) Why is *sends* spelled <sends>? Because the underlying form is /send-/.
3) Has the *-d-* in <sends> never been pronounced?

The generativist is bound to say No, because the systematic rule in English has been latent or potential all through the existence of what we call the English language: it was systemically, underlyingly potential in OE when these 3rd pers. sing. forms of the verb were realizations of another underlying form: [sendeþ], [sendþ], [sentþ], [sentt], [sent]. And when we come to the Middle English period and [sendəs] begins to vie with [sent] and [sendəþ] for systemic supremacy, the systematic rule must be warming up for the final battle with the inherited rule so that the moment when ə is dropped in /sendəz/ in the early part of the 15th century, the new systematic rule so affects the continued underlying representation of the verb that /send-/ is realized phonetically as [senz] under specifiable conditions, *sends*. The absurdity of the generativist's position is epitomized by the above question: was *d* in *sends* never pronounced? The absurdity, however, only comes from the synchronic approach: URs can only be set up on the basis of historical knowledge or facts, and Kenstowicz's simplicity reflections above on the choice of URs will boil down to historical reflections, just as Harris (p. 120) is bound to admit that synchronic (*stable, distributional*) alternants must be understood as sounds of previous stages of the language in question: the synchronic system is created[53].

The empirical relevance of what the synchronic approach puts between slants is precarious: be it traditional phonemes, generativist morphemes (URs) or other systemic elements, or for that matter targets whose bull's eyes the speaker aims to hit when he speaks (Hockett 1958:440-411; Strang 1970:8; Samuels 1972:9). In any case, the criterion of the empirical relevance of signs that have such concepts as their contents must be the existence of denotata, so let's ask the simple question: what is the denotatum of the sign *the Danish language*, the denotatum of *dialect X*, the denotatum of the modern English word *a man* and of modern English /p/?

ModEng *a man* denotes a specific person when used by Primus at T1(P1), a different person at T2(P1), and when Secundus uses the word at T1(P2), *a man* denotes another person. On all three occasions one English denotator, the same word, was used. What does *the same word* denote?

The sign *the same word* must denote the sign *a man*, used three times. The sign was not **used** three times, three spoken utterances were produced [mæn] of T1(P1), [mæ:n] of T2(P1), and [man] of T1(P2): the utterances represented by the bracket-notation manifest or

---

to earlier *h*-less morphemes of the words; how to explain this historical process: {SR of T1 → /autor-/ → [autor]} > {SR of T2 → /auþor-/ → [auþor]}? – For a syntactic parallel see Bessermann (1998:114), who transposes a modern syntactic rule to Chaucer's language. Bessermann commits the error that will be dealt with under figure 1.3.5 below.

53 This is of course a trivial statement, but the point is that the generativist does not draw the consequences of his historical presuppositions.

realize the denotatum of *the same word*, *in casu* and now put between slants, /a man/. Thus we have a manifestatio-metaphysics and a subsumptive relationship, represented by the vertical syntheses in 1.3.2:

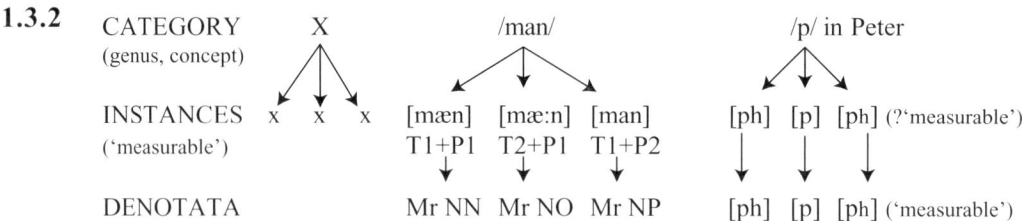

1.3.2   CATEGORY   X   /man/   /p/ in Peter
        (genus, concept)

        INSTANCES   x x x   [mæn] [mæ:n] [man]   [ph] [p] [ph] (?'measurable')
        ('measurable')   T1+P1   T2+P1 T1+P2

        DENOTATA   Mr NN Mr NO Mr NP   [ph] [p] [ph] ('measurable')

The diagram in 1.3.2 indicates that systemic elements contract various relationships with the level of denotata. As physical entities at the level of denotata the allophones will be denoted by the entities of the preceding level, which therefore will be signs in their own right: /[ph]/, /[p]/, /[ph]/, signs, however, denoted by /p/. This illustrates how the everyday language itself makes for dualist conceptions of itself, how ontological dualisms and modern subsumption- or denotation-theories or, to use an ideologically neutral term, the principle of classification are intimately related, and how, in particular, phoneme theory is the hotbed of idealization.

A category is a sign and categorization implies denotation; as a sign it may denote, and the phonemic notation must therefore be changed so that /p/ appears as a sign, */p/*: for /p/ to become a sign it must be understood as a sign function; here /p/ will be the expression and the definition (definiens) that the phonologist assigns to it will be the content:

1.3.3   /p/ r C(ontent = a definiens)

        [ph]   [p]   [ph]

But */p/* is no ordinary expression (word) of the English language, as it has a man-made definiens for its content. Thus its sign-functional structure is not a simple one, ErC, but the following: (/p/ =)ErC(ErC), as the definiens is a complex sign in its own right (a sentence). What has become of */p/*? It no longer denotes an empirical entity – as *a man* does – but is **made** to subsume/denote a number of entities under the scope of its man-made definiens – **one** of its definienses. It is not something 'underlying' or 'behind' its denotata; it is not some entity (an ogre; cf. Bernsen 1978:192) common to all the instances in question, a supraindividual, a social phenomenon that controls *its* (phonetic) realization and variation in a given substance; it is not a transsubjective entity in the above Paul-Goldmann sense. It is a shorthand symbol that is made to denote in accordance with a convention expounded by its definiens. In fact */p/* is no word of the English language and could be replaced by any physical symbol, because it merges completely with its content. It is this definiens-utterance that denotes *parole*-phenomena. A phonological theory is not a network of symbols, let alone phonemes *qua* entities in their own right, but of sentences couched in the words of the everyday language. Conversely, the English word *a father* could be given a man-made definiens – "a male parent" – without being replaceable in the English everyday language by that definition (Hansen 1985:85-89), and such a definiens would in no way exhaust the 'nature' or meaning of the empirical phenomenon, *a man*.

The word-phoneme /p/ is a category which is a pure(ly arbitrary) expression plus a definiens; to this extent Gregersen (1991:1.13;19[54]) is right in saying in accordance with nominalism that a theoretical description is the sole conveyor of meaning, whereas (theoretical) categories that are also words of the everyday language have this sign-functional structure: (ErC)ErC(ErC). The Glossematic technical term *analysis* has two contents, its definiens as a theoretical construct and its historical content – which cannot be shed – from its existence as a word of the object language. The complex sign function reads, with *analysis* as our example, as follows: the word of the everyday language, *analysis*, denotes its definiens, which is made to denote the former so that it denotes by means of the word of the everyday language, *analysis*, a given scientific process. *In the historical narrative, the ideal situation is such that the theoretical content of a technical term merges with the word's own inherent meaning*. This will be elaborated under 1.3.5 below and chapter 4.

The transcendence problem can be seen as an analogy of the everyday language's denotative-existential function: *an empirical theory contracts the same relationship with its data as the everyday language we use in our historical existence does with its denotata; the everyday language and the empirical theory are empirical facts of man's historical existence. As the empirical theory is meant to tell us something about (a section of) reality, so – in this sense – the everyday language also carries existence postulates.* Such parallelism will safeguard the theory against idealizing tendencies; the everyday language cannot be separated from its speaker and reality, concrete existential situation; sign-functionally, both the theory and the everyday language are constituents of their data and denotata, respectively. Finally, we see again the relevance of Hermann Paul's warning against, *in casu*, the assignment of any realistic-existential content to systemic entities so that they transcend their sign-functional existence and the realization- or manifestation-relation above becomes a realistic and temporal process. As far as I can see – with apologies to Ernest Gellner – no mainstream philosophical trend since Plato and Aristotle has – willy-nilly – avoided a dualistic fundamentum, which more or less invariably leads to an idealism whose antidote, nominalism, creates the transcendence problem which, in turn, – at any rate in linguistics – has ended in a synchronic absolutism that is virtually impossible to distinguish from Medieval Realism. Therefore it is a curious fact that the serialism of genetic-oriented theories has met with massive criticism for not being explanatory, let alone scientific, whereas hypostatization-like positions are acceptable: the Cartesian *cogito* or *ego* is not an individual entity, but a generalized one (Beck 1967:204; §2), the ideal speaker/hearer of generative grammar is an idealized entity (Givón 1990:6-7), and the synchronic descriptions of an everyday language's grammatical system that we find in various grammar books are also meant to represent a phenomenon that various groups of people share or have in common; modern structuralism also becomes idealizing since it eliminates the individual (Hjelmslev 1947:125-126); Samuels (1972:29; §6.2) disappears the individual speaker pretty quickly, and so does Milroy (1992:77). Even Hermann Paul must be criticized for not creating a stringently individual historical theory.

---

54 But we are not told where the theory aquires its meaning from.

Paul (1968:5) made the science of language belong to the cultural sciences, which entails that the constant interaction, as we saw above, between *Einzelgeistern* (p. 12) becomes the basic entity of linguistics; so in spite of the fact that each language phenomenon (utterance) is the product (*Werk*) of an individual person (p. 18), language development is contact-based, based on the interplay between at least two persons (cf. Samuels *loc. cit.*) and the mutual influence that two persons exert (*ausüben*) on each other (Paul, p. 12 note 1): language development (a cultural process and product) is therefore both *eine Arbeitsteilung* and *eine Arbeitsvereinigung* (p. 18) within a social totality (p. 13). This view leaves no room for the individual's historical sense, the intrapersonal aspect of language development; Paul ignores the fact that not only are 'to forget' and 'to remember' individual processes[55], but man's historical existence is also the individual's constant creation of the present from the present; that this existence is interpersonal is no reason for disregarding both the dynamic element in the individual utterance and the fact that the individual utterance is more than the creation of static variants whose continued existence is dependent on them being imitated, borrowed, or 'processed' by another speaker (p. 18). Paul's horizon is not consistently temporal or intrapersonal, and Gregersen (1991:I.348; 259) both correctly and not correctly remarks that Hjelmslev does not have a theory of successivity or of the period[56]: it is an elemental corollary of the synchronic point of view that the period as a dynamic-historical reality cannot be incorporated into either a point-in-time or achronic theory, but at the same time this point-in-time theory turns the (concept of the) identical object into that which makes successive phenomena cohere in an orderly manner (Nelson 1970:3), of which Descartes's wax experiment is the best example[57].

In order to avoid an idealizing historiography, the stuff of history must be grounded on the concept of the individual's activity without excluding (Goldmann-Paul) transsubjective processes and facts in man's existence (Paul, p. 33) and on the purportedly prescientific fact that at least one so-called contingent statement about reality is unquestionably true. The cognitive value and function of the everyday language cannot be denied[58]. With the anthropo-

---

55 What is remembered is carried into the present, thereby constituting part of the horizon against which the present is both understood and created. What is forgotten may at best play a similar role through its possible subconcious influence on the present.
56 The generativist's trajectory (cf. Harris 1994:120) is not a theory of language development; the same goes for Lass and Anderson's lenition and strengthening trajectories (1975:ch.5).
57 Thus synchronic linguistics is saddled with problems similar to what confronts the 'savage mind' (Lévi-Strauss 1962:314), namely how to unite the opposition between the period and the point ('a-period'): once the synchronic mind tries to integrate the material variants into its formal system, the successivity-coherences underlying the variants surface, and the synchronic linguist, as Paul said, steps on to the territory of history. The synchronic linguist must therefore conquer the opposition between the aperiodic nature of his formal system and the periodic nature of what the system purports to explain, empirical phenomena.
58 Bernsen (1978:189; see chapter 1.4. below) argues that it is absurd to accept that we have an everyday language and that we do not have knowledge of reality. Thus the generative programme is not only too restrictive, it has also not proved that the everyday language as a whole, let alone the language of *parole* (performance) is not applicable to cognitive situations, and the irony here is that – following Paul (p. 31) – the cognitive inadequacy of the everyday language may lie in the synchronic classification systems of the traditional – synchronic – grammatical categories; *e.g.*, noun – verb, subject – predicate/object, noun phrase – verb phrase. In this respect it is worth remembering that both Bloomfield and Hjelmslev recognized the

1.3. The system at work: synchronic processes

centric fundamentum of linguistics, the Glossematic connotative function, CrE(ErC), is not a corollary of the system, analytically reducible to the simple sign function, ErC, but its fundamentum. Thus a historical structure is made *une chose sans l'homme* by a theoretical sleight of hand which eliminates the connotatum: {Cr^{con}E(ErC)}> (ErC); it will always carry *la trace de l'homme*; and Paul (pp. 33;28) put it like this: *wie verhält sich der Sprachusus zur individuellen Sprechtätigkeit?*, a question Milroy (1992:77-78) also poses against a synchronic horizon of *social interaction*.

The synchronic grammarian, including the generativist, does not see that the dynamic *raison d'être* of *parole*-phenomena, the variants, is that they constitute ongoing historical processes – sound change, sound continuation, lexical diffusion –, thereby contributing to the continued existence of the dialect in question; therefore it is a theoretical fallacy to stop this process by having a number of variants picked up by a synchronic system's state so that these variants – at a given point in time – assume properties different from what they had when they were empirically relevant stages in a historical (*geschichtlich*) process; now they become realizations of a system's phonemes, underlying or lexical representations. However, neither Paul nor modern views envision the possibility that allophones, alternants or variants enter into both intrapersonal and interpersonal sequential and temporal relationships, relationships that testify to man's historical sense. Itkonen (1978:123-124) seems to be denying this possibility, saying that there can be no *conceptual relations (...) between purely spatio-temporal entities* (events). The objection to Itkonen is twofold: his point of departure is biased: he works within the confines of a reductive-dualist setting, where actions merely exemplify a given action concept, events being instances of a generic concept. Secondly, I shall demonstrate that it is possible to establish historical links between empirical entities; Itkonen should have used the word *perceptual* instead of *conceptual*, because the historical links are not percepta, but something to be comprehended.

The objective of the present essay will therefore be to demonstrate that intrapersonal spatio-temporal entities, *parole*-phenomena, enter into significant functions, testifying to what they are, the product of man's sense of history. This means that genetic-historical theories have no immediate metaphysical reckoning outstanding, whereas dualist-based synchronic theories do. On the other hand, the former type of theory has a data-constitution problem that nominalism reduces to a theoretical question (the *Selbstverstehen* of the theory). This metaphysical reckoning as well as data-constitution problem is an aspect of the notion of **empirical relevance**, a concept that is taken for granted in the literature or reduces to questions of verification or falsification. I shall attempt a negative sign-functional determination of the concept:

---

danger of generalizing linguistic categories of one particular language, thereby heeding Paul's memento. In a case-language like Old English, the object function, with three case-forms, must be defined differently from the object-function in its present-day successor. *E.g.*, are the <u>objects</u> to be defined identically in: *oþþæt hie <u>hine</u> ofslægenne hæfdon* :: until they had killed <u>him</u>. – The verb *beniman* may evoke an <u>accusative</u> case-form or a <u>dative</u> case-form as – in the modern sense – the object: *he <u>hine</u> his rices benam // he heora feorh <u>him</u> benam*, in so far as these forms are objects and not the words in the genitive, *rices*, and accusative, *feorh*, respectively. See Schibsbye 1977 vol. 3:30.

1.3.4 Empirical relevance is not linguisticism, **linguisticism** being a mental process that turns the E(xpression), the C(ontent) or the definiens of a sign into an independent, empirically relevant entity (Realism). Linguisticism has entities exist only in so far as 'we talk about them' or only as linguistic entities (Nominalism). Linguisticism is the assignment of (independently existing) properties to entities that are only established through the everyday language (Idealism) so that an entity X *qua* the carrier of properties becomes a denotator or a phenomenon of empirical reality (Essentialism).

A theory based on (the independence of) contents, expressions or definienses infringes the criterion of empirical relevance. Both the everyday language and the theory create sign functions in which they participate themselves, this being the indispensable lesson of Glossematics; but Glossematics did not draw the inescapable consequence of this position: theory and data do not exist in absolute separation of each other, owing to the sign-functional interdependence between content and expression.

The everyday language is always time- and place-specific – again owing to the nature of the sign function. The historical theory as well as the historical narrative is time- and place-specific (Dray 1957:104). They are historical products, each a part of reality in their own way, able to enter into the horizontal synthesis of Kausalnexus (Paul 1968:24) and to change according to the Critico-Philological Method (see 1.4.5): a theory or a narrative may be improved/rejected; the everyday language exists. Hjelmslev's Glossematics follows the same line of thought, having the definition of *diachrony* mirror that of *metachrony* (Hjelmslev 1975:123-125; deff. 210, 211), namely *continuation* between *metasemiotics* and *semiotics that are not metasemiotics*, respectively[59]. However, the Glossematic version leaves a lot to be desired because it has no dynamic or processual notion of historical processes.

The historical theory or narrative that describes the events of a specific period, T1+T2, is bound to change as this period develops into T3, thereby becoming a predecessor; secondly, in so far as the historical theory and the everyday language merge, the theory will also change; thirdly, the historian's transcendence problem contains a translation element.

A historical phenomenon, created – and destroyed – by cooperating people is *significatif* ErC(ErC), because the process involves the everyday language, and since the historian's narrative will be couched in the everyday language of the (his) present moment (E(ErC)rC), so the longer the period to be described, the greater the difference between the two languages. The gap will be bridged by the narrative's representation of the gradual and coherent development (existence) of the events from their relative (chosen) 'start' (=T1) to their near-merger with the present moment. The Figure in 1.3.5 below may help to visualize the parallelism (to be elaborated in chapter 4):

---

59 Gregersen (1991:2.31; see chapter 1.4. below) does not see this unifying aspect in Hjelmslev's theory, where metachrony is continuation between semiotics that are not metasemiotics, while diachrony is – not as Gregersen has it, the study of *parole*- or substance-changes, but – continuation between metasemiotics. A semiotic cannot be restricted to the form-changes that *a language* becomes subject to.

1.3.5

	1789	1789	1789	1850	1875	1900	1950	1999
Level 0/3 = Er	La Bastille stands	being destroyed	La Bastille destroyed	(… other Parisian events …)				
Level 1/2 = C	C(ErC)	C(ErC)	C(ErC)	C(ErC)	C(ErC)	C(ErC)	C(ErC)	C(ErC)
Level 2/1 = Er	E(ErC)	E(ErC)	E(ErC)	E(ErC)	E(ErC)	E(ErC)	E(ErC)	E(ErC)
Level 3/0 = C	Speaker NN T1	Speaker NN T2	Speaker NN T3	Speaker NN T4	Speaker NN T5	Speaker NN T6	Speaker NN T7	Speaker NN T8

Historian

The language of the historian's narrative at T8 will be different from the language of T1, the language that is a contributing factor in this particular historical event of T1. The historian will bridge the gap between the T1 and T8 on condition that he can reconstruct the language of T1, changing into the language of T2, etc. The modern narrative at T8 of T1 needs to be translated into the language of T1, and the narrative will cover the period from T1 to T7, because the historian's is no Archimedean vantage point; he cannot step outside his own everyday language to incorporate the present moment into his narrative.

What I have described here is the first step in the historical theory to be developed and is a generalized sign-functional exposition of this well-known type of problem in historiography: for us to understand, say, the English feudal society in King Ælfred's days, we must be able to reconstruct the precise meaning (or the denotata) of such words as *ealdormonn*, *wita*, *witena gemot*, as well as seemingly less technical terms (underlined) as King Ælfred's *eall sio gioguþ þe nu is on Angelcynne friora monna*; the same goes for William Jones's concepts of *affinity* and *a common origin*[60]. Secondly, it is also a comment on the uniformitarian principle, and an answer to the questions why the interpretation of the past through the glasses of the present can only provide us with a biased narrative of the past[61], and how to remedy this: through knowledge of how the everyday language develops.

Consequences of this theory are: can the history of the English language be described in

---

60 This means that my interpretation above of the Jones-quote is no attempt at retrieving what Jones *really* meant by these crucial concepts.
61 Uniformitarianism (Labov) is a reversal of the savage mind's 'stubborn faithfulness' to its past as a 'timeless model', which justifies the present in virtue of one single maxim: *les ancêtres nous l'ont appris.* (Lévi-Strauss, p. 313). While the savage mind supposes the past as an absolutum, Uniformitarianism takes the present as its absolutum: the present teaches us what was possible in the past. To exist in the present is a quality that so distinguishes its holder that the present can be severed from its (immediate) past, but at the same time it coheres with and shapes the past in its own image.

the Danish language? Can France's history be described by a nonspeaker of the French language? And: can a synchronic description of modern French be written in English? Finally, is there such a discipline as universal linguistics, achronic or panchronic lingustics, which must necessarily be implemented by a time- and place-specific everyday language?

I.4. **On genetic explanation. 'What must a language theory be able to do?' Empirical relevance is denotation. The Critico-Philological Method. An analysis of 'All A>B'. An aitional interpretation of 'to come from' and 'to develop into'. The Phenomenal Error. The meaning of *ex*istence**

Explaining in virtue of origin is regarded as the antithesis of analytical explanation in terms of subsumptive parameters (Riedel 1986:4). Dualist explanation is all that genetic explanation is not, and *vice versa*. In modern linguistics, synchronic explanation *is* explanation, genetic explanation non-explanation. Science as the search for a constancy *qua* a form with ontological pretensions leads invariably to the dichotomization of the object of the theory, and in linguistics the epitome of this dichotomization is the *langue-parole* dualism, which creates a transcendence problem and as a corollary hereof, what Gregersen (1991:2.233) calls the 'empirical problem'.

To specify the paradox: genetic explanation does yield knowledge the empirical relevance of which nobody denies (Gellner 1969:20). A dualist-based explanation yields knowledge, too; but this knowledge is of a special kind because the empirical relevance of one part of the dualism is not immediately relevant; to repeat: what empirical phenomenon does the *langue* – with its definiens – of synchronic linguistics correspond to (Hjelmslev 1972a:25-26; 1975:N 17; def. 38)? What is the empirical correlate of the *eidos* of Plato or Aristotle – or what is left after accidental and specific differences have been subtracted from a given phenomenon? What is left after the specific/differentiating differences in Kenstowicz's eight variants in chapter 1.3.1 above have been removed from the eight empirical phenomena? What does a Glossematic 'cleansing-process' of *parole* (Gregersen, *loc.cit.*) lead to – apart from the word *langue*? The sign referring to the residue, be it an *eidos*, a formal trait, an *essentia*, is not a priori a denotator, and the Socratean-Aristotelian definiens that is intended to denote this essence, is not a priori empirically relevant. And it is yet to be demonstrated what empirical phenomenon the generativist's *language module* denotes in man's brain and the Glossematic term *language* (see diagram 1.4.4), a member of the class of sign-systems, denotes when it comes to the analysis of, say, modern English[62], and how a dematerialized *langue* concept yields a typology of *langue*-structures, whose immanent-internal movements will explain history; and if the changes envisaged by the Glossematic programme must be cataclysmic (Gregersen, p. 235; cf. chapter 3.4. below), it is odd that Gregersen does not conclude that the programme is empirically moribund: where do we find just one such change? One must ask the simple question – of both Gregersen and Rasmussen (1992) as

---

62 Gregersen's examples (pp. 233-234) very significantly presuppose what they should prove. I think that Gregersen will not disagree as to the relevance of this question: what point in time does Togeby's description of the immanent structure *de la langue française* refer to?

1.4. On genetic explanation

well as of Glossematic theory itself: if the ultimate objective of synchronic theorizing is to explain history, why is it necesary to make this enormous theoretical detour, which furthermore disregards language change completely – why not set out from the facts of change and create a theory of history which may also find room for the static aspects of the everyday language[63]? The irony of this is that this essay, preserving the 'ontological unity' of the everyday language implements Gregersen's programme (see preceding footnote) through the creation of a historical theory, which does not accept a dichotomy within the historical sciences of the traditional type: historical :: ahistorical (see Gregersen's little boxes on p. 259): and to answer Gregersen's question (p. 252): what must a theory of language be able to do? A theory of language must at least be able to explain the simple isolative change, *e.g.* OE *a:* > *ɔ:*, on the basis of a theory of history, *in casu*, sound change.

In the modern conflict between historians and synchronic linguists, there seems to be this confusion in the basic arguments: what is scientific analysis and what is the nature of the everyday language? If the latter is a historical phenomenon, it cannot be subjected to conventional generalizing subsumptive analysis; however, this crucial question has been answered axiomatically (Hjelmslev 1928; 1972a[64]; Sørensen 1958:26[65]): the everyday language *is* basically synchronic, a conclusion reinforced by the chosen methodology, according to which the empirical world is populated by objects, not by processes; the historical process has been reified. Hjelmslev's view of reality only includes *objects*[66], which remains one of the metaphysical terms of the theory. The nature of scientific analysis has therefore been determined with no regard for the historical domain (Dray 1957:103-104). The closest we get to an argument is the postulate that for a change to occur, it must change in relation to something stable (cf. Gregersen 1991 vol. 2:31), while neither Paul's abovementioned view of the necessity of the historical point of view nor Coseriu's argument to the effect that the synchrony-history antinomy (1958 = 1974:9[67];93;95; repeated by Winters 1992:504) be-

---

63 Gregersen (p. 238) also disagrees with Hjelmslev's extreme reductionism, but his sociolinguistic paradigm (pp. 252-254) does not reinstate history as a science in its own right, seeing that this paradigm (*sprogvidenskab*) is to unite the classical antinomy between 'history' and 'structure' (p. 252). Gregersen does not go beyond the standard position according to which 'history has become sociology, language history language theory …' (p. 250), thereby overlooking the obvious fact that 'history should again be(come) history' and 'language theory a theory of history' (Paul!), for the simple reason that the everyday language is a historical phenomenon. A theory of variation, a theory of *parole* (*usus*) is not the exclusive domain of a social approach. Gregersen does not see that Labov's sociolinguistics does not cope with the inception problem, as the variant's social content is something that comes to it in the course of its existence (Labov 1994). – In his, admittedly, very summary illustration of his sociolinguistics, Gregersen (pp. 260-262) does not mention this inception problem.
64 In neither of these works does Hjelmslev's argumentation rise above the level of postulates.
65 Sørensen is here so candid as to state this apriorism explicitly: *What (synchronic) properties has a sign? To us a sign has only those properties which we have attributed to it (…)*. And on p. 47 we are told that *lexicography* as *a systematic science* is *synchronic lexicography*.
66 In the eyes of the Glossematician, reality consists of two subclasses of objects, the class of semiotic (object)s and the class of non-semiotic (object)s (1975:9). And significantly enough, the processual term par excellence, *analysis,* has been stripped of any dynamic content, it being a specific type of *description*, another indefinable.
67 The theory to be developed in chapter 6 does not accept Coseriu's sharp distinction between synchrony and history.

longs to the level of logic (*Forschung*), not phenomena (*erforschter Realität*), and that the synchronic attempt at explaining or justifying *der reale Wandel* by means of *der abstrakten Sprache* turns the issue upside down: abstract language should be justified (*begründet*) by the phenomena of historical development; change is not in need of being justified by an abstractum (cf. Paul's Realism).

Hjelmslev (1963:8) emphasizes the received position as follows: if linguistics (= general (if not universal) linguistics (Gregersen *loc.cit.*)) is to be a science, then its object must be subjected analytically to a specific set of scientific principles. The *Eigennatur* of the object plays no role; the everyday language *qua* fluctuation is reduced to being a model for the theory *qua* a constancy. Diderichsen (1966:21), accepting the historical nature of the everyday language albeit within a synchronic two-state conception, requires of a theory that it creates coherent knowledge of its object; with this we are back to our fundamental issue: while Itkonen denied the reality of spatio-temporal relations among data, Paul argued that because of the interconnectedness of cultural phenomena (cf. *die Wechselwirkung der Individuen* (1968:5-8)) the principles that we apply to data must come from (*sich ergeben*) an investigation of, *in casu*, the nature of historical development. And if we add Coseriu's view (1974:95) that a (mechanical) cause or causes are not what creates the empirical coherence of language development, the general state of affairs may be summed up as follows: the fact of change is denied by nobody; empirical change does imply empirical coherences[68]; the problem (*aporia*) of change is a theory-created issue; this problem is not to be solved by what creates it, a static, synchronic theory, but by a historical theory: if this is possible, the problem vanishes.

Apart from the question of the nature of the everyday language, the historical point of view has been criticized for operating with value-loaded, in principle and fact unsubstantiable concepts such as progress or regress, progressive deterioration or amelioration, as well as survival of the fittest (communicatively most efficient and functional), a teleological drive towards an optimum, balance and symmetry – all of which concepts remain, in the Glossematic sense of the word, metaphysical or entail logical flaws in the theory at hand (Gellner 1963:3-20; Hawkins 1979:641-642); and if we again add something to this traditional position, namely the conventional *aporia*-views mentioned above, it is obvious that genetic explanation has been carrying a load of heavy-weight problems that no indefatigable defender of history can hope to cope with, and the historian appears to have this alternative: continue to work within and with the concepts that synchrony imposes on his or her field – as Hopper and Traugott (see below) do, despite themselves; or: take the synchronic and other impositions for what they are and regard them – with Coseriu – as irrational solutions to a rational problem, and therefore (try to) reject the synchronic views by demonstrating their inadequacy, empirical irrelevance and construct a historical theory. If such an enterprise is feasible, then linguistics has a rational basis on which to compare the results of synchronic theorizing and a historical theory. To underline: historical explanation must be stripped of any overtones due to empirically irrelevant concepts; this obliges me to try to determine the often-used concept of empirical relevance against the horizon of the domain of the empirical phenomenon of language change:

---

68 It is the theory *qua* a theory of history or our sense of history that creates coherence; coherence is to be comprehended.

1.4. On genetic explanation

1.4.1 An empirically relevant concept is a sign that denotes and is denotable; or which is made to denote by another sign at least one empirical phenomenon or which contributes to the denotation of an empirical phenomenon. Such a sign is always time- and place-specific, is participative, and a relate in a connotation. – Cf. chapter 1.3.4.

This definition entails that an empirically relevant sign is not a potential denotator[69]. A statement about the history of the sounds of English or about the French Revolution is empirical, whereas fictional statements such as the Babel-Tower explanation of language change or Charles Dickens's *A Tale of Two Cities* are not. A historical statement is minimally relevant, if at least one of its signs denotes an empirical phenomenon; the dualism will fulfil this requirement if at least either of its relates denotes. Consequently, falsification through modus tollens is too sweeping: the statement *it is false that all swans are white* may only negate (a subset of) the quantifier of the A-statement *All swans are white*, some may be white: of all OE *a:* some *a:* do develop into ME *ɔ:*, and it is the type of connection between the successor and the predecessor of this unfalsifiable fact that is to be explained in an empirically relevant manner by the historical theory[70].

The more empirically relevant terms a theory contains, the better it is; progress will therefore belong to the man-made level of logic. An historical explanation is evaluated in virtue of (its degree of) empirical relevance; it is rejected (falsified) completely if all its concepts are empirically irrelevant. This latter possibility is hard to envisage, and its lack of relevance may be regarded as a reason why tradition, continuation at the logical level is possible. Hjelmslev's view of himself/Glossematics as a (revolutionary) linguist/theory with no predecessor – not even Saussure[71] – can only be excused because Hjelmslev had no historical theory by which to substantiate his claim (see 1970: *Prefatory Remarks*); but even cataclysmic change occurs against or with a predecessor.

These remarks will now be combined with the exposition in 1.3. (cf. 1.3.5). The ideal historical narrative will consist of this type of signs: an empirically relevant sign is denotable within a historical theory and enters into a horizontal synthesis to form a narrative *qua* a narrative that denotes (cf. 1.4.15). The theory (of language (change)) makes the modern language of the narrative in question cohere with the language of the events that it describes. The truth of Narrative T8, describing the events of T1, hinges on the translation of the language of the narrative into the language of T1.

The historical theory behind a narrative containing the sign *Napoleon won the battle of Waterloo* will demonstrate that this sign will be a potential denotator in the (French) lan-

---

69 Signs such as *a centaur, a horse with fifteen noses*, or *love has a beautiful colour* are potential denotators, but they are not empirically relevant except in their **ex**istential function.
70 The binary causal concept that Gregersen (1991:2.268) adopts is not placed within a general theory of historical development; that we may say *post factum* that 'Y as our present successor would not have been the case if X had not been the previous present succesor (its predecessor), thereby implying a determinism that is modified by the principle's second element: X as a present successor does not imply the necessary creation of Y as its successor', reduces historical development to a mere two-state affair. The binary nature of this causal concept, which presupposes the continuousness of historical development (see 1.4.7 below), will be elaborated below in chapter 1.5.
71 This is virtually the picture that is evoked in Gregersen (1991:2.1-60).

guage of 1815 (18 June)[72], because it is not denotable by the theory. Another way of putting this is that our 'false' sentence does not, cannot enter into a larger narrative synthesis[73].

On the other hand, denotation is primitive knowledge (cf. Sørensen 1958:15), which the language-acquiring child knows all about. Successful denotation *qua* communication is hard work, learnt experientially: all knowledge begins with experience, and human experience is co-extensive and merges inseparably with denotation[74]: *experience and denotation are man-innate faculties which presuppose the fact that reality has the 'innate' capability of being (made) amenable to experience and denotation, to being experienced and denoted in one fell swoop*. It is impossible for us to state simultaneously that we perceive (hear a sound, see a tree, etc.) or that we know something and that we do not have an everyday language. Bernsen (1978:189) deals with a similar cognitive problem: it is impossible for us to say both that we *possess language* and that we have no *knowledge of reality at all*. From the point of view of the everyday language we say: we cannot deny that we possess an everyday language and at the same time maintain that we have knowledge of reality: no knowledge without language. This means that in spite of the fact that we have no experience without the everyday language, all our knowledge does not come from experience. *Experience arises through use of the everyday language*[75]. Through this denoting function, the everyday language creates a distance to data, while at the same time it bridges this gap. The question is: why is it not experience that withdraws, leaving the scene to the everyday language, but the everyday language that withdraws, leaving the experienced as a mental phenomenon, since the converse relationship is equilogical: *Use of the everyday language arises through experience*?

Hjelmslev tried to answer the question by saying that it is the nature of the everyday language to be ignored; we tend to move away from the means of cognition to the cognized; but since the everyday language is not only the means of cognition, this answer is not the whole truth: again we must return to Hermann Paul's Realist warning: the everyday language does not cognize for us; if the experienced or cognized retreated, reality would be language-

---

72 That more narratives are needed to describe this historic event is more than a theoretical possibility in view of the British and Prussian participants. Paul's *Arbeitsteilung* and *Arbeitsvereinigung* now assume a more general meaning than Paul envisaged.

73 Gregersen (*loc.cit.*) turns this into an aesthetic, if not hedonistic factor: *jo længere årsagskæder jo større nydelse* (the longer causal chains the greater the pleasure).

74 For an exposition of this type of merger (syncretism) see Løgstrup (1984:19-20); while Løgstrup seems to work with sense-perception independent of the everyday language in that he makes our awareness of both our bodily-ness and our understanding (of or in the sense-perception) separate the mental process from the concrete sense-perceptum, my point of departure is that sense-perception cannot be separated from the everyday language nor can the latter be separated from understanding. This merger is conditioned by the existential function of the everyday language, and in historical analysis the everyday language 'dethrones' sense-perception. – Cf. Hauge 1985:665; Hansen 1993:159-160.

75 This directionality is not cogent. A consistent sign-functional understanding of denotation implies that the expression and the content of a sign are defined oppositively. Man's creative faculty lies in the fact that something can be *made* a content or expression; this may be another way of putting Løgstrup's: *Tegn bliver ordet først, når det bruges i udsagnet* (1976:53; 'a word becomes a sign, when it is used in *parole*'); although I do not subscribe to Løgstrup's *tegn-ord* (sign-word) dualism, I accept the structural dynamics that Løgstrup's view implies. I shall say that a sign function arises in *parole*, because the sign function is not the exclusive property of words, and the sign completes its meaning in *parole*.

1.4. On genetic explanation

created and end in fiction: it is not a question of either-or: Realism or nominalism, language-use (denotation) or experience/cognition, but a both-and that is embodied in the existential function of the everyday language.

The Glossematic sign function epitomizes this view of denotation in the relationship between denoting man and denotable reality: just as *the semiotic schema(ta) and usage(s) which we designate as the linguistic physiognomy N.N. (...) are expression for the real physiognomy N.N (that person)* (Hjelmslev 1963:118-119[76]), so N.N.'s everyday language becomes the expression for N.N.'s specific reality *qua* denotata. Thus reality is constantly being created and continued sign-functionally – and participatively: neither the individual person nor his everyday language can be separated from a given section of reality; and I shall generalize Sartre's position above (1.3.): reality will always carry *la trace de l'homme* and the everyday language[77]. To this extent man is a necessary part of the reality that he creates and continues himself, and this may be a way of escaping the either-or'ness of the nominalist fallacy (Løgstrup 1976:52). I shall represent the participative relationship as follows: 'a historical phenomenon(formed substance **r** maker) **r** maker' = 'a historical phenomenon is (trans)individual and *significatif*; it is a substance formed by the everyday language of its maker(s) and is as such never *sans la trace de l'homme*'.

How does this view relate to the standard one concerning the transcendence problem? The simple view of the problem is that of generalizing classification: language (a theory) subsumes a number of entities *qua* denotata under its scope, with the corollary that scientific analysis is deductive and truth becomes a theory-internal affair once the first inductive step has been made[78]. I shall illustrate my point with reference to the Classical A-Syllogism:

1.4.2a All human beings are mortal
Socrates is a human being
Socrates is mortal

1.4.2b Iron is composed of H_2O
Water is iron
Water is composed of H_2O

A theorem *qua* the conclusion of an argument such as *Socrates is mortal* acquires its denotative power or empirical relevance through the truth of its premises, as a false statement cannot be deduced from true premises. The denotative power of the premises is presupposed. However, this premised empirical relevance is only hypothetical. Secondly, a true conclusion can be deduced from false premises (1.4.2b) – this being one of C.I. Lewis's startling theorems (Lewis 1912; see Hansen 1983:459): neither does the denotative power of a conclusion make its premises true (denote) nor can premises or abductive guess-work endow a conclusion or Law, respectively, with denotative power.

---

[76] This characteristic aspect of Glossematics is curiously absent from Gregersen's extensive treatment of Hjelmslev's dematerialization of the everyday language, this being a recurrent theme in Gregersen 1991; on the other hand it seems to tie up with Gregersen's narrow understanding of *metachrony* (see note 14 in 1.3.).

[77] I am not saying that the existence of, say, the Alps or the Mississippi is impossible without language-wielding man.

[78] This will be elaborated below, but reference could here be made to Gregersen's correct observation concerning Hjelmslev's *naivrealistisk*[e] [videnskabs]*praksis* (Gregersen 1991:2.11-12).

A theory constructed as a logico-deductive argument like Glossematics or for that matter on the basis of universal quantification are *made to* denote, and what I shall call the Metahistorical Precondition (or generalization) in 1.4.3 will take care of this:

1.4.3 To date there is nothing to suggest that there has been, is, or will be at least one person/one everyday language (one entity p) that is not mortal/static (q). Or: to date there is nothing to suggest that this is not what happened in P1 at T1.

Hermann Paul's view of the priority of a historical methodology has been so generalized as to include the level of theory[79].

The Metahistorical Precondition is the sentence under which the statements of a generalizing argument exist structurally. It equips both laws or generalizations and particular statements with their respective denotative power, empirical relevance and truth value. This conventional type of scientific procedure has a before- and an after-reckoning. Solving the transcendence problem by *positing* or *posing* the theory or its theorems among possible/contingent data is no solution. All we can say is that a given theory apparently 'functions, is viable' in respect of a finite set of data. This projection-transcendence does not say anything about the generalization value of the theory. What necessarily came before the deductive argument, the Metahistorical Precondition and the necessity of contingency has not been considered by the theory/theoretician[80].

The Metahistorical Precondition is historical because of its overtly serial and temporal content; it permits quantification over historical processes such as sound change and historical events at the level of data so that we need not turn the scientific data into classes (Glossematic functives), *i.e.* generalized and generalizing entities.

Apparently, 'history' permeates man's existence. Paul demonstrated how the 'synchronic' linguist must turn historian once he begins to ask serious questions about his object; above it was demonstrated how scientific analysis hinges on an historical premise. The all-pervasiveness of history in man's existence may seem to prevent us from establishing a vantage point from which to examine history and to determine what has come to be a pejorative term, *the historical approach to ...* or *historical methodology*. However, we may again turn to Hjelmslev for a clue; in his *Résumé* Hjelmslev establishes a mirror-connection between development at the object level and the metalevel, a fact that seems to have been overlooked by Hjelmslev analysts (Rasmussen 1992; Gregersen 1991). In order to make my point clear a diagram may not be inappropriate; but first Hjelmslev's definitions (1975:124-125): *A diachrony is a continuation between metasemiotics. A metachrony is a continuation between semiotics that are not metasemiotics; continuation is semiotic change.* The Glossematic an-

---

79 Hjelmslev's theoretical starting point presupposes the Metahistorical Precondition. And with reference to Gregersen's criticism of Hjelmslev I shall specify Hjelmslev's theoretical naivety: his arbitrary selection of language-properties that the theory fixes by definition; the way the theory gets started is inconsistent with the cognitively exclusive function of the theory. The latter point refers to Gregersen's apperception concept, which I shall rephrase as follows: for Hjelmslev to get started, he must rely on the unquestionable truth of a single contingent statement, say, *This language is static*, *This language has this property* in order that he can generalize such contingent, inductive-based statements into *All languages are static*, *all languages have this property*. The cognitive function of the everyday language has been accepted.
80 It is not inappropriate to remind my reader of this essay's Sixth Guideline, the opening lines of Kant's *Kritik*.

1.4. On genetic explanation 133

alysis of objects looks like this (1975:9-10); the domain of diachrony has been underlined, that of metachrony bold-faced:

1.4.4

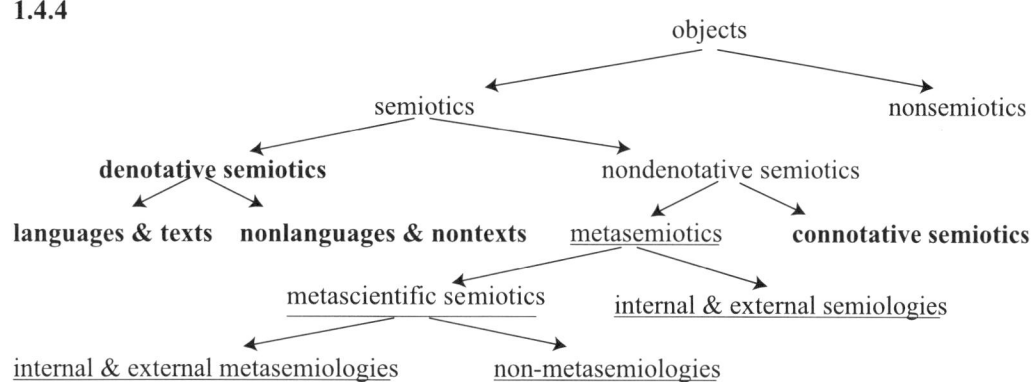

All change that hits a semiotic domain is defined in exactly the same way. Similarly, I shall say that being a historical product a theory will change according to the **Critico-Philological Method** (1.4.5), the dynamics of which mirror my general statement of historical change (1.4.6):

1.4.5 The Critico-Philological Method creates a constructive synthesis on the basis of critical analysis (of the inherited) and innovation.

1.4.6 For something to happen, *i.e.* to become the case *qua* a successor in the present, something else must have so happened prior to it that this exists simultaneously with the successor.

The presentic phenomenon in the historical process is a structural and man-made synthesis that consists of a new element (the successor) *and* what has been inherited from the past (the predecessor) into the present. The 'cause' creates something new *and* continues something against the horizon of change. This causal principle is **biuniformal**, the cause(s) **participative**.

For a new synthesis *qua* a theory to come into being, anterior processes (critical analysis) must have been performed by the theoretician, and the new theory will consist of something new and what can be 'utilized' of the old.

The historical process in 1.4.6 is an elaboration of the commonplace notion that for a language to change, it must be used (Samuels 1972:2; Saussure does not disagree, cf. Gregersen 1991:2.56). – This is partially true of a theory. However, the converse is also a theoretical possibility that will be examined: for a language to be used, it must change. This may sound paradoxical but it is seen as an obvious corollary of another linguistic commonplace notion, that of constant change. The product of the synthesizing process that is involved in both 1.4.5. and 1.4.6 is a *sign-functional structural entity*, a historical phenomenon, whose 'meaning' will appear from the structural interaction of its parts.

The critical and constructive procedures include normal historical processes; a concept is rejected (not continued), another one is added, the contents of other concepts are extended or narrowed. Falsification of the inherited does not imply the construction of something new, and it does not entail total absence of the inherited in the present. Critical analysis 'analyzes', separating out that of the inherited material which can be applied in the continued process of theory-building, involving the combination of new concepts with the inherited ones.

The Kretzschmar-Luick interpretation of the *First Germanic Consonant Shift* involves a reordering of, say, Streitberg's view of the same processes plus the addition of a new concept, the push-chain mechanism, which, in turn, entailed the dropping of the concept of atomism from historical analysis. Otto Jespersen's rejection of Luick's explanation of the *Great Vowel Shift* involves the continuation of the notion of historical coherence and the substitution of a drag-chain mechanism for a push-chain one: Jespersen's solution to (the start of) the 'Great Vowel Shift' is a chronological reordering of the same material within a new theory. Finally, if the lowering of Middle English short vowels is placed after vowel-lengthening, falsification of the ensuing narrative of this particular stretch of language history involves neither of the two sound laws as such, only their relative chronology. This means that change at the level of theory is both conservative (Quine 1972:2-3) and constructive: theoretical change and continuation are anchored in our sense of reality, and this sense has the sense of history as an indispensable part.

Just as we do not make a break between what becomes the past and the new present we create when we exist, the Critico-Philological Method has it that the construction of a new theory does not entail the complete falsification of the old: continuation between the old and the new, the past and the present has been established, and this continuation is aitional at all levels:

a new theory, but not only if complete falsification of the old
a change (A > B), but not only if complete loss of the old (A)
existence qua development into the present, but not only if complete loss of the previous present

As can be seen from the aitional relationships the point of view is that of the present, and dynamically we say – or rather know that the present develops into the present, this being an innate form of our historical sense; another innate part of it will be our knowledge of the present as coming from something, in traditional terms, a/its past. Lastly, the aitional formulation highlights the fact that history occurs in a continuous space, with discontinuities stemming from man, and the formulae's implicit continuation concept tallies with Gadamer's distinction between the *geschictliche Wirklichkeit* of the historical phenomenon and the theoretical process that turns it into a *historische Zufälligkeit*.

This leaves us with two versions of the traditional sound law A>B, with A developing into B and B coming from A. The objective of the following analysis is to demonstrate the ambiguity[81] of and the precise meaning of the (A>B) sound-law concept seen against the horizon of the abovementioned binary type of cause-effect parameter, a causal pattern that emphasizes the synthetic nature of historical understanding. Couched in the diction of the everyday

---

81 It will be dealt with below in terms of the formula's quantification.

1.4. On genetic explanation

language, A>B says that 'OE *a:* develops into ME *ɔ:*' with its corollary that 'ME *ɔ:* comes from OE *a:*'. The determinism of the binary causal principle is interpreted as a logical dependence, which, however, is incompatible with our knowledge of historical development.

	Aitional formula	*implies*	Conditional formula[82]
1.4.7i	ME /ɔ:/ ('comes from'), but not only if OE /a:/	⊃	OE /a:/ ⊃ ME /ɔ:/
1.4.7ii	OE /a:/ ('develops into'), but not only if ME /ɔ:/	⊃	ME /ɔ:/ ⊃ OE /a:/

1.4.7i says that as the present successor of the present successor OE *a:*, ME *ɔ:* may have come from this particular predecessor. 1.4.7ii says that as the present successor (= predecessor) of the present successor ME *ɔ:*, OE *a:* may develop into this particular successor. Again the alternative developments are virtually legion, and the determinism between *a:* and *ɔ:* appears from a logical interpretation of the aition.

Otherwise no determinism is involved immediately, seeing that the ME sound could have come from any other sound, and the OE sound could have developed into any other sound. The question is: how to find the necessity in the seeming freedom of historical data? This necessity will come from the 'historical reality' into which the two-state conception of historical development in 1.4.7 must be placed, and a theory of history will be constructed that reinstates the process of the Neogrammarian formula A>B into its *geschichtliche Wirklichkeit*. This implies that despite its synthetic nature the two-state conception of historical development only expounds a *processual Zufälligkeit*[83].

The development in 1.4.7 will now be put into the structural framework that was developed in the Prologue above in order to illustrate the **biuniformal** and **participative** causal principle that my theory entails:

1.4.8i   OE a: > ME ɔ:
           ↘ ↓ ↑
           (OE) a:

1.4.8ii   The present successor *a:* of T1   >   the present successor *ɔ:* of T2
          A necessary and causal factor   A necessary and causal factor
          in the rise of *ɔ:*; in this process   in the continuation of *a:* under
          a: becomes a predecessor   the aegis of sign-functional change
                                                 (T1 *a:* ≠ T2 *a:*).

---

82 To be read as: If OE /a:/, then it develops into ME /ɔ:/. And: If ME /ɔ:/, then it comes from OE /a:/, respectively. The determinism is too strict: 'not OE *a:* without ME *ɔ:*' and 'not ME *ɔ:* without OE *a:*'. The long *a*-sound does not necessarily develop into the open ME *o*-sound (in the Southumbrian dialects); it also merges with OE *æ* and *a* (see Luick 1964 vol 1:362). Under certain conditions it is correct that all ME *ɔ:* come from the OE sound. So here the determinsim must be modified by precise historical place- and time-circumstances, seeing that some short *o*-sounds were lengthened into a long open *o*-sound. Cf. also: Not all ME /i:/ come from OE /i:/, in that OE /i:/ and /y:/ merged in ME. So the 'meaning' of the A>B-fomula is highly polysemous, dependent on the quantification of the relates.

83 The juxtaposition of two synchronic states or entities – *historische Zufälligkeiten* – does not make a *geschichtliche Wirklichkeit*.

It is impossible to read a unidirectional dependency-relation into the process; and of our two processual concepts, 'to develop into' has an empirical immediacy, nay, existential immediacy different from 'to come from', which is purely mental, but no less empirically relevant.

The freedom of the historical process was understood in the light of isolative change: OE *a:* could have developed into a fronted sound ME **æ:*; but the aitional relationship also captures conditioned change *qua* exceptions. Not all OE *a:* developed into ME *ɔ:*, some were shortened; and the problem is aggravated by dialectal developments: is the Northumbrian continuation of the long *a* an exception to the Southumbrian change or the latter an exception to the former process? Since the latter question is a corollary of the empirical scope of the synchronic system in question, I shall not deal with this, but approach the problem from the traditional angle of the *langue-parole* parameter and quantification, and thereby demonstrate the abovementioned aspect of ambiguity

1.4.9 OE /a:/   [a:]   >   ME /ɔ:/   OE /i:/   >   ME /i:/
                 [ɔ:]   OE /y:/   >   ME /i:/
                 [a]   >   ME /a/
       OE /a/   >   ME /a/
       OE /æ/   ME /a/

Summary comments:
1) All A > All B   is not correct, since the quantification of OE /a:/ > ME /ɔ:/ is
2) **Some A > All B**.- OE *stan, an* > ME *stone, one*; *an* (the indefinite article).
3) All A > All B   is not correct, since the quantification of OE /y:/ > /i:/ is
4) **All A > Some B**.- OE *lys, mys* > ME *lice, mice*; OE *cild, min* > ME *child, mine*.
5) **Some A > Some B**   is correct, because the quantification applies to the OE *a*-sounds.
6) All A > All B   seems not to be empirically relevant at all.

To repeat: the meaning of the Neogrammarian sound law is contingent on the precise statement of the processes the single law is applied to. Furthermore the applied methodology may complicate this situation. I shall illustrate how a strictly serial point of view straightens out the problems: let the shortening of OE *a:* occur before the raising of *a:*, then nothing prevents us from implying an 'all' before *a:* in OE *a:* > ME *ɔ:*, just as the rise of this *ɔ:* occurs before the lengthening of some short *o* later into the Middle English period. On the other hand, dialectal spacing salvages the partial quantification in 2) seeing that *a:* used to occur south and north of the Humber, but later on was split into two: some OE *a:* > all ME *ɔ:* and some OE *a:* > all ME *a:*. Thus there are no restrictions on the possibilities of quantification, and the discussion about the exceptionlessness of the sound laws will be replaced by historical description proper. The exception-rule parameter so essential in synchronic grammar books, is theoretically specious.

My point is that an exception is no exception at all. The concept is created by the applied theory and not least the commonplace notion that regularity is significant, irregularity nonsignificant as well as insignificant[84]. Thus the Chomsky-Halle version of this Neogram-

---

84 In an influential book on English phonology (Chomsky and Halle 1968:ix) we read the following: *(...) citation of exceptions is in itself of very little interest*. This position implies that the scientifically absolute

1.4. On genetic explanation 137

marian parameter must be characterized as unacceptable, it being a complete contravention of Karl Verner's scientific position: no exception without an explanation. The purported exception is neither exceptional, paradoxical nor mysterious in virtue of itself, only in respect of something else: the rule 'Some A become C' is exceptional in respect of the (false) rule: 'All A become B', expressing an expected regularity. The regularity-irregularity-, significance-nonsignificance- and the rule-exception parameters vanish if they are placed within a continuum of 'many-few' or 'more-less', with each type of datum entitled to theoretical explanation (as in Samuels 1972:18-27), which leaves us with two equilogical and equiempirical statements (rules)[85].

The preceding criticism of the prevalent rule-exception paradigm of synchronic linguistics will be summed up in the principle I shall call the **Phenomenal Error.** The principle establishes a link with the principle in 1.1.2.

1.4.10 To commit the Phenomenal Error is to explain or decribe either of two equilogical and equiempirical phenomena on the basis of or in terms of the other phenomenon.

It follows from 1.4.10 that (mechanical) causal descriptions *qua* theories commit the Phenomenal Error, confusing constituent factors of the explanandum with a cause *qua* a theory: *I*-Umlaut is no theory of why not all PreOE *u* become OE *u*, and no theory of why **kuning-* is *kyning* in Old English. In so far as push- or drag-chain mechanisms are empirically relevant they are no theories, either, because they are precise descriptions of the purported movement of entities which change articulatorily. What then is a historical theory? *A historical theory is created in the image of the existential and cognitive functions of the everyday language.*

As the everyday language is not falsifiable, a theory/narrative is not falsifiable; theoretical statements are, as the statement *Napoleon won the battle of Waterloo*, either true or false. Theoretical statements are Janus-faced. As a historical narrative they constitute a whole in relation to their empirical relevance: data unifies and constitutes another whole in relation to a theory: theory unifies. A theoretical statement – like an utterance – completes its meaning or itself contextually, *i.e.* in *parole*. Statements do not change or develop, they being either true/historically accurate or false/historically inaccurate; an utterance does not change, it vanishes at the moment of it being uttered. The everyday language changes and develops in accordance with 1.4.6; and so does a theory in accordance with 1.4.5.

---

concepts of 'significance' and 'generality' may be compared (pp. viii-ix). Chomsky and Halle do not see that unless an attempt is made to regard the so-called exceptions as scientifically significant data, the empirical relevance of, *in casu*, a generalizing rule is a priori doubtful. The point is that the exception does not materialize until the theory has been constructed, and a theory that equips empirical data with the property 'exceptional, irregular' is a contradiction in terms. Bruce Mittchell's words in 1992 are universally valid: *They are there* – the seemingly mysterious and chaos-creating philological details.

85 As early as 1875 (p. xxviii) Leskien saw the theoretical intricacies of the parameter, reducing it to the dynamic interaction of two rules (*wo ... eine Regel die andre durchkreuzt*), and he added these significant words: *Lässt man aber beliebige zufällige, unter einander in keinen Zusammenhang zu bringende Abweichungen zu, so erklärt man im Grunde damit, dass das Object der Untersuchung, die Sprache, der wissenschaftlichen Erkenntnis nicht zugänglich ist*, a scientific position that antedates Hjelmslev's in *Prolegomena* by some 70 years.

1.4.11

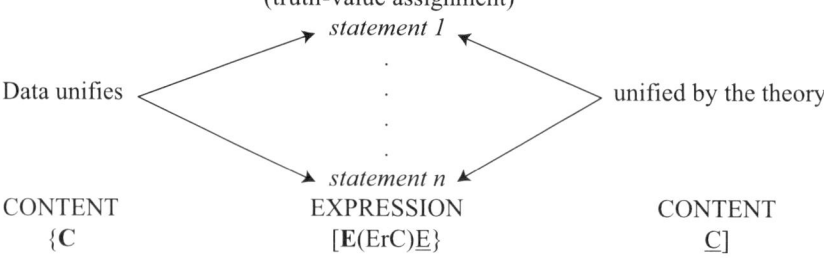

Since the theory is a system of definitions, there is still room for the everyday language, as the signs of the *parole*-level are not all defined by the theory. The construction of a theory is a totalizing enterprise and a *sine qua non* for the existence of scientific statements; the totality and the individual word live at the mercy of statements. A theory is therefore always *en ergeiai* in two ways: it makes a narrative, the class of statements *qua erga*, cohere. The rejection or modification of a theoretical word does not alter that word's status in the everyday language. Otherwise we could not speak of Geocentrism today. The theory like the everyday language exists through its statements/utterances and a statement/utterance exists 'under' a theory/the everyday language. Hence data *qua* denotata exists 'under' both a statement and a theory and the latter through data (empirical relevance). This view steers a middle course between the nominalist alternative[86]. The concept of meaning completion in *parole* acknowledges the contribution of data as well as the contribution of the theory. The everyday language is bound to embody a view of reality, not all of reality (Diderichsen 1966:21), and in conjunction with a theory it yields an exhaustive description of a section of reality. The sign-functional interpretation of the transcendence in 1.4.11 makes data neither potential nor an unknowable x: the historical phenomenon exists in the original sense of this word. That which exists 'comes from' something else, a something on which it is somehow contingent: in our case, the everyday language. The historical phenomenon is *significatif*.

Man's immediate awareness of (his own) existence emphasizes the successor- and effect-nature of historical phenomena, hence serialism *qua* the concept of the horizontal synthesis springs from the very nature of existence[87]: in our immediate understanding of historical reality, we apply our historical sense by supplying a cause-predecessor concept: the analytical mind is unable to grasp historical reality, unable to insert its perceptum into its historical reality.

What exists in a non-idealizing manner 'exists out of' something else that also exists. The empirical meaning of existence (German *geschehen*) implies genetic relations, and is not a katadynamic process (Hansen 1985:76) implementing a dated *manifestatio*-metaphysics. The modern, dualistic meaning of existence is static and idealizing. The dualizing element

---

86 Løgstrup's two views of nominalism apply to these extremes: language (theory) is a pure form whose meaning comes from denotata (reality), data's contribution being everything. Language (theory) is a pure form with definitions which become the meaning of data, data's contribution being nothing.

87 Cicero (*De Fato* IX.18.19) pointed out how the conclusion of a logical argument exists out of its premisses.

1.4. On genetic explanation

comes from the (implicit) contrast between existence and *essentia*, the *esse* of *entia*, the things 'in' or 'behind' which their *esse* is. The essence is that 'out of' which *existentia* (Nietzsche might have said *viventia*) exist or on which they depend katadynamically. The word *existence* means an 'out-of-something'-process (transcendence), and not an entity, a thing. Under this empirically relevant interpretation *essence* assumes the characteristics of a predecessor, and a successor *qua* a predecessor becomes the essence of its own successor in the historical process *qua* existence, the latter being constantly εν εργειαι.

Our Western Civilization has emphasized the static and objective (reifying) aspect of existence; it has turned the serial ενεργεια of the historical process into a timeless and dematerialized εργον so that motion becomes at best repeatable and history generic.

1.5.
Cause and the existence of the everyday language. Cause and datum merge: cause is not a theory. The basic definitions of the theory: simultaneity, historical and the spoken chain's. The affinity between the developmental series and *parole*. The Neogrammarian Paradigm is reinterpreted in the light of the advanced definitions

The framework in 1.4.11 leaves the questions: where to place the concept of cause and how to implement the theoretical concept? Following the Neogrammarian Paradigm – and Karl Verner – causes explain the exceptions to a given isolative change; following the linguistic tradition causes constitute a theory of change. As Coseriu (1974:ch.1) sees this scenario, the latter identification of theory and cause confuses the levels of logic and phenomena, and causes do not explain the 'problem' of the everyday language's changeability (p. 57).

Samuels (1972:2[88]) follows Coseriu (1974:56), turning part of the causal issue into a conditional one: the conditions under which (certain) changes occur. More importantly, however, Coseriu (pp. 91-93; 56-57) demonstrates the absurdity of the WHY-question: why does the everyday language change?, in that it implies that the everyday language should not change.

The implication is absurd because it negates/denies what cannot be denied/negated. The WHY question implies that we are to prove that change occurs, which will make up a circular argument: the linguist wants to prove what is presupposed; the futility of which objective Aristotle saw clearly: to prove that nature exists is absurd; and if we follow Aristotle (*Physics* II.1) a bit further, the synchronic conception of change parallels the physicist who wants to 'prove what is obvious by what is not', in our case, the language system; as nature is an inner principle of change (Aristotle[89]), so the everyday language is constantly εν ερ–

---

[88] Samuels distinguishes between expected and unpredictable change; if this distinction refers to intrasystemic change and extrasystemic (contact-based) change, respectively, I shall accept it, too. But then, to talk about *expected change* is a tautology, as it still implies a narrow view of language development, a limitation that inheres in Samuels's use of *change* as a technical term. In Samuels's theory there is no place for continuation: A>A, this being no change. Thus Samuels reduces the question of the existence of the everyday language to Coseriu's *historische*[s] *Problem* (1974: 56).

[89] Coseriu matches this quote with his reference to Plato's static use-conception, speaking being the use of the words of a 'preestablished' norm (*nómos*), in *Kratylos* (378b-388d).

γειαι. This energy is not to be reduced to a causal factor as Samuels (p. 2) does in his metacondition: 'for a language to change, it must be used', whose alternative: 'for the everyday language to be used, it must change', was envisaged above. In order not to turn this alternative into logical hair-splitting, I shall continue Coseriu's line of reasoning and generalize to this effect: it follows from the existential function of the everyday language that it is absurd to ask why the everyday language is not static, because the WHY question is tantamount to asking: why do I – we – humankind – exist: if we assume that 'I' do not develop into the next moment, T2, but that 'my' existence can be fixed or frozen at T1, the absurdity is evident: we turn what we are to examine and comprehend into something which it is not. The person who wants to carry out this experiment does not deny that the object to be examined also exists at T2, thus he confuses, in Coseriu's imagery, the moving train with a picture of it: he turns life into a still-life. Needless to say, but I do it for the sake of emphasis: Saussure's tree-comparison (1972:125) commits the same fallacy.

It is no condition for any scientific process that the research object be turned into a 'dead' entity; the scientific problems that a section of reality constitutes are merely methodological problems: no linguist has proved that it is impossible to create or construct a historical theory: so if the synchronic conception with concomitant reduction of the existence of the everyday language is to become *the* theory of language, then it must first have proved the impossibility of a theory of history. Apriori assertion and maintenance of the synchronic point of view on the pretext that its negation results in nihilism is scientific poverty and inertia, resulting in the minimalism of the generative programme (Chomsky and Halle 1968:ix; Chomsky 1988) as well as of Glossematics.

On the other hand, Coseriu's tripartite division of the study of language development (pp. 56-57) needs a few more words. First of all, the very focus on *Sprachwandel* suggests that Coseriu has a less generalized and less dynamic conception of the everyday language than is implied in the concept of the **ex**istence of the everyday language. This I infer from the naturalistic imagery by which he amplifies his three approaches to language development (see next paragraph), and from the fact that Coseriu sees [*die innere*] *Mechanik des Sprachwandels* against the static horizon of *Norm und System* (1975:96) *qua* in a broad sense *frühere Muster* and potentiality (1974:92), so that change (*Wandel*) arises in the system('s pressure points (1975:100)).

The *rational* problem is as irrational as this why-question: 'why do people die?' implying: *Warum sind sie nicht unsterblich?* The *general* problem is amplified by: what are the conditions under which people die – age, illnesses, etc. The third problem is the *historical* (note: *historische*, not *geschichtliche*) one: why did Mr NN die? – Why did OE *a:* develop into ME *ɔ:*?

The first problem asks for the rationality of human mortality or *Sprachwandel*, but the imagery suggests differently, namely language death, and that is an entirely different thing. The alternative that Coseriu may have in mind is: is it possible to imagine the existence of language as something different from what it is? And the only rational answer to this is NO! It is (humanly) impossible to keep value-loaded, progressive/regressive, amelioration/deterioration, 'decay'-based overtones out of the mortality-immortality parameter. To put it differently: a human life from birth to death is characterized by functions that differ naturally and expectedly in accordance with man's biological (inherited-necessary) and social (imposed-acquired) clocks; the existence of the everyday language is always in order as it is (Wittgen-

1.5. Cause and the existence of the everyday language 141

stein), and so is man's life; but the function(s) of the everyday language at T1 (in P1) is(are) exactly the same as it(they) is(are) at any other time or place.

I agree with Coseriu that the second and third problems depend on the first, the rational-theoretical one. It involves man's self-awareness or *Selbstverstehen*, and it cannot be reduced to the *general* and *historical* problems, as Samuels does. The rational question tries to come to grips with the empirical fact that *die Sprache* <u>wird</u> (…) *durch* (…) *"Sprachwandel"* (p. 91); but I disagree with Coseriu as his generalization is not sweeping enough: he does not say that the study of language is the study of *das Werden der Sprache*, but subsumes the latter under the study of language *Wandlungen* (p. 91) within the abovementioned system-norm context[90]. This restriction – opening for a static view of language – is repeated in his general *Sprache-Sprechen* conception of the everyday language (p. 92). The former relate of this dualism is recreated (*wiedergeschaffen*) by the latter, because *Sprechen* is based on previous (language) patterns and <u>is</u> *Sprechen-und-Verstehen*; and it is constantly renewed (*erneuert*), because *das Verstehen* transcends that which becomes (is already) known through the *Sprache*, which precedes (exists prior to) the concrete speech-act or utterance; and Coseriu concludes (first): *Die wirkliche Sprache ist geschichtlich und dynamisch, weil die Sprechtätigkeit nicht darin besteht,* <u>eine Sprache zu sprechen und verstehen</u>, *sondern darin,* <u>vermittelst einer Sprache etwas Neues zu sprechen und verstehen</u> (p. 92). Secondly, the continuation of *die Sprache* is an adjustment process in which *die Sprache* adapts to the needs and requirements (*Ausdrucksnotwendigkeiten*) of the speaker, and Coseriu – in my view correctly – in contradistinction to Saussure (1972:109; cf. below) makes continuation dependent on change: no continuation without change, but this view is not compatible with the above restriction: the study of language investigates a phenomenon that continues and changes: *das Werden der Sprache*. The problem seems to lie in Coseriu's perhaps seeming distinction between *Wandlung* (*Wandel*) and *Werden* as well as in the (seeming) fact that all through his exposition the becoming of the everyday language implies a contrast to a (static) *Seins*-concept that surfaces on p. 91, where he says that change (*Wandel*) belongs *zum Wesen der Sprache* – what else – what ?other properties – belong to this *Wesen*? And on p. 25 we read: *Das ganze Sein der Sprache bewegt sich notwendigerweise in diesem Zirkel*: *das Sprechen* (…) [*existiert*] *durch die Sprache*, and the latter only when spoken (cf. also p. 23), but on p. 94 *die Seinsweise der Sprache* seems very synonymous with the everyday language's *Wesen* (cf. Coseriu 1975:11-101).

My main point of criticism of Coseriu refers to his basically Archimedean and therefore somewhat static conception of the everyday language. Quoting Martinét, that *the ever changing and conflicting needs of their users are* <u>permanently</u> *at work silently shaping,* <u>out of the language of today, the language of tomorrow</u> (Coseriu 1974:91 note 63; emphasis added), Coseriu draws the obvious analogy that these same *Ausdrucksnotwendigkeiten* also shape (*ausbau*en) *auf der Grundlage der Sprache von gestern die Sprache von heute*. The two points of view add up to the traditional tripartite view of time with the time line divided into the past (period), a present (point) and the future. Of these two positions Martinét's is empirically irrelevant, because the future, let alone a future language(state), have no empir-

---

[90] As a corollary of this Coseriu sees *sprechen* as *eine Sprache sprechen* (1974:25), where speaking is also 'to exist', a necessary condition for man's historical existence.

ical relevance, both being states of mind of the present[91]. Both points of view suffer from a monolinear conception of historical development, albeit both accept the simultaneity of the relates of, in Coseriu's terms, the *Sprache-Sprechen* dualism (cf. Coseriu's circle above), and Coseriu does not implement his circle-view of the everyday language and the concrete utterance (cf. chapter 3.4. below).

Finally I shall add a few comments on Coseriu's *neu-alt* contrast (cf. 1974:67). It is difficult not to read value-loaded connotations into such an opposition, which, in the final analysis, will reduce to a simple present-past contrast[92]. The old is *frühere Muster* on which Sprechen is conditional: how can the present's utterance be dependent on something that does not belong to the present? Any concrete utterance involves the simultaneous presence of the speaker's understanding of his everyday language. And since what is permanent (= 'what does not change') is not *geschichtlich* (p. 92), and since the everyday language is the latter, there is no such thing as a former or previous language pattern being continued unchanged into the present. As a corollary of this Coseriu's *wiedergeschaffen*[93] is also spurious if it implies a *manifestatio* dualism, and not 'continuation under the aegis of constant change'. This latter concept makes his *überwunden* empty, as there is nothing to conquer: the present utterance (*das Neue*) does not conquer a thing of the past and it does not in any significant sense renew the everyday language[94]. To conclude: the new is synonymous with the present, the old is incompatible with the new and constant change. To a certain extent I am certain that Coseriu's diction belies his intended meaning; on the other hand his choice of words is compatible with the abovementioned (limiting) focus on *Wandel*, not on the everyday language's existence, a fact that is corroborated by his determination of *eine Neuerung* (p. 67) and *Wandel* as the spread and generalization of an innovation (p. 68); and when he states explicitly that *die Sprache* precedes the utterance of the present, he is being at best ambiguous: are we dealing with a logical or a temporal *Vorausgehen*? In conclusion, despite his neutral view (p. 95) of *Wandel* as *Geschaffenwerden* – no *wieder-*, this constant *Geschaffenwerden* is closely correlated with change *qua* innovation (A>B), because the general factor that creates the everyday language, the speaker's *Sprachfreiheit*, is to be understood in terms of the conditions under which *die Sprachfreiheit die Sprache zu erneuern pflegt* (p. 95). Coseriu's view will be generalized as follows in that I accept the concept of freedom (cf. Bréal's *la volonté humaine* quoted by Coseriu, p. 93)[95]: the everyday language exists (= is

---

91 There is a constructive-synthetic alternative to both Coseriu and Martinét: the present creates the past and/or the present creates the present thereby creating the past, *i.e.* my existential view of the everyday language. Martinét and Coseriu share the traditional view of a previous phenomenon that determines its successor, thereby ignoring the just-mentioned alternative that it is what becomes a successor that 'shapes' that which becomes a predecessor.
92 It will appear below that the simplicity of the past/old-present/new opposition is far from simple, neither so static nor so equiempirical as Coseriu implies.
93 On p. 95 Coseriu omits the *Wieder*-aspect and only says that the everyday language *wird immer geschaffen*.
94 This type of transcendence – transcendence of preestablished patterns or structures – also appears in Sartre's view of the dialectics of existence (chapter 3.2. below).
95 There is a discrepancy in Coseriu's over-all conception: the speaker's total freedom that he implies in (1974:56): *Die Sprache [wird geschaffen]*, nay, *[ist] unaufhörliches Schaffen*, and *[ist] nicht fertig* (p.24), is restricted by the controlling force of *die Funktion der Sprache* (p. 24) in that *die Sprechtätigkeit*, which is both *Schöpfung und Wiederholung (Wieder-Schöpfung)*, takes place within the limits *vom System*

1.5. Cause and the existence of the everyday language

continued from the present to the present) against a horizon of freedom (to be determined) and the conditions for this freedom will involve the empirically relevant concept of man's historical sense, which will be implemented by a historical theory.

Thus Geneticism must escape the pitfalls of transcendentalist dualisms (essentialism), value-loaded determinations[96], and be based on an appropriate definition of *to exist*, *i.e.* couched in the language of history (time), so that the historian's statements are not spatially dominated. The same applies to the dialogue-monologue contrast[97].

To put a phenomenon in a developmental series is not – say critics of Geneticism – to explain that phenomenon. But the placing of a phenomenon in a developmental series is a precondition for it being explained in an empirically and structurally relevant manner. A phenomenon placed in its historico-temporal context is distinguishable from fictional entities. So this first methodological step is simply a corollary of structuralism: a phenomenon acquires (part of) its meaning from other phenomena, *in casu*, its predecessor and/or successor. Consequently the autonomy requirement of synchronic linguistics is structurally a contradiction in terms. An immanent structural description can only become a part-description; its object being not describable as a successor or as a successor-turned-into-a-predecessor. This autonomy requirement follows from the applied theoretical standpoint: scientific description is subsumption, classification of data, under a concept by means of analysis *qua* a deductive movement. Historical linguistics will also avail itself of both definition and subsumption, but the difference is that the subsumed is not regarded – in Coseriu's words – as an *ergon*, but as something 'dynamic and historical'. Unfortunately, Coseriu did not see his *energeia* against the horizon of *die geschichtliche Wirklichkeit* of the developmental series.

The terms – incorporating a causal concept – of the historical theory will be defined in this and the following sections; they constitute an attempt at creating a unified historical theory with no exclusively synchronic basis[98]. By *unified* I mean the task of combining into a coherent whole time- and space-, historical and synchronic, dynamic and static as well as

---

gebotenen Möglickeiten (1975:94). Coseriu 'does not solve the perennial dilemma of 'freedom, *voluntas*' and 'determinism', man's craving for freedom and novelty and his craving for certainty and predictability. The argument of this essay is that the latter need is not to be fulfilled by a transcendence into the quietism of an ideal haven of (various) systems.

96 Cp. various functionalist views containing such concepts as conflict (Martinét), greater clarity (Horn), communicative efficiency, and such indefinables as tendency, disposition (Hjelmslev), economy (Coseriu 1974:67-68; Samuels). Note also the emotional approaches to the fact of language change that were adduced in the *Prologue*.

97 Coseriu (1974:60) makes a *non-sequitur* conclusion, inferring the primacy of the dialogue and of the hearer from the assumption that *zu bedeuten* (the speaker signifies) presupposes *zu deuten* (the hearer understands). The soliloquy or prayer involves both processes – and so does the historian's (writing of his) narrative. In spite of its irreducibility the speaker-hearer situation (Coseriu 1974:65, 67, 70, 79; Keller 1985:233-236) assumes neither empirical nor theoretical precedence over the monologue, and the function of *die Einzelsprache* – (*die Sprache* or *sprachliche Wissen* (Coseriu, p. 65) differs from person to person) – is not only intersubjective *Mitteilung* seen against the horizon of the above-mentioned *etwas Neues zu sprechen und verstehen*, but *eine Bedingung* in man's individual and social existence.

98 Both Samuels 1972 and Kanngiesser 1972 work on overtly synchronic premises such as the *langue-parole* dualism, and as suggested above Coseriu (1974:67) is also influenced by this dualism, and he has not succeeded in demonstrating the empirical relevance of the system as a *sui generis* entity and his conclusion (p. 246) revives a *manifestatio* metaphysics.

simultaneity- and succession concepts. The definitions will be constructed with a view to what is traditionally called conditioned change (exceptions to isolative changes), changes that cannot be explained immediately by the interaction of 'historical freedom' and 'social-systemic necessity' and which all contain an overt causal element. The possibility of setting up such definitions will demonstrate why a theory of change cannot be equated with a causal explanation, the cause being a part of the effectum. So causes belong to the level of data, not that of logic or theory.

The first definition, with one (two) undefined concept(s) (*non*)*fullbloodedness*, is a partial definition of the developmental series[99]:

1.5.1 A developmental series consists of non-accidental horizontal syntheses each of which creates an empirical vertical synthesis, and it contains at least one full-blooded element.

Hence a developmental series does not consist of only nonfullblooded elements and a synthesis is a process, not only the result of a (perfective) process (cf. def. 11 and chapter 6.).

Def. 1 A horizontal synthesis is a synthesis that consists of at least two equiempirical elements.

Def. 2 An accidental horizontal synthesis is a horizontal synthesis that is purely man-made (= arbitrary).

A Def. 2 entity is a Glossematic class or hierarchy and as such an entity that can be subjected to Glossematic analysis; or is a strictly nominal entity, which follows from the theory in question (cf. Quine 1971:1-2); an accidental horizontal synthesis cannot be a part of a developmental series, a consequence that rejects Hjelmslev's hierarchical view of semiotic change (1975:dff. 203, 208) and which observes Paul's *Weg mit allen Abstraktionen!* – Fictional processes are also Def. 2 syntheses.

Def. 3 A non-accidental horizontal synthesis is a horizontal synthesis that is not purely man-made (= non-arbitrary).

A Def. 3 synthesis arises out of the two processes in 1.4.6 and 1.4.5; and both definitions 2 and 3 underline the anthropocentric factor in all historical processes, at both the theoretical and phenomenal levels. Furthermore a Def. 3 process is not katadynamic formation of unformed matter into a formed substance[100]. The elements of the developmental series are all formed, just as – at the theoretical level – the construction of a new theory is always based

---

99 At this stage in the theory construction *process, synthesis* and *series* are used almost interchangeably.
100 The Glossematic triad of form, unformed substance and formed substance is no more empirically relevant than Anaximander's *to ápeiron* 'the boundless, limitless' (Jones 1970:11-12); both are linguistically motivated constructs.

1.5. Cause and the existence of the everyday language

on a previous formed one[101]. No historical theory is purely man-made; modern formal logic (the propositional calculus) is therefore not a historical theory. An element in a developmental series has an Eigennatur, which stems from its substance.

Def. 4   A non-accidental horizontal synthesis is a horizontal synthesis in which endogenous motivation and exogenous motivation occur, and metagenous motivation does not occur.

Definition 4, amplifying Def. 3, brings up the subject of causation (motivation), and it states that the 'cause' participates in the elements of the horizontal synthesis. I shall also argue that this definition unites historical and *parole* processes, and that the causal concept involved is both participative and biuniformal.

Def. 5   Exogenous motivation (= exogeneity) is simultaneous motivation that originates from the Eigennatur of any one preceding and/or succeeding, and/or simultaneously existing phenomenon.

The concept of Eigennatur is intended to emphasize that the elements of a developmental series are not mere raw-material; it follows from the fact that the structural context-tenet (cf. above and below) is sooner or later bound to entail the question: where does that meaning come from which arises in the interaction of the elements of a structural totality? – Definitions 5 to 10 are seen from the process's, not the effectum's perspective.

Examples of a Def. 5 process: Assimilations such as OE *gieldan*; *cyning*; *ceald* < PreOE *geldan*; *kuning*; *kæld*. – OE *cild* [i:] < [i]. – ME *home* < OE *ham*. – ME *they* (as a loan word). – *Passenger*; *captain* < *passager*, *capitain*.

Def. 6   Endogenous motivation (= endogeneity) is simultaneous motivation that cannot originate from the Eigennatur of a succeeding phenomenon.

Endogenous motivation is also involved in exogenous motivation, which means that there is no absolute difference between isolative and conditioned change.

Def. 7   External endogenous motivation is endogenous motivation through which the synthesis that the endogenously motivating factors and the effectum constitute, is characterized by distance *qua* attachment and distance *qua* separation.

---

101 If, as Gregersen (1991:2.17) argues, Hjelmslev saw himself as having no predecessor, then such an Athene-like conception of a theory demonstrates a complete lack of a historical sense, a lack that reminds us of Descartes's Third Rule: '(…) our inquiries should be directed, not to what others have thought, nor to what we ourselves conjecture, but to what we can clearly and perspicuously behold and with certainty deduce; for knowledge is not won in any other way (…).' (Translated by W.T. Jones (1969:157)). And since Descartes only admits two mental operations, intuition and deduction, in the search for knowledge, it is obvious that the mental operation we call our historical sense has been excluded a priori and arbitrarily.

Specification: +Attachment *and* +Separation.

External endogenous motivation is the basic historical process in which a phenomenon exists from T1 to T2. The separation concept suggests the difference between a successor and what becomes its predecessor, and the attachment concept suggests the participation of the cause in its effect.

Def. 8   Internal endogenous motivation is endogenous motivation through which the synthesis that the endogenously motivating factors and the effectum constitute is characterized by distance *qua* attachment and not by distance *qua* separation

Specification: +Attachment *and* -Separation.

A Def. 8 process refers to the role of suprasegmentals such as accent and intonation in the historical process, and the specification suggests the interrelatedness of suprasegmentals and segmentals in the spoken chain: neither is present, if the other is absent. Suprasegmentals[102] (accent) are involved in the abovementioned PreOE *I*-Umlaut, the rising of OE [a:] in *ham*, as well as the lengthening of [i] in OE *cild*. – I shall be dealing with analogy and ME *thei* below.

The Glossematic connotative function, in which the speaker (connotatum) becomes the content of his or her own language (CrE(ErC)), is internal endogenous motivation owing to the originally inalienable connection between a person and his everyday language; it is also involved in the concept of loan words: as a national language connotes its nation, a dialect connotes its group of speakers, so a loan word connotes its origin, its connotatum. This means two things: the simple sign function, ErC, is not empirically relevant; any sign will always connote a specific everyday language: CrE(ErC); no synchronic systems-explanation is capable of determining which is and which is not a loan word: at T1 the dialect of the poet of *Havelock the Dane* contained no *þei*, 3pl. pers. pron.; at T2 the Scandinavian loan appears so that his dialect has this variation[103]:

> þanne micte <u>chapmen</u> fare (...) Oueral þer <u>he</u> willen dwellen, 'Then could merchants travel (...) Everywhere there they will stay'
> þei hidden hem alle, and helden hem stille, 'They hid themselves all, and held themselves quiet'.

This illustrates a fundamental problem with the-synchronic-cut notion of immanent syn-

---

102 I do not enter into a discussion about stress, accent, etc. I am relying on the early discoveries within the field of Experimental Phonetics made by the German Functionalists Horn (1930, 1936) and Luick (1912, 1922, 1923, 1927, 1930, 1930b, 1930/31,1931), Lira (1976), Parmenter *et al.* (1933), Peterson (1961); Ladefoged (1972:1-49). See also Saussure 1972:103. – Recently, the explanatory power of such findings have been resumed by Samuels (1972; cf. Back 1979) and with reference to this theory I shall make do with this determination of accent (stress): a certain degree of pulmonary (subglottal) force. It is a sign-functional interpretation of such data that has and will be utilized in this essay.

103 From Bruce Dickins & R.M. Wilson, *Early Middle English Texts*. (Fifth impression). London: Bowes & Bowes, 1961: ll. 51-54 and 69, respectively.

1.5. Cause and the existence of the everyday language 147

chronic linguistics: no correct synchronic description can be made of the T2 dialect: the variation can be registered in statistic terms. Secondly, the loan word *þei, thei*, etc. was structurally isomorphic with the inherited conjunction *þei, thei* (< OE *þeah* 'though'). At T2 there is nothing to suggest immanently that one of these forms makes the dialect less homogeneous than it was before the cut was made. This simple fact explodes one of the die-hards of the synchronic myth: synchronic analysis presupposes one and only one language state. Not even the Glossematic requirement of structural homogeneity (Hjelmslev 1963:115) can be fulfilled without the necessary presence of a horizontal synthesis of two language systems of which one is bound to be the simultaneously existing system *qua* the predecessor system, the donor system, of a given language state, the recipient of that loan word: the connotata in these syntheses: (*þei* rC)ErconC (= Scandinavian) :: (*he* rC)ErconC (= English) do not appear from the structural interplay between the two signs and the other signs in this dialect at T2. – *Mutatis mutandis* analogy has a similar processual structure[104].

Def. 9   External exogenous motivation is exogenous motivation through which the synthesis that the exogenously motivating factors and the effectum constitute is characterized by distance *qua* separation and not by distance *qua* attachment.

Specification: -Attachment + Separation.

The PreOE *I*-mutation belongs here due to the sound intervening between the affectum and the *i/j* sounds. Similarly, the notion of distance involved in the (contact-based) synthesis constituted by a loan word/a donor language and a native word/a recipient language is maximal and the loan process therefore belongs here, as analogical processes do; here the expanding paradigm assumes a donor function. – The type of change involved in *passenger* (ME *passager*) seems to belong here, while the loss of *-i-* in *captain* seems a moot question.

Def. 10   Internal exogenous motivation is exogenous motivation through which the synthesis that the exogenously motivating factors and the effectum constitute is characterized not by distance *qua* separation and not by distance *qua* attachment.

Specification: -Separation and -Attachment.

Taken in a broad sense assimilation- and dissimilation-processes which involve adjoining sounds belong here. Examples: PreOE *geldan* > *gieldan, cild* > *ci:ld, feld* > *fe:ld, wifman* > *wimman*, PreOE *æld* > *eald/ald,* *cyþde* > *cydde,* *þeofþ* > *þeoft,* *ciesiþ* > *he ciest* (inf. *ceosan*), inf. *cweþan* 'to say': *cwiþ-is þu* > *cwiþs-þ þu* > *cwisst* > OE *þu cwist*.

These four definitions (7-10) do not define four categorically different changes; on the contrary a given change is a processual synthesis in which the principle of multiple condi-

---

104   In *Sir Orfeo* we read: Of al þinges þat men seþ (l. 11), 'Of all things that one sees'; and l. 4-5: Layes þat ben in harping / Ben yfounde of ferli þing 'Songs that are for the harp / are composed out of wonderful things'. From J.A. Burrow and Thorlac Turville-Petre, *A Book of Middle English*. Oxford: Blackwell. 1992:112.

tioning will be effective; and the definitions therefore tally with an aitional definition of the concept of multiple conditioning: A * B: A, an effectum, but not only if B, (a specific motivating factor). Secondly, the amenability of the affectum (its Eigennatur) – to being affected by a cause – will be considered by external endogenous motivation. If all change were a segment's adaptation to synchronic environment, (external) endogeneity would be superfluous.

The notion of distance has been introduced as a technical term, defined by the four primitive terms +/-Attachment and +/-Separation. Synchronic linguistics presupposes distance at various stages in their methodology; the synchronic cut and the consequent notion of diachronic links (cf. Samuels 1972:10-11) presuppose it; the immanent language state presupposes it in two ways: the unanalyzed object is a priori at a distance of all other objects, and distance appears as a side-product of the analysis in that the resultants of any analysis presuppose a distance between, be they NPs and VPs, Subjects and Predicates, or various sententials; any formation process implies it. Likewise, the past and the future presuppose some distance between them, and so do the constituents of the speech situation. The theory establishes four types of distance, to be defined below in terms of directional concepts: [+Attachment & -Separation], [+Attachment & +Separation], [-Attachment & -Separation], and [-Attachment & +Separation], and as will appear from the preceding exposition, these definitions do not make a sharp cut between synchrony and history.

The concept of exceptions now finds its theoretical place: an exception implies the concept of a phenomenon's Eigennatur; the *-i-* in PreOE **cild* was not amenable to exogenous motivation by *-ld*, and the consonant cluster was not strong enough to bring about 'breaking' (Campbell 1964:§148): the ('breaking') consonant cluster *-ld-* (in PreOE), but not only if 'Breaking'[105].

In historical processes the motivating factor becomes a part of the resulting effect, and effect and process will thus merge, and I shall demonstrate how the effect becomes a structural phenomenon in relation to its origin, the affectum. When such a phenomenon is an element in a developmental series, it will be called an *intersection*. This merger of process and effect (and affectum) is the reason why it is impossible to maintain a separate parallelism between (the language of) time and (the language of) space, as Jakobson (1962:651) implies. Time

---

105 Lass's criticism of teleology and function in his 1980-work (ch. 3) for not being able to predict and for making for unordered *post-factum* explanation is per se relevant criticism of certain theories of language change, and it seems to apply to my principle. But Lass does not consider the fact that amenability does explain why, say, **cild* did not become **ciuld* in PreOE, because he focuses on the cause-concept, thereby committing the Phenomenal Error; our consonant cluster is no more a causal factor in relation to the *i*-sound than the latter is in relation to the former. Explanation in terms of overt causes should not conceal the fact of covert conditioning (here: not represented in writing) and therefore other conditioning factors, in our case progressive assimilation instead of the expected, regressive assimilation: if I may be excused for using the slant-bracket notation, I can explain myself briefly: the variant [ld] of /ld/ in PreOE *cild* is not the same variant as we have in PreOE *æld*. Lass might expostulate that explanatory principles should have well-defined, restricted scopes; correct, but theoretical understanding of phenomena should not be restricted by (1) a narrow (arbitrary) view of what explains (*i.e.* causes) and (2) a narrow (preconceived) view of what is to be explained: the continuation of PreOE *i* in our sample word deserves, nay, is in need of as much explaining as PreOE *æ* > *ea* and *æ*.

1.5. Cause and the existence of the everyday language

and space will merge in the historical phenomenon, of which our definition system can define three (cf. def. 31):

Def. 11 A historically genetic process is an intersection that is so produced through **internal** and **external endogenous** motivation as well as **exogenous** motivation that the intersection becomes a constituent of an empirically relevant vertical synthesis which involves a change process. – Cf. Def. 14.

A Def. 11 process involves an element of theory-controlled freedom, which a Def. 13 process does not, where the plural ending could have been *-en* or *-e*. In chapter 5 the difference between development with change of substance (A>B) and development with no change of substance (A>A) will be defined. Extrasystemic change, such as the adoption of a loan word, say ME *hie > þai*, will be defined as seen from the point of view of the receiving language and the loan itself:

Def. 12 A loan process is an intersection that is so produced through both **internal endogenous** motivation and **external exogenous** motivation, and **not** through external endogenous motivation, that the intersection becomes a constituent of an empirically relevant vertical synthesis which involves a change proces. – Cf. Def. 15.

A loan process is not a historically genetic process because the intersection (the loan) does not involve external endogeneity. The two-state model, ME *hie > þai*, is only appropriate as seen from the points of view of content: 'they > they' and of completed spread with the ouster of the form *hie*.

Analogy is structurally related to the loan process in that the spreading grammatical element corresponds to the loan word and the resulting intersection becomes the start of a new developmental series:

Def. 13 An analogical process is an intersection that is so produced through both **internal endogenous** motivation and **external exogenous** motivation, and through **external endogenous** motivation, that the intersection becomes a constituent of an empirically relevant vertical synthesis which involves a change process. – Cf. Def. 15.

Analogy is a historically genetic process and differs from a Def. 12 process in the absence of external endogeneity. Both processes presuppose that the affected language is not *densely continuous* (cf. below). – The *þing > þinges* process does not follow from the past history of the OE plural of *þing*.

Illustration: Def. 11 process   Def. 12 process   Def. 13 process

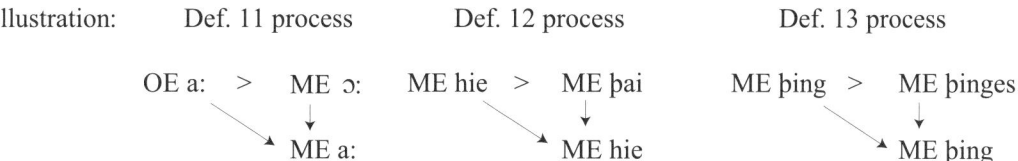

The figures illustrate two things: the concept of simultaneity, which is therefore not the exclusive property of synchrony, and that of linearity, which is not monolinearity. The simultaneity concept in the above definitions differs from the one in synchronic dualist theories. Here it is looked upon in two ways: the monolinearity of the Saussurean signans excluded concurrent elements and if such a syntagmatic segment occurred, the process was to be interpreted as one of superimposition; this point was aptly criticized by Jakobson (1962:636), but within a static conception of *syntagmatic context*; his *concurrent factors* did not include anticipation and 'postcipation' *qua* dynamic processes. Secondly, simultaneity refers to the elements of the system in the sense that all systemic elements may in principle appear, pace the concatenation rules controlling the linearity of *parole*, in the same *parole* slot at any one given point in time because *parole* is mere contingency. Furthermore, these systemic elements all exist simultaneously at one specific point in time in a world different from their manifesting world; this appears most clearly from the way Glossematics defines its process, by an undefined both-and function which involves no temporal succession (cf. Gregersen above); the same goes for its definition of the system, defined by an equally undefined either-or function[106]. From this it also follows that *parole* (the process[107]) is looked upon as a monolinear entity. I shall show how the historical process and *parole* are related, having the concept of simultaneity in common (see the end of this chapter).

The spoken chain is both linear and simultaneous. Synchronically, the spoken chain is defined by means of points *qua* the product of (mutually) preceding and succeeding segments (syntagmatic relations). This is an arbitrary definition seeing that a point *qua* an element may be defined by the lines that intersect it. Thus the point or the intersection is not per se *the* scientific unit of language analysis, and syntagmatic succession does not necessarily reduce to being combinations controlled by the paradigmatic point of view's paradigms *qua* classes of simultaneously preexisting elements. The synchronic problem hinges on the fact that the linearity of the spoken chain cannot be upheld non-circularly, since the synchronic element may be regarded as an intersection, the product of lines running through it. Thus the elements of the pardigm/calculus, far from being the defining terms, now become the defined, namely by 'the lines of *parole*'. The line and the point are equilogical entities.

My diction is deceptive. There *is* no point to be defined by intersecting lines, the intersection process produces the point: when a number of lines (processes) intersect, they produce a point (a language element) called an intersection. The figure in 1.5.2 below underlines the fact that my definitions are not categorical ones, as it implies that a given 'line' – type of motivation – is not effective to the exclusion of other lines. Lines are *en ergeiai* simultaneously, thereby illustrating the multi-layered and multi-conditional and aitional nature of historical phenomena or existence. An intersection is a bundle of lines *qua* motivating factors (cf. 2.5.1).

---

106 Similarly Jakobson (1962:636) also makes a sharp distinction between simultaneity and successiveness.
107 I am not equating the vague Saussurean *parole* concept with Hjelmslev's process.

1.5.2[108]

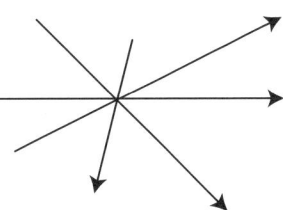

I have used the spatial metaphor of a line, but that is not tantamount to accepting a purely spatial conception; simultaneity (synchrony) and succession (history) are not mutually exclusive.

Continuity is difficult to define historically; it seems superfluous to try to do this, since it is virtually synonymous with existence. So Saussure's use of it in his determination of change (*altération*) (see Coseriu 1974:92) explains nothing, and Coseriu's reversal of this point of view (*loc. cit.*) becomes a more tractable and attractive alternative: continuity rests on change, since no change implies permanence and therefore absence of *Geschichtlichkeit*. However, Coseriu's view is too simple, as his determination of change (*Wandel, Veränderung*) is too narrow (cf. above). Continuity includes both a static and dynamic aspect. Continuity *is* (Coseriu, Saussure), it is created (Coseriu), and it is broken, arrrested, or interrupted, and so is discontinuity. This obliges me to determine these two seemingly correlative concepts in relation to a Def. 11 process: the concept of continuous development into the present (existence) implies both a separation-from process and an attachment-to process. Hence the intersection of the present moment *qua* the intersections of the spoken chain and the developmental series has both a *cutting* and a *cohering* function. An intersection cuts a line into two subsections without creating an absolute gap. Despite their separateness, there is always some kind of attachment between a successor and what becomes its predecessor. Again my diction is not dynamically appropriate; but my objective is not to reject all static concepts; to be precise: the cutting element is always the here-and-now intersection constituting the actual present, this being a structural phenomenon, which has appeared from the vertical syntheses of the Defs. 11, 12, 13; but at the same time its over-all structural nature disallows the autonomy of the cutting element.

The continuity concept will now be determined from the point of view of historical theory, not from that of the mechanisms of change as was suggested above in chapter 1.4., and I shall now set up a number of definitions expounding the concept with reference to the spoken chain, while its historical nature will be defined in chapter 6, where the theory of the developmental series will be constructed.

The spoken chain does not cohere densely; loan words penetrate into all languages, and sounds are added/lost[109]. In spite of the fact that the Anglo-Normans imposed on the English language their indelible stamp, we cannot find any discontinuity in the English language in the 11th, 12th and 13th centuries. The continuity concept also underlines the rejection of the

---

108 Cf. Ottosson's (1996:49) reference to Bruns's 'definition' of hermeneutical, and note 140 in the *Prologue* below.
109 In English: *sound, thumb, numb, dumb, against, whilst; the Thames, Anthony, author.*

mechanical (Neogrammarian) view that change is remedial, following functional or communicative break-downs, that change is '... from function to non-function to function ...'. On the contrary the existence of any everyday language tells us unambiguously that it or rather its elements are discontinuous against the horizon of continuity. In this sense language development is always gradual and discontinuities are to be explained in respect of continuity: the everyday language never loses its collective orientation or existential function.

The spoken chain is no proper – spatial – line, although no element of it is not without extension. An utterance is an ordered succession of intersections existing in the same place at different moments, and intersections may exist simultaneously, *i.e.* concur. So a more descriptively adequate image of *parole* would be *a palimpsest*. Tempo is therefore an essential element in speech, and not merely a paralinguistic feature: one segment cannot occupy its allotted position for too long (cf. Løgstrup 1976:11); it is not only in our meditative moments that we may let our imagination fly, the 'flight' of our speech also contributes to the meaning of an utterance. Each element of it must disappear ('be erased') in order that its successor may be produced. As the understanding of an utterance seems to be predominantly holistic, function-directed speech (Samuels 1972:28-29) cannot be generalized; the merging totality character of suprasegmentals and segmentals (internal endogeneity) and anti- and postcipation processes (exogeneity) points in the same direction. The palimpsest metaphor indicates a type of discontinuity that is seen against the horizon of continuity.

Two lines contributing to the production of an intersection are kept separate by our synchronic tradition; they are different, but they have so basic similarities that they can be seen from a unifying stance, and neither can be left out of consideration when we come to the question of how the spoken chain is understood by the hearer (cf. Def. 14.); synchronically, the separation is due to the fact that of the two – in Jakobson's terms – co-ordinates of language – the axes of simultaneity and successiveness – the latter axis is polysemous; it means either synchronic-static extension *qua* syntagmatic concatenation, this being atemporal because of *parole*'s rule-controlled (predictable) nature, or temporal succession, *i.e.* a non-synchronic term. This distinction is not empirically relevant, a fact to be reflected in the following definitions:

Def. 14 *The developmental series*: a developmental series does not intersect itself; an intersection of a developmental series **ex**ists **out of** the process of external endogeneity. – Cf. Def. 11; Def. 23.

Def 15 *The spoken chain (an utterance)*: an utterance intersects itself; an intersection of an utterance **ex**ists **out of** the processes of exogeneity and may not exist out of endogeneity[110]. – Cf. Defs. 12 and 13; Def. 23.

In the spoken chain two intersections do not so cohere that (i) either of them cannot disappear, (ii) no intersection can be added between them, (iii) either cannot change, and (iv) no

---

110 An utterance <u>may</u> contain loan words. – It is difficult, if not impossible, to envisage relevant conditions under which an utterance in English (by an English speaking person) will not be motivated by endogeneity. This would imply that the speaker could produce an utterance in French entirely in accordance with the endogenous and exogenous processes that characterize French.

1.5. Cause and the existence of the everyday language

intersection can appear before or after them or concur with them. Since these four facts are empirically irrefutable, a synchronic explanation of them to the effect that (a) *parole* is contingent, non-necessary, subjective and motivated by speaker idiosyncrasy, and that (b) *langue* is potential, dealing with only categorical and significant possibilities, is inadequate.

Exogenous motivation creates separation as well as coherence through both assimilation and dissimilation (cf. Hjemslev 1970:46; Samuels 1972:41) against the horizon of continuity, and I shall demonstrate that (1) exogenous motivation, so-called subjective *parole* variation[111], and endogenous motivation are *organizing facts of the everyday language, hence principles of linguistics; both processes are united by the concept of the present's development into the present (existence)* and that (2) (convergent and divergent) developmental processes are *gradual, neither saltatory nor spontaneous, neither regular or irregular nor destructive or paradoxical, but always in order*[112].

In synchrony with its exclusive emphasis on analysis the all-important concept is that of the absolute break; the notion of the *apriori* scientific legitimacy of the synchronic cut precludes any thematization of that concept. From the level of logic – with its ongoing, often contradictory, division of all objects into binary subclasses – to the level of data – with its immanent and autonomous language state – the break-concept is presupposed. The following will be such a thematization, and the exposition will also introduce a set of primitive notions to describe a fact that the above linearity-discussion did not include, namely directionality.

Metacondition 1: All empirically relevant entities (existents) are produced with extension.

It follows that all empirically relevant entities have/create a BEFORE part and an AFTER part; the nature of such a sequence will be modified by the following metaconditions.

Metacondition 2: No cutting intersection is so produced that it produces a historical synthesis with an absolute **first** intersection ([+First] and/or with an absolute **last** intersection ([+Last]. – |→←|.

It follows both that the figure |→←| ([+First] & [+Last]) is impossible as an illustration of a historical synthesis or the synchronic element (an immanent and autonomous state or system) and that a break will occur within a line; if the figure should occur, it does not characterize horizontal, only vertical syntheses.

The spoken chain does not consist of segment a and segment b so that either implies and only implies the other. This type of coherence will be called **dense cohesion**. The spoken chain does not consist of two and only two elements, which means that the Glossematic both-and definition of the *text* must be defined aitionally: 'a * b & b * a', this being identical with the conjunction 'a & b'.

Our historical existence is continual, apart from birth and death the individual cannot experience any absolute, non-continuous break in his or her life. The individual's existence has

---

111 Jakobson (*op. cit.*:638) correctly emphasizes the *decisive role* of *the subjective impression of the listener* in verbal communication and therefore its importance to linguistics, but he overlooks the correlative role of the 'subjective expression of the speaker'.
112 This is of course dependent on how these concepts are defined.

an absolute beginning and end, all are born into a continuing and continued world, which – in structural diction – continually influences us and we do it. Existing, the individual lives with the past (in postcipation) and in anticipation of continuation; the directionality of existence is always characterized by '[+First] & [-Last]' in the sense that the present (moment) develops into the (next) present (moment), thereby being First in relation to the next present, simultaneously becoming non-Last and non-First:

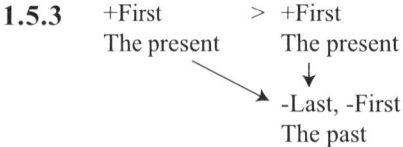

1.5.3   +First   >   +First
       The present   The present
                       ↓
                  -Last, -First
                  The past

The present is the 'cause' of continued existence and the past. Thus with the help of Metacondition 1 cosmological questions have been ruled out. Logically, the axiom is irrefutable because talk about the limits of reality will always take place in terms of the 'existents' of that same world which is being denied: the great void – be it Hjelmslev's unformed substance or Anaximenes's boundless – exists contingently on that which exists. Therefore we shall be committing the Phenomenal Error, if we introduce the concept of a historical nothingness or for that matter the concept of the future. In logical terms, the implementation of logical contradiction, A & nonA, is participative: the contradictory of A involves A (cf. Favrholdt 1965:174-176).

Such developments as ME *an ewt > a newt; soun > sound* (Lat. *sonus*), (Pre)OE *-es þu > -est þu* (2nd pers.sing.pres.ind.) tell against the rejection of Metacondition 2. Similarly, no synchronic theory can decide when the denotatum of *the English language/Old English* came into being *qua* an empirical entity that did not exist at T1, but at T1+ ... . The absolute end of a language is also conditional on our defining its demise. So *the demise of Old English* (Nielsen 1998) refers to no empirical entity, only to the linguist's arbitrary decision, at best to his definition of the sign at the level of logic. Hence the actuation problem (Weinreich *et al.* 1968:102; Milroy 1992:76) is a theory-made pseudo-problem that is tied up with eschatology-like notions such as a change's start, progress and end as well as the synchronic cut.

Metacondition 3:   No cutting intersection is so produced that it produces a section with a last intersection [+Last] (←|) and a section with a first intersection [+First] (|→).

This axiom makes absolute discontinuity impossible and the cutting of a line into three subsections possible. ME *passager* (> *passenger*) was not characterized by dense cohesion and so illustrates the fact that no gap is so absolute that no third element may become a 'gap-filler':

1.5.4   ab > a # x # b   (←| |→)

Thus discontinuity occurs on the ground of continuity or continuity participates in discon-

1.5. Cause and the existence of the everyday language

tinuity. The axiom illustrates a discrepancy between the binary principle of analysis as a logical operation and its empirical relevance: how can the separate categories of, say, noun phrase (NP: *John*) and verb phrase (VP, predicate: *left*), or subjectival and verbal, capture the coherence between the corresponding *parole* phenomena in the utterance *John left*?

Metacondition 4: No cutting intersection is so produced that it produces itself as a predecessor.

It follows that a predecessor implies a successor. Metacondition 4 can be elaborated as follows:

Metacondition 5: No cutting intersection is so produced that it produces an intersection as a successor.

The Eigennatur of any cutting intersection is that of being produced as a successor of a successor (cf. below).

Metacondition 6: A cutting intersection is so produced that it makes all other intersections (its) predecessors.

By *all other intersections* in Metacondition 6 I mean either an ordered class of intersections producing a line *qua* a developmental series or an ordered class of intersections producing a line *qua* a spoken chain.

From metaconditions 4 and 5 it follows that all produced intersections are successors, and the successor intersection of the present moment that produces the present intersection of the (next) present moment is a successor-intersection that will be called *a here-and-now intersection*. The here-and-now intersection never turns itself into the predecessor of its own successor *qua* a development into the future from the present. The future is not an empirically relevant concept; it can only be established on what is not the future, *i.e.* inconsistently. But owing to the nature of the spoken chain a here-and now intersection is regarded as a predecessor in respect of the possibility of (the) exogenous motivation (of form-directed speech): [y] in PreOE *kyning* is a predecessor in respect of [i]. The structural nature of historical existence can be stated preliminarily as follows: to be a predecessor is an acquired property, a structural property, while to be a successor is both acquired and innate: what characterizes all historical phenomena *qua* their Eigennatur:

Def. 16 A here-and-now intersection is regarded as an intersection in the spoken chain if it can be motivated in a vertical synthesis by an intersection that follows and/or precedes it; if it cannot be motivated by a following intersection, then the here-and-now intersection is regarded as an intersection in a developmental series.

From Def. 16 it follows that to the temporal linearity of the spoken chain – and the developmental series – must be added the concept of simultaneity.

Two endogenous lines contribute to the production of a here-and-now intersection. How can an exogenous line, which is yet to be or has been produced affect a here-and-now intersection? From where does it exist as a causal agent? It *exists* in its developmental series in what will be called the L-continuum (see below chapter 1.6.1): against the existence of a here-and-now intersection (= b) what has become its predecessor (=a) exists no more and what (= c) will succeed it does not exist yet *qua* elements *equiempirical* with it. Here I need the qualification *equiempirical* – to be determined below – in order to anticipate criticism for reintroducing astructural monolinearity: a here-and-now intersection is not a so-called totality sui generis; it exists as a part of a totality for which it itself is a necessary condition – which it motivates and is motivated by (see 1.5.3 and 1.5.8).

The spoken chain is not produced by selection – from the store of a finite calculus – of discrete and irreducible, monad-like systemic entities. Interestingly enough, the Glossematic view of variants (Hjelmslev 1963:82) is compatible with Samuels's form-directed speech in connection with exogenous motivation as a general dynamic factor in the production of an utterance. The traditional conception of language being made up of a set of smallest possible entities – the ultimate entities – is yet to be proved as empirically relevant; and Jakobson's coupling of this type of objective for linguistic analysis with the *differential aspect* of language phenomena is arbitrary (Jakobson 1962:636-637; and with Morris Halle: *op.cit.*:465; see also Hjelmslev 1963:60-75; 81-84). Secondly, the view still has to demonstrate that a sign-functional phenomenon like the everyday language can consist of non-signs! Thirdly, I shall demonstrate that this differential aspect can be generalized – not only as our synchronic tradition does in its view of 'constant variation' - to language change and continuation. But first a summary of the cutting and cohering function of the here-and-now intersection.

Dynamically, the here-and-now intersection creates different types of coherence in order to bridge the separation that follows naturally from its endogenous nature: it is different from its predecessor[113]. Since it has no successor, it is characterized by [+First]. Continuation *qua* the present's development into the present is bound to relativize [+First]-elements so that any successor will both turn its predecessor into a [-First]-element and imply its own [-First]-property.

The static and geometrical view of continuity implies no historical directionality; it is strictly monolinear and simultaneous only from the point of view of its calculus; historical directionality is multitiered and simultaneous. A line as a finished product may always be divided arbitrarily. In 1.5.5 below any one element (point), say f, may be regarded as the [+Last]-element of a predecessor section and its predecessor as a [-Last]-element. Accordingly, g is the [+First]-element of the successor section and its successors will therefore be – remain – [-First]-elements. The structural dynamics of a static line are a foregone conclusion with predictable distribution of the above four properties, once the (synchronic) cut has been made:

1.5.5   abcdefghijk → abcdef] [ghijk

The written text is the closest we get to a geometrical line; but this does not imply that the point-line conception describes the spoken (and written) chain as a dynamic phenomenon.

---

113 More correctly: 'it becomes different from the intersection that it makes its predecessor'.

1.5. Cause and the existence of the everyday language

A film reel is not the moving film we watch on a cinema screen. The moving pictures are not the representation of the film reel's static row (calculus) of pictures *qua* the representatum or manifestatum. On the contrary the film reel is a poor representation of the story of the film we see on the screen. The film reel completes its meaning in cinematic *parole*[114].

The possibilities of the static system behind the line in 1.5.5 will be enumerated in 1.5.6, where A is the cutting element:

1.5.6	B	A		B	A
	Last	First		First	Last
1)	[+Last]	[+First]	5)	[+First]	[+Last]
2)	[-Last]	[+First]	6)	[-First]	[+Last]
3)	[+Last]	[-First]	7)	[+First]	[-Last]
4)	[-Last]	[-First]	8)	[-First]	[-Last]

1) This line contains an **absolute gap** (= ← B] [A →) and is in need of a bridging factor. This is the kind of gap that the synchronic two-state model must presuppose. 1) is historically impossible.
2) This line contains a **partial coherence** between the two elements in that [-Last] implies that some element may appear between B and A (= B → A →). – The addition of sounds presupposes this type of gap: OE *eald, gieldan*; ME *newt, sound, passenger*.
3) This line also contains a partial coherence in that [-First] implies the possibility of an element before it: ← B ← A.
4) This line contains a partial coherence, stronger than 2) and 3), seeing that both [-Last] implies the possibility of a following element and [-First] implies the possibility of a preceding element: B → ← A.
5) This line contains no gap and is characterized by **dense cohesion**. While 4) does not imply a two-element synthesis, 5) does. Synchronically, it could represent the autonomous state, torn out of its structural context. Like 1) this possibility is impossible
6) This line contains a partial coherence: ← B ← A.
7) This line contains a partial coherence: B → A →.
8) This line contains a separation, ← B A →, but different from the absolute gap in 1).

The logic of the system is such that [+Last] &[+First] (in 1)) and [-First] &[-Last] (in 8)) are not synonymous. While [+Last] implies that at least one element precedes it and no element succeeds it (←]), [-First] only implies that at least one element precedes it (←). – These logical possibilities will be further restricted by the general theory in chapter 6.

In conjunction with the processes in 1.5.7 below, definition 17 introduces the simultaneity concept and reflects the affinity of the history of an element with the production of an ut-

---

114 At the theoretical level it is a commonplace that a theory is always underdetermined for meaning in respect of data; why does our tradition find it so difficult to accept the parallel – at the level of phenomena – that dualistic views of a phenomenon suffer from the same weakness: the system-behind, the calculus, does not provide all the meaning of the spoken chain. Or: we do not understand an utterance with reference to its underlying system.

terance; it is a superdefinition of the here-and-now intersection of the spoken chain and the developmental series. Differences, due to the sign-functional nature (meaning) of existence, will be elaborated below.

Def. 17 A cutting intersection is so produced that it becomes an intersection with a [+First]-property, and its predecessor becomes a [-Last] & [-First] intersection with no [+Last] and [+First] properties.

The dynamics of the cutting intersection of the developmental series will now be described. The [+First]-property implies that the here-and-now intersection acquires the dynamic predecessor properties [-First] and [-Last]. While the former points to a predecessor, the latter indicates that there is no absolute break with the successor. Thus the here-and-now intersection changes its predecessor, this process being the reason why change is not from unformed to formed, but from formed (matter), which is de-constructed – not destroyed – to formed (matter). The fact that the here-and-now intersection becomes a predecessor because of its existence conditions does not make it unformed. What becomes a predecessor involves coherence with its successor because the present implies itself. The traditional view to the effect that there is no present without the past from which it develops is a static view of existence that turns the past into a preestablished fact on a par with the present: the past > the present, an entity in its own right. Rather, implying itself, the present implies itself both as the present and as its past. Thus the implication is an aition: the present * the present[115].

The innate coherence between a successor and its successor (existence) will be called **participative coherence** (external endogeneity). With the present intersection being defined by [+First, -Last], its development into the present creates a predecessor defined by [-First, -Last], which participates in the new present:

1.5.7   A   >   B   >   C
        [-Last, +First] > [-Last, +First] > [-Last, +First]
                            [-Last, -First]

The processes in 1.5.7 illustrate the process in 1.4.6 above and suggest the inadequacy of the monolinearity of the static two-state model.

The psychological reality of what grown up people regard as *the* past – conceived as a resoluble class as one – asserts itself in the language-acquiring infant. To her all she remembers has just taken place. Yesterday holds all her memories. The notion of many yesterdays or yesterevents as temporally and spatially ordered series is yet to be developed[116]. She can only come to master her past as she goes on learning more and more language. So when we

---

115 This in contrast to the standard view of self-identity, the tautology: p ⊃ p. Here we find the epitome of immanence and autonomy, while the aition: p * p, is a contingent statement to the effect that it is absurd to claim that existence is possible if non-p; a historical phenomenon is never identical with itself – whatever that means. It is curious that this concept – self-identity – so crucial to the autonomy and immanence of the Glossematic semiotic, and for that matter to the standard synchronic view of change: something is stable through time(s) of change, is not thematized in synchronic (and Glossematic: Gregersen 1991; Rasmussen 1992) literature.

talk about the past in the English language as well as in Danish *fortiden* and German *die Vorzeit*, the available words are very appropriate: one word agreeing with its verb in the singular (not so in the plural noun in Latin *ges gestæ*).The simultaneity concept of the vertical synthesis of the developmental series is also empirically relevant to the child: what happened yesterday, a moment ago, plays an important role as she goes about creating her present through the gradual ordering (= resolution of the class as one) of the factual syntheses that the events of her yestermoments, yestermonths, yesteryears come to make up. The person who denies the psychological reality of his past, *i.e.* his historical sense, has no realistic grasp of his present.

The uniqueness of the present moment is indicated by the [+First] property of the here-and-now intersection; it is a synchronic commonplace that this uniqueness is attached to the individual person (speaker) and that it can only become a scientific object if it either contains a generalizable element (dualism) or is subsumeable under a concept. In other words the *proprial* nature of the successor (see Dray 1957:104) is to be removed; but then we shall not be able to place our phenomena (data) in a chronological series (Luick); so proprialization will be defined as follows:

Def. 18. Proprialization is the creation of a definite time- and place-position by the historical process that we exist by.

It is a corollary of this definition that the objects of physical nature do not contribute to our sense of history; they do not create definite time- and place-positions. Existing, man makes (physical) reality sign-functionally conformable to the preconditions for his own existence, his sense of history. In the objects of nature the present and history do not merge.

The predecessor is nonproprializable. Our historical theory aims at making the predecessor as determinate as possible. But to become a predecessor has certain costs: we shall never be able to say that this, X, is what occurred at T1-1, seeing that the place X occupied then is occupied by Y at T1 and X will have changed (cf. 1.4.6).

A successor is both an effectum and a necessary condition for its predecessor, the latter being a part of the former. In this sense the present creates the past and makes it continue.

---

116 Jakobson's referral (1962:649) to and acceptance of Henri Wallon's views are not empirically relevant. Nothing empirical substantiates the claim that the language-acquiring infant is capable of handling such abstract analytical mental operations as the logical law of noncontradiction; such mental functions grow as the secondary processes begin to control and check the primary processes (see Jacobsen 1976:27; 17ff.). On the other hand, I tend to agree with Wallon that *La pensée n'existe que par les structures qu'elle introduit dans les choses* (…), because he seems to follow Goldmann's view of the sign-functional nature of man-made reality. But there is no inevitable transition from this to the conclusion that all thinking, even the infant baby's rudimentary understanding of her surroundings, hinges on the prior existence – and application – of *binary* logical oppositions. Such oppositions, being linguistically motivated, develop with the child's growing grasp of her everyday language. Above I demonstrated how what is traditionally considered a simple binary contrast contains a complex relationship: the categorical statement (the if-then proposition): *All lions are mammals*; *John is a smith*, is only a valid factor in our logical systems if we imply the condition that Not all mammals are lions, etc. I therefore venture to rephrase the Wallon quote as follows: the everyday language exists only through the utterances we produce when we **ex**ist. – Reality is not only a class of *choses*.

There is a gap between the present and its predecesor, when the present develops into the present. However, there is no discontinuity between the relates of the traditional notation OE *ham* > ME *home*, and the deceptiveness of this notation is reinforced by its Parmenides-like interpretation: when ME *home* was 'there', OE *ham* could not be 'there', too, and vice versa. This static conception of the historical process is due to its monolinearity, which cannot accomodate the structural nature of the historical phenomenon: *when* home *is there,* ham *cannot be there, too, but is there notwithstanding*: OE *ham* is also present with ME *home* as a structural part of the latter under the aegis of change and continuation:

1.5.8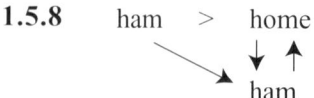

Two questions are outstanding: how to define a successor as against its predecessor; what is meant by 'a vertical synthesis'? As regards the first question, the synchronic literature takes it for granted: we know beforehand which state of the two-state model's two states is the predecessor, which one the successor. This is entirely inappropriate, because given two states, either may just as well have developed from the other: the below early modern English system of long monophthongs could just as well be the predecessor of the system of long sounds that we find about 1400:

1.5.9   Early Modern English   1400
      ɪi   ʊu   >   i: u: LateME *ride, house*
      i: u: e: o: *meed, mood*
      e: o: ɛ: ɔ: *meat, home*
      ɛ: a: *name*

And the Middle English substitution of *findes* for (Southumbrian) *findeð* could easily be reversed. Thus the Glossematic attempt at doing away with temporal directionality (Gregersen 1991:1.32) will merely result in an unrealistic understanding of the facts and events of reality. If Glossematics, as Gregersen puts it, 'liquidates' time as a theoretical parameter, the Glossematician must also say that Glossematics could have been constructed by Plato or Thales (cf. chapter 2.5.4). – The vertical synthesis will be determined by definition 19:

Def. 19 An intersection is so produced that the process creates two subsections of which one has a [+First]-element, the other a [+Last]-element.

The definition specifies item 5) in 1.5.6 above as a vertical synthesis, characterized by dense cohesion, and elaborates the distance concept of internal endogeneity, while at the same time the extension of the intersection is maintained. No third entity can appear between the two subsections constituting a merger of cause and effect. Definition 19 describes a part of a historical phenomenon, the synchronic (phonetic) aspect of a spoken here-and-now intersection. This means that something must make such an intersection[117] cohere with other intersections. That something is its historical Eigennatur and amenability to influencing and being influenced by other spoken-chain intersections.

1.5. Cause and the existence of the everyday language

Glossematics eliminated not only time, but also the linearity of the sign's *signifiant*; the latter is possible because of the palimpsest nature of the spoken chain, impossible because of the temporal directionality of the utterance; furthermore, if the sign's directionality is eliminated, regressive and progressive assimilation would merge. The remainder of the section will consist in an analysis of the directionality in exogenous motivation, implement the definition system above, demonstrate ideally how the concept of cause belongs to the level of data, and determine the function of change.

Historical succession (**ex**istence) is not monolinear: A>B, consisting of two fixed elements: [-Last] (A) & [+First] (B); it is multi-linear in so far as its elements are equiempirical:

1.5.9   A   >   B
        [+First] [-Last]   [+First] [-Last]

The absurdity of this static version of the **historical process** is obvious and will be resolved by its dynamic and multi-layered version in 1.5.10:

1.5.10   A   >   B   >   C
         [+First] [-Last]   [+First] [-Last]   [+First] [-Last]
                                [-First] [-Last]   [-First] [-Last]
                                A   B

where the successor so influences its predecessor that the latter 'points back' to its origin/predecessor through the acquired [-First]-property[118]. The processes in the **spoken chain** mirror those in 1.5.10:

1.5.11   A   +   B   +   C
         [+First] [-Last]   [+First] [-Last]   [+First] [-Last]
         [-First] [-Last]   [-First] [-Last]

where the replacement of A by B turns [+First] of A into [-First]. The 1.5.11 process is ideal, devoid of exogenous motivation. Both processes underline the simultaneity aspect: for A (in 1.5.11) to be influenced by its successor, the two must exist simultaneously. This will therefore be a minimal determination of form-directed speech in that – in deference to the diction of our tradition – selection into *parole* will not be successive selection of single (systemic) elements. *Parole*-simultaneity in the non-ideal utterance is an intersection's ability to be continued under the aegis of exogeneity.

The coherence that simultaneity in **internal exogenous** motivation implies will be called **cohesion** and in **external exogenous** motivation **complex cohesion**. The two assimilation

---

117 This vertical synthesis is only a part of the whole intersection; def. 19 subsumes the vertical synthesis in 1.5.8 under its simultaneity concept.
118 How the previous successor, A, influences its successor requires a sign functional interpretation.

processes, **regressive** and **progressive** will be described in 1.5.12 with its definition in Def. 20, and in 1.5.13, defined in Def. 21.

1.5.12i PreOE hælp   >   OE healp 'helped' (*the verb* helpan)
            ea + lp   =   [-First] [-Last] + [+First] [-Last]
            lp   =   [+First] [-Last]

The appearance of the 'broken' vowel presupposes the simultaneous existence of the 'breaking' consonants, which must be continued into their own overt appearance. Anticipating and influenced by its successor, the intersection *ea* does not acquire the [+First]-property, and as such it cannot alter the [+First]-property of its predecessor, *h-*. The *energeia* of the *parole*-processes that produces our word is described as follows:

1.5.12ii A [h]   +   B [ea]   +   C [lp]
              [+First] [-Last]   [-First] [-Last]   [+First] [-Last]
              [-First] [-Last]   ←   [+First] [-Last] ←→
                                      [lp]

Regressive assimilation will be defined as follows:

Def. 20   An intersection[119] is so produced as to create no predecessor subsection and a successor subsection (itself) with [+First]- and [-Last]-properties. – A.
            An intersection is so produced as to create a predecessor subsection with no [-First]-property and a successor subsection (itself) with no [+First]-property. – B.
            An intersection is so produced as to create a predecessor subsection with a [-First]-property, thereby making the predecessor of its predecessor a predecessor with the [-First]-property, *and* a successor subsection with [+First]- and [-Last] -properties. – C.

Progressive assimilation will be defined on the basis of this *energeia*:

1.5.13i   PreOE geldan > OE gieldan 'to yield'
              g + ie   =   [+First] [-Last]   +   [-First] [-Last]
              g   =   [+First] [-Last]

1.5.13ii A [g]   +   B [ie]   +   C [l]
              [-First] [-Last]   [-First] [-Last]   [+First] [-Last]
              [+First] [-Last]   ←→   [+First] [-Last] ←
                                       [g]

---

119 *I.e.* a here-and-now intersection.

1.5. Cause and the existence of the everyday language

Def. 21 An intersection is so produced as to create no predecessor subsection and a successor subsection (itself) with [+First]- and [-Last]-properties. – A.
An intersection is so produced as to create a predecessor subsection with no [-First]-property and a successor subsection (itself) with no [+First]-property. – B.
An intersection is so produced as to create a predecessor subsection with no [-First]-property, thereby making the predecessor of its predecessor a predecessor with the [-First]-property, *and* a successor subsection with [+First]- and [-Last]-properties. – C.

In both assimilations the effectum does not acquire a [+First]-property, which must be shared with the motivating intersection, which – in 1.5.13ii – preserves this property while its effectum appears, and which – in 1.5.12ii – acquires this property while its effectum appears. Logically, the two processes are similar and as such Hjelmslev's suspension of temporal directionality may seem appropriate, but the biuniformal concept of cause will make the difference clear: *what causes something to change is made to continue* (cf. ⟷ in 1.5.12ii); *what is made to continue makes something change* (cf. ⟷ in 1.5.13ii). Thus the Glossematic position can only be upheld if we disregard the dynamics and simultaneity of the vertical synthesis in exogenous motivation.

Progressive assimilation has a certain affinity with the historical process; but here, it will be demonstrated, the motivating processes are sign-functionally different.

As regards **external exogenous** motivation I shall make do with a **regressive** example, the PreOE *I*-mutation, as in **kuning-* > *kyning-* (> OE *cyning*).

1.5.14 A [k]   +   B [y]   +   C [n]   +   D [i]
         [+First] [-Last]   [-First] [-Last]   [+First] [-Last]   [+First] [-Last]
         [-First] [-Last]   [-First] [-Last]
                         ⟵── [+First] [-Last] ──⟶
                              [i]

Def. 22 An intersection is so produced as to create no predecessor subsection and a successor subsection (itself) with [+First]- and [-Last]-properties. – A.
An intersection is so produced as to create a predecessor subsection with no [-First]-property and a successor subsection (itself) with no [+First]-property. – B.
An intersection is so produced as to create no predecessor subsection with a [-First]-property, and a successor subsection (itself) with [+First]- and [-Last]-properties. – C.[120]
An intersection is so produced as to create a predecessor subsection C with a [-First]-property, a predecessor B with a [-First]-property, and a predecessor A with a [-First]-property *and* a successor (itself) with the properties [+First] [-Last] – D.

The coherences of the ideal spoken chain constitute a gradual monotonic movement in which a single intersection replaces its preceding successor; the dynamics will be predictable:

---

[120] This intersection [n] does not turn the first predecessor [k] into a [-First]-element; this was done by the anticipated [i]-intersection.

1.5.15

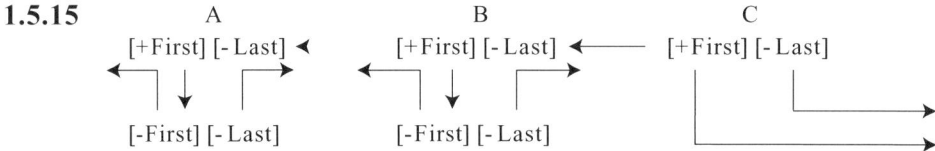

The properties of the intersections in this homogeneous process will be interpreted as follows: the here-and-now intersection points strongly towards its successor and will be connected with its predecessor through its structural influence on it (Metacondition 6); the predecessor points (*i.e.* is made to point) to both its successor and predecessor. In a word, even the ideal spoken chain is a phenomenon in which *tout se tient*, but also a phenomenon that is not an autonomous entity, only relatively autonomous, since the acquired [-First]-property of any predecessor points to something preceding and the innate [+First]-property of the successor points to a new presentic phenomenon: the spoken chain is part of an ongoing (communicative) process.

It is obvious from the above diagram that if **structural** simultaneity was not an essential factor in the constitution of the utterance, the final intersection (the here-and-now intersection) would not cohere with the rest of the utterance. On the other hand, it is equally obvious that the basic structural principle, that of context-based meaning assignment (cf. chapter 3), cannot establish sufficient differences between the intersections unless each of them has an Eigennatur. While the first intersection (A) is defined differently from its successors – except the here-and-now intersection – it assumes the same structural properties ([+First], [-First], [-Last],) as the other intersections and I shall say that like the other intersections it **completes its meaning** in relation to these, the first intersection will therefore not infringe Metaconditions 2 and 4. But these processes merely establish one difference-relation: that between the here-and-now intersection and its predecessors, a degree of differentiality that is hardly sufficient in meaning-conveying functions (Bloomfield's non-significant differences): the monotonicity of the ideal utterance defies the existential, including communicative, functions of the everyday language. This entails the universality of exogeneity: no empirically relevant conception of the everyday language is without it, which in turn means that **exogeneity is not merely a function of human imperfection**.

Four more consequences: 1) function-directed speech belongs to the production of the ideal utterance (1.5.15); 2) only form-directed speech is empirically relevant, and must constitute a continuum of degrees; 3) the monad-like quality of systemic elements is incompatible with form-directed speech; 4) the notion of constant change has been given a formal determination.

The here-and-now intersection is left partially dangling in the utterance as its coherence with its predecessor is weak; it will be called **gap cohesion**: in the ideal utterance its predecessor is *made to* cohere with it, but this is only from a synchronic stance, since each *parole* intersection is also a part of a developmental series.

Gap cohesion explains the ever-present possibility of an element appearing/disappearing before or after another element. The participative coherence of the developmental series does not allow such 'intrusions' (or losses); this distinction only appears if we accept exogenous motivation. Otherwise the dynamics of the ideal utterance correspond exactly to those of the developmental series: this makes 'A develops into B' and 'C replaces B' synonymous, which appears from Def. 23:

1.5. Cause and the existence of the everyday language

Def. 23 A here-and-now intersection is so produced that it creates two subsections such that the predecessor acquires the properties [-First] and [-Last] and the here-and-now intersection, itself, the [+First] and [-Last] properties in a vertical synthesis.

This definition clashes with definitions 14 and 15 above, a clash that will be redressed when exogeneity is allowed to play its universal role in the constant change (existence) of the everyday language.

What happens when an intersection is exogenously motivated? That such an intersection acquires [-First]- and [-Last]-properties indicates that it becomes better integrated into the spoken chain – in deference to the tradition, as a compensation for its inherent novelty. The net-result is greater internal coherence in that these [-First]- and [-Last]-properties make the effectum (here: B by C) establish coherence with both its predecessor and successor. It also appears from the diagram that *parole* is intersected by itself through the anticipated existence of C[121].

1.5.16

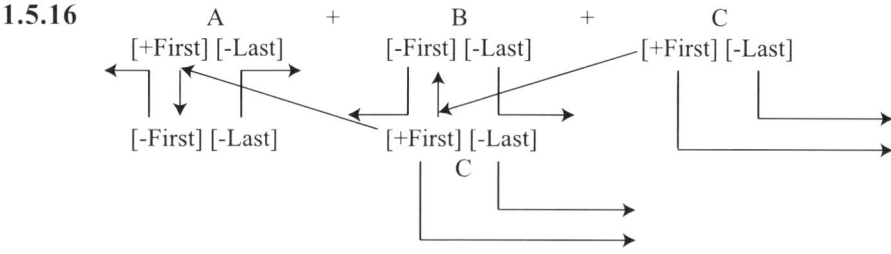

How do we know which intersection is the effectum, which one the efficiens in 'conditioned' change[122]? Our theory singles out B because of its properties[123]; but change is more than exogenous change. The answer will be given in chapters 5 and 6, where the contribution of internal and external endogeneity will be demonstrated. Secondly the over-all integration of the here-and-now intersection (= C) of the spoken chain as well as the other intersections into the everyday language will be demonstrated in 1.6. This leads to the first determination of change: *change makes for coherence; it ensures that the elements of the utterance are not understood analytically as autonomous entities. Consistently conscious and deliberate selection of the so-called smallest, systemic entities of the everyday language does not produce intelligible language; such speech production (function-directed speech) is astructural; therefore the model in 1.5.15 (with no exogenous change) is an idealized view of the spoken chain in which the single element's self-identity threatens to defeat the necessary differences*

---

121 In accordance with the palimpsest nature of the spoken chain the horizontality of the diagrams should be replaced by columns in that the horizontality image represents a view of time as being extended in space.

122 This question is analogous to the synchronic systems-problem: which one of two allophones is the phoneme, – to be put between slants? From (Pre)OE we can mention fricatives /f/ [v], [f]; in modern English, what is the phoneme controlling dark and clear 'l', and in *Newsweek* (April 6, 1998:37) we find this headline: *Our Buddy, The Prez*. How do we know synchronically that *Prez* is a variant of /president/?

123 Note that the synchronic point of view has no means of telling us whether [hælp]/[geldan] or [healp]/[gieldan] in PreOE is the changed form.

of the spoken chain[124]. Speech tempo sees to the spoken chain's intelligibility through its emphasis on the differences in identity (an element's self-difference), and therefore to necessary (exogenous) change and coherence. Change is not the contradictory of nonchange; change and nonchange are two different modes of historical existence, therefore subsumeable under the same theory; it follows (cf. 1.5.16) that change as well as nonchange (to be demonstrated below) *is a structural, relational phenomenon, with an aetiology of its own.*

The discontinuities that a given theory will inevitably produce are part of our scientific heuristics: each of them must be bridged. No historico-structural phenomenon is a relatively autonomous entity from the scientific (logical) point of view. The affinity, expounded by the search for missing links and gap-fillers, of progress in empirical research with other types of historical existence is clear: both processes are continuations developing into the present and each level contributes to the creation of the present.

An immediate corollary of this partial determination of change is that the traditional synchronic view, change being necessarily from something stable and fixed, is untenable (see Gregersen 1991:1.31)[125]: existence is the present's **constant** development into the present, and it consists of equiempirical elements.

1.6. Vertical syntheses. Four ideologies. Two dualist theories compared. 'To be history' and 'to have history'. *Langue* and *parole* are equiempirical in the L-continuum. The everyday language is always *en ergeiai*. More criticism of the phonemic principle: assimilation, loss and addition. Katalysis

All dualisms imply a mental space in which the vertical synthesis exists, seeing that the two relates of any dualism exist simultaneously – in place P at time T. The concept has been implied in the advanced view of **ex**istence and change (1.4.6), endogeneity and exogeneity as well as in the constructive synthesizing of the Critico-Philological Method (1.4.7). Dualist theories, however, do not enter into non-accidental horizontal syntheses and cannot constitute a *Kausalnexus*, a fact that Coseriu's diction above and Gregersen's 'empirical problem' testify to[126]. The definition system of the vertical synthesis will now be set up.

Def. 24 A vertical synthesis is a synthesis that consists of non-equiempirical phenomena.

The concepts 'non-equiempirical' and 'equiempirical involve no contradiction, but different empirical elements (cf. Def. 28); this is due to the fact that for two empirically relevant entities to exist simultaneously in a specific place both cannot be of the same empirical nature.

---

124 The reason why an infant baby learns a language is precisely such differences; consequently the ideal language of 1.5.15 cannot be learnt. The generativist's postulate about language-learning has now been rejected theoretically.
125 A widespread version of this is seen in various synchronic handbooks on phonetics or phonemics and grammar (derivation and compounding), see Davidsen-Nielsen 1994:96. – I shall be dealing with this below.
126 Gregersen (1991:2.233) does not generalize it to include its historical aspect, the step from variant to change.

1.6. Vertical syntheses 167

Def. 25 A non-accidental vertical synthesis is a vertical synthesis that is not purely man-made and not metagenously motivated.

Def. 25 implies the existence of data; data are not potential (Def. 29).

Def. 26 An accidental vertical synthesis is a vertical synthesis that is purely man-made and linguistically motivated.

Def. 26 defines the ideal type of vertical synthesis, namely the deductive view that an objective theory is to be constructed independently of data. In practice the ideal never transcends the level of logic and it makes data potential and contingent. It implies a view of man as omniscient, able to find an Archimedean chair and desk or to come at some distance of himself in order to hear himself speaking his everyday language.

Def. 27 An empirical vertical synthesis is a vertical synthesis that is not purely man-made, is katagenously motivated, not metagenously motivated, and is characterized by participative coherence.

A Def. 27 synthesis is a part of a horizontal synthesis and has become a part of a totality. The vertical synthesis of the developmental series is empirical, consisting of the here-and-now intersection (internal endogeneity) and the simultaneously existing phenomenon that motivates/motivated it through external endogeneity; the former entity of this structural phenomenon (totality) is an *existent*, the latter acquires an empirical status different from that of what becomes its successor; it has changed. For it to change something else must have occurred, namely the present's development into the present[127].

Def. 28 Paragenous motivation is simultaneous motivation in a vertical synthesis; it stems from a phenomenon whose Eigennatur is not of the same kind as the Eigennatur of the motivated phenomenon (effectum).

Def. 28 expands Def. 24 and includes the transcendence problem, where the motivated element may be the object of description in a denotation, or a theory controlled by data. Its relevance will appear in Def. 32.

Def. 29 Metagenous motivation is non-potential paragenous motivation so that the vertical synthesis that the motivating factor and the effectum constitute is characterized by EITHER distance *qua* [-Separation], [+Attachment], [-Attachment] OR distance *qua* [-Attachment], [+Separation], [-Separation].

Def. 30 Linguistic motivation is potential paragenous motivation so that the (potential) vertical synthesis that the motivating factor and the (potential) effectum constitute is characterized by distance *qua* [+Separation], [+Attachment], [-Attachment].

---

[127] The change process occurs through katagenous motivation, to be defined below in 2.1.

The syntheses of Def. 29 may be described by a relation of inherence if we are adherents of an Aristotle-like conception of the *langue-parole* dualism, or by a relation of participation if we are of a Plato-like persuasion. As such Def. 29 is a metaphysical statement of ontological dualisms, and in the final analysis[128] these two vertical syntheses will share the form (idea) *qua* motivating factor that we find in the vertical synthesis of Def. 30, this being linguistic or in more prosaic terms, a word from an everyday language. On the other hand, Def. 30 refers to the nominalist view of the transcendence problem and includes the discussion of existence postulates: 'the process of putting a phenomenon, *in so far as it exists,* under a category or of classifying it as a member of a class – the category or class being something the phenomenon cannot be without' creates a vertical synthesis that is not immediately empirically relevant, because the theory has only been validated internally by *e.g.* the Empirical Principle of Glossematics. Denotation *qua* subsumption complies with the epistemological requirement that we *cognize at a distance of that which is cognized,* and the correlative requirement that *what is a precondition for something cannot be a part of that something.* Note how Def.s 29 and 30 observe the affinity between modern nominalism and Platonic idealism: the gap between data and theory is maximal.

These definitions enable us to amplify the distinctions between the various ideologies referred to here and in chapter 1.1., namely (1) (Aristotle-like) immanent idealism, (2) (Plato-like) transcendent idealism, (3) linguistic immanence (Glossematic nominalism), to which I shall add (4) historiographical investigation:

1.6.1

(1)	(2)	(3)	(4)
[+Attachment]	[-Attachment]	[-Attachment]	[+Attachment]
[-Separation]	[+Separation]	[+Separation]	[-Separation]
[-Attachment]	[-Separation]	[+Attachment]	[+Separation]

The reason for the mutual affinities is that in the final analysis each 'ideology' is dependent on the everyday language[129], and of these (3) drew the full consequences, while (4) has not realized the extent to which the everyday language plays a role in our historical existence, and none of the ideologies has grasped the existential nature of the everyday language; or: to what degree the everyday language makes an absolute distinction between linguistic and nonlinguistic reality impossible.

Ad (1)   The combination [+Attachment] [-Separation] suggests a dualism of inherence and immanence: linguistic and nonlinguistic reality are mutually dependent. The feature [-Attachment] indicates the difference between the dualism's relates.

---

128 As ontological statements Def. 29 is regarded as dated metaphysics.
129 Neither Plato's world of ideas nor Aristotle's form-concept can be grasped, let alone created without the everyday language, and as such the everyday language assumes an ontological status, in modern diction, enunciates existence postulates. Hjelmslev's Glossematics, on the other hand, is explicit about this: all objects – semiotics and non-semiotics – are components of semiotics (1963:127), and all objects will therefore be *illuminated* and *viewed* from the *key position of linguistic structure.* – In my view, this is only true of the everyday language, *i.e.* Hjelmslev's *linguistic theory* and *semiotic structure* is/are also illuminated or viewed from the point of view of the everyday language.

1.6. Vertical syntheses

Ad (4) The combination [+Attachment] [-Separation] is interpreted as in (1), and the [+Separation] suggests a sharper distinction between linguistic and nonlinguistic reality.

Ad (2) The combination [-Attachment] [+Separation] suggests a complete gap between linguistic and nonlinguistic reality (transcendence), while [-Separation] recognizes that there is a connection between the two relates, in our case, one of participation.

Ad (3) The combination [-Attachment] [+Separation] also suggests a complete retreat from empirical reality and as such the theory's affinity with (2), linguistics being a 'Platonic idea'. On the other hand the feature [+Attachment] indicates that transcendence into nonlingustic reality is not other-worldly; the form lies immanently in this world (Hjelmslev 1943:127)

The synchronic *langue-parole* dualism saddles synchronic linguistics with a metaphysical reckoning in respect of Def. 29; since this definition represents dated metaphysics, we are left with Def. 30, in accordance with which the *langue-parole* dualism is a linguistic, hence theoretical representation of the spoken (and written) chain of a given everyday language – in so far as the latter exists.

The dualist conception must be taken seriously for the simple reason that it presents us with the important historical notions of simultaneity, continuity and change; on the other hand since synchronic linguistics has not provided the historian with a consistent and exhaustive as well as simple description of one language system, let alone the systems of two states of the same language, it is impossible for the historian to evaluate the purported synchronic basis of historical linguistics and language development. However, two theories are available whose conceptual bases are sufficiently explicit so that they can become subject to critical scrutiny. Paradoxically speaking, they are also so different that a comparison will reveal more differences and hardly any similarities.

Glossematics places the dualism at the level of logic, thereby substituting a theoretical dualism for an ontological one. The system-process dualism represents two different points of view of *the same entity*, and the articulation of the system-process concepts ends with *a language* and *a text* (Hjelmslev 1975:def.s 38 and 39). These two classes are components of a denotative semiotic, a semiotic none of whose planes is a semiotic (p. 9; def.s 24, 26). *The same entity* above refers to the class of variables constituting the manifestant of a manifestation (def.s 28, 31, 32 and Note 17). Apart from the definitorial fixing of the object of linguistic analysis, we know nothing about the empirical relevance of these Glossematic entities, since the substance is the variable in a manifestation, where a language is the constant: a language does not presuppose the presence of a manifestant. A language (*langue*) as a specific type of correlational hierarchy, subsumeable under the systems-concept is of no avail, either, it being merely a potentiality of the theory. And, to repeat, nominalism and historiography are an odd couple: I fail to see how history can be regarded as the potential manifestation of a finite class of elements (functives) definitorially characterized by properties deducible from two sets of indefinables: *to be a necessary condition for* and *not to be a necessary condition for*, as well as *either-or* (the undefined fundamentum of the system) and *both-and* (the undefined fundamentum of the process). Secondly, the historical definitions in *A Résumé* (def.s 203-213) are anything but adequate as descriptors of the historical process of one language or dialect's development into its successor. These few definitions boil down to the following statement of language change, *i.e.* metachrony: Metachrony describes continuation between connotative semiotics, *e.g.* (CrE(ErC)), and denotative semiotics (ErC) and loan-contact. *Metachrony* is *a continuation* or *semiotic change* between *two non-scientific semiotics*, of which one is *a Fore-semiotic* and the other one *its After-semiotic;* such semiotic change *is a contact*, which is *a relation* (a both-and function) *between semiotics* (and *derivates* of different semiotics). And with Gregersen's correct analysis of Hjelmslev's view of change[130], the poverty of the theory speaks for itself. However, the crowning absurdity of Hjelmslev's view of language change is that his theory of language change is not Glossematic! In his 1934-lectures at Aarhus University (= Hjelmslev 1972) we are met with a host of undefined technical terms such as *norm, artikulationsforskydning, skifter udtale, systemskifte*, sprogets *bevægelser* (p. 45); *strukturen ... står (ikke) i optimum* (p. 73[131]), *et relativt optimum* (p. 75), *det absolutte optimum* (p. 106), *midlertidig nødforanstaltning* (p. 87), *det sygelige stadium* (p. 89), *disposition til reduktioner* (p. 90), *simplifikation kan give bagslag, en katastrofe kan indtræde* (p. 113), *ligevægt* (p. 122). There is no reason to rub more salt into this Glossematic wound: the absurdity consists in the fact that Hjelmslev does not observe his own scientific requirements and strictures (cf. Hjelmslev 1963) when it comes to language history: a theory of language is *metaphysical* if its terms are not defined. Nowhere does Hjelmslev define such terms as the above ones. Hjelmslev's view of language change is metaphysical; in his own words Hjelmslev subjected historical events to *poetic, subjective, aesthetice, aprioristic treatment*, this being the contradictory alternative to *scientific treatment* (1963:9-10).

Following Samuels (1972:139-142; 28-29), the *langue-parole* dualism represents the two empirically relevant entities that constitute the everyday language. As such his position implies an ontological dualism[132]. The equiempirical nature of the respective denotata of *the spoken chain* and *the system* is suggested by the relation linking them, a relation of *selection*[133]. However Samuels does not escape the fallacy of a definition-29 metaphysics, because of his idealization of the selection process: who or what selects what? And the system and its entities are not equiempirical with the spoken chain in that what appears in the spoken chain is not what has been selected.

---

130 It is curious that Gregersen – in his otherwise pertinent criticism of the Glossematic view of history – does not attack Hjelmslev's theory of language development from the point of view of the relevant Glossematic definitions; the same criticism must be levelled at Rasmussen 1992, who, however, merely echoes the words *du maître*. Note that none of the articles in Thrane 1980 deal critically with Hjelmslev's view of history from Glossematics's own point of view; and only two articles, Jørgen Rischel's (p. 65) and Søren Egerod's (p. 115) include *A Résumé* in their respective bibliographies. – See the following note.
131 Here *optimum* means a one-for-one correspondence between expression and content (= ErC; cf. p. 121); elsewhere it means the complete disappearance of a grammatical category (pp. 73; 74); and what do the following statements mean: "En form dør i reglen ved at dens betydning trænges op i det abstrakte plan" (p. 73), "sprogsystemet kunde give sig til at fuldbyrde de forandringer, som det maaske længe havde haft dispositioner til" (p. 74)?
132 This phenomenal dualism will also be reflected at the level of theory; thus chapters 2 and 4 (in his 1972-book) and chapters 3 and 5 describe *parole* phenomena and *langue* phenomena, respectively.
133 I am not considering the historical function of selection, in which a variant becomes a systemic element (p. 140).

1.6. Vertical syntheses

What Samuels and Hjelmslev have in common is a katadynamic dualism in which something manifests a preexisting entity: Glossematically a linguistic usage (a substance) manifests a linguistic schema (a form) in a manifestation in which the form is determined by the substance, where Samuels makes do with an adaptation of the Saussurean dualism: the system, being a group of opposition-based *norms*, which are realized as variants (of themselves) in the spoken chain. The other Glossematic dualism, the one between system and process, which apparently makes up a linguistic schema, is less clear, because the definition system establishes a connection between a correlational semiotic and relational semiotics; all Hjelmslev (1963:39; cf. 1975:74, N 35) says is that the process determines (selects) the system. However, this formal statement is in contrast to Hjelmslev's otherwise vague diction: it is only a *thesis* that there is a *correspondence*-connection between a system and a process (1963:9-11), and this correspondence is elsewhere described as a system *underlying* a process (p. 10), a system *belonging to* a process (p. 63), or as a system *lying* or *being present behind* a process (p. 39). Finally, *the existence of a system does not presuppose the existence of a process* (p. 39), from which postulate it does not follow – unless circularly – that the process presupposes the system[134]. This relation of presupposition does not follow from the Glossematic articulation of the class of objects, but is based on an analogy from the transcendence problem, *i.e.* Hjelmslev's axiomatic view of what constitutes scientific analysis. The issue must be divided into an empirical and a logical part: empirically, there is no data to suggest that there is no process – being *the more immediately accessible for observation* [than the system](1963:39; cf. 1975:74, N 35) – without a system; empirically, there is no data to suggest that a system has been 'observed' to exist without a process. Hence, it is impossible to infer a unilateral dependency relation between the system and the process. Logically, knowledge of the everyday language is a Platonic idea, hence the form is logically prior to any other datum: there is nothing either before or after the theory; the theory is even before its (future) object (predictability). And in this sense, it is correct to claim that *[a] process is unimaginable – because it would be in an absolute and irrevocable sense inexplicable – without a system* (1963:39). What Hjelmslev is saying here is trivial: knowledge of X presupposes a cognitive system, or: no object is a scientific object without a theory to establish it – the former – as a scientific object[135]. It becomes non-trivial if Hjelmslev is trying to obliterate the distinction between empirical object and scientific object, thereby implying an ontological dualism of the *nihil-est-sine-ratione* type: the empirical object has its existence from its scientific existence (in a world of ideas): *the process comes into existence by virtue of a system's being present behind it* (p. 39). There is a systematic ambiguity here, which can be traced back to the basics of Hjelmslev's epistemological stance: in *Prolegomena* we read as follows: the science of language searches for *a* (linguistic) *constancy* to be *projected on* a (linguistic) datum (p. 8); this constancy is a *system* through which *the phenomena* at hand are *interpreted*[136]. Next we have our thesis of there being *a system* for every *process* (pp. 9; 9/10), and finally on p. 10 the Glossematic theory has it that *a process has an underlying*

---

134 See Sørensen (1958:§§14.3, 25.3) for a pertinent criticism of the concept of presupposition.
135 Sign-functionally, a scientific object, having no meaning in itself, is the expression (**E**) of a sign function in which the theory is the content (C): **ErC(ErC)** *i.e.* the theory conveys meaning to the datum (**E**).
136 The function of such a system is to be *aggregating, integrating, generalizing, exact* and *systematizing*.

system – a fluctuation an underlying constancy, so that we end up with the following synonymous dualisms: a constancy and fluctuating/changing reality, a system and phenomena, a system and a process, and a metaphysics we have encountered before: the linguistic theory foresees all *events qua combinations of [the] elements [of the calculus]* and establishes *the conditions for their realization* (p. 9). On pp. 10 and 11 the words *theory* and *object* are used, while a theory must *agree with so-called (actual or presumed) empirical data*. How do *theory* and *object* correspond to the above-mentioned pairs?

The systematic ambiguity may now be expounded as follows: the word *process* belongs either to the logical level, and is therefore a technical term, or to the level of data, and is therefore no technical term, referring to the empirical data, realized phenomena in fluctuating, changing reality.

A solution may be seen in Hjelmslev (1941:103-104): here experience is defined by knowledge obtained *by the application of* the theory (method): there is no experience of any object before or after the application of the theory or *without the use of scientific methodology*. Any other definition of experience or empiricism creates *metaphysical* knowledge, and to be metaphysical is to work with undefined technical terms; in short, *theory* is synonymous with *procedure* and *scientific methodology*, it being the *constancy* in our experience; *a constancy* is implemented by the dualism, *system* and *process*, so that the theory contains (specifies) this dualism plus the *realization conditions* for this finite calculus of *elements* and their *combinations*. However, this produces another ambiguity as regards the notion of realization conditions which may be represented in a diagram (cf. 1.2.)[137]:

1.6.2	the level of theory constancy	projection or application relation	level of data fluctuation; substance
(1)	realization conditions system ⟶ process	is projected on to is applied to	[a physical, physiological, (psycho)logical entity]
(2)	system & process ⟵ realization conditions	realization conditions ⟶	[data]
(3)	system, *langue* ⟵ realization conditions	realization conditions ⟶	process, *parole*, spoken chain
(4)	system	denotes	data

---

137 To put it differently: what does it mean to say that *a process has an underlying system – a fluctuation an underlying constancy*? Unfortunately, the following pages in *Prolegomena* are of no help either: on p. 10 we read that *behind the textual process* there are a number of systems, and on pp. 12-13 *a system* is to be order[ed] to *the as yet unanalyzed text in its undivided and absolute integrity*; does *text* belong to the level of data or theory? Note that 1.6.2(3) is a monoplane conception that is often found in synchronic grammarbooks.

1.6. Vertical syntheses

If 1.6.2(1) is the Glossematic version of the transcendence problem, then Glossematics has the insurmountable task of how to erect a methodological construction which is to provide the meaning ((ErC)C) of the actual or presumed data (E) of the phenomenal level – with no recourse to this level. As we shall see Hjemslev could not do this and I shall show how Glossematics rests on an inconsistency, because a theory constructed only on the premises that *its object* (1963:10) requires is bound to place this crucial word, *its object*, in one of the boxes above and then pandemonium – epistemological, cognitive, ontological – will break loose (cf. also *text* in note 137).

Before proceeding with my comparison I shall let the cat out of my theoretical bag: there is only one type of experience that is both before and after its object, and that is the experience that the everyday language embodies: there is no experience either before or after the 'application' of the everyday language: the everyday language, experience and existence are co-incident.

We may now ask about the historical relevance of Hjelmslev's synchronic views. The system, we are told, *governs and determines* the process *in its possible development* (1963:39). In so far as this is a historical statement, the following analogy will be a corollary: the linguistic schema governs and determines the development of its linguistic usage, and the general implication for history of Hjelmslev's position (1963:127) is this *requirement*: for a given historical reality such as the everyday language, history proper *qua a fluctuation* to be treated scientifically and not subjectively or aesthetically, a constancy *qua* a historical schema must be established and *projected on the "reality" outside* the historical schema (Hjelmslev: *language* (1963: 79)), *i.e.* the historical usage (pp. 8; 81). The consequence of the analogy results in this absurdity: just as the process is virtual and the system may exist without a process (pp. 39-40), so the historian has to envisage a historical system with unrealized historical phenomena or events, since the existence of these can be foreseen as *a possibility*.

Clearly, Samuels's systems concept is a reified mental entity that does not escape an element of idealization (undefined properties have been underlined): the system is *the total of accepted and intelligible norms, established by opposition, used by a specific group of speakers for a specific period* of time. Historically, a norm is *a variant misunderstood as an acceptable form* (1972:10), and as such Samuels follows the negative view of language development (see *The Prologue*); secondly, while his theory presupposes a theory of how a norm is continued unchanged for some time, Samuels must be credited for having worked out a relatively detailed relationship between the communicative function of the *stable* systemic units (p. 28) and the spoken substance's dominant influence on the system, because the theory considers the sign-functional origin of change. The weakness is the lack of a theory of continuation and the idealization of the systemic unit, which forces Samuels to work with the circularly defined abruptness concept of the equally undefined synchronic state concept.

Hjelmslev's theory is well-defined and explicit with no existence postulates; his dualism is a thesis. Samuels's theory is less explicit and his dualism is an (ontological) axiom. Glossematics provides us with a dualist description of one substance entity (in so far as it exists), Samuels sets up one description of a dual entity. Hjelmslev's potential data unifies the two components of the analysis, Samuels's theory unifies data (1.4.10), bringing order to chaos. Hjelmslev takes a too simplistic view of the transcendence problem, which Samuels does not consider.

An example of the undefined state concept

In ME long open *o* and long close *o*, /ɔ:/ (< OE /a:/) and /o:/, had been continued for a couple of centuries; as two norms they expounded intelligible and accepted oppositions as in ME *boat* 'boat' and *boot* 'boot'. They belong to one system as long as they remain stable and unchanged. This period is contingent on the length of the period during which the variant [u:] in *boot* has *not* been *misunderstood as acceptable ... by the hearer*. When the variant [u:] is accepted into *langue,* the system has changed: S1 > S2 /ɔ:/ :: /u:/. Owing to the historical nature of the system, the process cannot be explained by what it creates, the system, seeing that S2 is the successor or the effectum of S1 and the process, respectively. On the other hand, if we do not regard this type of isolative change as systemic, a theory of continuation is needed to account for the continuation of S1 during the processes when a variant arises, is spread and accepted into the same system from which another element has been ousted.

1.6.3

S1					
T1		T2		T3	
system	parole	system	parole	system	parole
/o:/	[o:]	/o:/	[o:]	/u:/	[u:]
			[u:]		
/ɔ:/	[ɔ:]	/ɔ:/	[ɔ:]	/ɔ:/	[ɔ:]

The selection of [u:] into *langue* may be an understandable process, but it makes no sense to say that [ɔ:] continues to be selected into langue – for what reasons or under what conditions? Secondly, it makes no sense, either, to say that at T2 a selection process creates the new variant [u:], *i.e.* a speaker selects something that does not exist!

Change is abrupt: either /ɔ:/ :: /o:/ (S1) or /ɔ:/ :: /u:/ (S2/S1) exists, and in between we have a diachronic link (Samuels, p. 11) implemented by the gradual spread of the variant [u:]. What about the notion of the synchronic cut? Let's assume that the synchronic historian makes his cut at T2, then it is impossible to know whether the process into T3 is one of continuation or of change. Thus we are back to the basics of the synchronic method: how to set up a systemic element /?/ on the basis of variants, variants controlled by this 'most wanted' element /?/. In other words, the synchronic method is a search for a historical fact: at T2 the two variants [u:] and [o:] are the synchronic data coupled with the assumption that a systemic element /?/ exists at T2 which also existed at T1, the point in time when /?/ existed and controlled its own manifestation: the systemic element is thus a predecessor phenomenon in relation to its manifestations (T1 > T2) and a successor phenomenon (T3) in relation to its manifestation (T2 > T3).

Furthermore, the whole picture will become extremely muddled when more changes are involved: it is a commonplace that English changed during the 15th c from S1 /ɔ:/ :: /o:/ to S2 /ɔ:/ :: /u:/ and that S2 was continued until /ɔ:/ began to be raised; but was S2 changed when some [u:] (*blood, flood*) were so shortened as to participate in the development of ME /u/ into /ə/, *i.e* when the contrast /u:/ (< ME /o:/ *boot*) :: /u/ (< ME /u/ *but*) changed into /u:/ :: /ə/? Or: how many states will the synchronic linguist have to set up in order to cover what Luick called *die grosse Vokalverschiebung* and Jespersen *one great linguistic movement*?

1.6. Vertical syntheses

The project ends in absurdity, and it is not surprising that no linguist has ventured to write the history of any everyday language on the basis of the existence of states and systems, but adopts make-shift solutions in order to salvage the systems concept.

Following Samuels (p. 28), *the system is the essential complement of the substance of communication*: how can two different systems, say that of the *younger generation* (S2) and that of the *older generation* (S1) (p. 112), constitute the common ground of a communicative situation? Samuels adds these epicycles to his system: S1-speakers will understand the forms of S2-speakers: but then communication does not presuppose a system (S1) through which the spoken chain is interpreted, let alone the existence of a common supraindividual system; or: code- or style-switching by S-2 speakers ensure communicative efficiency: but then communication does not presuppose a supraindividual system, S2-speakers will communicate by means of forms (variants) that belong to a system that is not theirs; or: both parties may use synonymy in terms of *redundant forms*[138].

To sum up: with Hjelmslev's non-cogent argument above, the conclusion that the system is not necessary whereas the spoken substance is, is not hard to come by, and I think that Samuels's definitions (p. 139) of the spoken chain and the system suggest this kind of independence on the part of the former *vis-à-vis* the system, and his basic assumption that for a language to change, it must be used, also points in the direction of an autonomous spoken chain. And I shall generalize these facts into this Hermann Paul-like position: for an everyday language to exist, it must change, from which the notion of constant change follows. The system – in the singular – is bound to be idealized, if not hypostatized, from the moment it appears in *its* minimal manifestation, as something used by two interlocutors or a group of *habitual interlocutors* (pp. 89; 140). If the system remains a theoretical construct, we may ask: do we need the systems-concept – do we need the notion of 'the step from *parole* into *langue*' – in order to understand the existence of the everyday language? Following Samuels (p. 177) *variation ... consists of departures from existing norms,* which translated into the language of time means:

> the historical process that creates a (synchronic) variant at T2 is deviation [u:] from a norm /o:/ existing at T1 and T2; this variant may become a norm /u:/. The norm is the historical product of a historical process: a historical product deviates from a historical product.

There is no non-historical reason to uphold the formulation: 'the variant deviates from the norm', at the expense of: 'the norm (/o:/) comes to deviate from the variant ([u:])'. Both formulations boil down to the simple historical fact that somehow one sound 'develops into' another sound, and this process, I shall argue, is an empirical process that cannot be observed, only comprehended. To this end we need a theory of historical existence. After these

---

138 Samuels creates an inconsistency: on p. 139 we read that *two interlocutors* will share a system: how do we know that? Secondly, this position cannot be generalized, as we have just seen. The only conclusion is, given two interlocutors the synchronic linguist will not know whether they share the same system or they have two different systems, and since they may and may not share a system, *the* system is not a necessary condition for the substance of communication.

critical remarks on the nature of the synchronic system (systems concept), the remainder of the chapter will be constructive.

The equiempirical status of both *langue* and *parole* may be upheld (Samuels and Hjelmslev). What is immediately empirical is the spoken chain (Hjelmslev), and therefore the empirical nature of *langue* and *parole* will be mediate; this means that *langue*- and *parole*-elements may constitute vertical syntheses with the spoken chain. The distinction between *langue* and *parole* is not an absolute one, and both types of elements constitute a continuum, an L-continuum, from which selection into the spoken chain occurs. This selection process involves no manifestation, but will be shown to be a historical process in which longer or shorter, in Samuels's diction, stretches of speech will be produced. This selection process is the continued existence of the everyday language, *i.e.* the present moment's creation of the present, and is not restricted to occur from a finite group of preexisting entities constituting a *homogeneous* system. Loan words and on the whole intersystemic contact are as fundamental processes as intrasystemic ones.

Selection is creative, continuative and existential, and is determined by greater or less speaker awareness of the everyday language, – **creative** because an utterance is the product of elements that have no preexistence, – **continuative** because an utterance is produced out of preexisting elements, – **existential** because an utterance contributes to our historical existence. The speaker's awareness of the everyday language is expounded by his knowledge of it *qua* a historical theory, this being ideally identical with his historical sense: as such the everyday language is a cognitive system. The speaker's ability to create the present moment requires knowledge of how to order and set up the phenomena of the present in horizontal (temporal) and vertical (simultaneous) syntheses.

The elements[139] of the L-continuum can enter into non-accidental syntheses and developmental series. The continuum is ordered by endogenous and exogenous motivation; external endogeneity ensures the structural basis of internally endogenously motivated entities. The elements of the L-continuum become predecessors (katagenously motivated by the utterance) in relation to the spoken chain, which will also become an existent. Exogeneity will accompany the other factors according to speaker awareness[140]. The longer the selected stretch is, the more likely it is that external exogeneity will occur. Assimilation, regressive and progressive, involves the selection of stretches, not single elements. But it is difficult to determine which is which, and cases may not be as immediately clear-cut as we see in such OE examples as *gieldan, healp, cyning;* modern English *mi[ʋ]lk* as well as clear and dark *l*. The so-called exceptions, *e.g.* OE *cild, helpan* could suggest progressive assimilation, and the LateOE and ME weakening of posttonic vowels/syllables lends itself to three explanations: OE *stanas* > ME *stan[ə]s* is progressive assimilation if the weak stress of the inflectional syllable is the partial product of the strong stem-stress, on condition that the word is the selected stretch; it is regressive assimilation if the weak stress is produced in anticipation of a following word with stem-stress; perhaps more likely the weakening of these ME inflec-

---

139 These are all stretches of speech. All elements of the spoken chain have extension; the extension of the spoken chain does not come from elements without extension.

140 *I.e.* form- and function-directed speech; while Samuels (1972:28-29) regards the two processes as virtually categorically different, I shall regard the difference as one of degree.

1.6. Vertical syntheses

tional syllables may be due to multiple conditioning in the spoken chain in the sense that a larger stretch of speech than the word is selected: Cv́Cv̀ + Cv́Cv̀.

It is a phonetic commonplace and a corollary of the fact of constant change that sounds in the spoken chain are produced under the (assimilating) influence of both a following (regressive) and a preceding (progressive) sound; thus exogeneity characterizes all the elements of an utterance: the *p/t* in *pet* is/are different from the *p/t* in *pat* and likewise the vowels are variants that are characteristic of this particular type of context different from the respective vowel variants in *bed/bad, ate* [et]/*at* (see Hjelmslev 1963:60-75). These facts go to show that the minimal unit of language analysis is a synthesis; the historian and the synchronic linguist agree on this. They differ as to the interpretation of the exogenously motivated elements. Synchronically an assimilated element is explained by the system, say the elbow-room of the phoneme in systemic space, the historical analysis will make it a part of a developmental series in accordance with the constant-change universal. But it is difficult for the synchronic phonologist to keep his perspective consistent. In Davidsen-Nielsen (1994:148-149) the modern English assimilation product *happen* [hæpm] is regarded as phonemic because a phoneme boundary has been crossed and we have two different *phoneme sequences*: [hæpn] :: [hæpm]; in other words this [-m] is a variant of the phoneme /m/, while the [-n] is a variant of the phoneme /n/. The consequence is that /m/ controls the [-m]-pronunciation, and /n/ the [-n]-pronunciation. Where does this leave the phonemic (invariant) description of the sign *to happen*? If this [-m] is a variant of /m/, how does the phonologist know that an assimilation process has taken place? On the other hand, the fronted/centralized [u:]-sound in *music* is a *non-phonemic* assimilation product due to the preceding [j]-sound, because this fronted pronunciation is regarded as a variant of the long back monophthong /u:/. This is also a historical statement: how does the phonologist know that a process of centralization/fronting has occurred? For all we know (Luick 1964:§§ 489-490), the sound in this French loan word was a front sound [ü:] that developed regularly into [ju:], so that there is no reason to assume a fronting process here; on the contrary, an assimilation process in the opposite direction is perhaps more relevant, if the *u*-sound is (ever) pronounced as a back vowel[141].

The elements of the L-continuum may be divided into *langue*- and *parole*-elements, a non-categorical partitioning. *Langue*-stretches have historical existence in that they have all become predecessors of their respective successors, as they create the present out of the present. *Parole*-stretches have historical existence in so far as they are parts of a developmental series. The spoken chain has no history – except perhaps interjections (Diderichsen 1970:139). When the present is created, the relevant L-stretches contract empirical vertical syntheses in developmental series with their respective intersections of the spoken chain, and the L-element out of which a here-and-now intersection exists becomes historical in that it becomes a predecessor. An example:

PreOE **kyning* was a hapax legomenon the first time, T2, it appeared; at T2 the front sound cannot be re-duced to PreOE /u/. And it is of no avail to the synchronic linguist to find

---

141 Again we are reminded about Paul's words about the synchronic linguist who is a closet historian: once we begin to look into the logical geography of synchronic diction, we move into historical terrain. Davidsen-Nielsen is here saddled with the synchronic dilemma: which of two or more sounds to elevate to systemic or invariant status.

word forms at T2 such as *kuning of P1, *kyning of P2, *kuning of P3 and *kuning of P4. In order to explain the variant, he would need a historical theory that provides a diachronic link between the forms, and the moment appeal is made to the language at T1, when no *kyning existed, his possible classification of *kyning as a variant of /kuning/ at T2 makes him a historian. But without a proper historical theory, which of the two states is the predecessor state cannot be decided (see chapter 5; Luick 1964:7). The historical theory must consider the idiolect and the theory describe the process out of which our hapax legomenon exists: this process occurs when Primus so creates *kyning as a here-and-now intersection in the spoken chain that it enters into an empirical vertical synthesis with what becomes its predecessor *kuning, the L-element[142].

The discontinuities of the L-continuum will be defined dynamically as follows. Discreteness is determined by the factors motivating the L-element in question. The smallest element of the L-continuum is created by most motivating factors, internal and external exogeneity and endogeneity. The longest element is created by the fewest motivating factors: a historical phenomenon is a phenomenon that has a history determined by a certain number of motivating factors. Therefore a sentence like *None of this means that McDonald's is shaking in its shoes* (*Newsweek* December 26, 1994:72) is not an element of the L-continuum nor has it a history. Most of its motivating factors cannot be separated from the factors that motivate its constituents. A totality has no history, it does not enter into a developmental series, but it has the potential for doing so; otherwise we could not talk about the denotata of the word (not a sentence) *a sentence, e.g. Colourless green ideas sleep furiously* is a sentence-like construct that may be said to have entered the L-continuum of the English language with word-like status. Similarly, Bache and Davidsen Nielsen's discussion of ambiguity (1997:23) does not transcend the metalingual level: the person who said: *Flying planes can be dangerous*, was not in doubt as to the meaning, and the listener – if there ever was one – would not be in doubt either: the context would certainly contribute to the precise meaning. – For a very instructive example of how so-called ambiguity may work, see Le Carré *The Honourable Schoolboy* (1977:135. Hodder and Stoughton): "'I want to open an account, Frostie,' said Jerry (…). … 'From all you told me last night you haven't the resources to open a bloody piggy bank. (…).' … "I didn't mean we'd open *my* account exactly, sport. I meant someone else's,' Jerry explained." The purported ambiguity need not inhere in the language, but in the interlocutor's ignorance of Jerry's historical circumstances.

The preceding exposition may be represented as in 1.6.4, which implies no ontological dualism, and as such no absolute distinction between *langue* and *parole*. All elements of the L-continuum are signs and the sign function ErC must, as Ingerid Dal points out (see Willems 1995:168-170) be generalized to all language elements.

---

142 The theory to be developed will demonstrate how we can see which of the elements of this vertical synthesis is the successor; anticipating this, I shall here specify 'meaning-completion': the here-and-now intersection completes its meaning in the spoken chain as a potential historical phenomenon, the L-element as a historical phenomenon. We need this difference seeing that the mental space of the vertical synthesis implies non-equiempirical phenomena.

1.6. Vertical syntheses

1.6.4 Creative and continuative selection: Form-governed or Function-governed
the speaker selects from the L-continuum, producing an utterance. This is not katadynamic manifestation, but historical, continuing the everyday language as a historical and potential historical phenomenon.

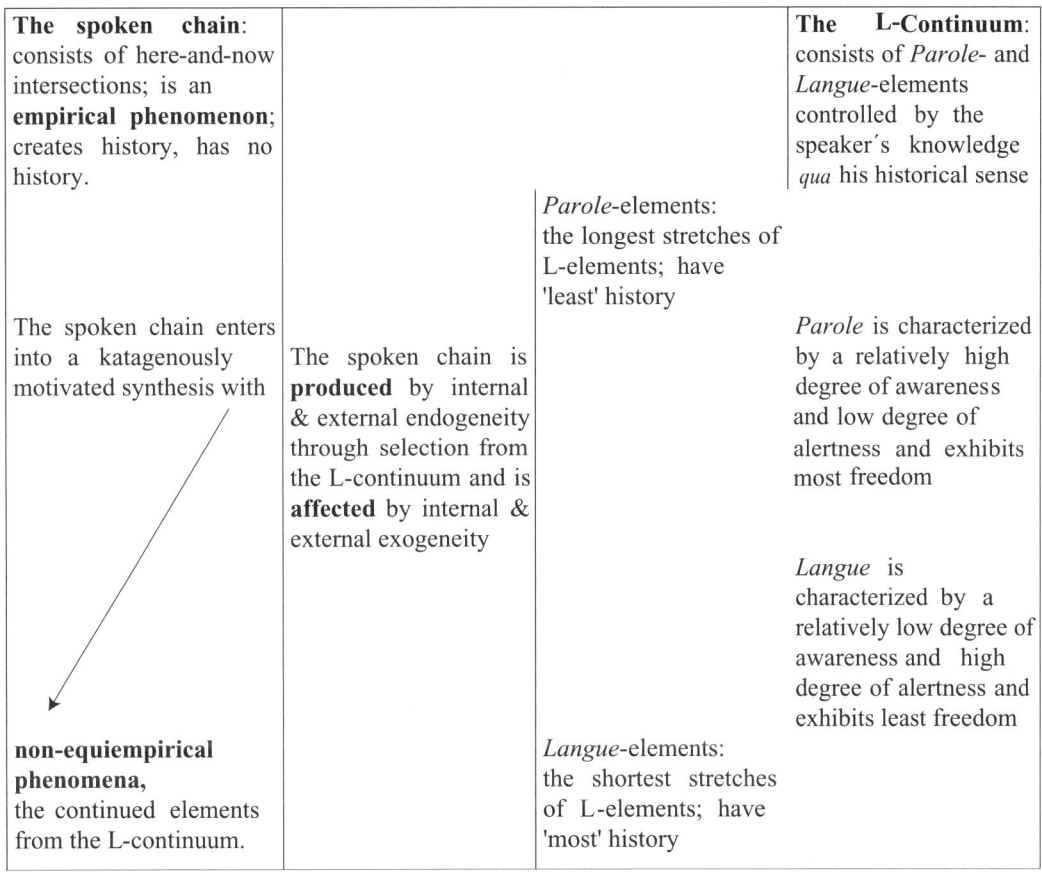

The diagram in 1.6.4 is to be elaborated from the perspective of the everyday language's existential function.

When a spoken chain is produced, the relevant L-elements develop into the present, creating a structural phenomenon (through katageneity) of which the endo- and exogenously motivated here-and-now intersections become the concrete, empirical utterance, while the L-elements, becoming the respective predecessors of the utterance, are continued under the aegis of change. The utterance is a complex entity consisting of non-equiempirical phenomena. The concrete here-and-now intersections *exist out of* the L-continuum's elements, which thereby become historical (predecessors), and are synchronic in the sense that existing here-and-now, they vanish.

The non-equiempirical part of the historical phenomenon is not an objectum, and the distinction between the short-lived existence of the speaker's utterance and the continued existence of its historical elements is connected with the two selection-processes. The historical parts of the utterance and the elements of the L-continuum are entities 'in our *self-awareness*'[143]. Vanishing and fading out of a speaker's (and hearer's) mental focus, the utterance as a totality makes room for its meaning or denotatum. While such objective totalities do not continue to exist in our self-awareness, those of their constituents that exist in developmental series will enter into structural relationships with their respective predecessors, and may then develop into a new present.

The smallest L-elements owe their continued existence to the greatest degree of alertness on the part of the speaker; they are the elements we are least aware of: hence for such an element to be selected it requires a selection process of the greatest degree of alertness, while the longest L-elements are those we are most aware of, hence their selection does not require a particularly great degree of alertness: what is continued of a given utterance are those elements that we are most alert to; hence an utterance is not continued – is *per se* not 'continuable'[144]. The utterance itself – being a totality – does not develop into the present; it vanishes, it does not become a predecessor, but becomes a constituent of the present moment *qua* a part of non-linguistic reality, namely the present moment. Thus when we speak, a section of the L-continuum will be *en ergeiai*, while the remainder will function as the over-all structural background of the spoken chain; the existence of the everyday language and human existence are hardly to be separated[145] (see chapter 4).

The mentalist basis of the advanced theory of our historical sense does not involve a simple causal statement: the L-continuum is not the controlling cause of the spoken chain *qua* an effect. The synchronic dualism, on the other hand, is both a causal and a theoretical statement of the system's manifestation in *parole*. The system controls the directionality of a given change in virtue of such systemic factors as phonological space, functional yield, symmetry or optimum. Unfortunately, all such concepts are difficult, if not impossible to equip with tractable definitions. Although the speaker is extremely alert to the elements ('phonemes') that expound those concepts, it is yet to be proved that any native speaker becomes either alert to or aware of such purportedly mental configurations, and the same goes for the dualism's (mental) reality. And in my ongoing criticism of this ontological hypothesis

---

143 The diction is far too spatial and static, but I keep it in deference to tradition.
144 We are particularly alert to the word and sounds, while it is doubtful that the syllable and sentence structure can attract a similar degree of alertness. (Certain) rules of word order and synchronic grammatical rules may assume a middle position. The fact that we use such phrases as *to pronounce a word correctly* or *wrongly, to use the wrong* or *correct word,* while we hardly ever hear *He used a wrong intonation, sentence pattern* or *syllable,* or *produced a wrong p* seems to suggest something about the mental status of the various types of L-elements. As for grammatical rules it should be noted that all such rules have been created by historical processes and no rule will survive permanently. This tells us that we are not particularly alert to such phenomena, although we are very aware of them in a sign-functional context (see Jenny Cheshire's analysis of *e.g.* the present tense *-es* inflection in non-standard speech (1982:31-43); and the following examples also tell about the precarious existence of grammatical rules: *"This is a total heartbreak for my family and I."* And: *When Paul and Linda's own wedding was announced back in 1969, […]. (Newsweek* May 4, 1998:39).
145 The distinction between linguistic and nonlinguistic reality is therefore not appropriate, because what is real to us is sign-functional. The diagram in 1.6.4 is to be placed in a sign-functional context, and it will be demonstrated how the smallest L-elements are signs constituting what will be called an **attitudinal** system and how **ex**istence involves **synonymy** and **polysemy**.

1.6. Vertical syntheses

I shall now focus on two types of change that make systemic explanation absurd, which means that the dualism in 1.6.4 is not a reformulation of the *langue-parole* dualism:

1.6.5   /invariant/   [variant]
             [variant]

This model does not account for the rise of such words as these: ME *passenger*, *nightingale*, *sound*, (*thumb*), *thimble*, *thunder*, **n**ewt; -*apron*, *cap-tain*; and in the modern language we have words subject to elision: *carefully*, *preferable*, *chancellor*; *gorilla*, *police* (see Davidsen-Nielsen 1994:150; Gimson 1994:213-215; 261-262). First of all: how does the phonologist know that something has been added or lost? The answer is: we know historically that the expanded or elided forms are/were later than the respective shorter and longer forms.

The involved phonemes can hardly be said to be the controlling factor in either process. It is absurd to claim that the (structure of the) phonological system or underlying lexical representations – ME /kaptain/ or /kapitain/ etc. – govern the distribution of their variants in the spoken chain:

1.6.5i   ME /i/   [i]   *capitain*
                         [Ø]   *captain*

1.6.5ii   ME /Ø/   [Ø]   *passager*
                         [n]   *passanger* (see *OED* sub *passenger*)

1.6.5iii PreOE /u/   [u]
                         [y]   (*I-Umlaut*)

1.6.5iv   PE /n/   [n]
                         [m]   (*to happen*)

What makes sense is the application of our historical sense: one form is earlier and older than another younger form, and from this structural knowledge we can infer the fact of synonymy. The synchronic linguist may counter that derivation and compounding and alternating forms may reveal which form is 'original' or basic: the synchronic problem emerges again: the synchronic cut will only yield simultaneously existing forms, and it is impossible to decide which one is the synchronically 'original', *i.e.* systemic one. A final example:

In modern English vowels in unaccented syllables are [ə, ʊ, ɪ] and the phonologist (Davidsen-Nielsen 1994:95-96) makes these phonemes *reduced forms of the other English vowels* in the processual sense that *e.g.* /ə/ in **a**tomic is the *weakening of the /æ/ in* **a**tom. The question is: what is the /ə/ in *atom* a weakening of? The /ɔ/ in *atomic*? A consistent representation of this view of the dualism will look like this:

1.6.5v PE /æ/   [æ]   **a**tom/at**o**mic   [ɔ]   /ɔ/
               [ə]   **a**tomic/at**o**m   [ə]

an interpretation that makes it impossible to elevate either of the words to being the underlyingly systemic representation or establishes this absurdity:

1.6.5vi /ætɔm/ or /ətəm/   [ætəm]   conditions A
                                   [ətɔmɪk]   conditions B

The synchronic dualism reduces to a historical basis (cf. Paul): for a sound to be the weakened form of a 'full' sound, the two sounds must be placed in a chronological order (cf. Luick) and shown to be connected with a diachronic link (Samuels 1972:10-11): the full vowel is the origin of the weakened one. Secondly, based on the simultaneous existence of two equiempirical forms, it is synchronically impossible to make either of them more basic, etc. than the other without committing the Phenomenal Error.

The dualism must (1) specify how **absent** (in *parole*) an element may be without being (systemically) absent, *i.e.* still be **present** in *langue*? OR: how **present** must a systemic element be in order to be **present**[146]? (2) demonstrate that the systemic elements and 'their realizations' (variants) do constitute a vertical-simultaneous synthesis and not a dynamic-temporal synthesis. Katalysis is not the answer to the first problem, as Gregersen (1991 vol. 2:232) indirectly points out: the loss of <-e> [-ə] as a verbal ending in Middle English *ich gife* (< OE *ic giefe*) 'I give; ich gebe' becomes a historical fact that at some time overrules any theoretical attempt at encatalyzing an ending: the grammatical system will not be able to enforce its previous *Zwang*.

The linguist's task will be to describe the speaker's L-continuum, how it appears as an entity in its own right, different from all other everyday languages. In chapters 5 and 6 below a dynamic theory of a section of an L-continuum, namely (long) monophthongs, will be set up, and in chapter 4 an argument will be presented for the necessity of two language universals that synchronically are regarded as non-systemic *parole*-phenomena: synonymy and polysemy. Secondly, it will be demonstrated how these two universals follow from another universal, that of constant change, and how they constitute the necessary basis of the sign-functional theory of history.

---

146 That the four processes in 1.6.5 are/were potentially present in the system, because as realized they did not infringe any word- or syllable-rule, is a circular explanation.

CHAPTER 2

A constructive view of Geneticism and a critical view of the synchronic dualism's two-state model

―――――

> *Perspicuum enim est attendenti ad temporis naturam, eadem plane vi et actione opus esse ad rem quamlibet singulis momentis (quibus durat) conservandam, qua opus esset ad eandem de novo creandam, si nondum existeret; adeo ut conservationem sola ratione a creatione differre, sit etiam unum ex iis quæ lumine naturali manifesta sunt.* Descartes *Meditationes* III.49.

2.1. **More definitions. The structure of the historical phenomenon: synchrony and history unified in the spoken substance. Aristotle's four causes. The dualism makes 'variation' static and synonymous with 'difference'. The two-state-model is basically no theory, but expounds a certain view of (the mechanism of) change**

THE CONCEPT OF the horizontal synthesis agrees with the two-state model, both involving a predecessor (a before-state) and a successor (an after-state) linked by a process. The historical notation A>B implies a two-state model, and I have yet to show the historical inadequacy of it. The concept of simultaneous motivation was an indication hereof, and so was my interpretation of the line metaphor[1]; the inadequacy of the synchronico-dualist basis of historical analysis was also demonstrated: Plato's ontological dualism was an attempt at combining the empiricism of Heraclitos and rationalism of Parmenides into a change-state compromise, whereas the modern introduction into linguistics of the dualism was a negative reaction to the previous century's historical conception of the everyday language. It was no attempt at establishing the position of history within the spectrum of other sciences. The *langue-parole* dualism with its focus on analysis creates division and separation, and so does the other Saussurean dichotomy, history and synchrony[2]. There is no division of labour against the horizon of cooperation between equal partners, historians and synchronists.

―――――

1 Lass (1997:32-43) only focusses on the positive aspects of the metaphor's scientific relevance; he does not subject the basic metaphors of our trade to any analysis, *e.g.* that of the arrow of time; history is therefore monolinear.
2 The adoption of the opening line of Cicero's 1st speech against Catilina: *Quo usque tandem abutere, Catilina, patientia nostra?* as the linguistic banner under which Viëtor and Otto Jespersen and others launched their language teaching reform is telling (Juul 1995:56); and if I may refer my reader to the opening paragraph of Cicero's second speech against Catilina, nobody can doubt the legitimacy of eradicating historical knowledge from language teaching.

What is the empirical relevance of the two-state model – what does the underlined string of words above denote? What is a state – where does the former end, where does the latter begin? How is the predecessor or successor identified? How does the process connect the two states – is it a phenomenon outside the states[3]?

The developmental series is a complex synthesis (1.5.1) which defines what makes the everyday language cohere multiconditionally against the horizon of discontinuity; the coherence is based on **ex**istence. Existence – historical development – implies simultaneity, and the latter makes the historical phenomenon a transsynchronic structural totality. That the spoken chain is 'more' than the sum of the L-continuum's elements follows from the concept of 'meaning completion in *parole*' (cf. also my brief discussion of ambiguity above).

The present is an effectum, a successor in a developmental series, which is defined in definition 31:

Def. 31a   In a historical process that phenomenon which is stable in the vertical synthesis of the process is its successor (the here-and-now intersection; the new), and the phenomenon that changes is the predecessor (the old, the inherited)[4]. – Cf. Def. 31b below.

Thus the monolinear historical process represented by the changes in 2.1.1 formalized by A>B

2.1.1   OE ham   > ME home // PreOE * kuning > *kyning // PreOE *kæld > *keald
ME passager > passenger // ME capitain   > captain

represents a chronological truth, and is theoretically defined – only in virtue of difference – by the involved mechanisms (causes) of change. Therefore the empirical effectum (successor) is explained by something which is part of itself (cf. the Phenomenal Error). The structural dynamics of the vertical synthesis (def. 31) presuppose the simultaneous existence of both phenomena, so that 'to exist simultaneously' will mean 'to exist out of the same phenomenon as a whole' in relation to a third phenomenon (cf. chapter 1.5.7 and the Jones Hypothesis).

A successor acquires part of its meaning from the horizontal synthesis it contracts with its predecessor, and its meaning is therefore a product of external and internal endogeneity. Conversely, the predecessor will be motivated in a vertical synthesis by its successor and will as such change – the successor will not change, because there is nothing to (be) change(d); it is stable in time. The seeming tautology, *For something to change, something else must have happened*, will be rephrased as follows: *for an endogenously motivated process to occur, a katagenously motivated process must also occur*. Change is therefore aitional (1.4.6): change, but not only if change. – A more elaborate definition of the change concept will be given in chapter 6.

A predecessor changes; otherwise it would be no predecessor (Metacondition 3). In dynamic and more empirically relevant terms: the present that develops into the present changes,

---

3   The following sections will contain a systematic ambiguity. I shall be dealing with elements mostly, not totalities. Thus the A>B notation will represent both sound changes and state-transitions; synchronic linguistics leaves language, state and system undefined. Note Saussure's uninformative equation: *la langue = le langage – la parole*, which implies: *la parole = le langage – la langue* and *le langage = la parole + la langue*.

2.1. More definitions

becoming the predecessor of the new present. The process is not monolinear, because when A *is*, B cannot possibly *be*, and vice versa. What makes A and B cohere in A>B is no physical entity, but *man's totality-creating sense, his sense of history*.

In the present's constant development into the present two ordered phenomena contract historical links. Change is to be defined on the basis of a theory devoid of metagenous and linguistic motivation, *i.e.* hypostases, finalism and essentialism. Change is not to be defined in terms of two entities' exchange of properties (two states), let alone one entity's exchange of properties (essentialism). Change is relational and structural and occurs within the totality of a phenomenon[5]. The French Revolution is what it was and did not change, neither did it become the new property of an ideal carrier entity that somehow exchanged properties[6]. Change is a structural process involving in addition to endogenous and exogenous processes what will be called **katagenous** motivation, which occurs in the vertical synthesis of the developmental series:

Def. 32 Katagenous motivation is paragenous motivation that originates with the Eigennatur of the stable element (in a vertical synthesis) in an endogenously motivated process.

Katageneity is the (structural) energy of the vertical synthesis in which a merger occurs differently from the mergers of internal endogeneity and exogeneity. The specification of katageneity will illustrate the dynamics of the process. – The 2.1.2-diagrams illustrate both the non-monolinearity of the historical process and its dynamics and the definition of part B of the here-and-now intersection:

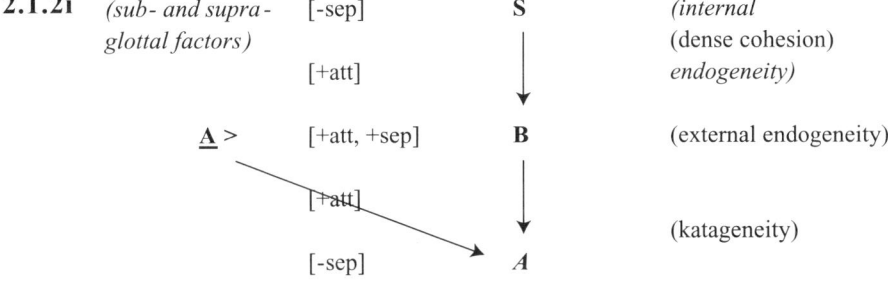

---

4   Nietzsche was right in maintaining that the 'is' which becomes the 'was' does not turn into *ein Nichts*.
5   Thus my theory is a return to structuralism, but not the static structuralism that Saussure and Hjelmslev introduced. Structuralism is dynamic and therefore basically a historical point of view; I shall demonstrate in chapter 3 that Structuralism's determinants imply the concept of the period and cannot be kept within the purportedly static confines of the notion of a point-in-time. Hence it is a pleonastic way of speaking when we say: to give history a structural interpretation. Thus the irony of this is that when we analyze the everyday language structurally a historical aspect is necessarily implied. Structuralism and synchrony cannot be synonymous. Hermann Paul's words about the linguist who does not know that he is being historical have been relevant since the advent on the scientific stage of the contradiction in terms: synchronic structuralism = synchronic history. The moment we deal with man-made or partly man-made phenomena, we shall have adopted *eine geschichtliche Betrachtung* (Paul 1966:20; Hansen 1982a; Keller 1985:220-221).
6   The demolition of *La Bastille* is a historical event and is an integral part of the continued history and existence of what we call *France*. Whether the process was destruction or otherwise is a value-loaded issue.

Def. 33 The here-and-now intersection (= B) is an existent, both determined and determining. It is at the same time a potential historical product and an active constituent of the present moment; as such it makes **A** katagenously determined differ from **A** endogenously determining – with A completing its historical potential.

The process A-into-A, in which A is made to differ from it itself, requires the addition of the property [-Attachment] (not indicated in the figure). The difference between the two mergers ([-sep] [+att]) will appear in 2.1.2ii.

As a totality the present intersection is aitional in that 'B * B'; B exists out of its predecessor and other factors so that the principle of multiple conditioning asserts itself. Definition 33 emphasizes the structural whole of the present intersection, while the diagram in 2.1.2ii is the dynamic counterpart: endogenous motivation produces a complex merger; internal endogeneity allows no structural interplay, while the mental space of katageneity does. B makes its predecesor **A** a historical phenomenon: **A** > *A*, and is separated from its predecessor in the proces of external endogeneity, while still constituting a merger with *A*. The property complex **[+sep, -sep, +att]** (cf. 1.6.1(4)) refers to the potential history of the intersection. The here-and-now intersection appears as a perceptum inseparable from its sub- and supraglottal motivation, but separable from its origin, **A**.

While A is continued (= [-sep]), differing from A through the acquisition of the property [-att], B continues [+att] to constitute a structural merger with its (changed) predecessor: the property complex *[+att, -sep, -att]* indicates the dual nature of the historical phenomenon and this duality is no ontological dualism of the Aristotelian type (1.6.1.(1)).

2.1.2ii

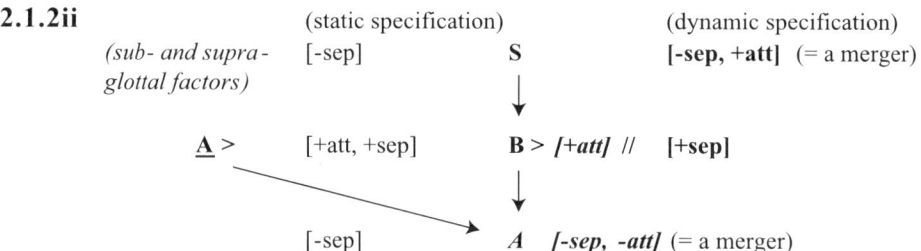

The diagram reads as follows: the (synchronic/static) merger of internal endogeneity [-sep, +att] remains, and merging with external endogeneity [+att, +sep], it creates the here-and-now intersection with these properties **[-sep, +att, +sep]**[7]. The here-and-now intersection becomes discontinuous in relation to its origin, **[+sep]**, but constitutes a merger with its continued predecessor under the aegis of change with the italicized properties. The historical identity of *A* and **A** is symbolized by [+att, -sep], properties that are changed katagenously into [-att, -sep], whereby the difference between *A* and **A** is safeguarded. The katagenous merger of B and *A* is indicated partly by [-sep] (from **A**) and [+att] (from B). The [-att]-property of *A* is to be seen in relation to **A**, and katageneity will be specified by these properties: *[+att, -sep, -att]*. Finally, the [+sep] property of B and the [-att] property of *A* symbolize the discontinous development of **A** (the present) into the present.

---

7 The figures in 2.1.2 need a sign-functional interpretation in order to describe a fullfledged historical process.

2.1. More definitions

The latter figure exhibits the affinity of immanence-inherence (of the ontological dualism of Aristotle (1.6.1(1)) with history; together – following the Critico-Philological Method – they constitute a new synthesis, namely dynamic historiography, which takes its point of departure from the notion of the present's development into the present. Since this notion also defines existence, the historical narrative that applies this dynamic point of view will also express existence postulates – as against the static point of view in 1.6.1(4)[8]. The figures in 2.1.2 represent the unification of synchrony and history, static and dynamic processes; they indicate how analytic reasoning will not be able to grasp historical totalities, a fact to be emphasized when the above type of entity is seen as a sign-functional phenomenon.

The inadequacy of the monolinearity of the two-state model also appears from its causal concept: *-i-* in PreOE **kuning* is the cause of the *-y-* in (Pre)OE *cyning*, with the main stress on the root of the word being a contributing factor:

2.1.3 IF   Cause   ?Ø
                    a following *i/j* + stress   a following *i/j* + stress
                    ↓

         THEN   *-u-*   >   *-y-*
                    Affectum   Effectum
                    State 1 of T1   State 2 of T1+1

The diagram shows how a cause loses its causal property[9], owing to the Eigennatur (amenability) of the (purported) effectum. The loss of such a property implies the question: how did the sounds acquire this causal property? When the mutated form was produced, the affectum is not the material on which the cause or causes worked. On the contrary, what the synchronic point of view can do is to register the simultaneous existence of two different – mutated and unmutated – forms, produced by two speakers between whom no causal connection can be established. The simultaneous presence of the cause with either of the two forms or with both forms is just as untenable, indeed, synchronically contradictory. How will the synchronic linguist explain that in State 2 certain sounds sometimes function as causes? A synchronic theory will have to be transsynchronic!

This so-called mechanical causal principle, dominant in historical linguistics, is an Aristotelian *Nachlass*. Both Pheidias's marble block existing simultaneously with his *Athena Parthenos* and Aristotle's four causes are easily recognizable: (a) the material cause, (b) the efficient cause, (c) the final cause, and (d) the formal cause (2.1.4; Jones 1970:224; Lear 1988:28-42). While Pheidias's original marble block is describable as unformed, it makes no sense to assume that PreOE **kuning* is an affectum *qua* unformed material, and PreOE **kuning* did not exist independently of the group of preOE speakers, while no marble block is dependent on any person; furthermore, the limited number of causes invite criticism. Why not 5 or more (cf. Cicero *De Fato*: p. 166)?

---

8   This implies a redefinition of the Uniformitarian principle as it is applied by Labov (1994) and Lass (1997). To explain the present or any successor through its past (predecessor) does not capture how we comprehend the historical process called **ex**istence.

9   The cause *is altered in fulfilment.* (T.S. Eliot, *Little Gidding*).

2.1.4   (b) speaker/Pheidias   ↔   (c) design/function of end result
         (plus '**tools**')   or effectum

↓

         (a) *kuning* /marble block   ↔   (d) form of end result (effectum)[10]
         affectum   *cyning* / statue

We could ask: where is Pericles as a motivating factor, who commissioned through his everyday language Pheidias to design the Parthenon Marbles? What is more significant is that the final – the 'for-the-sake-of-which' cause – is a statement in the everyday language, and so is the formal cause of the effectum. The form is no non-linguistic essence inhering in the material cause. In sum, the efficient cause is a language-wielding agent, the form and the final cause are sign-functional phenomena, possessed by the agent *in his soul* (Lear 1988:62). Thus we are back to the principle of participation: the everyday language is an essential part of man's existence as well as a precondition for it. Let us labour the differences between the two types of causality; the structure of the language example(s) looks like this

2.1.5   internal endogeneity + external exogeneity   internal endogeneity

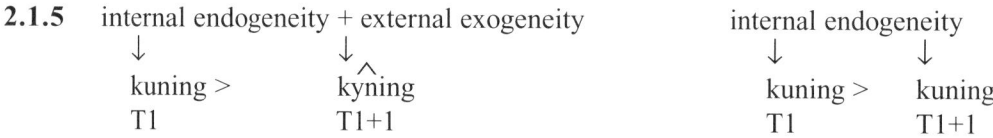

         kuning >   kyning   kuning >   kuning
         T1   T1+1   T1   T1+1

The exogenous and endogenous causal factors do not affect **kuning*; they are simply constituent factors of the words in question. When Pheidias finished his work, the marble blocks did not exist any longer, while our unmutated form existed simultaneously with the mutated form, at any rate for a period. Our language data enables us to understand simultaneously existing variant forms historically, chronology being (part of) the code by which we understand them. This chronology is retrievable by the grammar of the language in question (man's historical sense) or by a historical theory. Similarly, but presumeably through different processes[11], we are able to understand the man-made nature of the Parthenon sculptures by placing the finished product in (transsynchronic) horizontal syntheses. This comparison leads to an elaboration of the definition of the historical phenomenon (Def. 11):

Def. 31b   A historical process is the creation of a phenomenon through exogenous, endogenous and katagenous motivation; it constitutes a horizontal synthesis and an empirical vertical synthesis. It creates a here-and-now intersection, which becomes part of two mergers in two vertical syntheses characterized by internal endogeneity and katageneity. The phenomenon completes its meaning in *parole* and man's existence, if it is an everyday language phenomenon, and in man's existence and

---

10   The form is an ideal cause in the sense that it expounds a sign function in which the (ideal) form is the content and the particular concrete entity (one of) its expression(s).
11   The medium of language is sound, hence language phenomena are more fleeting than *e.g.* a statue.

2.1. More definitions

in an everyday language if it is not an everyday language phenomenon[12] (a statue; the destruction of *La Bastille*). This historical phenomenon is a proprial whole. – Cf. 2.5.2 and Def. 31a.

Finally, I shall describe the above PreOE example in terms of the *langue-parole* dualism:

2.1.6
CAUSE → /kuning/
[kuning] ← formal, function-governed style
[kyning] ← informal, form-governed style

The systemic element is the 'material' (cause) that efficient causes so work on that the two variants are 'realized'. The formed entity /kuning/ is neither amorphous nor equiempirical with the variants, it being a controlling, norm-setting entity, a bull's eye that does not fit in with the Aristotelian system of causes. That the [kyning]-variant should be the first step in a teleological drive or a function of the over-all PreOE system is an unsubstantiated hypothesis.

As has been indicated more than once, synchronic linguistics has a data constitution problem and a theoretical reckoning in the light of definitions 26 and 27; is the system a supraindividual affectum of the synchronic selection process and is its change governed by supraindividual formal and final causes? That the change of OE /ha:m/ into ME /hɔ:m/ is implied *formally* or *finally* by OE /ha:m/ is yet to be proved. And the variant [hɔ:m] is synchronically impossible if it is to be understood by the hearer with reference to a system that has /a:/ and not /ɔ:/. The solution proposed by Weinreich *et al*. (1968:122; against Saussure 1972:140) in terms of ordered heterogeneity explains nothing because they have no theory of how variants arise[13]. The idealist supposition of the synchronic *langue-parole* dualism may now be summed up: *the systemic slant notation of the theory implies a* given *datum whose properties follow from the definitions of the theory. The systemic object is an expression of the theory's activity, not that of an empirical object. The theory cannot predict the development of its data; systemic-based prediction is impossible, unless the prediction has been presupposed* (literally 'said beforehand'). *A given object means 'given' by the theory, not by or in a Kausalnexus*. The nominalist projects this theory on contingent data; the Realist claims that the theory expounds the nature or essence of data. The dualism confuses the two relational concepts 'to vary' and 'to differ'.

The variation-relation of the dualism has it that a variant differs from (= varies in relation

---

12 Curiously, Gregersen (1991:2.255) arrives at a seemingly similar conclusion: sociolinguistics studies the 'life of a language in [its] society' and 'the life of a society in [its] language'. Gregersen is more right than he dares to believe himself, therefore he has to put his postulate in disarming diction. I shall argue with reference to Saussure (below in chapter 3.4.) that the object of such a sociolinguistics is made *viable*, not *vivant* because Gregersen misses the temporal dimension, in Saussure's words *la loi de la tradition*. The moment a sociolinguistics adopts the time dimension it becomes historical. It is no counteranswer that time is implied in 'life': society and time must be connected overtly and inseparably, and then the sociolinguist is a historian.

13 They seem to conflate the problem of actuation and that of origin; at any rate they do not consider the latter.

to) its systemic element, which means that it is not a perfect manifestation of the latter. This does not prove the stability of the system, nor will the definition of the systemic element explain what creates a variant (2.1.6). However, the property 'to be a (synchronic) variant' invariably implies the existence of at least one other variant, otherwise there would be no difference between the systemic element and its manifestation:

2.1.7 T1 (a simultaneous relationship)
/X/   [x1] in P1   Primus
      [x2] in P2[14]   Secundus

The dualism makes the two (different) variants identical in relation to their manifestatum and this manifestation sees no significant connection between the two differing variants. As was said above the dualism implies an unspecified manifestation ontology couched in the atemporal language of space so that the scenario in 2.1.7 can be both grasped in one single act of the mind and produced instantaneously.

It will be demonstrated that the historical phenomenon (the bracketed entities) presupposes the language of time so that a variant varies in relation to another empirical entity in time, thereby making the slant-entities superfluous. In chapter 6 it will be illustrated how this variation is seen in relation to the origin of the involved phenomena (variants) and how – in accordance with the definitions established so far – the variants enter into a significant simultaneous relationship. To sum up: the static point of view has it that existing at the same time, T1, in different places, P1 and P2, two equiempirical entities 'come from' the same 'source' (2.1.7). The historical point of view says that existing in the same place, P1, two empirical entities exist out of their common origin from T1 to T2. The above exposition was the first step towards the two points of view's unification, a unification that the mechanical cause-effect interpretation of the two-state model cannot effect.

2.2. An interlude: further criticism of the slant-bracket convention. Contingency. The synchronization of the speech situation. A summary of the essay's argument. What makes a system static? An alternative

Synchronic linguistics puts the empirical differences of the speech situation on the Procrustean Bed of identity through the /X/; this slant-entity is used to suggest that the two examples of the slant-bracket convention in 2.1.7 and 2.1.6 reduce to the general schema in 1.6.3, and that the speaker-hearer situation is reducible to one person. The process of near-mergers (Labov 1994) indicates that the speech situation is not reducible, and the notion that change occurs through some kind of hearer-imperfection – be it when the infant baby is learning her mother tongue or when 'misunderstandings' (faulty imitation) arise in the speaker-hearer situation (Samuels 1972:10) – also entails a lack of symmetry in speech production and speech reception.

---

14 The irreducibility of the speaker-hearer situation follows from the *langue-parole* dualism. To be elaborated in 2.2.

2.2. An interlude: further criticism of the slant-bracket convention

The synchronization of the speech-situation implies that at any one given point in time the empirical data[15] available to the linguist's analysis will be distributed minimally over two persons' connotations within a time- and space-continuum where no speech-situation can possibly arise:

2.2.1 Primus produces   [x1]   in P1 at T1 $C^PrE(Er^{den}C)$
Secundus produces [x2]   in P2 at T1 $C^SrE(Er^{den}C)$

In spite of empirical differences the two utterances are similar and therefore reduce to the systemic entity /X/ through some identity criterion or other (cf. 0.3.4). Glossematically, we say that substitution is operative, commutation is not[16]. The two variants are not only identical in relation to /X/, they *are what they are* because of /X/, and it is with reference to /X/ that the hearer understands what the speaker says [x1]. It follows that the systemic entity cannot belong to either person and it must therefore be a subjectless, supraindividual entity, *un fait social,* devoid of any trace of connotative subjectivity. And the totality of systemic elements is an instrument, a superstructure, for language-users, *end-users of historically evolved systems* (Lass 1997:xviii).

Despite the general acceptance of the latter type of characterization questions about origin and other dynamic concepts are irrelevant, and so are the questions: what is the empirical relevance of such a systems concept – how can the individual language user participate in supraindividual entities, speaker/hearer transcendent causes of the spoken chain, *shared by you and me* (Eco 1972:57)?

The synchronic relationship contracted by /X/, [x1] and [x2] and constituted by the mutation test (Hjelmslev 1963:73-74) suppresses these original relations: (1) time and place have been subtracted, (2) the communicative-anthropocentric relations between Primus and Secundus on the one hand and on the other between Primus and [x1] as well as Secundus and [x2] (the connotation) have been eliminated, and (3) there is no immediate and linguistically relevant relationship between variants, either idiolectal (from T1 to T2 in P1) or between speakers (from P1 to P2 at T1), a relationship of transindividual and transsynchronic cooperation.

Obviously, such an extreme immanent stance goes against our simplest language experience, and it is untenable *vis-à-vis* the universal of constant change. It is incompatible with the fact that change arises in *parole* (Samuels 1972:2; and Gregersen's empirical problem) and with the time-factor involved in any speech-situation: the supraindividual system must

---

15 Constituting his data, Hoenigswald (1966:2) also sets out from the idiolect, but moves on to the class-concept, suspending the period by means of this contingency argument: what speaker A says at Time 2, he might just as well have said at T1.

16 The synchronic linguist needs an identity criterion; this problem may be *an unnecessary complication* (Hjelmslev 1963:61 note 13) in a linguistically motivated theory; but the step from 'x and z can be reduced to X' to 'x and z are identical in respect of X' is hard not to make. Furthermore, since Hjelmslev (pp. 64-65) must rely on the undefined, if not undefinable notion of *differentiations of intellectual meaning* the Glossematic reduction principle implies this negative definition of identity: "differentiation not based on intellectual meaning", a criterion that is hardly more informative than 'significant or non-significant contrasts' (cf. 0.5. above). – Hoenigswald (*loc.cit.*) displaces the problem of identity from lingustics proper to what he calls *the same* or *different life-situations* or denotata.

exist unchanged through time (for a period; cf. Samuels 1972:28). The contingency argument leads to this absurd situation: synchronically, any individual speaker's historical existence is irrelevant, because it is assumed that all of his everyday language, which he is going to manifest through a whole life could have been produced at any one given point in time: what I do today might just as well have been done yesterday or be done tomorrow – or might not have been done at all. Not only does contingency suspend interpersonal communication, it also makes the notion of history self-defeating and absurd; certainly a truly nihilistic consequence. But this nihilism is so extreme that it negates human existence, and unless humanhood is completely immaterial, the consequence cannot even be said![17]

So the *contingency-of-life-situations* assumption backfires: it implies *parole* in the shape of the latter being pronounced; and in its less extreme version it implies that the historical notions of *earlier* and *later stages* (Hoenigswald 1966:3) are impossible to work with. An example: the appearance of *thei* 'they' in Middle English could have happened at any other point in time than when it really occurred. As any sequentially occurring phenomena could have occurred at any other point in time than the one at which they actually occur, the ordered linearity of *parole* becomes superfluous: /#ls-/ would be legitimate in English as well as /#sl-/ (Hoenigswald, p. 1). But since we do find restrictions on how a language sequence must be built synchronically, it is not unreasonable to assume that there is some kind of necessity involved in the appearance of historical events and phenomena. And the question is: in what sense are restrictions *qua* language rules necessary or merely historical accidents? The paradoxical answer is: as synchronic entities, they are completely arbitrary, as 'historically evolved 'entities', they are controlled by historical accidents. But as historically evolved entities they are neither accidental nor arbitrary on condition 1) we have a theory and 2) historical evolution is not accidental. Another, but related aspect of this dilemma is: following Paul the synchronic linguist is a closet historian, if the theoretical relevance of the speech situation and ordered linearity is accepted. Or: if the synchronic theory makes language phenomena ideal through the unmediated step from the individual (variant) to the group (system), the theory's basis is empirically irrelevant.

As I am not arguing for an either-or position, we must find some common ground for the historian and the synchronic linguist. They have in common, first of all, *parole*; consequently they share a variant concept both through the concepts of endogeneity and exogeneity and the fact that an empirically relevant connection (diachronic link) between two variants can be established in katagenous relationships. Secondly, history and synchrony share a simultaneity concept. I shall sum up my position in the following items:

1) a dynamic relationship can be established between variants [x1] and [x2];
2) the relationship, both historical and static, will not involve the hypostatized entity, /X/;
3) if it exists, the state with its system is a historical product[18];

---

17 While Jordan in Jeanette Winterson's *Sexing the Cherry* (Vintage 1989:10) may *record* the journeys he *might have made* and not the ones he did make, no synchronic linguist has described the grammatical system that might have been the case instead of the actual one.
18 Nobody seems to deny this fact (Kanngiesser 1972:71;66): generativists are historical, Saussure is historical, and Hjelmslev had to include history in his system of definitions, and recently Lass's 1997-view of the system is historical, although he does not relate this to his 1980-postulate: *No change is ever necessary* (p. 131).

2.2. *An interlude: further criticism of the slant-bracket convention*   193

4) the speech-situation is irreducible and with *parole* necessary:
5) the empirical relevance of superconcepts such as national language, dialect, etc. should be substantiated;
6) 'small differences are significant': change as the gradual and ordered occurrence, not snow-ball-like accumulation, of more or less 'imperceptibly' small variants is empirically relevant (cf. Coseriu 1975:228);
7) the everyday language has never been seen to lose its individual and collective orientation, and man's historical sense is an integral part of his over-all sense of reality;
8) a theory of history cannot be equated with the mechanisms of change, but merges ideally with our sense of history.

The concept of the collective orientation (item 7) is a corollary of the fact that no everyday language develops irrationally, randomly or individually; any 'idiolectal utterance' can be translated in the sense that it becomes a listener's property or the speaker can translate it into another utterance of his (synonymy), and a Danish text can be translated into any other everyday language[19]. This collective orientation will be grounded theoretically on man's historical sense *qua* a theory of development: man's knowledge of his own language (Coseriu 1974:49-51;56) is a precondition for his (understanding of his historical) existence[20].

The imposition (items 2 and 3) of the *langue-parole* dualism on the two-state model ends in this absurd situation: synchronically, a spoken chain belongs to one state or system, L1, and the step from variant to change, whatever its intricacies (Samuels 1972:10), occurs from one moment to the next:

2.2.2   T1   T2
       /A/   > /B/
       PreOE */kuning/   > */kyning/ 'king'
       OE /ha:m/   > ME /hɔ:m/ 'home'

       L1   > L2
       /X1/ [x1]
            [x2]   > /X2/

How can [x2] all of a sudden lose its dependence on /X1/ and itself become that which controls and determines itself, *i.e.* its own manifestations [x2], [x3]? If the criterion is that loss of its conditioning factors entails (possible) independent, continued existence, then its dependence on its systemic 'counterpart' /X/ must also go by the board[21]: in a word, the sys-

---

19 Hjelmslev (1970:134-135) provides a reason for the theoretical possibility that the idiolect is linguistically significant and the speech-situation therefore irreducible: each speaker (*physiognomy*) has his own *distinctive stamp* and may therefore have a *linguistic structure* different from other physiognomies.
20 I disagree with Coseriu's reliance on the systems concept (Coseriu, pp. 50-51) as that which expounds our language knowledge, because the historical explanation is just as cogent. To say that in Modern English the word **klim* is a potential word and that **mkli* is an impossible word implies a calculus view of change that does not dispel the mystery of change (Coseriu, p. 56): change is therefore reduced to being a function of potentialities (dynamis).
21 The role of function is curiously absent from such discussions (see section 2.6. below).

temic element is superfluous. In conditioned change loss of conditioning factors is (may be) identical with the property of systemshood (to be a manifestatum); but what about isolative change? Here we have no conditioning factors (*pace* Samuels), except the sound's systemic affiliation: ME [ɔ:] loses its manifestatum /a:/, and assumes – in the very same moment – a new manifestatum -/ɔ:/ – itself.

The two-state model is indifferent to the directionality of change (item 6): which state or form is the predecessor, which one the successor? That certain types of change are more 'natural' than others is both a commonplace in the literature[22] and an unsubstantiable one. That, say, dissimilation PreOE **kyning > *kuning* is less likely than the well-known assimilation process is not an inviolable law. So the type of directionality in 2.2.2 is an instance of question-begging: the historian wants to prove that L2 is the successor of L1 or to establish that the Vietnam War was later than World War 2; and he does this by his prior knowledge of L1 and L2, etc. Furthermore, the synchronic point of view cannot explain continuations, and this also applies to the sociolinguistic model of Weinreich *et al.* (1968:149-151), where the basic notion of *orderly differentiation* excludes continuations[23]. Finally, how many languages and/or states and/or systems (cf. item 5), how many steps from variant to change, are needed to explain the history of what we call the English language from 1000 to 1700?

Let's return to Plato. The world of forms was postulated in order that Plato could find a place for the changing world of sensuous experience, whose reality it was too absurd to deny. The world of forms does not explain (modern) reality, hence Plato's philosophy has not carried the day (Rescher 1973:44-45) unchanged – outside of modern synchronic linguistics.

As the state- and *langue*-concepts are intended to explain change, the justification of the two-state model – as well as the system (cf. Kiparsky above) – through the incidence of change is circular; within the context of what is called the same language *more than one synchronic state can only be established if systemic change is presupposed as an empirical fact*. A language state is a man-made construct, a section of a language arbitrarily delimited and cut out of a processual whole. Either it is an extensional entity and as such its empirical relevance must be established; in Glossematic jargon it is a resoluble class as many. Or it is a metaphysical entity (a class as one) which has been turned into a scientific object. Furthermore, we are confronted with a systematically synchronic equivocation: in *linguistic state (état de langue)* the concepts *state* and *system* merge if we follow the Saussurean dichotomy: *la langue* and *la parole* constitute *le langage*. Similarly in Hjelmslev (1972b), here the system is embedded in the state and the concept of systematicity is imperceptibly reified into an entity, the system. And when we read the following programmatic enunciation: *(…) language changes being investigated must be viewed as embedded in the linguistic system as a whole* (Weinreich *et al.*, p. 185), Hermann Paul's warning springs to mind and one cannot help asking: what does the complex sign *the linguistic system as a whole* mean?

What synchronic linguistics as a historical theory is in need of are empirically relevant definitions of its crucial concepts:

---

22 See the discussion on 'Naturalness' in Thrane *et all.* (1980: ch. 3). Here Wolfgang Dressler (p. 77) makes 'naturalness' an explanans of stability, a position criticized by Roger Lass (*op.cit.*, p. 93). Is the continuation of Northumbrian *a:* more natural than the Southumbrian change of *a:* in Old and Middle English?

23 It will be an irony of this essay that the advanced theory of history will generalize this sociolinguistic commonplace of orderly differentiation or heterogeneity.

2.2.3   A synchronic state/system is an X which is stable, static, unchanging.
      A stable X is an X which is Q
      A synchronic state/system is an X which is Q.

If this elemental programme cannot be fulfilled, the position in 2.2.4 is to be replaced by that in 2.2.5

2.2.4   A system S1 controls and determines the communicative efficiency of its language state and the latter's whole existence **despite** the universal fact of S1 changing into S2.

2.2.5   The systematicity of language is the cause of the existential function of language and the existential function of language presupposes change.

For the everyday language to be systematic it must change, because change or the present's development into the present is not random, irrational, but an individual process performed against the transsynchronic – and transindividual – horizon of cooperation.

2.3.   **Serial relationships vs. generic history. The two-state-model presupposes directionality and prediction is impossible. The meaning of history: historical phenomena are sign-functional. Change involves a three-state-model. More remarks on the concept of cause. The future is a linguistically motivated concept. The everyday language is always *en ergeiai*: *ein unaufhörliches Schaffen***

External endogeneity lends itself to a dualist and transcendentalist interpretation if the motivating factor – like immaterial entelechies or *teloi* – inheres in or exists outside the material substance of the elements of the developmental series, whose dynamic principle is the interaction of endogeneity, exogeneity and katageneity with a speaker as agent.

Simultaneous motivation and the concept of distance as *both* attachment *and* separation prevent idealist forces from being introduced[24]. Developing into the present, the everyday language is maintained through the dynamics of its historical grammar *qua* a historical theory.

---

24 Jakobson's *telos*-concept (1962:1;652-653) is purely man-made; it is not a question of being afraid of purposiveness, as Jakobson claims linguistics is, but a question of the theory in question being able to determine purposiveness rationally. The goal-directed behaviour, the corollary of the adoption of a teleological criterion, entails a change concept that is delimited by *the start* and *the finish* of a change, both being equally undefined. This teleological function is too value-loaded and invites its own immediate defeat the moment a language function (in this word's full ambiguous sense) cannot be seen as purposive: what is the purpose of first creating rounded front vowels in PreOE and then destroy them some 4-500 years later? As Kant (1923:18) put it: *ein Organ, das nicht gebraucht werden soll, eine Anordnung, die ihren Zweck nicht erreicht, ist ein Widerspruch in der teleologischen Naturlehre*. The teleological criterion must distinguish between the purposiveness of the origin of a form; the purposiveness of the continuation of the form, the purposiveness of the (fossilized) form at a given point in time. Lass (1997:ch. 2; pp. 309-318) enumerates

To put a phenomenon in a developmental series is, according to critics of serialism, a methodological specification of the series itself, and the registration of a multiple state chronicle serves only as a means towards the general readability of the sum of historical events. Once the mobile principle of a given series, which is a point-in-time factor, has been stated, the series with its individual phenomena is redundant. As indicated, the mobile unit is a set of causal factors motivating the transition from State 1 (the start) to State 2 (the end), and this is an adequate and sufficient historical explanation of discontinuous change. Samuels's 1972-theory is a case in point; chapters 2, 3, 4, 5 and 6 of this work embody mobile principles (factors) that are supposed to explain the development of a given language – perhaps of a specific ('synthetic') nature. However, Samuels's use of the sign *the cyclic swing*, *i.e.* from synthesis to analysis, suggests that his theory is intended to cope with the full swing, and not only with the transition from synthesis to analysis (pp. 80, 82-83). If this is so, Samuels's theory assumes that history is generic, that historical processes are essentially identical or recurrent, a view that is also presupposed in current grammaticalization studies. The developmental series can – by the linguist's *synchronic cuts* – be divided into states of which the transition from State 1 to State 2 is essentially identical with the transition from State 2/3 to State 3/4, etc., 'essentially identical' meaning 'explained by the same principles'.

The before- and after-states of such a two-state model of change are to be registered as such *immediately*, and no serialisation is needed beyond this. The model does not consider the full endogenous nature of historical phenomena, assuming that exhaustive description of 'X of L2 in P1 at T2' is adequate and sufficient without recourse to X's historico-structural nature. In so-called synchronic descriptions (of, say, modern English[25]) it is scientifically legitimate to describe, say, the definite article, the class of concord- or aspect-phenomena, as if they were not interdependent parts of a structural whole – *où tout se tient*[26]. A particularly gross consequence of this method of isolation is the linguist who does not use *authentic, unamended* data, because *the unedited (...) example usually contains at least two or three more grammatical problems than the one which the student is trying to focus on* (Preisler, p. 6). However, this method is just one extreme of a continuum whose other extreme is the Glossematic attempt at isolating the denotative language (ErC), and in between we have the notion of a homogeneous language. To put this differently: in so far as a class of phenomena can be subsumed under a law- or rule-like concept (definition), that phenomenon will have been explained in a scientifically adequate manner. But this 'monograph'-approach is virtually always motivated by predecessor descriptions of the same phenomenon, and the paradox is this: the historical nature of scientific pro-gress is accepted (cf. Hjelmslev 1963:7), while the historical nature of scientific description is denied – or separated from the description.

---

much data whose purposiveness is hard to see; and the imperfection view of language development does not agree with purposiveness. – The rise of the 3rd person singular flexive in the present tense, *-es*, in Middle English may have fulfilled a grammatical purpose, but its purposiveness today has certainly changed to a predominantly socio-regional 'stylistic' function.

25 Whatever this term means (cf. Bache & Davidsen-Nielsen:1997:2-3, Greenbaum 1991:2-4; Preisler 1992:5-6;19); in Vestergaard 1993 no attempt is made to even state the language described.
26 Toynbee (1963:1-16) criticizes this type of scientific treatise, in particular the monograph-type.

2.3. Serial relationships vs. generic history

The structure of two-state explanation is reminiscent of explanation through the vertical syntheses of subsumption. The causal factors take on a function that is superordinate in relation to the phenomena involved, the affectum and effectum being mere raw-material of change, manifesting the workings of generalizable factors: the concrete change of L1 into L2 is *one instance of a generic history* (Rescher & Urquhart 1971:155). It is worth a thought that Gellner (1969:1-2) suggests that socio-historical recurrence is a pathological event; otherwise, this causal view commits the Phenomenal Error: phenomena of an empirical whole are elevated to the status of causal-dynamic factors, while other phenomena are 'downgraded'. Let's be constructive.

The historical alternative is this: the fact that X of L2 appeared in P1 at T2 is empirically significant in respect of both its predecessor and (eventual) successor; thus the three are mutually dependent and temporal succession implies (at least) a three-state model. The specification of the historical series as a chronicle of events is informative, telling us about the route taken (trajectories and evolutionary tendencies), contributes to our understanding of *why something happened and why something else did not happen*, enables us to set up developmental directionality, and last but not least it will invariably include continuation (A>A).

2.3.1 Historical explanation explains through an empirically relevant theory of development (ideally = the process itself) the route taken by an element in a developmental series, the mechanics of both stability and change, and variation and continuation in so far as variation and change as well as stability and continuation are different processes. – (Stability is created).

The refutation of the two-state model in favour of at least a three-state model hinges partly on the answers to such questions as

Q1   To what extent is the fact that X1 develops into X2 instrumental in X2's development into X3?

OE a:> ME ɔ: > 1500- o: X1 > X2 > X3

Q2   To what extent is the fact that X1 develops into X2 instrumental in X1's Eigennatur?
Or: To what extent is the fact that X2 having developed from X1 develops into X3 instrumental in X1's Eigennatur?

X1 > X2 > X3

Answers will be given indirectly in my continued investigation into the logical geography of the two-state model, the inadequacies of which model make series-specification more than a historian's desideratum; it becomes a language-theoretical imperative. I shall now take another look at the change from synthesis to analysis in order to substantiate my argument against the 'verticalized', generic two-state model.

The generic assumption is that the transition from synthesis to analysis can be explained in a way that is not type-wise different from the transition from analysis to synthesis (cf. 2.3.2 below). Thus in the change, L2 > L3

NB.
1) the loss of a (synthetic) flexive, say, a dative -*e*,
2) the rise of a(n analytic) grammatical element, say, a preposition, *to*
3) the development of free word-order into fixed word order

are not in principle different from the processes that in the change of L1 into L2

NB.
1) had created a (synthetic) grammatical flexive (where none was previously)
2) had changed a lexical element into a grammatical flexive
3) had created free word order (where fixed word order existed previously).

If this is the case, the respective processes will not be dependent on the available material (raw-material). As far as I know, no linguist is prepared to accept these consequences of the generic view of change. This means that Stockwell and Minkova's 1988-rejection of the unity of the Great Vowel Shift on the basis of a postulate to the effect that the internal phonological laws implemented in 'the Great Vowel Shift' are recurrent processes must be taken with more than a grain of salt.

Since the two processual types in 2.3.2 require different causal factors, there is no cogent reason to infer that the differences are due to the respective sets of causal factors and not to the different types of material (affecta); and since the causal factors turn out to be constituents of the very processes, the notion of the historical phenomenon's *concrete* Eigennatur has been asserted; historical phenomena are not mere raw material; and the historical theory in question cannot make do with formal features *qua* explanantes.

2.3.2

	Synthesis		Analysis	#	Analysis		Synthesis
	L2	>	L3		L1	>	L2
	-e	>	Ø		lexeme, Ø	>	flexive
	lexeme	>	preposition		?	>	lexeme
	free	>	fixed word-order		fixed	>	free word-order[27]

What unifies and explains this long-term change must be a theory of history that will honour the specific serial Eigenart of historical processes[28].

Let's consider the concept of long-term change more closely from the perspective of a non-linguistic example. Is it totally unlikely that modern events have as part of their causes events of the not-so-near past? Our answer is No! The French Revolution and Hitler's War have repercussions in modern society in the sense that certain facts today cannot be ex-

---

27 As far as I know, no grammaticalization study describes how free word order arises.
28 It is a curious paradox that proponents of such long term changes as circular (or chain) shifts do not draw the consequences of their acceptance of the unity of these changes, namely that the application of a two-state model conceals the nature (unity) of these changes, reducing them to being a sum of state transitions, not different from the simple isolative change (OE /a:/ > /ME /ɔ:/).

2.3. Serial relationships vs. generic history

plained without us or historians having recourse to the two past events. I am not implying that the actual beheading of Louis XVI or the Nazi Army's 1939-invasion of Poland is the mechanical cause of events of today. What may live on from the past – beyond a given modern *effectum* resulting from the processes of a continued mechanical cause-effect chain – is the **meaning** of the past, *i.e.* meaning as a constituent part of a historical phenomenon[29]. The past, *i.e.* part of it, lives on in the everyday language. In this sense, the meaning of our two sample events above, whose meaning would never have come into existence if the events had not occurred, may be a modern motivating factor. But then the idea of well-defined, self-contained historical states (like categories) must be abandoned[30].

The notion of the meaning – not meaningfulness – of a historical phenomenon is easy to illustrate. The Fall of the Bastille in Paris in 1789 was not only a transindividual historical event in the eyes and minds of those who actually participated in or witnessed the process. To those Frenchmen living in Marseille on the 14th of July that year and Parisians who *heard* about the demolition of the Bastille in 1791, the event was a historical fact. The reason for this cannot be the (perceptually) mechanical aspects of the event, only its meaning element. The meaning of the everyday language (the C in the sign function ErC) lends structure or form to concrete, physical events, so that a historical phenomenon *qua* a sign is created. A historical event with no meaning is a *contradictio in terminis* (cf. Itkonen 1978:122-123). The French language of 1791 embodies the facts of 1789; it was both a contributing and describing factor in these particular events. For a description to be produced, there must be something to be described; and this something must be describable. It is describable in so far as it is compatible with the describing medium. The meaning attached to a heap of rubble (the destroyed Bastille) provides these material entities with a form that makes it graspable by the everyday language[31]. The everyday language carries existence postulates with it, because it creates existence.

Linguistically, what is called the past arises from the present's constant development into

---

[29] Definition 31 does not consider the content-constituent of historical phenomena. This will be developed in chapter 4, but few remarks may not be inappropriate here. Extralinguistic change is change which, springing from non-linguistic reality – the invention of the sewing machine – affects the everyday language. This process occurs in a sign function contracted by an utterance *qua* an expression (= E(ErC)) and the concrete *denotatum qua* the content (= C), *i.e.* of the former complex sign. This direction of influence may go both ways: What Robespierre and his fellow countrymen then put in their native language became part of (or influenced) the meaning heritage of the French language. They may have transmitted a correct or wrong conception of, *in casu*, the French Revolution (= what happened in their days). This alternative is one of the *raison d'êtres* of historiography *qua* the search for historical accuracy.

[30] *Forum Romanum* is an integral part of today's Rome; it is both a necessary condition for it and a part of it. Thus this non-significant part of modern Rome cannot be (thought) removed without changing Rome significantly. The absence of *Forum Romanum* is both the sufficient and a necessary condition for the change of Rome. There certainly are such things as 'small' differences – they have all the properties that other differences have: distinctive, contrastive, significant, and most importantly they are empirical.

[31] Recently we have witnessed a furious debate started by the German novelist Martin Walser over the Holocaust's influence on modern Germany: *postwar Germans have rebuilt their lives in the shadow of the Nazi past*. (*Newsweek* December 21 (1998):23). History and structure are inseparable here; the present cannot explain the past. The weaknesses of Labov's Uniformitarianism are exposed and the slogan 'history in synchrony' must be stripped of its spatial little-boxes overtones in that 'history' is an integral constituent of any present.

the present against the horizon of sign-functional processes. Therefore Coseriu's view *that Jeder Sprachzustand zum grossen Teil Wiederherstellung eines früheren ist* (1974:22;59) will be interpreted to the effect that the present consists of the part of the past that has been continued under the aegis of change; and this is a sign-functional process, otherwise we would not be able to say that part of the present 'points to' or signifies the past. As such the everyday language is eminently suitable for the analysis of the past (cf. Uniformitarianism): it is its own theory. However, this also means that a synchronic state description that does not consider this aspect of its object cannot come up with an exhaustive description: *a synchronic state description is only exhaustive because it does not ask certain questions, in particular, ones that open with a WHY or which ask for the AGENT or the speech situation of a concrete language act. The exhaustiveness that a synchronic theory provides is seen in relation to the theory itself*[32]*,* which a priori excludes the Wiederherstellung *property.*

While the present has access to the past, it has no access to the future. The future is no historical phenomenon, it being pure meaning or potentiality, as such linguistically motivated. Historical prediction is meaningless. Kanngiesser (1972:8;75) puts it like this: *von der Möglichkeit einer Struktur kann nicht auf ihre Existenz geschlossen werden* (cf. Coseriu 1974:50-51), and the Glossematic requirement that linguistic theory should be able to foresee all possible languages is meaningless when it comes to the historical dimension, not only the typological one. Perhaps a theory could predict Old English and Modern English as possible language types, but no theory could either foresee the possibility that an Old English language would come into being or predict on the basis of the potentialities of Old English what the modern language would look like; typologically it is *hard to imagine two languages more different from one another than Anglo-Saxon and Modern English* (Comrie 1989:203; cf. Popper 1974:105-119). The metaphor, the arrow of time, will therefore be reduced to two points, the present and the past[33], and most importantly, the purpose of the present is not to be equated with a drive *ad non esse* (as Augustine has it; see Løgstrup 1978:11; 11-17[34]) seeing that a drive *ad existere* is the logical alternative which can be implemented by a theory of history.

---

32 Coseriu (1974:20) correctly remarks that 'synchronicity' *gehört zum Sein der Beschreibung und nicht zum Sein der Sprache.*

33 The nature of the future as a linguistic tense category as against the preterite and the present tenses is revealing. The future seems always to be a *præsens de futuris*-tense, and its varying histories in the Indo-European dialects suggest that it was not a category equilogical with the present; to a less degree this also applies to the formation of the younger – non-ablauting – preterite forms; but the conceptual affinity between the present and the past seems closer than that between the present and the future (I am thinking of the original ablaut-system in the Indo-European verb). The very notion of *præsens de præteritis* and *præsens de futuris* suggests that the equilogical division of the arrow of time into the past, the present and the future is not empirically relevant (Coseriu 1974:132-151).

34 Following Løgstrup's exposition of Augustine's analyses of time, I venture to criticize Augustine for his monolinear view of existence and Løgstrup (1978:31) for grounding time on destruction, the passing of time being grounded on irreversibility, and the latter on destruction. History is irreversible, I agree; but history is not destruction unless under a value-loaded interpretation or in the case of human (biological) existence through the displacing of a process from the future to the present – destruction is a future act, anticipated by analogy. History is irreversible, irreversibility is grounded on the present's development into the present and concomitant creation of the past; ergo history is … . Here I must prefer Nietzsche's *vivo* to Augustine's *moriar*!

2.3. Serial relationships vs. generic history

The synchronic-based two-state model idealizes its causal factors (as *e.g.* in Comrie (1989:225)). They become generalizable and recurrent – properties that are essential to the generic process. The alternative, a one-state model in which the present state is always the cause of the following, is logically tenuous: in L1 *qua* an autonomous entity the cause and the **af**fectum merge; and not only that; the **ef**fectum (L2) *qua* unrealized potentiality participates in the merger, too. Thus the effectum-to-be, L2, – the successor of the predecessor L1 – is the cause of itself as well as of that which is otherwise the cause of *it*, its autonomous predecessor L1: the potential structure of L2 is an upholding part of the structure of L1 (cf. Question 1 above); a successor (L2), a contributing cause of its predecessor L1, is part of the cause (L1) which is the cause of L2 *qua* an effectum. In a word, the future state is the cause of itself. Why do we need L1? L1 is the container of L2, which therefore contains L3, which therefore contains L4. This is absurd; that the Old English language should contain all the immanent states that develop from it is meaningless. Secondly, that a successor state L1 should contain the successor state L3 makes L2 a predecessor before it has become a successor and L1 a predecessor simultaneously with it being a successor; this infringes Metacondition 4 above[35].

To conclude: cause or 'to be a cause' is not an innate property of the historical phenomenon; it is a structurally acquired property that arises from the latter's existence conditions, endogeneity and exogeneity as well as katageneity: developing into the present, the present turns the present into the past. The causal potential of the historical phenomenon (*qua* an effectum) is not to be seen against the empirically irrelevant horizon of the future, but against the past, and dynamically against the present. The past as the cause of the present is therefore an imprecise way of saying that the moment the present becomes the past, it acquires the causal property.

The 2.3.3-figure represents an intersection's historical existence, a proprial entity becoming a historical phenomenon. The dotted lines represent the present's development into the present against the horizon of the palimpsest metaphor.

The synchronic view operates with a Cartesian view of the time line, that of absolute discontinuity: what upholds an entity at T1 is the same as upholds it at T2 (see (7) in 2.4. below): actuation- and transition problems are irrelevant to such a conception of continuation. Consequently, if an entity changes from T1 to T2, the upholding factors must differ; the two problems are no longer irrelevant, but in either case a period is a sum of contingently connected points whose respective entities preserve their self-identity despite change. This view of change makes do with the upper part of the figure in 2.3.3, and we can see how Uniformitarianism can be observed: the past is not different from the present, or the present tells us how the past is.

The dynamic view of history outlined so far considers the everyday language not as *ein einmalig Geschaffenes, sondern (...) ein unaufhörliches Schaffen* (Coseriu 1974:56; 21-22), whereby the transition- and actuation-problems, the start- and finish-problems vanish. The concept of the present – here I agree with Uniformitarianism – is our starting point; but the

---

35 This criticism also applies to the the one-state, calculus point of view, argued for by Hjelmslev 1963 from the scientific point of view and by Hjelmslev 1928 from the point of view of data (a universal category table; cf. 2.6. below).

present is dynamically effective and is to be comprehended against the horizon of its development into the present. Therefore the nature of the present cannot be generalized so as to be imposed on the past: the past is 'essentially' different from the present (cf. the lower part of the diagram; against Uniformitarianism). The figure also highlights the structural nature of historical linguistics: as a historical phenomenon it completes its potential when it becomes the past in the structural whole made up of the horizontal and vertical syntheses of endogeneity and katageneity.

2.3.3i

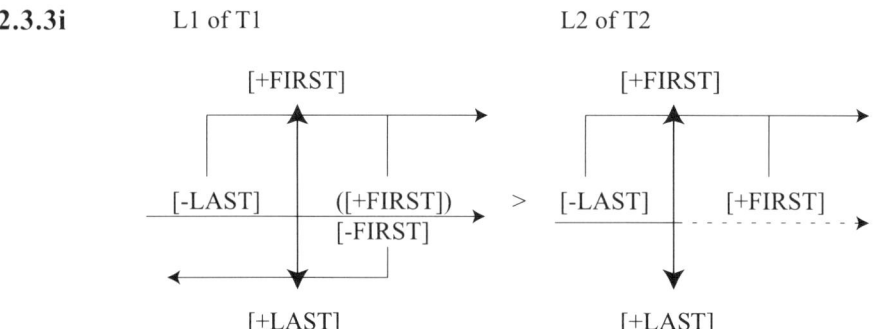

This figure is a static picture of the two-state model, which cannot differentiate between the <u>forms</u> of the predecessor and the successor, L1 and L2 being identical in relation to the synchronic state conception. The figure in 2.3.3ii is a dynamic interpretation of the former, where we can see the <u>formal</u> difference between L1 and L2, these being different in relation to their chronological placement in a developmental series in that L1 loses its [+First] property historically (cf. Figures 2.1.2 above):

2.3.3ii

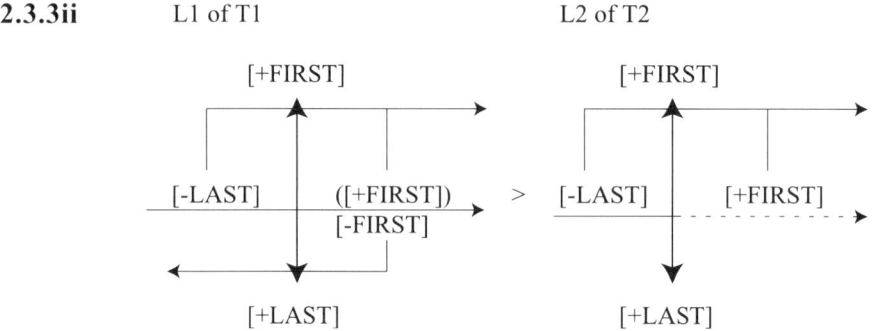

ein unaufhörliches Schaffen of the present and the past.

2.4. The two-state-model and the generic mechanical cause-effect parameter. Its logic implies the concepts of multiple conditioning and a phenomenon's *Eigennatur*. Historical causes are biuniformal, constitutive, effective and non-effective. Answers to Questions 1 and 2 above. The concept of cause is redundant: Aitken's Law

Three elements are involved in the mechanical cause-effect chain as a definition of change, an affectum, an effectum and a cause (cf. Coseriu 1974:94ff.): the cause affects the affectum, effecting an effectum. The predecessor, be it a state, system, a single phenomenon, is the affectum, the successor the effectum. The cause is external to the affectum, and the definitions of predecessors and successors are independent of the definition of causes. This causal chain precludes causes making for stability and continuation; by implication, stability and continuation require no cause. Synchronically, to be an effectum/a successor and an affectum/a predecessor are acquired properties. The following seven items are postulates underpinning the mechanical cause-effect parameter:

1) The same cause produces the same (type of) effect. – The deductive-nomological model's cause, *i.e.* a *modus-ponens* type: $((C \supset E) \& C) \supset E$.
2) Two similar effects imply the workings of the same cause: $((E \supset C) \& E) \supset C^{36}$.
3) A cause is not inherently attached to an object.
4) A cause is recurrent, non-unique and supraindividual. An effectum is an individual member of a plurality of likes.
5) The spatio-temporal setting of a causal chain is immaterial. An effect could have happened anyplace else.
6) There is never more in the effect than (is (contained)) in the cause.
7) The continued existence of an effectum does not presuppose the continued presence of its cause.

The postulates are epistemological diehards in our Western Civilization and are inseparably tied up with the transcendence problem[37]. Like an explanans or theory, this concept of cause operates at a distance of its effects (Hoenigswald 1966:v), and the following scenarios will expound the causal mechanics implied by the two-state model in accordance with the postulates. The Figure 2.4.1i represents a developmental series (a) in a certain place P1:

---

[36] The deductive-nomological method's condition for predictability; this predictability is incompatible with cause being the sufficient motivation in a process as in (1). The logical intricacies will be dealt with in this essay's sequel.

[37] Cause and effect; explanans/theory and explanandum/datum; eidos/form and individual; ground/reason and consequence. The identification of historical knowledge with causal knowledge as well as cause with theory commits the Phenomenal Error and idealizes historical linguistics into a generic-subsumptive discipline, where processes become instances of causes.

2.4.1i (a) > X > A > B > C > D >
           T1   T2   T3   T4

(b) in P1   C1
             ↓
             A > B

(c) in P2   C1
             ↓
             G > B

(d) in P1   C1   C1
             ↓   ↓
             A > B   G > B
             T1 T2   T8 T9

Scenarios (b) and (c) indicate that the same cause affects different phenomena, producing the same effect at the same time but in different places. Scenario (d) is the purely temporal counterpart: the same cause is operative in the same place (language) at different points in time.

The 'histories' are all generic and lead to the dualism, with the process defined by the cause, C1: /Generic process/ [A > B] = [G > B], despite differences. As the model, however, contains an implicit simultaneity-asssumption, it entails the following Parmenides-inspired problems:

1) where the affectum is, the effectum is not;
2) where the effectum is, the affectum is not;
3) where the cause is, the effectum is not, because if the effectum has come into being, the causal process has stopped; if the cause was present together with its effectum, the cause should again be effective – <u>unless</u> the effectum is so effective as to somehow neutralize the strength of the cause;
4) where the cause is, the affectum is not, because the effective cause would have produced an effectum – <u>unless</u> the affectum is so effective as to neutralize the strength of the cause.
5) Hence where the cause is, the affectum is not, because an effectum should be there; where the effectum is, the cause is not, because an affectum should be there. Cause as a dynamic factor, affectum and effectum are mutually separate concepts[38].

---

38 The alternative is that the properties of being affectum and cause are acquired: the (unfrozen) lake of T1 <u>became</u> an affectum after a temperature of -1 centigrade had made its water freeze at T2; this lowering of the temperature <u>became</u> a causal factor in relation to the water of the lake, which <u>became</u> an affectum/effectum *due* to its amenability (to being influenced by the cause). After the lake <u>has become</u> an effect, the temperature is no longer an effective cause, but <u>has become</u> upholding. – Another well-known example: the explanation of the cracking up of Hempel's automobile radiator on a cold night does not make the empirical

2.4. The two-state-model and the generic mechanical cause-effect parameter

The <u>unless</u>-codas in 3) and 4) foresee the concept of the material, concrete Eigennatur of the phenomena as a determining factor in the causal process. If the cause is upholding, too, it is difficult not to regard it as a constituent part of the phenomenon. This dual function makes the cause **biuniformal**, it being effective, upholding and continuing in relation to its effectum. Let's be more explicit.

The here-and-now intersection of the developmental series is the effect of something and the cause of it, too, *i.e.* the latter's continuation. Or: what is the cause of a here-and-now intersection becomes a part of the latter. These mutually related processes take place when the present develops into the present. Therefore Coseriu's view of the existential process is too simple: *Diskontinuitätsfaktor in bezug auf die Vergangenheit ist der 'Wandel' zugleich Kontinuitätsfaktor in bezug auf die Zukunft* (1974:22). In the first place, the concept of the future is empirically irrelevant and that of time is not monolinear; secondly, Coseriu's process does not bridge the gap between the past and the present (*pace* the quote in 2.3.), and why should the continuity-element only be seen against the horizon of a non-existing future. Thirdly, it is doubtful whether Coseriu makes his *Wandel* create the present. I shall therefore elaborate Coseriu's view as follows: *discontinuity-factor in relation to the past is the process of* Wandel (= *the present's development into the present) at the same time as it is continuity-factor in relation to the past*[39]. Next I shall implement the developmental series in 2.4.1.

A historical cause Z (Civil Dissatisfaction or Heightened Subglottal Pressure[40]) changes A into B, and C into D, but not B into C

2.4.1ii

$$\begin{array}{ccccccccc} & & & & & Z & & & Z & \\ & & & & & \downarrow & & & \downarrow & \\ > & X & > & A & > & B(E) & > & C & > & D(E) & > \\ & & & T1 & & T2 & & T3 & & T4 & \end{array}$$

B (= revolutionized A // [e] from [ɛ]) is effected by Z and so is D (= revolutionized C // [i] from [e]). Hence B and D are identical in spite of the fact that they are different (assumptions (6) and (1)) and the affecta are different.

The mutual difference and similarity of B and D imply multiple conditioning; Z cannot be the sole cause of B and D, although it is responsible for their common core of identity. Another supraindividual cause must be introduced, which solves nothing; or the notion of the Eigennatur of the respective affecta must be accepted as somehow active material causes

---

process a part of a historical process proper (Hempel 1942:346); and the effective cause is hardly turned into an upholding one in this case, the broken radiator remains broken in spite of the future weather conditions! My point here is also that we need a distinction between the historical (man-made) domain and physical nature in contradistinction to the so-called Chomskyans' (and Lass's 1997-)attempt at making (historical) linguistics a natural-science discipline (see Anttila 1992:323; and cf. Swiggers 1992; Keesing 1992; Macnamara 1992; Winters 1992).

39 Thus I have replaced Coseriu's change-concept with a more general one, namely that of existence, in that Coseriu follows Jakobson's teleological conception of the everyday language's over-all dynamic function (Coseriu, p. 23). *Wandel* and existence cannot be separated, indeed, they merge. – Coseriu is also critical of the mechanical cause-effect parameter.

40 The causes have been chosen with a view to my historical examples, the French Revolution (cf. also the 1848- and 1917-revolutions in France and Russia) and the raising and/or fronting of vowels.

(rehabilitation of the material substance). The common cause will then account for the relational identity (essence) of B and D, their respective predecessors for their differences. But only the introduction of essentialism, the essence-accidence parameter, salvages the logic of postulates (1) and (6). But which is the more important – the 'nearest' *genus*, the cause Z, or the specifying differences between A and C?

What cause sees to it that Z hits A and C, and not B and D? The only answer which escapes a causal infinite regress and the notion of historical coincidence is that – in contravention of postulate 3 – A and C are amenable to the workings of Z, while B and D are not. This means that before Z can become operative, something else must have happened, which made Z operative *and/or* A and C amenable to being affected by Z: the same cause Y produced A and C out of X and B, respectively. This analysis establishes the following concepts:

1) the principle of multiple conditioning as a necessary aspect of the cause-effect parameter;
2) the concept of a phenomenon's Eigennatur as a necessary factor in change. Aristotle's designation, the material cause, is not appropriate, because it underlines the ergon-nature of objects, whereas Eigennatur implies energeia, compatible with the dynamics of structural processes;
3) the concept of amenability (the rehabilitation of the concrete);
4) the complex nature of change, epitomized in the dynamic principle: for something to occur, something else must (have) happen(ed);
5) the period as an empirically relevant phenomenon.

I shall elaborate items 1) and 5). I suggested above that Cause Y and effect B are responsible for the fact that C becomes amenable to the workings of cause Z, and that therefore Y and affectum X must be responsible for the rise of A, which evokes Z. Thus when it comes to the change of A into B, cause Z and affectum A are responsible for the fact that B is affected by Y; from this it follows that D must attract to it cause Y, which couple will produce affectum E, which will be hit by cause Z, etc. This is absurd and the involved recurrence will be broken by the introduction of the notion of multiple conditioning in connection with that of a phenomenon's Eigennatur.

The upshot of the latter argument is that this idealistic mechanical cause-effect parameter and the two-state model are incompatible. The synchronic, the well-defined and self-contained state concept with its immanent and autonomous object is not empirically relevant. I have now answered Question 1 above: the fact that B (X1) develops into C (X2) causes C (X2) to develop into D (X3). This is not the answer that I am going to pin my argument on, though. No historian is likely to subscribe to the determinism involved. But it goes to demonstrate how the two-state-model as the product of two autonomous states is self-defeating; at least three states will be needed, owing to the peculiar dynamics of the applied causal concept.

That history is generic is the demonstrandum, not a necessary presupposition for the scientific analysis of history (cf. chapter 2.5.). Historical causes are not ideal entities. Civil Dissatisfaction may indeed have effected the 1789- and 1848-revolutions in France; but as an ideal cause, it must be made operative by another (ancillary?) cause[41].

A historical cause is **participative**: it participates in its effect; it is **biuniformal**: it makes

2.4. The two-state-model and the generic mechanical cause-effect parameter

something occur, while something else occurs. Both properties are closely tied up with a phenomenon's Eigennatur. The biuniformal cause makes for change, changing one phenomenon, and does not make for change, making another phenomenon continue. The scope of the principle embraces both the system that develops like a calculus (recombination of elements)[42], and the sequential nature of the spoken chain and developmental series. As such it can be incorporated into my continuum reinterpretation of the *langue-parole* parameter in chapter 1.6. above, whereby the categorical difference between the two types of elements are removed.

Aitken's Law will here be used as illustrative material, this law being paradoxical (Lass 1976:54-56) because it seems to be made up of apparently two disparate processes within one and the same system. In Scots short vowels were lengthened conditionally and long ones shortened conditionally. How to reconcile these two contradictory quantitative processes? Heightened subglottal pressure (= [+Stress]) causes short vowels to lengthen before /r, v, d, z, #/. The same suprasegmental pressure is a realization condition for long vowels, while lowered subglottal pressure (=[-Stress]) is a realization condition for (continues) other short vowels and causes long vowels to shorten outside the above environment[43].

Advocating a finalistic explanation with two components, agents (the two rules) and telos *qua* both an antecedently existing cause and an effect, Lass has the effect of and the motivation of the agents merge in the same concept. Lass's explanation suffers from a number of arbitrary assumptions and he commits the Phenomenal Error; in the first place he reduces one empirical process to that of another in that the production of short vowels is a straightforward manifestation of the underlying system (*a single underlying system* of short vowels), while the manifestation of long vowels depends on a lengthening rule (p. 55; and 2.4.2 below). Secondly, his explanation is a return to an Aristotelian teleology with four (arbitrary, cf. above) causes: (a) the predecessor system of the Scots dialect (the material cause); (b) the two agents, lengthening and shortening rules (the efficient cause); (c) the intended successor system *qua* a system of a specific form (the formal cause); (d) the telos is the final cause. Thirdly, it is absurd to have the obtained result merge with what motivates its rise, *i.e.* the telos and agents.

Lass's manipulation of the data appears from what should be the historical starting point of the explanation: four equiempirical processes with four equilogical causes (V = short vowel; VV = long vowel; R = lengthening environment):

2.4.2	System 1	>	System 2	Process type
	V + Condition 1	>	V	continuation
	V + Condition 2	>	VV	change
	VV + Condition 3	>	VV	continuation
	VV + Condition 4	>	V	change

---

41 Note that an *i:*-sound (= Cardinal 1) is not amenable to the raising effect of heightened subglottal pressure.
42 In Hansen (1987) I have demonstrated this causal principle in an explanation of the *First* and *Second Germanic Consonant shifts*.
43 I have simplified this 17th century data (cf. Collinge 1985:3). – Cf. note 44 below and chapters 5 and 6 for the logic of the style contents.

The diagrams in 2.4.3 explain Aitken's Law in terms of the concept of biuniformal and participative causality on the basis of 2.4.2. Multiple conditioning explains why the FS variant [VV + R] was selected, and not [VV+non-R], the long vowels (bold-faced) being equally suitable as forceful-style products. Internal exogeneity contributes to the selection of the former variant, while the principle of least effort may account for the selection of the RS variant [V+non-R] of /VV/. However, like all systems-based explanations, this explanation also contains an arbitrary element, a fact that is due to the scenario's two-state conception and lack of a historical theory. The 2.4.3i-diagram represents the basis for the selection processes in the change within the dualism's framework, showing how the two sets of **[VV]+R** variants and [V]+nonR, respectively, are not sign-functionally incompatible:

2.4.3i /V/   **[VV] + R**   and   HSP → FS
              [V] + nonR   and   -HSP → NonFS (includes NS and RS)
              *[VV] + nonR   and   HSP → FS
        /VV/ [V] + nonR   and   LSP → RS
              [VV] + R   and   **-LSP → NonRS (includes NS and FS)**
              *[V] + R   and   LSP = RS

The conditions – the R-environment and HSP – realizing a short vowel /V/ as the sign function '[VV]rFS' participate in the production of some long vowels as [VV] as nonRS variants, and neither variant's style content – **long vowels in an R-environment** with the contents FS and nonRS[44] – is incompatible with the other's. The subglottal presure that produces change in a short vowel (> [VV]) is the normal realization condition for a long vowel and will as such be correlated with normal style here /VV/. But since some /VV/ have been shortened, internal endogeneity will here create a sign with a relaxed style content: '[V]rRS', which in turn will structurally entail the realization of other /VV/ as signs with a NonRS content. – The figure below illustrates the motivating processes.

2.4.3ii

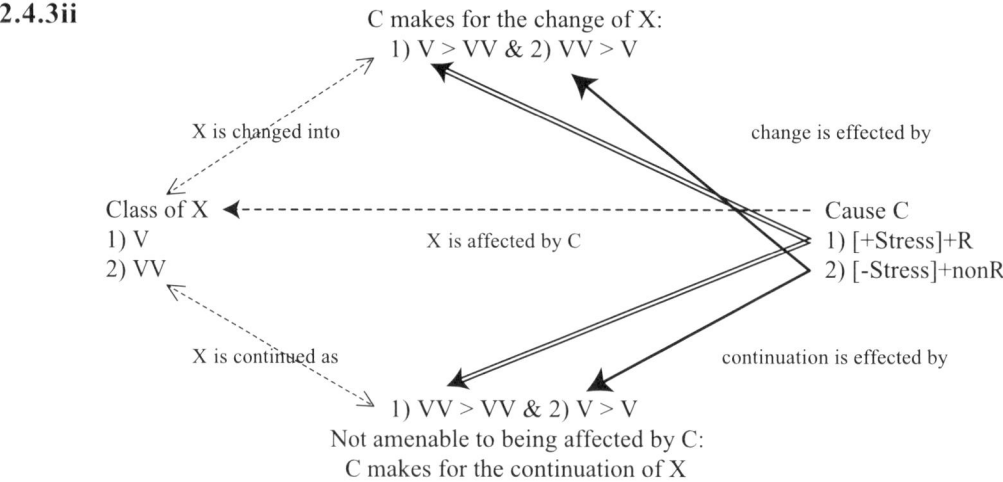

---

44 It will be shown below that in the attitudinal system consisting of three attitudinal sign functions RS, FS and NS NonRS is polysemous in respect of normal style (NS) and forceful style (FS) and NonFS in respect of normal style and relaxed style (RS).

2.5. On the validation question

An apparent consequence of the biuniformal and participative concept of change is that the word *cause* should perhaps be struck from the vocabulary of history and replaced by *realization condition* and *constituting factor*, or simply by *process*. Cause, motivation, development and change (and variation) are all concepts that are embodied in the word – the *par excellence* empirically relevant concept – **existence**.

Arguing *e silentio* from the fact that X is not changed by C, to the conclusion that C is not effective in the constitution of X, is empirically noncogent. Obviously, our standard logic (embodied in the law of non-contradiction 'not both A and nonA') cannot cope with the empirical facts of change with, in our case, a 'cause' making for both change (A) and non-change (nonA). What unites the absurdity is the concept of constituting factor: C participates in two phenomena such that one appears to change and the other to be continued.

The logic of the latter is as follows: if C is effective in the constitution of a phenomenon X, C is certainly a constituent of X, irrespectively of X being changed or not. On the other hand, if C changes a phenomenon X, then C must also be a constituting part of X; but here there is no alternative whatever. To be effective is the larger concept of the two: *to be effective * to cause change*, as such we do not need the latter; if necessary, it can be derived from the former. The following sums up the biuniformal and participative concept of cause:

1) X_1 develops into X_2 and Y continues. C is part of the realization conditions for Y. – This has been exemplified by Aitken's Law, where the motivating factors are internal endogeneity and exogenous motivation.
2) X_1 develops into X_2 and X_1 is continued under the aegis of change. – This is exemplified by so-called isolative change, where the motivating factors are external endogeneity and katageneity. – (To be developed in chapters 5 and 6).

Change and continuation, A > B and A > A, have been combined by the notion of '**being effective in the existence of**'.

2.5. On the validation question. The immanent state is self-validating, as a historical state it is validated transcendently. The state concept is analyzed in analogy with the phenomenon; it implies the concept of breaks. The state seen in the light of contingency and potentiality; contingency leads to absurdity: the necessity of the contingent fact

Geneticism is criticized for validating the historical state externally or transcendently[45]; the historical phenomenon is not independent (cf. chapter 2.3. above). Such criticism presupposes the empirical relevance of the autonomous state concept. The historical state fulfils its existence in relation to something outside itself (p * p), while the immanent state is what it is because it preserves and contains its own self-identity (p ⊃ p): it is not defined in relation to the fact that it is bound to change. While synchronic linguistics does not define its state

---

45 This section is written with Glossematics as its horizon and silent opponent. Nowhere does Hjelmslev define the state-concept, and his view of historical linguistics is dominated by reconstruction.

concept, but takes it for granted, it is my purpose to show that an argument can be set up that demonstrates the untenability of the concept of the immanent state, an argument which underlines the fact that the synchronic state is an *explanandum,* not an empirical given.

Synchronically, the historical state finds its raison d'être in the historical chain, preferably both as the predecessor and successor of some other state. As a successor it finds its essence in the fact that it 'exists as the heir to a past out of which it has grown' (Nietzsche 1966:ch.3), and this knowledge is the sufficient and necessary justification for its present existence. Or: each state is validated by being a stepping stone in a Grand Evolutionary Scheme with some ultimate goal – positive-progressive – or for that matter, negative-regressive – as its end-station[46].

The synchronic objection to the historical point of view rests on a specific definition strategy (2.5.1 below) and therefore mistakes a methodological convention for an empirical fact, a form for its content (Popper 1974:53; Nietzsche 1966:232). This fact was noted by Coseriu (1974:20-21) in connection with his criticism of Saussure; synchrony belongs to the *Sein der Beschreibung* and not to the *Sein der Sprache*[47]:

2.5.1 1) a scientific phenomenon is an extensionally defined class to be analyzed deductively into its 'members';
2) a class is therefore not validated transcendently, but by its elements;
3) a historical phenomenon is validated transcendently;
4) the historical phenomenon is therefore not a class, but an element of a class,

hence 5) the historical point of view is scientific if and only if it observes the principles of a theory constructed with a view to dealing with phenomena as (immanent) classes or as classes of classes;

hence 6) the historical stage must be regarded as a (synchronic) class, historical analysis requires two such temporally consecutive stages, and consists in the finding of differences between two such stages (Hoenigswald 1966:v; Labov 1994[48]).

---

46 See Olsen (1979) for a summary of various conceptions of the historical process. – Lass and Anderson (1975:167-183; 188-205) and Lass (1976:51-56; 57-72) view certain so-called major sound changes as unitary entities with temporal extension so that the time it takes for the process to be completed involves the notion of transitory states *qua* ancillary steps, which fill the gap between the start and the finish of the whole process. So synchronic linguistics saddles itself with the further problem of defining the transitory state as against the state that is not transitory.

47 Otherwise Coseriu is too kind to Saussure, saying that Saussure was only interested in methodology, not *Ontologie,* in being able to distinguish the two points of view, synchrony and diachrony (p. 21). But Saussure's conception of the sign is not pure methodology, it carries empirical assumptions with it (cf. chapter 3 below).

48 This involves what Labov (p. 21) calls the Historical Paradox: the historian studies differences between the present and the past, but we do not know how different the past was from the present. I have two objections to this narrow view of historical linguistics: it can be based on no full-fledged historical theory in that it cannot accommodate continuations; and with reference to Nietzsche, the limitation is necessary: Labov does not see that the very fact that we 'forget' something of the past is a fundamental constituent in our historical existence. Historical analysis, let alone the historical process, is not what Lass (1997:111) calls *replication.* If the present was mere replication of itself (its past), there would be no difference between the present and the past, no history. Lass is caught up in the web of his own biological metaphors and concepts, continuing the imperfection view of history.

2.5. On the validation question

The fixing of scientific phenomena as classes and the consequent fixing of the scientific horizon by the state concept eliminate the stage property of the state; the absurd corollary is that this property is assigned to the state, when we come to the historical analysis. What the symbolism 'A > B' represents cannot be regarded as a complex empirically relevant unit, but regarded as two independent objects with '>' an unnecessary complicating factor, as the change from L1 to L2 is realization of the potential of L1 through laws internal to L1.

A method's internal validation is an aspect of the transcendence problem and the historian may counter: the (often transcendent) use of formal theories requires validation, while the historical phenomenon is self-validating because of the very nature of historical **ex**istence[49]. Finalistic or telos-based validation may thusly be eliminated from historical analysis: the concept of the future state plays no role in historical existence; the present state is not validated by the fact that it may become the predecessor of a future state. On the contrary, the present (*qua* a here-and-now intersection) is the meeting ground of the present and the past, with the past being validated by the fact that the present exists out of it (external endogeneity), and with the present by the fact that it continues and creates the past (katageneity). The historical phenomenon is validated as it becomes *a necessary constituent of the present* (cf. 2.5.3iii below). From a global point of view, the historical phenomenon is validated externally by the fact that it contributes to man's historical present, *it being a constituent of it and a precondition for it*. This leads to the following implementation of the figure in 1.5.2 (cf. chapter 2.1. and Def.s 31):

2.5.2 A historical phenomenon **ex**isting into the present (1) embraciates its surroundings so that they become both a precondition for its actual existence and part of it[50], and (2) exbraciates – through its *gegenständliche* nature – its sourroundings so that it can be regarded as existing at a distance of these surroundings.

With the sign function (ErC) as a precondition for both processes, a historical phenomenon's surroundings are: what becomes its past (external endogeneity; katageneity), its progenitor (connotation; internal endogeneity; exogeneity), the meaning that arises when it completes it in *parole*.

The spoken chain like the destruction of *La Bastille* (the creation of a pile of rubble) or the making of a marble statue contributes to the creation of the present, and the linguistic tradition's characterization of the spoken chain as individual implies an anthropocentric point of view. The spoken chain is 'mine', not the group's; the tearing down of *La Bastille* was the activity of cooperating individual persons that led to a transindividual fact (through the division of labour), and the statue is the sculptor's. The generality of the sign-functional connotation cannot be doubted. Dynamically, the spoken chain is an *existent*, it exists out of 'me' and the history of the L-continuum's elements. The synchronic linguist regards *parole* as a material existent in relation to its underlying system or form. This system is necessari-

---

49 Attempts at justifying existence seems to end in religious belief or moral arguments that carry no cogency. Hjelmslev's belief in the Empirical Principle's three requirements as the ultimate scientific arbiter is also a postulate: consistency is left undefined, exhaustiveness is undefined; simplicity may be a quantifiable requirement, but the notion of the simplest analysis presupposes that the relevant analyses are exhaustive.

ly collective, social or general, its manifestation or products – when *en ergeiai* – are necessarily neither collective, social nor general. This is the Synchronic Paradox and synchronic linguistics has no empirically relevant logic to maintain both assumptions, social and nonsocial. Hence, like the stage-property, *parole* is discarded as a scientific object, and with this Gregersen's Empirical Problem vanishes – from the theory, not as an empirically relevant problem.

In what sense does the spoken chain become transindividual? The spoken chain is both individual ([+Attachment]) and non-individual ([+Separation]) in relation to its 'origin'. Thus 'my' utterance is also 'not mine', but my interlocutor's. With a slight alteration of Coseriu's *Wandel*-concept: the spoken chain <u>becomes</u> *Diskontinuitätsfaktor* (exbraciation) in relation to the speaker and *Kontinuitätsfaktor* (embraciation) in relation to the listener. Not until then has the spoken utterance completed its – I hesitate to say purpose or function – transindividual nature or itself, and Paul's view of how language becomes what it is through both a division of labour and cooperation has been asserted (cf. also Goldmann). Let's complete the analogy: there may be a form behind or plan with the finished work of art, but hardly a system *qua* a collective calculus of discrete entities; the same goes for the Parisians' concerted effort that led to the July-events in 1789.

Saussure did attempt to validate the synchronic state through the purported psychologcial reality of the simultaneous structural interplay between A and B in the synchronic functions, A::B, constituting the state's paradigms (*e.g.* the past tense :: the present tense), while the historical formula A > B had no such reality because it did not enter into psychologically real, structural relationships. I shall demonstrate that this is not correct. Saussure's point is not irrelevant, but depends on the definition of the historical formula[51]. But Saussure's argument backfires as it transcends the borders of the immanent state. Psychological reality is transstat-ic; any speaker of a language is aware of, even alert to, social- and geographical differences that enter into functional relationships with elements in his own (immanent) system: ModEng *you* (the singular and plural) vs. *y'all, youse*; the singular *-es* of the verbs in Southern British English also contracts a socio-dialectal function with Ø. – Cf. also Labov's Bill Peters example.

The extensionally determined synchronic state, self-contained and autonomous, presupposes the concept of **a break**. When the synchronic linguist makes his cut, he implies that there is **a break before** the state he wants to deal with: **[L2**, and a **break after** the state he does not want to deal with: **L1]**; and by analogy it is assumed that there is a break after L2, **[L2]**, as well as a break before its predecessor: **[L1]**. In dynamic terms, a **break-after** will evolve when what has become a predecessor is no longer the successor of the present, and at the same time a **break-before** will evolve when a new successor has been created.

The two-state model presupposes two such well-defined breaks separating L2] from [L3; but the break-after concept is somewhat spurious as it does not rise in the dynamics of the present moment: how can there be a break-after the language state of the actual present

---

50 Pheidias did not make his sculptures as unmotivated brain children of his.
51 I agree with Weinreich *et al.*'s criticism of Saussure, but do not accept their consequences, as they merely replace one synchronic conception with another one (1968:120-122), in fact the same synchronicity as Labov continues in his 1994-work – conducted against the general horizon of orderly differentiation and heterogeneity.

2.5. On the validation question

moment? What will succeed that break? It would create the end of this present successor – the end of existence. Or: how could a successor break into this successor-turned-predecessor in order to create historical coherence? That this break-after should be created in respect of the future makes no sense. The future (state) does not exist. *The synchronic state par excellence* is no well-defined object; it is bound to be open-ended. The irony of this is that the break-after is created by the developing language: when a state develops, it creates a break before itself and a break after its predecessor; developing further it acquires a break after because of its successor's break-before. The logic of the two-state-model will be summed up as follows:

2.5.3i A predecessor state, L1, and its successor state, L2, must each have a set of phenomena that constitutes a break-after, defined by being the last (elements) of L1 ([+Last]) and by a set of phenomena that constitutes a break-before, defined by being the first (elements) of L2 ([+First]). By analogy, L1 will also have a set of beginning elements, [+First], and L2 a set of ending elements, [+Last].

This scenario infringes Metaconditions 2 and 3 in chapter 1.5. What will fill the absolute gap between State 1 and State 2? There is no communicative situation without the everyday language. In the history of any everyday language there is no gap without language. Human existence in this absolute gap is impossible: no existence without language. Secondly, with such a gap between two synchronic states, there would be space for another language to move into and occupy, but with no external endogenous connections; there would still be a gap. The notion of the well-defined synchronic state is absurd[52]:

2.5.3.ii ... > [L1] > [L2] → ... > [L1] # [Lx]# > [L2]

The objections to the synchronic scenario in 2.5.3 mirror the standard criticism of the mechanical (Neogrammarian) conception of communicative break-downs due to the mechanical wear-and-tear of a language with the break-downs being redressed subsequently. However, the notion of pressure-points, imbalances and the like, communicative break-downs, even of a languageless situation will always be seen against the horizon of the everyday language.

The immanent state, L, necessarily with a break around it and unaffected by the fact of change, is naturally characterized by the features of internal endogeneity [+First] & [+Last]. We can therefore describe its *monolinear* development into L2 in greater detail as follows. L2 will have no [+First] feature in accordance with metacondition 2. The two states will cohere through the [-First] feature of L2, which cannot acquire the [+Last] feature and [+First] feature in order to become equiempirical with its predecessor. Correspondingly, the predecessor of L, L1, cannot have the [+First] and [+Last] features, but must be characterized by the [-Last]-feature:

---

[52] The closest we may get to such a situation is when a language experiences an extreme extrasystemic contact situation (*B-contact* (Samuels 1972:92)). If we take the imposition of Anglo-Norman French and later on Central French on Middle English as an example, we must still say that this extreme contact situation created no gap in the existence of English. On the contrary, the imposition occurred against the horizon of a receding and continuing native language.

2.5.3iii   L1 [-Last] > L [+First] & [+Last] > L2 [-First]

Historical coherence is established by L1 and L2 in relation to their successor/predecessor, which preserves its autonomy. This scenario is inconsistent as all three languages must be equiempircal, and if we equip L1 and L2 with the obvious features of [+First] and [+Last], respectively, the model becomes absurd, as L can be neither [+First] nor [+Last]:

2.5.3iv   L1 [-Last] & **[+First]** > L **[+First]** & *[+Last]* > L2 [-First] & *[+Last]*

The alternative is that the autonomous state acquires properties that are consistent with its successor and predecessor, and it does so if it possesses changeability and if historical development is not monolinear; changeability will therefore be as essential to the immanent state as stability:

2.5.3.v

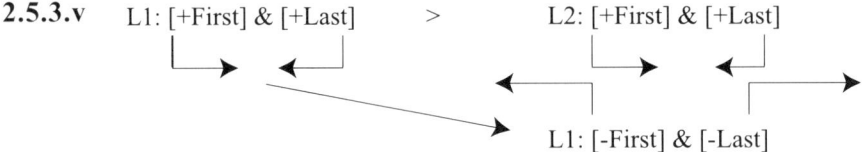

It is the developing state (L1; the original successor) that establishes the coherence of the present with the past and which equips the present with its dynamic element ([-Last]). Another curious thing is that the present state does not *complete itself* unless it has become a predecessor[53]. Furthermore, a language state is a structural whole in which a historical element is necessary; this element is neither an expression of heterogeneity nor an archaic layer (Weinreich *et all.* 1968:149), but a necessary and structurally integrated part of the state *qua* an empirical whole. In this sense, *under the aegis of change*, the synchronic state *wiederherstellt* its predecessor (Coseriu 1974:22).

The concept of totality has been seen partly in analogy to its parts, and I am not implying that the concept has the same empirical relevance as its parts. I have tried to answer the question which the synchronic linguist should have answered: what are the synchronic limits of the state and how is the latter to be constituted? While a sign or the spoken chain appears to exist objectively, in separation from the speaker, the state has no such objective existence, and the synchronic linguist still has to answer the question: is the totality a metaphysical statement disguised as a scientific supposition or is its existence purely nominal, to be defined extensionally by its denotata (cf. 2.5.1 above) – which, to anticipate chapters 4.1. and 4.2. below, will always be single phenomena conceived as parts?

However, if the just mentioned objectivity, which agrees perfectly with the epistemological requirement that we cognize at a distance of the cognized, is so generalized as to exclude questions about the relations between two consecutive states, all that will be left to the historian is the notion of the systemic potentiality of the calculus point of view: history is made into the study of potential developments. This is absurd – or fiction.

---

53 The figures are to be understood in connection with 1.5.6 and 1.5.11 above.

2.5. On the validation question

In chapter 1 we saw how our Western Civilization from its awakening has tried to dress the world of sensuous experience in the straitjacket of staticness, and in particular to exclude immanent historiography from the domain of scientific objects. In the diction of modern nominalism empirical data is contingent and accidental; historical events are therefore not the proper object of scientific analysis, at best raw material for a plunge into real reality. Being contingent, the empirical data is not necessary. This means two things: data is dependent on something else (a system) and it need not have existed where and when it actually did. It was merely successful in the competion with other possible data which was not realized at a given point in time. Its realization was in fact due to historical accidents – over which the scientist holds no sway! So the proper scientific object is the system that holds the class of potentialities *qua* possible realizations and impossible realizations.

With the three concepts – contingency (accidence), potentiality, objectivity[54], which underpin the immanent, autonomous and well-limited state-concept – the synchronic linguist may shunt man's historical sense and the developmental series into the museum of curiosities[55]. Potentiality means systems-controlled possibilities; contingency makes one possibility as relevant as any other possibility; *gegenständlich* objectivity makes the history of the system unnecessary. I shall draw some of the consequences of this position and demonstrate its absurdity.

For cognizing man 'to get started', there must be at least one experience that kick-starts him into his ideal h(e)aven – if that is where he wants to go; that this experience could have been different from what it was does not matter: *what is empirically possible presupposes the necessity of at least one realized possibility*. This does not mean that there is a cogent, predictable reason why B and not anything else appeared in L2, and why A of L1 did not develop into C in L2. It means that now that B is the case, nothing else can be the case here and now. B has no alternative, despite the fact that C could have been the case; the fact that C is a possible alternative does not imply that anything could have been the alternative of A. This inexorable fact is not 'blind' to being treated by a theory, a theory of history, a theory that describes the structural-historical nature of the present's development: the theory will therefore reflect the constraint that **existence** puts on its own possibilities. Thus what is to be rejected is the statement of contingency in 2.5.4:

2.5.4 If anything, say X3 and not X2, could have become the case in L2, there is no reason to assume that (i) the fact that X1 of L1 does develop into X2 determines X2's development into X3, and that (ii) X2 as the successor of X1 should in any way influence X1.

Let the synchronic state L1 be determined by all the phenomena that may appear in P1 at

---

54 In the above *gegenständliche* sense.
55 Irrespectively of whether we shall ever obtain adequate knowledge of how the human brain works, there is no *apriori* reason for the assumption that our mind functions ahistorically, nay, antihistorically. There is no noncontingent reason for the supposition that for man to obtain a sensible grasp of reality, his mind or language must perform in the ahistorical mode. That, on the contrary, our mind is historically structured in compliance with our historical existence is not unlikely (Jacobsen 1976:27).

T1. L1 is then positively and negatively defined by all the phenomena that may and may not appear in P1 at T1, respectively.

What enables a phenomenon A to appear in P1 at T1 and prevents (a phenomenon of the class of) nonA from appearing there and then? Nothing (in the) past (of A) is so effective; besides the past is irrelevant to the present. Hence (any phenomenon of) nonA might just as well have appeared in P1 at T1; this is extreme anything-goes contingency. Thus (the) chaos (view of the empirical world) follows from contingency. If contingency is rejected, the chaos view must go, too.

The synchronic state is not completely contingent and it is static. What are the mechanics of such stability (2.2.3)? What makes A fill a position (P at T) so that nonA does not fill that position? The mechanics of stability are such that A in P at T is upheld by a cause and nonA prevented from occurring there and then. The synchronic state has an element of necessity to it. The issue is not why French is not spoken in Britain, and English in France – this might have been the case; but why the English language consists not of words and sounds from French and English, but of only English words and sounds. Why should a state lose this necessity, when it 'becomes' a stage?

I shall formulate the alternative below in the Non-Contingency Requirement (= A) and its contradictory (Requirement B):

REQUIREMENT A
> If A, the class of phenomena constituting L, is such that it cannot appear in any other place than P at T thereby excluding all other classes from P at T, the mechanics of stability must explain what makes A time- and place-bound.

No synchronic theory can tell us this, which leaves us with

REQUIREMENT B
> If A, the class of phenomena constituting L, is such that it can appear in any P at any T, the mechanics of stability must explain what makes A time- and place-free, and what makes A and no other phenomenon appear in P at T in spite of the fact that A is so time- and place-free that it could have appeared anywhere else.

No synchronic theory can tell us this. So where does this leave synchronic theorizing – how are we to understand its contingency- and potentiality-concepts? Saying that A is *a* possibility whose realization is contingent on historical accidents is a statement of what should be proved. The fact that we can so think of the empirical world does not legitimize the transference of mental possibility to empirical property or fact

The dynamic dimension fares no better: if A of L1 may appear in P1 at T1 as well as in P1 at T2, and B of L2 may appear in P1 at T1 as well as in P1 at T2, what will the mechanics of the change from L1 to L2 be like? There will be two sets of motivating factors (internal laws), one changing A into B, if and only if A appears in P1 at T1, another set changing B into A, if and only if B appears in P1 at T1. Furthermore, B of T2 (< A of T1) could just as well not have appeared at T2, but at T3: A of T1 > ? of T2 > B of T3. The synchronic linguist cannot answer why A and B appear; he must resort to historical accidents, which paradoxically serves his purpose: history is no scientific object; but only because the argument is circular: history is not scientific/rational because it is not theory-*fähig* – history is not theory-*fähig*, because it is not scientific/rational. The three purportedly validating *properties* of the dualism of the synchronic point of view:

the state/system	is	*potential,*	*necesessary,*	*objective*
the material of the state	is	realized possibility,	contingent,	individual

lead to absurdity: the synchronic linguist describes possibilities that need not be there where he places them, unless he can demonstrate that the possibilities are there because of some necessity. And once he tries to answer this question, he becomes a historian.

In the light of such problems it is understandable why Glossematics relies on the *Selbstverstehen* of strict nominalism, a method by which any – past, present, future; realized or unrealized – language can be described, a method that enunciates no existence postulates, a method validated by a logical principle: consistency, by a relative principle: exhaustiveness, and by an aesthetic principle: simplicity

With reference to Kant's First Proposition (1923:18): a phenomenon that does not fulfil its purpose is a teleological contradiction, we may conclude that from a communicative-functional as well as the more basic, the existential point of view, a language possibility that does not fulfil a sign-functional function is an empirical contradiction.

2.6. On the logical geography of the A>B relationship of the two-state model. Historical necessity and criticism of contingency. Synchrony defeats the two-state-model. Cause is an acquired property. The category, which is not stable, makes for synchronic stability. The Commutation Test and a definition of change, an epistemic concept

Contingency and immanence make relations between two states immaterial to the nature of the state[56]. In spite of this synchronic theory operates with different states of the same language, an extremely polysemous concept which reduces the *temporal* and *spatial* differences between L1 and L2 to identity in respect of what the undefined sign *the same language* denotes (cf. chapter 3.2. below).

Glossematics (Hjelmslev 1970:136-137) provides a formal statement of the relations between a mother language and its daughter languages within the confines of the two-state model: the logical relation between a predecessor – be it a state or a sound – and its successor is one of unidirectional presupposition (= 'L2 ⊃ L1' in 2.6.1):

---

[56] In the early days of the synchronic point of view Hjelmslev could even claim that there is no causal connection between a change and the establishment of a system (1928:227). This is either nonsense or a superfluous statement in so far as 'to establish a system' belongs to the logical level of methodology: if the linguist has made his synchronic cut, separating System 1 from its successor, then the synchronic linguist is bound to disregard what preceded System 2.

2.6.1   If L2, then L1 = Not a successor without its predecessor
   or   If L2 develops into L3, then L1 develops into L2[57].
   If L3 comes from L2, then L2 comes from L1.

The very dependency is immaterial to the descriptions of L1 and L2. Therefore the logical formalization of the relation by the logical conditional is appropriate[58]. The historical properties of the state, *to be a predecessor, a successor; to be a cause, an effect; to be a mother language, a daughter language; to be a stage* are state-external. And the synchronic point of view does not escape a certain degree of inconsistency: from where does an immutable state acquire its stage-property? If no change is involved in the constitution of L2, then the fact that L1 develops into L2 is not causal (instrumental) in the rise of L1's stage-property. An answer may lie in the following.

The peculiar nature of the function contracted by L1 and L2 is one of correlation, *i.e.* the Glossematically undefined either-or-function: the mother language and the daughter language do not exist at the same time, an assumption that safeguards the linguist against regarding A>B as an independent unit: despite the historical relation between L1 and L2, L1 and L2 can be analyzed independently of each other, and since L1 does not exist at the same time as L2, then no causal relation will exist between them. Hjelmslev adopts one interpretation of the Jones Hypothesis[59].

In the domains of logic, geometry or mathematics the irrelevance of Requirement A above and the relevance of Requirement B may be evident, but once we put them in bodily form, the weakness of the dominant view will immediately catch the eye of Doubting Thomas.

ad A   *Existence is time- and place-bound*
   Could Kepler have done what he did without his predecessors? Is his achievement of such a nature that it could not have been achieved either previously to or later than its actual occurrence? – Could the French Revolution/the destruction of *La Bastille* have taken place irrespectively of what events happened before it? – Could these events not have taken place previously to or later than their actual occurrence? – Is the development of OE /a:/ into ME /ɔ:/ time-and place-bound? Could Primus not have produced /ɔ:/ or [ɔ:] without him or others having produced /a:/ or [a:]?

---

57 It is note-worthy how different this dynamic-synthetic version is from the analytic-static one; its content is far more deterministic. *E.g.*, is it the process between L1 and L2 that is the sufficient condition for the process between L2 and L3, which *makes L2 develop (into L3)*? Or must we make do with a virtually vacuous interpretation to the effect that 'the fact that L1 has changed is the sufficient condition for the fact that L2 *may* change'. Note again the meaningfulness of the everyday language; is this a statement of empirical necessity to the effect that the actual present state is bound to develop – to create a new present or must we appeal to the generalization of the contingency-principle: the fact that L1 did develop into L2 does not imply that L2 will develop into L3. And if we apply the contingency of the actual present state to its predecessor, then we are back to square one: historical accidents made L1 develop into L2, change being unnecessary, which entails this absurdity: the present's development into the present is unnecessary. And the phonemic version: if ME /ɔ:/ develops (into ?), then OE /a:/ develops into ME /ɔ:/. Is this the meaning of the form: ME /ɔ:/ ⊃ OE /a:/?
58 The logical constant ⊃ attributes no meaning to what it connects, p and q: p ⊃ q.
59 If the past does not exist, the notion of the synchronic cut is redundant!

2.6. On the logical geography of the A>B relationship of the two-state model

ad B *Existence is time- and place-free*
Could what Kepler did just as well have taken place in the days of Thales or Plato or today? – Could the French Revolution/the destruction of *La Bastille* just as well have taken place in the days of Thales or Plato or today? – Or could all three events just as well *not* have taken place? – Is the development of OE /a:/ into ME /ɔ:/ time- and place-free? – Could /ɔ:/ have been produced at any other time?

These are not unlikely answers:
(i) Kepler presupposed the work of his predecessors;
(ii) his scientific achievement could have been reached at any other time without the presence of any other astronomical theory;
(iii) if Kepler had lived at any other time, he could have done then what he did in the first decades of the 17th century (*pace* (i)),
(iv) Kepler's Laws could just as well not have been discovered then.

However, what must bother the contingency supposition is that Kepler did what he did then, neither Thales nor Plato anticipated his work, and nothing else happened then than what actually did happen.

The four answers show how we tend to discard the anthropocentric basis (the connotation, *la trace de l'homme*) of historical achievement and how intermingled historical activity and physical nature are. Answers (ii-iv) are true because of the object under investigation, recurrent or stable objects of nature.

This, however, is not the real issue, but it shows how historical events may have got their chaos-making properties, contingent, accidental and non-necessary. Scientific discoveries are historical processes, and we must take a brief look at a commonplace in history, the purported ambiguity of the word 'history'. Our tradition has it that the word has two senses (cf. Lass 1997:16-18); it has three: (1) it refers to historical events, what happens (*geschieht/ist geschehen*); (2) it refers to the product of the scientific activity of the historian, a history or an exposition of some historical events; (3) it refers to the (type of) activity of (2), too. This scientific activity contributes to the present's development into the present, just as the destruction of *La Bastille* did. As Kepler's Laws could have been discovered before they actually were, *La Bastille* could have been destroyed before 1789 or not have been destroyed at all. Similarly, the sound change above could have occurred in the 9th century, and could not have occurred at all – one can *imagine*.

For the destruction of *La Bastille* to take place and for the OE change to occur, that building and OE /a:/ must exist. Similarly, for Kepler to discover his laws, he must falsify ('destroy') Geocentrism – or must he? Did Kepler falsify Geocentrism or did he discover his Laws?

In discovering his Laws, Kepler falsified Geocentrism.
In making /ɔ:/ appear, speakers of OE/ME 'destroyed' OE /a:/.
In making a certain pile of rubble, Parisians destroyed *La Bastille*.
(In writing their grammar of modern English, Bache and Davidsen-Nielsen ... did what?
In writing his history of the French Revolution, X ... did what?)

Development into the present is more than the seeming monolinearity of percepta (cf. chapter 1.4.6). However, the problem has now been moved to the successor: why did the successor, say, not occur before it did? – and we are back to the A/B-questions above. The issue may be stated as follows: **what is the connection between the rise of something – including the present, and the change of something else – the present?** My answer is: irrespectively of whether we can establish a connection of some kind, the 'connection' is to be **comprehended**: comprehension is the name of the game. And we comprehend by means of a theory – of history. And the next step in this process is to formulate a general and necessary relation between the processes: **for something to occur, something else must occur or have occurred.** This dynamic principle defines the notion of empirical necessity which is the relation between two phenomena constituting a non-accidental horizontal synthesis. The relation is such that it excludes the possibility that *any* thing may succeed a given successor or that *any* thing could have preceded a given successor. Contingency is rejected and a kind of determinism is now a property of existence which is not the static determinism of *modus ponens*[60] (cf. Jones 1969b:170).

If all data is contingent and the substance of historical phenomena, its *physical substratum*, is inert, mere raw-material *qua* malleable, adaptable and form-able matter, unformed until an action-constituting *pattern* (Itkonen) is correlated with it, it is incomprehensible why Thales did not form historical preconditions *qua* the norm that controls the social behaviour that was to result in the setting up of Kepler's Laws. Thus we have two theses: that of still-to-be-proved contingency (Thales could have done so) or *empirical necessity*: the destruction of *La Bastille* was time- and place-bound.

The so-called raw material of historical development – and of the manifestations of forms – is necessary beause it is *formed*. Historical existence is from form to form occurring in structural processes, not a Glossematic katadynamic triad (Hjelmslev 1963:49-52; Hansen 1985). In his presentation of reality as different from what it used to be, Kepler's sense of history must have played a significant part: Kepler must have known that change is possible – that something need not be what it is (said to be – by his predecessors); but for reality to have this element of contingency, something <u>must</u> in the first place be or have been the case; secondly, for Kepler to meaningfully reject geocentrism, he must have supposed that the received world picture must be replaced by another (not any other) theory, which received part of its meaning from the old theory. The notion of existence *qua* formed matter emanating arbitrarily from unrealized possibility *qua* unformed matter has no empirical relevance over and above the empirical necessity of historical existence *qua* always formed existence. The same goes for the speaker of an everyday language: he or she must have 'innate' knowledge of the fact that he or she is not obliged to replicate or continue the past faithfully, but may change it.

As regards the French Revolution – or L2 – I think few synchronists or logicians will care

---

[60] It is virtually vacuous to say the following: on the assumption that a daughter language unilaterally presupposes a mother language, then if a language is a daughter language, there also *was* a mother language. – $((p \supset q) \& p) \supset q$. Hjelmslev would say: the mother language's former presence is *a necessary condition* (*i.e.* the constant functive in the genetic function) *for* the subsequent presence of the daughter language, whose presence is *not a necessary condition* (*i.e.* the variable functive) *for* the former presence of its mother language. – The further investigation of the determinism of formal logic will be dealt with in a sequel to this essay.

2.6. On the logical geography of the A>B relationship of the two-state model

to present an argument for this event (and L2) to be time- and place-free. The next question would be: what about the predecessor events (L1) and successor events (L3)? Would they establish an equally meaningful historical process : L1 > (LX >) L3, as L1 > L2 > L3? Let's adopt a middle-of-the-road course.

Something else could have occurred instead of the French Revolution; this contingency will be controlled by empirical possibility; and empirical possibility is in turn controlled by empirical necessity: Y could have happened instead of X, but not anything but X, *i.e.* any member of the class of nonX. This means that the French Revolution and the French language spoken in those days were both one possibility of the restricted class of possibilities that could have been the present's development into the present in those days; the same is true of ME /ɔ:/: all were realistic successor possibilities in their respective days that could not have occurred at *any* other time or place.

We are now back to the perennial starting point, *i.e.* how to unite (an aspect of) idealism and the concrete: realization/manifestation (also in the dualist-katadynamic conception) and historical development/the present's constant development into the present (in the existential conception), with both processes being **necessary**? The present's development into the present is neither a deterministic-necessary process nor a random-accidental one, but is so necessary (rational) that it always keeps its collective orientation (cf. 2.5.2) and can be subjected to theoretical treatment. In Lévi-Strauss's and Luick's words and Hermann Paul's spirit, the theory of history will be formed in and by the code of history, **chronology** including a choice-factor, which makes it amenable to comprehending data.

Hjelmslev's stance, including the one in his 1975-work *A Résumé*, takes for granted what should be established: what is the predecessor, and what the successor, in genetic functions (*continuations*; cf. 2.6.1)? Only by presupposing a preestablished chronology can the synchronic linguist distribute his state-properties, predecessor (the constant functive) and successor (the variable functive). And only on condition that he sticks to the two-state-model will he be able to distinguish formally between the specific mother-language and *its* daughter language (2.6.2). On the other hand, if L2 of T2 develops into L3 of T3 (cf. 2.6.3), then Hjelmslev will be needing the language of time to uphold the distinctions; in other words, the very placing of phenomena in a temporal sequence is meaningful, significant and necessary in order to distinguish between the phenomena of the series:

2.6.2	L1 of T1 >	L2 of T2		
	mother language	daughter language		
	'presupposed'	'presupposing'		
	the constant	the variable		
2.6.3	L1 of T1 >	L2 of T2 >	L3 of T3 >	L4 of T4
	mother language	daughter language	daughter language	daughter language
	'presupposed'	'presupposing'	'presupposing'	'presupposing'
	a constant	**a variable**	*a variable*	<u>a variable</u>
		a constant	<u>a constant</u>	
		'presupposed'	'presupposed'	
		mother language	mother language	

There is no reason to labour the synchronic dilemma much further: the four-state model in 2.6.3 requires three formally different entities; as immanent entities L2 is indistinguishable from L3. Finally, the synchronic cut with its concomitant notions of a start and an end creates empirically irrelevant entities (L1 and L4) as against the transitory states, and where do we make the cut between L2 and L3 unless the respective material substances of L2 and L3 are known?

The weaknesses (2.6.3) of the formalizations in 2.6.1 and 2.6.2 could be redressed by making the (idealistic) concept of cause add directionality to the historical series. However, as no synchronic linguist – to my knowledge – has provided data that can implement the four-stage model in 2.6.3. I shall illustrate my point with our sample datum, OE /a:/ > ME /ɔ:/; the diagram in 2.6.4 will therefore also be a comment on my previous analysis of the cause-effect chain. The Glossematic model implies a determinism that cannot be validated; it is impossible to infer anything about the predecessor (OE /a:/) on the basis of the successor's (/ME ɔ:/'s) dependency on the former.

2.6.4[61]   OE   >   ME   C1 +HSP   C2 +LSP   C3
         C1   ↓   ↓   ↓
         C2 → E   /a:/ > /ɔ:/   /o:/ > /ɔ:/   /oʊ/ > /ɔ:/
         C3

The argument could be reversed, so that making the present (ME /ɔ:/) our start we should find the same lack of determinism: the successor of a successor cannot be predicted. Given this type of data, the linguist would not know how to order it genetically, unless he presupposed the chronology of history as a cognitive *Vorurteil*; and this is tantamount to saying that man's sense of history is a cognitive code.

The very logic of the two-state model ends in a multi-state developmental series, in which the identity of the states cannot be maintained. Hence the developmental series is not a function of a number of two-state transitions. The irreversibility of time (chronology) assigns properties to the historical phenomena that prevent the generalization of a two-state model. One such property is 'to become a cause': the present's development into the present creates a successor that becomes the cause of the previous present becoming its past, an outcome that is partially corroborated by the daughter languages L4 in 2.6.3 and L2 in 2.6.2. The Glossematic version of the two-state model does not tell us what change is or for that matter stability. As I am not following the standard view of the priority of stability[62] nor Coseriu's methodological restriction of the concept of synchronicity – and by implication, that of stability – to the domain of 'points of view', I shall now take a closer look at the concepts.

Empirically speaking, nobody has seen a language change, and nobody has seen a language static, and everybody agrees that all everyday languages, of whose class nobody has seen a single member as *a class as one* (= in its objective entirety), are equiempirical facts.

---

61 The example also illustrates how the affectum is more than raw-material: different causes create the same effects – because of the affectum.
62 In Hansen 1983 I demonstrated how there is no logical way of awarding the prize of priortiy to either change or stability.

2.6. On the logical geography of the A>B relationship of the two-state model

Apparently, the everyday language exists, but not only if it is seen to change (p) or to be static (nonp)[63].

Change and stability are concepts whose empirical relevance is not immediate. However, *not immediate* means *not perceptually immediate*, and since there is no reason to restrict the empirical to what is perceptually immediate, there is no a priori reason for the exclusion of change or stability from the empirical domain. The question is: are the two concepts equiempirical? Hjelmslev awarded priority to stability, and this priority is dominant in modern linguistics. I shall argue that both are empirical, but not equiempirical. Our two basic incompatible positions are these:

a) **we know** that all languages change, and **we infer** that no language is stable (nonA);
b) we have no immediate perception of (a) change, hence **we infer** that a language is stable (A).

The former assumption has empirical relevance, but not its corollary. The second assumption has empirical relevance, but not its corollary. I shall resolve the paradox as follows – in favour of the concept of change: *change is an empirical fact; it is not perceptually immediate at a synchronic point in time; it is not perceptually given in a mechanical cause-effect model's framework. Change is an innately epistemic concept which merges completely with the fact that the everyday language exists and with the latter's existential conditions including human existence. Change is not grounded on destruction, because that would imply an incomprehensible aporia in existence to be justified by a static value-loaded 'myth', nor on the finished product, because that would imply a less radical aporia to be justified by a static value-loaded telos. Historical existence is characterized by being always* en ergeiai, *hence anything that negates this is a contradiction. Our sense and grasp of reality involve our sense of history*. Whether stability is a similar superconcept is a question that its proponents must answer. But since I do not reject the empirical relevance of stability, I shall try to arrive at what is stable in the everyday language by way of an analysis of the (synchronic) category concept and commutation, two cornerstones in the synchronic construction.

The synchronic sign is immutable and unchanging (Saussure) and must be so for communicative reasons (the linguistic tradition). Hjelmslev (1928:66-67;77;78ff.) is more informative, as he implies that the concept of category makes for stability. A category is immutable and unchanging, but not unconditionally: the category is the servant of its theory, as the word (sign) serves the everyday language. Does the (grammatical) system serve the everyday language or the synchronic linguist's purpose?

If there is anything in our historical existence that is not stable, it is a category (a sign). From Thales onwards, our Western Civilization has been living proof of the mutability of categories. Certain categories are real diehards; eventually, they will succumb to change. In our historical existence stability may be a desideratum under certain conditions, but in so far as something is stable and durable, it exists at the mercy of change. In our scientific lives, falsifiability is incompatible with categorical stability. Secondly, the immutability of the class of signs (categories) is a synchronic tautology: existing at a point in time in a space with no extension, the sign cannot 'move'. This leaves us with an alternative: the sign's immutabil-

---

63 The everyday language is both A and nonA (cf. below chapter 3.4.).

ity determines the (synchronic) nature of the everyday language; the synchronic nature of the everyday language makes the sign immutable.

Hjelmslev (1928:66;67) may be right in saying that *la notion d'évolution a dissolu parfaitement des catégories,* because [d]*ans la diachronie, les catégories changent de sens et fonction*. But all he can do is to fix his position axiomatically (*à titre d'axiome* (p. 81)): *les catégories sont par définition liées à un système de stabilité* (p. 67)[64], in other words the historian must work on an axiomatic basis. That is historically untenable, because the ultimate consequence of the category view is that change is reduced to recombination of categories pre-existing as immutable entities in a universal table of categories (the panchronic point of view (pp. 107;268;293)) – which, incidentally, Hjelmslev was unable to set up here in 1928 (pp. 296ff.[65]).

The category accounts neither for stability nor for Hjelmslev's odd notion of *la causalité synchronique* (1928:229ff.[66]), it has the same purportedly limit-setting function (pp. 232; 238; 240; 242) as such systemic elements as the phoneme has in modern phonology.

If anything is stable, it must be what is created; at the moment of its creation, a man-made product is stable, and if it is durable, it may be subjected to some kind of change, and as such become a historical phenomenon. I am not concerned with the kind of durability that the Parthenon Marbles exhibit or for that matter the changes they have undergone; the spoken language cannot match such persistence. What is stable in the everyday language is the single individual utterance, irrespectively of how short-lived it is (perceptually). Due to its internal endogenous motivation it is stable. The *raison d'être* of its stability is its brief life (*Flüchtigkeit*), the latter being a precondition for communicative viability: stability is found in the spoken utterance[67]. If stability originates with the spoken utterance, then – in so far as the concept of the functional or systemic unit is communicatively relevant – the system is the place where constant changeability is to be found. This leaves us with an alternative: the unchanging system must be tied to a point-in-time conception (by definition; cf. Coseriu's methodology view of synchrony); the system has some empirical relevance because it is durable under the aegis of change. Thus what synchronic theorizing describes as categories, a system, is in a state of constant flux, change and development.

The single individually produced utterance *is* stable, it being a precondition for the listener's capability to subject it to correct interpretation and for the speaker to monitor his utterances. But from this we cannot infer that all we perceive is all of reality. No Old English listener was able to understand [bɔ:n] (the new variant of OE ban /a:/) through reference to the synchronic mechanisms *du langage*, grammatical function, government *rection* (Hjelmslev 1928:109;123;127), and (the) commutation (test).

---

64 This position is virtually repeated by Saumjan (1971:28-29), who does not question the synchronization of history, a synchronization brought about by *the methods of structural linguistics*'.

65 The class of pronouns (1928:324) created insurmountable problems.

66 This is an early example of Hjelmslev's attempt at depriving dynamic concepts of their dynamics and at removing the study of language from the world of temporal phenomena. Cf. his use of the word *history* in *Prolegomena* (1963:9), where he makes it change its meaning within the space of two paragraphs from '*a discursive form of presentation ...*' to the static and 'pre-given' movement of the analytico-deductive procedure of analytic reasoning.

67 Hardly a surprising conclusion, but far more tangible than the purported stability of an ideal entity such as a category.

2.6. On the logical geography of the A>B relationship of the two-state model

It is curious why the literature leaves out the possible role – in the step from variant to change – of commutation, the epitome of synchronic mechanisms of language. It would have been logical if the step from variant to change was finished when the new sound contracted a commutative function; instead loss of motivating factors accompanied by continued existence is the hallmark of the process. The reason is obvious: *this basic fundamentum of synchronic theorizing plays no role at all in the historical process*. This is no less than a contradiction in terms when seen against the synchronization of history. In OE *ban* 'a bone' and *bon* 'an ornament', *ham* 'a home' and *hom* '?'[68] (long *a* and long *o*) could be said to enter into a commutative function, but that minimal pair was non-existent in LateOE that had /bɔ:n/ (or /ɔ:/) correlate with another word (phoneme) in the sense that the test could decide when, say, *bon* /o:/ contracted a commutative function with /bɔ:n/ *ban, bone*, and no longer with /ba:n/. That substitution existed between [ba:n] and the new variant [bɔ:n] reveals nothing. Granted that the variant [bɔ:n] had achieved a codified status, so that both [ba:n] 'the old' and [bɔ:n] 'the new' contracted mutation with *bon* [bo:n], the listener – or the speaker – would not know whether it was the systemic entity /ɔ:/ or the systemic entity /a:/ that contracted commutation with /o:/.

The step from variant to change will then consist of three stages: at the first stage the negative result of a commutation test with OE /a:/ and */ɔ:/ is due to the fact that /ɔ:/ is non-existent; at the second stage, the result is mixed: at the level of words [ba:n] and [bɔ:n] contract a function of substitution and each contracts a function of commutation with LateOE *bon* [bo:n], but here /o:/ contracts no function of commutation with a systemic element as the synchronic point of view has no cogent synchronic means by which to elevate either [ɔ:] or [o:] to the systemic plane; at the third stage commutation occurs at both levels: EarlyME /ɔ:/ :: /o:/ and *bone* :: *bon*. This analysis is only a conventional registration of facts after the processes, not a theory of historical development.

On the other hand *if our Old Englander knew that [ɔ:] and [bɔ:n] were historical existents, then he also knew how to interpret the new variant, namely through reference to his historical sense (grammar): he would know the meaning potential of the possible variants of the particular element in his L-continuum*.

To prolong the pains of the synchronic system: knowledge of, say, a sound system cannot be generalized; its application will always be dependent on the material in question: *knowing that /x/ is a systemic element and what its systemic function is is not sufficient to ensure the generalization of this knowledge*. In modern English /ɔ:/ and /aʊ/ are commutatively related in the same way as /oʊ/ and /aʊ/ are: *core :: cow*; *bow :: bow*; but in *sauna* commutation is suspended between /ɔ:/ and /aʊ/, and examples where the significant contrast between /ʌ/ and /ɒ/ is suspended are plenty (Gimson 1994:105)[69]. Neutralisation explains nothing because one sound (not always predictable) will always be pronounced. My final word here concerns the inconsistent use of the phonemic principle: in Gimson (1994:182-183) the intensional description of the phoneme /l/ refers to none of the variants, clear or dark *l*. But

---

68 The meaning of this word is doubtful.
69 In Danish one could mention the forms of the verb *at rydde op* 'to clean, tidy up' where the intervocalic phoneme is either /-d-/ or /-ð-/, which otherwise contract commutation, as in *en gud* /guð/ 'a god' and *en gut* /gud/ 'a chap, fellow'. See *Den Store Danske Udtaleordbog* (Munksgaards Ordbøger. 1991. København: Munksgaard).

other phonemic descriptions describe an actually used sound as *e.g.* /f/ (p. 166). The theoretical implication is this: in the former type we have a proper (simplifying) description based on a common core found in the data; in the latter case the phoneme merges with some variants and differs from others[70].

To conclude: *the study of the spoken chain in its Eigenart is the study of something that is stable in virtue of its realization conditions, endogeneity, exogeneity and katageneity. The elements of the spoken chain are the mobile units of language development.* – Hermann Paul's view has not been refuted.

Stability *qua* non-changeability and non-durability (*Flüchtigkeit*) is a precondition for efficient communication, and stability *qua* changeability and durability (development) is a precondition for the communicative viability of the everyday language. – Hjelmslev (1963:46) does not disagree with the latter part.

The study of categories and systems is the study of something that is changeable in virtue of its existence conditions. A category and a system are historical superstructures[71]– Hermann Paul's view has been observed, and Lévi-Strauss's is to be modified: the word *a systemic element* or *langue element* should be replaced by *a predecessor* in order to underline the historical basis of the horizontal synthesis as it comes into existence when elements from the L-continuum are selected into the spoken chain.

The spoken chain is more[72] *than an immediate perceptum produced by internal endogeneity;* as such the structural point of view has been asserted.

---

70 In this connection the *Introduction* to the latest edition of *English Pronouncing Dictionary* (1997:v-xv) makes interesting reading. (1) *It is important to remember that the pronunciation of English words is not governed by a strict set of rules; most words have more than one pronunciation, and the speaker's choice depends on a wide set of factors* (p. vi); (2) *The use of phonemic transcription in works on pronunciation (including this one) has remained in the "realist" tradition established by Jones (...)* (p. viii). (3) And in *There are many places in present-day British and American English where the distinction between* /ɪ/ *and* /i:/ *is neutralised. For example, the final vowel of 'city' and 'seedy' seems to belong neither to the phoneme* /ɪ/ *nor* /i:/. *The symbol* /i/ *is used in this case (though it is not, strictly speaking, a phoneme symbol; there is no obvious way to choose suitable brackets for this symbol, but phoneme brackets* // *will be used for simplicity)* (p. xiv) a new sound, neither a phoneme nor an allophone, seems to have entered the English language.

71 This requires a determination of *a superstructure*: it is a transindividual entity which acquires its meaning from its constituent parts; it is always a class as many and can never enter into a *Kausalnexus*. It disappears the moment (a part of) its individual basis disappears (Goldmann 1966:152).

72 This '*more*' will be implemented by the theory of history.

CHAPTER 3

Structure and History are compatible – neither antinomical nor two different views of the same phenomenon

> *(...) science is (essentially) a unity. (...) history of science teaches us again and again how the extension of our knowledge may lead to the recognition of relations between formerly unconnected groups of phenomena, the harmonious synthesis of which demands a renewed revision of the presuppositions for the unambiguous application of even our most elementary concepts.* Bohr 1971:282

3.1. **What is structuralism? Sartre and the structuralist-existentialist controversy. Sartre's critique of the ahistorical nature of structuralism. Contingency is necessary and history is not paradox, but 'deconstruction' and construction of structures *qua* totalities**

IN THE 20TH CENTURY the word *structure* became the most meaningful and meaningless word of *le langage de la réflexion*[1], and in our century the word is still in great demand and we should therefore know what it means; but it is hardly inappropriate to claim that most treatises which use the word or apply structuralist approaches to a given subject matter take the logical geography of *structuralism* for given, providing hardly any specifying determinants of the -ism or *structure*[2].

Structuralism (cf. Boudon 1968; Piaget 1970) is an ideology about the nature (= structure) of reality and about how to approach reality. As such structuralism will already be presupposed before we acquire access to reality, and no non-circular definition seems possible.

---

1   In Williams (1976) *structuralism* occupies nearly 7 pages while the word *history* receives a 2-page description. – The polysemous use of the word can be seen in Bache and Davidsen-Nielsen (1998), where it refers to the very build-up of the grammar book in question (p. 1), to one of the functions of the order of sentence constituents, which is *to signal information structure* (p. 113) as well as to the particular build-up of *groups of words* (p. 2).
2   Culler's criticism (1982:20) is too sweeping, but essentially correct: [structuralists] *depart with nothing – no common method or theory – that they can point to and call 'structuralism'*. We can detect a common core of propositions (principles), but the irony is that neither have these been worked into a structural whole, *où tout se tient*, nor have their terms been defined (see chapter 3.2. below), to which must be added the fact that Culler makes no mention at all of the works of Hjelmslev and Togeby (1965), where we do see structural programmes in various degrees of completeness, or of Sartre's pertinent criticism to the effect that structuralists (*in casu*, Lévi-Strauss) only apply their method(s) to *des systèmes déjà constitués* (Sartre 1966:89).

This position is reminiscent of Hjelmslev (1971:110;1972a:34): *Il n'y a ni connaissance ni description scientifique possible d'un objet quelconque sans recours à un principe structural*, and may explain Topolski's vacuous statement: the structural approach is *dominated by a striving for structural interpretations of entirety* (1976:125). Structuralism even became a non-ideological ideology (Schiwy 1973:27; Leach 1982:37: *'a way of looking at things'*), and in the first decades of the previous century the concept so revolutionized the whole science of linguistics, not so much through concrete breakthroughs in language explanation as an intellectual reaction to the historical paradigm of the Neogrammarians (Lepschy 1972:21-25)[3]; later in that century it came under severe attack from certain linguistic quarters, in particular, in the United States (Lyons 1982:5-6) with the result that structuralism as an overt linguistic paradigm has been backgrounded in the linguistic literature, a fact that does not prevent generativists from using the word *structure ad nauseam* and polysemously[4].

Modern structuralism is Platonism in disguise, epitomizing dualist reasoning. Hjelmslev (1972b:86) called the instruments of scientific analysis a Platonic idea, and Glossematics is a structural approach to language. Togeby (1965:5-6;194-195) did not adopt the Glossematic view of structure (totality) and immanence (autonomy; independence) as being synonymous, but equated structure with *langue* in contradistinction to *parole* (style), and immanence with *fonction*, the immanent structure of the French language being the functional nature of its *langue* conceived as an autonomous totality. But both maintained an ahistorical conception of the language state, autonomous in respect of both succeeding and preceding states as well as of all extralinguistic phenomena, and both – Hjelmslev and Togeby – were as such in sharp opposition to Saussure's *vision historicisante de la langue* (see chapter 3.4. below).

Diderichsen (1966:169;188)[5] is dualistic, defining structure by the features the written and spoken manifestations (p. 169) of the same text have in common. However, he adds (p. 170) that any written rendition (notation) of a text is a structural analysis, the concept seems therefore rather devoid of meaning, being the essence of data and a short-hand term for analysis (division) of a text according to certain principles, respectively. This confusion testifies to the above-mentioned generality of the word; it refers to both the level of data (analysan-

---

3 We still need a comparison between, say, Henry Sweet's or Otto Jespersen's descriptions of English and recent grammar-books dealing with the same subject matter from a structural, be it transformational or otherwise, approach to demonstrate how the latter type is better ('truer', 'correcter') than, has surpassed Sweet (1968 (1891)) and Jespersen *MEG*. I am not thinking of the changes that English has undergone since then. In Denmark we could mention the conflict between the two contrasting views of Kr. Mikkelsen and H.-G. Wiwel (Diderichsen 1966:21-24).

4 As far as I know this School does not provide a definition of the word; so we only know what *structure* means referentially, that is with reference to what the word is used about as *e.g.* in Chomsky (1988:64-68).

5 The editors of this posthumous work have given it the peculiar, but interesting title *Helhed og Struktur* (Totality and Structure). Diderichsen's descriptive apparatus solves the riddle seeing that the former concept is a technical term. – Lass (1997:366) identifies structure with a form: a structure serves a function as *e.g. deictics locat[ing] participants in (…) a social context (…)*. In Kuiper and Allan (1996:3-4; 14) 'structure' is a property of language (pp. 3-4), a property of *a sequence of words* (p. 14), a *tree-diagram* and a *box-diagram* that appears after the application of an implicit analytical principle as well as a metaterm defined by *parts that are organised in some way* (p. 16), in other words a totality, a conception oddly reminiscent of Diderichsen's. However, since a bridge is also a structure (p. 16), Kuiper and Allan confuse the levels of object and description.

dum) and the level of theory (analysans). – In Hjelmslev (1972a:31,32,36; 1928:115;184-185) the meaning of *structure* is the spoken substance, the antithesis of what it means in his later works.

If there is a natural affinity between the concept of structure and (Plato-like) dualist thinking, structure and history are an odd couple, and the rise of the notion that structure and history are antinomical (*e.g.* Weinreich *et al.* 1968:98) is not to be wondered at – a commonplace in the intellectual debate of the nineteen-sixties in Europe.

The rise of structuralism went hand in hand with both the advent of a number of new successful sciences and revolutionizing breakthroughs in physics. Neither was linguistics left stranded with the irrefutable results of the 19th-century historical syntheses nor was it the passive recipient of the new scientific methodology. On the contrary, linguistics embraced the new methodology of the then dominant intellectual trend, logico-positivism *and* contributed significantly to the development of the movement before the Second World War (see Bloomfield 1971). No matter how dubious the value of the concept of the phoneme is to the scientific investigation of the man-made world, its impact on other sciences is not to be underrated[6]. Linguistics had seemingly beyond all doubt established the empirical relevance of a supraindividual fact *qua* an invariant that was the conceptual peer of the smallest/irreducible entities of the natural sciences[7]. Not only that; linguistics positioned itself right at the centre of Pre-World-War-Two thinking, which was quite natural in the light of the fundamental significance to human existence of the everyday language. However, symbolic logic became an end in itself, entrenching itself behind the seemingly evident characterization of human language as the medium that casts a veil of ambiguity over reality (cf. chapter 4.1-4.; Church 1956:1-9; 23-31; Lewis 1960:2-3; Alwood *et al.* 1977:164-167) so that the pure-form concept of logic with its syntactic transformation structures became the only viable means through which man could obtain access to reality in a non-subjective manner. While linguistics was open to the results of logic, neither logic nor the so-called meta-science, the logic or philosophy of science, subjected the everyday language, let alone the irrefutable language universal that no language does not change, to non-biassed analysis (Diderichsen 1962:252). The results of the 19th-century language historians as well as dialect geographers were *terra incognita* to the logico-positivists. Linguistics, however, is also to be blamed for the consequent state of the art. The Danish linguists Louis Hjelmslev and Hans Jørgen Uldall

---

6 Nearly all the articles in Robey 1982 use the notion of the 'phonological revolution' as an important opener. This emphasis appears to the present writer to be nearly contradictory to the non-critical, non-defining treatment that this important concept receives in *e.g.* John Lyons's wide-ranging 1968-work. To mention but one example: the commutation test, the corner stone of all structural treatments, is not discussed by Lyons.

7 Lévi-Strauss's 1945-article (pp. 35-37) is a virtual panegyric on the new phonemic concept as conceived by N. Trubetzkoy and R. Jakobson: *La phonologie ne peut manquer de jouer, vis-à-vis des sciences sociales, le même rôle que la physique nucléaire, par exemple, a joué pour l'ensemble des sciences exactes* (p. 35). In his 1962-work Lévi-Strauss (pp. 326;327;329) also hails the invariants as the identifying elements in *la diversité empirique des sociétés humaines* in order to establish *une humanité générale*. – It is interesting to note how little effect *cette révolution* has had today, when *e.g.* Lass (1997) feels obliged to turn to biology in order to gain new insight into the development of the everyday language; the same, *mutatis mutandis,* must be said about Anttila's return to hermeneutics (1989; 1992).

launched the ambitious but natural project to establish the science of language as *the* science whose methodology united the methodology of the natural sciences, including logic and mathematics, and the methods used in the humanities – in so far as the latter could exhibit any methodology (Uldall 1957). This notion, the possibility of creating a universal methodology with human language as its starting-point, did not catch on. Mainstream linguistics from the nineteen-fifties continued its methodological dependence on what other sciences could and could not do, and did not participate in the intellectual discussion that centred on the so-called structuralist-existentialist controversy as well as the earlier positivist-controversy in history; the science of language did not create through internal, immanent scrutiny novel concepts, but relied on more or less non-congenial concepts adopted from the logico-mathematical domains, hereby precluding other interesting, more language-congenial concepts that arose in the intellectual milieu of Post-War Europe[8].

The preceding remarks can only be a radical sketch of some positions in modern linguistics and are intended to emphasize the *possibility* that the study of language development – in contradistinction to the mainstream conception – is a non-dependent discipline with a theory of its own. It is also meant to suggest the line of reasoning that will be pursued in the following: structuralism primarily as a 'school' in linguistics is more than classification and the finding of patterns, and the meaning of *structure* is more than that of such words as *pattern, configuration, Gestalt, form, totality*. It is necessary a priori to acknowledge the universal nature of structuralism (Cassirer 1953:304-306), as it embraces the two most basic and fundamental *existence modi* of phenomena, namely that of coherence and attachment (synthesizing epitomized by *tout se tient*) and separation ('analyzing' epitomized by *tout est différence*), from which concepts those of transcendence, relation and immanence are inferred.

Structuralism is also more than the four Saussurean dichotomies, **substance** : **form**; **paradigmatic form** (group-patterns) : **syntagmatic form** (sequence-patterns); **langue** (form, stability, stasis, system, social) : **parole** (substance, fluctuation, style, individual, multiplicity); **diachrony** (history, non-structure, variation, chaos, disorder, contingency) : **synchrony** (states, systems, structure, order, identity, necessity) (cf. Lyons 1968 and 1982:11-15), and more than a principle of reinterpretation of (phonological) patterns when a sound change spreads (Labov 1994:302)[9]. Structuralism both as an -ism and in its linguistic sense shares

---

8   The return to hermeneutics as advocated by Anttila (1992; and Itkonen (1978) commits the Phenomenal Error. Lass's criticism and rejection of hermeneutic explication (1997:336ff.) is not based on the attempt at creating a non-dependent theory of history, but merely a battle between hermeneutics and biology, two transcendent views of language change; and in grammaticalization studies such as Heine *et al.* (1992:28) a cognitive approach is the use of the word *cognitive* instead of *grammatical* as in: *(...) we shall be concerned with strategies employed for the expression of a specific range of cognitive entities called "grammatical concepts."* Then they add six characteristics of these concepts, of which temporality or *temporal* structures are not a *characteristicum*, whereas *topological structures* are. This disregard for a given object's (perceptum's) temporal specification, a specification which makes the entity empirically relevant – in Saussure's diction *vivant* (see below) – is a permanent fixture of Structuralism's: Mepham's description (1982:122-123) of how objects are perceived (recognized) is a case in point: here *recognition* of objects involves no sense of time, but only establishment of the perceptum's external – *hic-et-nunc* – spatial relations. Note Mepham's peculiar, atemporal use of the word *recognition*.

9   In Lass (1997:366-369) structure is opposed to *use* and is reduced to merely referring to a grammatical phenomenon.

3.1. What is structuralism? Sartre

these features 1) man as a creative, a historical agent is decentred or eliminated. Man is at best reduced to being a relate or intersection in a structural network whose relations determine him exhaustively (Mepham 1982:123). This follows from the *immanence* requirement in that a totality *qua* a class or *ein Gegenstand*, being the fundamental descriptive unit, is not itself determinable by external determinants[10]; (2) man's *engagement* does not spring from his historical Eigennatur, but all that *l'homme structural* has to do is to discover meaning, patterns, in short, structures. From this it follows that man does not create structures; meaning and structures are *there before* man. *l'homme engagé* is passive, not really *en ergeiai*[11]; (3) a *connotative semiotic*, dealing with physiognomies (Hjelmslev 1963:114-119) and complex sign functions, does not surface until the linguist widens his perspective (to, say, phenomenologico-hermeneutical relations); the connotative function contracted by *the linguistic physiognomy N.N.* and *the real physiognomy N.N (that person)* is a function of denotative semiotics (= ErC): man does not involve himself in but *is* involved in reality; (4) the linguistic universal of *language change* is not in the Universal Component of Glossematics, let alone in the definition of a semiotic (Hjelmslev 1975:123-137; 3-56); and (5) *history* needs revision and redefining (Pingaud 1966:2-3). History is a succession of independent static forms existing at different points in time; each state is governed by internal laws. The timeline in general, and in particular the life of a language – like *parole* – is point-reducible. History's movement is that of a *Laterna magica*, not a cinematic one[12].

There is no denying that the structuralist controversy referred to above was vitiated by another conflict between Marxism and (Freudian) psycho-analysis; following Sartre, Marxism was the alternative to structuralism. I do not profess to know anything about either Marxism or Freudianism, but there is no reason to close one's eyes to the fact that Sartre levelled pertinent critical remarks at structuralism, and there is no ground for seeing the Structuralist Discussion exclusively against a horizon of philosophical isms. Such a circumscription of the field of vision is due to the fact that the philosopher-structuralists forgot to real-

---

10 The structural analysandum does not enter into horizontal (= temporal) syntheses. This epistemological requirement was also postulated by structural anthropology (Mepham 1982:110-114; 122-123), and Lévi-Strauss (1945:37) pays particular attention to the *internal* laws governing the development of the structural object: [les] *tendence[s] vers un but*. – Cf. also le Carré's protagonist in *The Russia House* (London: Hodder and Stoughton): [Barley] *thought of Katya again (...). She has no precedent and no sequel. She is not part of the familiar, well-thumbed series.* P. 185. Barley turns his mistress into *eine historische Zufälligkeit*.
11 Man is only active in the sense that he classifies (cf. Lévi-Strauss 1962); man was not a meaning-maker, only an uncoverer of meanings. There is a contradiction involved here: man finds the patterns, does not create them. But if the classification is based on a choice, perhaps controlled by a useful praxis as in Lévi-Strauss's savage mind, man must be active in making this choice. Thus we are back to the nominalist-Realist issue: if the classifier is a nominalist, he cannot be a mere recipient; if a Realist, classification and reality (nature) merge and he is passive.
12 Lepschy's definition of structuralism follows slightly different lines, but is not adequate to distinguish it as a particular scientific approach: he emphasizes the *model* aspect taken from mathematics or physics so that linguistic facts become analogous to non-linguistic data; *generality* or *abstraction* is another feature, which can hardly be said to characterize a structural approach; they are typical of any formal approach, yielding a simplified description of the object with such a description justified by the notion of *relevant features* versus *irrelevant features*. Thus we are back to the impossible task of defining what is significant or essential and their opposites non-circularly or non-arbitrarily. And with this it is not surprising that Lepschy ends in the *manifestatio*-metaphysics of the *langue-parole* dualism (1972:25-35; 30).

ize that the rise of the principles of modern synchronico-structural principles must be understood in the light of the developments and results of the science of language in the previous century (Hansen 1983:23-37)[13].

Structural linguistics may entail a constructive, but transcendent view of history, while as an ism structuralism was a reaction to other isms, Existentialism and Positivism. Therefore the history-structure antinomy only cropped up in structuralism when the latter was confronted with the possibility of historical development through internal laws. The development of a structuralist philosophy or view of reality was under no tacit obligation to consider the necessity of change. In either case, the crisis of history became a painfully relevant concept to historians[14], seeing that historical analysis was unnecessary and at best *in itself* unscientific (Sartre 1966:88).

Included in the following remarks[15] is the point that Sartre's criticism of structuralism relies on basic structural principles, which goes to confirm the general scope of structuralism. On the other hand, no ism or theory monopolizes a concept; this is a corollary of the Critico-Philogical Method: a concept may be modified, is subject to change and may be placed in different (theoretical) totalities, thereby creating something new. A concept is structural, – not ideological, but is made ideological – and is preserved against the horizon of development or existence: structural is therefore neither the exclusive property of synchrony, the defining feature of the latter, nor incompatible with a theory of history.

Structuralism is the rejection of history, *le refus de l'histoire* (pp. 87; 89). A structural history is a contradiction in terms[16], for it envisages not an archaeology, merely the geology of development: history being the sum of simultaneously existing layers in the same place, determined by different time dimensions (p. 87). Structural history is *le mouvement par une succession d'immobilités*, so that each *immobilité* (state/*état*) contributes to the formation of a series of *couches suc[c]essives qui forment notre 'sol'* (p. 87); structural history is no archaeology, because unlike the archaeologist the structural historian does not examine *comment chaque pensée est construite à partir de ces conditions, (...) comment les hommes passent d'une pensée à une autre. Il lui* (i.e. the structuralist) *faudrait pour cela fair intervenir la praxis, donc l'histoire, et c'est précisément ce qu'il refuse* (p. 87)[17]. The theme of this philosophical debate had been a commonplace in linguisitics for fifty years, and the notion of *l'immobilité d'une réflexion historique* must be modified seeing that a transcendent and dependent conception of language change was and is accepted by structural linguists; *but the*

---

13 Lepschy's treatment of structuralism also appears as a reaction against this; and the same applies both to Lévi-Strauss's 1945-exposition of (linguistic) structuralism and to Cassirer's (*op.cit.*; and 1945).
14 I am thinking of the discussion for and against Carl Hempel's Covering-Law model, a discussion that reminds us of the Glossematic postulate that 'to be scientific' presupposes the adoption of a particular point of view, with nihilism the alternative.
15 which are based on a 1966 Sartre interview in *L'Arc* 30. – All page references are to this article.
16 It is this contradiction that I shall refute, thereby demonstrating the contradictory nature of 'structuralism's rejection of history' (see chapter 6.).
17 The geology metaphor may not be quite linguistically appropriate unless supplemented with the palimpsest metaphor; even so its relevance is superficial, because the concept of *Wiederherstellung* (Coseriu) is not grounded on layers, but on structural integration (*e.g.* the Forum Romanum is an integral constituent of Rome). But the immobility-view is relevant to the state-concept, just as the *praxis-history* identification reminds us of *parole*.

3.1. What is structuralism? Sartre

historical dimension is not necessary in order to comprehend the nature of language. As Saussure (1972:105) put it: *Mais dire que la langue est un héritage n'explique rien*, adding the cryptic condition *si l'on ne va pas plus loin*.

Sartre and structural linguists agree that a 'praxis' with an acting subject is a necessary factor in (the explanation of) state transitions. No transition/transcendence is possible without *l'homme parlant*. The linguistic/philosophical consequence is that a systems concept is implied as that which is to be transcended (pp. 88-89). *S'il n'y a plus de praxis, il ne peut plus y avoir non plus de sujet* (p. 91). The speaker/hearer is not an external determinant of the everyday language, and the moment of the autonomous system is when it is regarded as an entity *déjà constituée, [une] chose sans l'homme* (p. 89; cf. below).

Sartre's crucial objection to structuralism is found in the next quote, which epitomizes the general position of synchronic linguistics, including that of Glossematics (Hjelmslev 1963:9-10; 125-127; 1972a:34):

> Renonçant à justifier les passages, on *(i.e. the structuralist)* opposera à l'histoire, domaine de l'incertitude, *(as well as the domain of* le non-sens *and* l'absurdité), l'analyse des *structures* qui, seule, permet la véritable investigation scientific (p. 88). (My additions).

Structurally, historical analysis is impossible; historical theorizing is *doxologique*. Only structural analysis is scientific. Historical analysis is not structural, the domain of history is not structured; historical analysis is non-scientific. Linguistically, scientific analysis is a principled analysis that establishes a constancy (a form) 'behind' multiplicity, fluctuation and variation; this constancy is to be implemented by the hypothesis that to each process there is a corresponding system. This programme applies to all humanistic objects. Accordingly, only and only if a constancy of a premised nature can be seen in the facts of change, shall historical phenomena be amenable to scientific investigation. Historical analysis has not been ruled out as an a priori impossibility; structuralism is not a priori incompatible with history, on condition that a Plato-like ideal constancy is compatible with historical existence.

A less stringent compromise is Lévi-Strauss's (1962:325-326): Lévi-Strauss regards the dialectical reason as a supplement to the analytical reason, the former being the effort the analytical reason must make in order to renew or transcend itself. Apparently, analysis and structure, system and synchrony take precedence over history, the dynamics of which are reduced to being an instrumental function in the existence – transcendence ?and construction – of structures and analyses. In my view this is a false opposition for two reasons[18]: (1) analytical reasoning may just as well be the effort that the dialectical reason makes when it contributes to **ex**istence, that is when the latter selects of the 'past' what is to become a part of the present, and the temporal gap that Lévi-Strauss envisages between them is no gap, but a simultaneous process in which it makes no sense to talk about priority. (2) Lévi-Strauss defines 'analysis' by 'to define, to distinguish, to classify, to contrast' (p. 325), but analysis can

---

18 Sartre (p. 90) also contrasts structuralism and dialectical reasoning, pointing out the structural weakness that structuralism (including Lévi-Strauss's) cannot cope with the rise and origin of structures – be they *"chaudes"* or *"froides"*.

in no way be regarded as prior to discursive processes: the process of distinguishing X from Y presupposes some kind of synthesis whose two parts are separated in order to be defined oppositively so that X and Y can be classified – brought together again – in the synthesis at hand, and for X and Y to establish contrast they must enter into a contrasting synthesis mutually or with other parts.

Sartre does not reject structuralism; on the contrary, *l'analyse structurale devrait déboucher sur une compréhension dialectique* (p. 89[19]). His argument runs as follows (pp. 88-89):

Sartre accepts the dualism: language is a synchronic *système d'oppositions,* a *totalité de relations (...) déjà constituée.* This is the immanent and *objective* nature of language. It follows that language is *inerte*, a dead, inactive substance[20]:

> C'est le moment de la structure, où la totalité apparait comme la chose sans l'homme, un réseau d'oppositions dans lequel chaque élément se définit par autre, où il n'y a pas de terme, mais seulement des rapports, des différences. *And Sartre continues:* Mais cette chose sans l'homme est en même temps matière ouvrée par l'homme, portant la trace de l'homme (p. 89).

Accordingly, the everyday language is janusfaced; it is *un élément du 'pratico-inerte'* (p. 89) and does not exist unless spoken (*parlé*) (p. 88). The word *pratico* refers to the individual's structural activity, while *inerte* refers to an inactive *objectum*, and not to man. The concept of the system being *déjà constitué* connects the two parts of language. This network of oppositions exists before the individual speaker, created antecedently to him; therefore it appears as an objectum. But the system does carry *la trace de l'homme*; in the final analysis it is a man-made phenomenon, *ouvrée par l'homme*[21].

Sartre's is also a succinct criticism of the naturalistic basis of structuralism. Language as an immanent object is determined by internal *qua* constitutive relations; in nature we only see external ones (p. 89; Cassirer 1953:306). The immanent state contracts no such relationships; secondly, objects of nature are not man-made, while language regarded as a system *déjà constituée* is not found *dans la nature* nor are *des oppositions telle que celles que décrit le linguiste* (p. 89)[22]:

---

19 Here we must add that reasoning in time presupposes the mental process that has selected X and Y as elements to be synthesized. So the opposition that Lévi-Strauss (1962:326) formulates as an absolute contrast between a science/a process that constitutes (creates) man and a science/a process that dissolves man is false: existence at all the three levels of the word *history* is a both-and. Lévi-Strauss works with a too sharp, if not absolute opposition between spatial simultaneity (co-existence) and temporal succession (see p. 329). Above I suggested that neither excludes the other.

20 This is reminiscent of Saussure (1972:112;113), who called language *sans le temps viable, et non vivante.*

21 Sartre's view may easily be aligned with Samuels's mechanical stance – with selection between the created system and the spoken chain – and Hjelmslev's connotation. This type of argument will also be found in the Durkheim discussion below.

22 Criticizing Radcliffe-Brown, Lévi-Strauss also rejected the biological interpretation of kinship terms (1945:52); the social character of kinship terms was a presupposition for the possibility of the 'new' phonology to influence anthropology. Therefore, Saussure's well-known tree-metaphor, which led to an analogy between the growth of a tree and language development, is inappropriate.

3.1. What is structuralism? Sartre

> La nature ne connait que l'indépendance des forces. Les éléments matériels sont liés les uns les autres, agissent les uns sur les autres. Mais le lien est toujours extérieur. Il ne s'agit pas de rapports internes comme celui qui pose le masculin par rapport au féminin, le pluriel par rapport au singulier, c'est-à-dire d'un système où l'existence de chaque élément conditionne celle de tous les autres (p. 89).

Thirdly, the functions of the everyday language cannot all be explained with reference to its system[23]:

> Dans le système du langage, il y a quelque chose que l'inerte ne peut pas donner seul, la trace d'une pratique. *And Sartre continues with a view to the imposing character of the system's* structure. La structure ne s'impose à nous que dans la mesure où elle est faite par d'autres. (...) il faut donc réintroduire la praxis, en tant que processus totalisateur (p. 89).

No object of nature imposes itself on us; *ein Gegenstand* may stand in our way – and we in its. The imposition of a language on all its speakers is necessarily an external relationship, when the infant baby learns her native language. In this process the system's structure is *déjà constituée* and is imposed on the baby by way of the spoken chain (praxis). This does not entail that the baby and the grown-ups cannot affect their native language – in fact they do affect it. Following Sartre, the everyday language exists at the mercy of somebody using it and in being used or imposed, it is changed by the user or the person exposed to it.

Both the 'inertial' and 'practical' properties of the everyday language point to its origin and to *le sujet*. Being *inerte* the system cannot be the cause of itself; external relations are called for to explain language. A more consistently structural conception of the everyday language implies that it is the terminus of necessarily external relations[24]; it is an intersection 'run through' by intersecting lines, among these the connotation, which makes the everyday language subject-hooked: *le langage n'existe que parlé, autrement dit en acte* (p. 89), in my diction when it completes itself in *parole*, when *en ergeiai* no structure exists *sans nous*; each structural entity, say the singular, points to or implies a totality (*un tout*), but this totality is nothing unless there is somebody to make it function, *i.e.* change: *Chaque élément du système renvoie à un tout, mais ce tout est mort si quelqu'un ne le reprend pas à son compte, ne le fait pas fonctionner* (p 89). This contradicts the autonomy of the everyday language, but Sartre defines the autonomy by the fact that the everyday language is – has been – *déjà constituée*, and thus he justifies the linguist's synchronic approach (cf. Coseriu's methodology argument). However, the property of being 'constituted' reveals *la trace d'une pratique*, and therefore *la trace de l'homme, i.e. l'homme structural*: in sum *l'homme structural* is a historical person who constructs his language through *un processus totalisateur* in *parole*. – This type of totalizing process will be elaborated in chapter 5.1. (from 5.1.7 onwards).

---

23 Referring to what he calls the *junk* of language, Lass (1997) sets great store by this fact (cf. note 38 below).
24 It is curious to see how Wang and Lien (1993:347) re-invent the *dialectical relation* between internal and external factors in sound change that Samuels (1972), using other words, worked into a theory of language evolution.

Sartre advances no theory of history, only certain key concepts which fit into a historical, non-static conception of man and his inalienable language. Knowledge of language is achieved not analytically or intuitively, but through reasoning in time (*une comprehension dialectique* (p. 89)), and as such there is perfect correspondence between mind and, *in casu*, the nature of the everyday language, which is not a naturalistic unit. Sartre emphasizes the active nature of man's inalienable relation to his language: man creates it, produces it and knows (how to use) it; man changes it as well as observes its rules. But Sartre follows the generalizing (and analytic) methodology[25] of the synchronic tradition. Talking about totalities and *le sujet*, Sartre is referring to generic entities; this implicit generalizing point of departure lands Sartre in a dilemma for ideological reasons: he does not know how to reconcile the two facts that the totality is anterior to man and the totality is man-made; he needs the autonomy of the static system: *(...) il y a des stases de l'histoire qui sont les structures* (p. 91), which the single person 'receives', and which the very same person revolts against (cf. below).

The dilemma is solved if we accept an *individual*-based and consistently dynamic point of view, and do not set out from a sociologized Cogito with its 'already developed' and static system (*stases de l'histoire*) as well as from the individual's group (Lévi-Strauss 1962:330). Secondly, Sartre focusses on only one aspect of history, namely change *qua* revolution; thirdly, like Lévi-Strauss he does not see how the analytical and dialectical interact without constituting an opposition.

The child is born into a *formed* world with a *formed* language, a reality *déjà constituée* the structure of which she is able to transcend as she **ex**ists[26]. When a structure is imposed on an individual, that person will already have *engagé lui-même dans l'histoire* and be involved in such a way that he cannot but destroy (*détruire*) the imposed structure[27] in order

---

25 Lévi-Strauss may be right in criticizing Sartre for being inconsistent: his dialectical method (or *comprehension*) presupposes what he is criticizing, the analytical *raison*. The former has it (1962:349) that *toute raison constituante suppose une raison constituée*, and the basis of this 'already constituted' reason is grounded on the (?)biological make-up of our mind, without which there would be no constituting (discursive) reason: the latter's 'initial' conditions must be *données, sous la forme d'une structure <u>objective</u> du psychisme et du <u>cerveau</u> à défaut de laquelle il n'y aurait ni praxis, ni pensée*. (My emphases). Admittedly, Lévi-Strauss's argument is logical, not historical; as such it is not cogent, and it does not disprove the possibility that man's mind (brain) is temporally structured. Secondly, his argument reminds us of the generativist notion of a language module in our brain and contradicts his argument against Radcliffe-Brown's biological foundation of the 'social' family (cf. note 22 above). – What Lévi-Strauss – and Sartre – perpetuate is the logical argument that for something to be built, constructed, to be said or written, even thought, this something must have existed somehow and somewhere before it is manifested or realized. It is this linguistically motivated aspect of idealism that has vitiated – let me confine myself to my immediate subject-matter – modern linguistic reasoning; the view entails a dated *manifestatio*-ontology, a static and katadynamic view of reality.

26 Hence use of the concept of 'preformed' is redundant, unless for strategic purposes, and the same goes for preexisting 'meaning', which Sartre's *l'homme parlant* does not possess (pp. 90-91).

27 If the structure is a mental one like the everyday language, it makes no sense to say that the infant baby destroys or changes the language of her parents: she creates her own language, which will become different from anybody else's. On the other hand, it is correct that when on that memorable day in July 1789 certain Parisians destroyed *La Bastille*, their *engagement* did destroy a social totality: but still the process is to be subsumed under the general historical process: the rise of something is accompanied by the change of something else.

3.1. What is structuralism? Sartre

to *en construire de nouvelles* (*i.e.* structures) *qui à leur tour le conditionneront* (p. 91)[28]. Being able to affect individuals, the structure must exist anterior to the affected individual: *l'homme reçoit les structures – et en ce sens on peut dire qu'elles le font* (p. 91). Thus structure and history are not really antinomical (perhaps for ideological reasons); against the horizon of the structural interaction of the two processes of transcendence and imposition they are two aspects of the same man-made process. A structured totality is a historical product: (i) it is created by *une praxis qui déborde ses agents*, (ii) it is *un moment du pratico-inerte* (p. 90), (iii) it is upheld, continued and changed by history, (iv) man cannot remain 'decentred'[29], and develop only in so far as the structures that contain him develop through their internal laws. Therefore history is not *l'ordre. C'est le désordre, (…) un désordre rationel*:

> Au moment même où elle maintient l'ordre, c'est-à-dire la structure, l'histoire est déjà en train de la défaire. (…) l'homme est (…) le produit de la structure, mais pour autant qu'il la dépasse (p. 90).

Sartre's view of history and structure is too narrow: history is not disorder and only order *qua* development through the laws of a structure; history is only disorder when seen against an a priori static horizon. Secondly, man is affected by a structure, *i.e.* the meaning of it, and it is the subject's task to discover and decipher it: *le sens advient à l'homme* (p. 91). In so far as man creates a structure, he also *fait le sens*. Sartre acknowledges the equiempirical nature of the two opposed processes, that of imposition, which makes man *parlé*, a passive, recipient or structural *terminus*, and that of transcendence, which makes man *parlant*, a constructive and synthesizing agent, but all within a context of the static horizon of the *langue-parole* dualism, which necessarily entails the two-state model's start-process-result conception.

Sartre's sketchy model of historical development is not dissimilar to the functional view of language change *qua* the maintenance of equilibra (Martinét and Samuels)[30]. His model finds a more detailed and significant role for man and the very process is not quite that of the monolinear two-state model. But he agrees with the functionalists in his view of the process as being somehow marked and as connecting two states: Sartre focusses on the *dépassement*-process so that the static *déjà-constituée* aspect of a structure becomes his point of departure (p. 93); hence change is a revolution, not evolution; and he does not consider how continuation (of structures) is brought about:

> il y a sujet ou subjectivité (…) dès l'instant où il y a effort pour dépasser en la conservant la situation donnée. Le vrai problème[31] est celui de ce dépassement. Il est de savoir

---

28 This reminds us of Kant's view of history *qua* the interaction between the individual and the created totality.
29 (…) *le décentrement du sujet est liée au discrédit de l'histoire* (p. 91).
30 Interestingly enough, Lévi-Strauss (1945:49) sees the dynamics of his social structures in terms of (dis)equilibra.
31 This task is turned into a passive and recipient process by the structuralists (Lacan in Mepham 1982:126; and Sartre, p. 91); in linguistics the process is brought about by 'loss of conditioning factors'; conceived as a restructuring (Hockett 1958:456); Saussure makes do with a statement to the effect that the problem exists, but in such a way that we can only *concevoir* the process (I do not disagree with Saussure, but for different, if not contrary reasons). On the other hand, there is no reason to make it the task of philosophy, and not history's, to try to *penser ce dépassement* (Sartre, p. 95).

comment le sujet ou le subjectivité se constitue sur une base qui lui est antérieure (…). *And* l'essentiel n'est pas ce qu'on a fait de l'homme, mais ce qu'il fait de ce qu'on a fait de lui (p. 95).

Historical development proceeding from something static through a process (of deconstruction) to something static, occurs as a person's reaction to something that has been done to (imposed on) him.

Owing to his ideological persuasion, Sartre commits the Phenomenal Error, turning one of two equiempirical facts into a hyperconstruct. What one individual revolts against was created by another (group of) individual(s), and what he himself creates will eventually be deconstructed. Historical development, **ex**istentially value-neutral, is a process of equiempirical phenomena constantly *en ergeiai*, not the destruction of one (undesirable) ergon and the consequent construction of the ideal ergon. History is both change and continuation. Sartre makes history an ideology[32], whereas our historical sense tells us that (the foundation of) history is non-ideology[33]; change is *per se* not marked in relation to unmarked stability.

Sartre distinguishes too sharply between structure and history, and cannot as such make structural man and historical man merge. The structural state is the spin-off of historical processes (*un moment du pratico-inerte* (p. 90)), and since man creates the structure and is determined by it, his own product, the relationship the two of them contract is necessarily participative. Sartre does not draw the consequences of this intimate relationship, perhaps because of his reckoning with Lévi-Strauss[34], but focusses on the negative aspect of the distance that man is able to create between himself and the structures: 'let's destroy the past in order to create a better reality'. In neutral terms, the individual is able to create a physical distance between himself and the inherited structures so that he can deconstruct them[35]; it is as if Sartre stands – in Løgstrup's diction (1984:17-23) – on the edge of reality.

Sartre is critical of the structuralist change concept, his target being Lévi-Strauss. His criticism is levelled at the fact that structural analysis is *only* applied to *des systèmes déjà constitués* (p. 89). The structural anthropologist – like the linguist (cf. 0.3. above and Lévi-Strauss 1945) – makes an arbitrary cut which prevents him from considering the fact that all societies – *froides* and *chaudes* – have been *élaboré, formé par les hommes* (p. 90); the methodology of structuralism cannot explain this evolution; its analytic reasoning is bound to make the seeming stability of a given society its essence (cf. Wilden 1980:7; Bukdahl

---

32 *Ce qu'on a fait de l'homme, ce sont les structures, les ensembles signifiants*: man reduced to *l'homme parlé* is studied as a sign-functional unit. *Ce qu'il fait, c'est l'histoire elle-même, le dépassement réel de ces structures dans une praxis totalisatrice* (p. 95). And further below: *Renoncer au marxisme, ce serait renoncer à comprendre le passage* (p. 95; Sartre's emphasis).

33 This is not to say that value-systems – in Lévi-Strauss's diction, systems of attitudes – do not exist simultaneously with other systems, in his case, the system of names or terms (*appelations* (1945:40-41;51); and I think that Lévi-Strauss is right in not regarding attitudinal systems as an extra layer to be added to the basic system or – in the synchronic jargon – as a function of the denotative system. – More on this below.

34 *Mais le structuralisme, tel que le conçoit et le pratique Lévi-Strauss, a beaucoup contribué au discrédit actuel l'histoire (…)* (p.89).

35 *Chaque génération prend, par rapport à ces structures, une autre distance, et c'est cette distance qui permet le changement des structures elles-mêmes* (pp. 93-94).

3.1. What is structuralism? Sartre

1979:291-300; 300-301)[36]. Sartre's criticism of the structuralist's view of history (evolution) is slightly ironical[37]: *Quand elle ne meurt pas sa belle mort, la structure succombe par accident. Mais ce ne sont jamais les hommes eux-mêmes qui la modifient, parce que ce ne sont pas eux qui la font: ils sont au contraire faits par elle* (p. 90). History is passive in its own history, existing at the mercy of the internal laws of the static structures as well as historical accidents. The problem is that – when we come to linguistics – the number of such exceptionless laws is small, if there is a single one that has not been broken[38]; in anthropology the internal law – following Sartre (p. 90) – reduces to an expression of what combination possibilities a given system's elements permit. As such structuralism entails reductionism (Leach 1982:49-51; Diderichsen 1966:169), and contingency is bound to follow from it, as the internal laws do not explain the history of a system exhaustively: given two different societies, the anthropologist reduces both to being variant manifestations of one and the same basic structure *qua* a(n ideally global) calculus consisting of a finite number of combinable (systemic) elements (cf. Lévi-Strauss 1945:46-47; and note 36 above). This system contains all possibilities, *in casu* unrealized and realized societies and language systems[39]. Sartre pertinently asks: *Pourqoi toutes ces possibilités ne se sont-elles pas réalisés?* and he repeats the structuralist's feeble answer: *Parce qu'il y a contingence*, while correctly denying that *l'histoire puisse se réduire à ce processus interne* (p.90).

The nature of structural man emerges from these quotes: *l'homme ne pense pas, il est pensé, comme il est parlé pour certains linguistes. Le sujet, dans ce processus, n'occupe plus une position centrale, il est un élément parmi d'autres, l'essentiel etant la 'couche', ou (...) la structure dans laquelle il est pris et qui le constitue* (p. 92), a programme implemented by Lévi-Strauss (1945:48-49): *L'erreur de la sociologie traditionelle, comme de la linguistique traditionelle, est d'avoir considéré les termes, et non les relations entre les termes*. His basic system of relations reduces to the following three *simultaneous* relations: a (horizontal) relation of *alliance* (marriage), whose terms (variables) are *wife* and *husband*; a (vertical) relation of *filiation* (genetic descent), whose terms are *parent* and *child*; a (horizontal) relation of *consanguinité* (consanguinity), whose terms are *brother-sister* relationships. Sartre's geology-metaphor is appropriate: the system consists of three identifiable layers, and its dynamics are those of generic history couched in teleological terms: *La parenté n'est pas un phénomène statique; elle n'est existe que pour se perpétuer* (p. 49).

---

36 Sartre could also have criticized Lévi-Strauss (1962:329) for being a reductionist and idealist: no less than *la vérité de l'homme* resides in the system's invariants which determine the system's different manifestations, societies separated in time and space. Sartre (p. 95) has an indirect comment on this 'synchronization' of truth: *La vérité est là, coupée de ses déterminations antérieures. (...) elle se donne d'emblée au présent, come si, entre le moment présent et le moment passé, il y avait une véritable coupure,* [que] *l'on n'explique pas, mais que l'on constate*.

37 With some justification; see also Lévi-Strauss's historical accidents in his 1945-article p. 51.

38 Not unlike the Glossematic position as expounded by Francis Whitfield (in Thrane 1980:48): a system has certain dispositions towards change that are either 'held in check' by unpredictable extrasystemic factors or 'released' by other unpredictable external factors; the internal inconsistency of the synchronic stance is evident: (a) dispositions have no connection whatever with the definition system of the theory and the theory is as such metaphysical, and (b) it tries to 'include' the universal of change on the basis of the 'excluding' approach of synchronic immanence.

39 This obvious parallelism between the 'new' phonology and anthropology is curiously absent from Lévi-Strauss's 1945-article (p. 51).

My exposition has focussed on the idealistic basis of structuralism, and it makes one wonder why structuralists are not concerned about the empirical data, the accidents, that wield such irresistible power over their structures. But even the staunchest structuralist-idealist must accept the empirical necessity of at least one contingent realization of a structure from where cognition begins[40]. The significant upshot of this is that since this cognition is bound to be discursive (dialectical) – no man can grasp a section of reality in one single (analytical) act of the mind – Lévi-Strauss's criticism of Sartre for presupposing what he is rejecting – the analytical mind – is perhaps logically correct, not empirically: analysis is always discursive, reasoning through time; reasoning within the single point in time does not occur in this world.

The everyday language as a structural socio-fact exists in an epistemological and empirical vacuum (Saussure does not really disagree!). It presupposes existence conditions that are at odds with experience; linguistics should not stop at the structures. *c'est (...) un scandale logique* (Sartre, p. 95): if we substitute *the synchronic cut* for *La vérité* in the quote in note 36 above, the words preserve their truth value and relevance.

Structuralism – the linguistic state-conception – should explain the gaps it creates; but the moment it tries to do so, it oversteps the bounds of its professed objective and becomes self-defeating. Only by a deliberately narrow definition strategy can structuralism (synchrony) maintain its strictly non-periodic stance and exclude history. But an empirical reckoning will remain in the form of such questions as: what makes for the static nature of the linguistic sign and its system(s)? What justifies the elimination of the human factor from man-made structures? Why is the *dépassement* problem, hence the empirical fact of change, relegated to the domain of contingency and non-significance or to the unknowable (Saussure 1972:105) or to as yet unknown internal laws? Why is change viewed as law-breaking, and not as law- or structure-preserving, let alone as structure- and systems-constituting?

In conclusion: I do not subscribe to Sartre's purple vision of the historical process, it being *une totalisation perpetuelle* with man as *l'homme totalisé et totalisateur*, so that the task of the historian is to comprehend *l'effort de l'homme totalisé pour ressaisir le sens de la totalisation*, which occurs in *une praxis totalisatrice*[41]. I do not adopt his view of the relationship between man and structures. Being only reactive, it is not dynamic enough; but I do accept his emphasis on the human agent, an emphasis that will be summed up in linguistically relevant terms as follows: in the Glossematic hierarchy of objects the speaker/hearer of the connotation is reduced to being both a terminus of a sign-functional relation subsumed under the the class of non-denotative semiotics and the object of an external semiology (Hjelmslev 1975:deff. 44, 49). Therefore there is a reckoning with the empirical fact that this terminus *qua parlé/pensé* changes or transcends its denotative semiotic in virtue of its *parlant/pensant* (acting) faculty. This transcendence occurs in the irreducible speech situation, where the speaker (speaking) produces a relationship in which the spoken chain: [abc], be-

---

40 The binary opposition between realized and unrealized turns out to be participative in that the potential is always seen against the horizon of the realized: thinking of the potential co-thinks the realized. Ingerid Dal's criticism of the Prague School's conception of the phoneme strikes a similar vein; see Willems (1995:168-169).

41 Sartre leaves this task to the philosopher. – *Praxis* is in its *mouvement une totalisation complète* (p. 95), *processus totalisateur* (p. 89).

3.2. Linguistic definitions of structure 241

comes the expression part (= E) of a synthesis which cannot be foreseen by some (semiological) calculus or other of preexisting systemic elements, and in which the speaker (spoken) becomes the content (= C): C r E[abc]. This synthesis or merger is resolved by the hearer as he identifies [abc] as: *That's what Primus said*, thereby creating a new connotation with himself as the new content-element in that he is able to separate these constituents, the speaker, the referred-to, the utterance.

3.2. **Linguistic definitions of *structure*. The idea of the static and supraindividual comes from the metalanguage. The idea of change comes from the everyday language, not the system(s) concept; a dynamic conception of language. Structuralism, man's sense of reality and history are correlative concepts. Analysis of the synchronic systems concept will eventually end in history. A quick glimpse at the language-learning infant. The Glossematic argument backfires. The structural programme is a metaphysical statement**

The epitome of linguistic structuralism is enunciated in the following statements – which appear as stereotypes, not as defining for the works in question (see Culler 1982[42] and chapter 3.1.):

3.2.1i [language is] an integrated whole, *un système où tout se tient,* that is to say a system in which each unit is defined by its place in the overall network of oppositions (Bynon 1977:76).

(…) dans un état de langue donné tout est systématique; une langue quelconque est constituée par des ensembles où tout se tient. (…) Qui dit système, dit ensemble cohérent: *si* tout se tient, chaque terme doit dépendre de tout autre (Brøndal 1943:15;93-97). – Italics added; cf. *hypothèse* in 3.2.1ii.

If Bynon's use of *defined* refers to the level of data so that a given unit is described in relation to the other units of such an integrated whole, the structural programme is impossible,

---

[42] And it will not be inappropriate to quote what Kant said in 1783 (1969:10-11), **both** in relation to Culler's criticism of structural practices. *Making plans is often the occupation of an opulent and boastful mind, which thus obtains the reputation of a creative genius by demanding what it cannot itself supply, by censuring what it cannot improve, and by proposing what it knows not where to find. And yet something more should belong to a sound plan of a general critique of pure reason than mere conjectures if this plan is to be other than usual declamations of pious aspirations,* **and** in relation to the 'novelty' of the structural programme itself: … *pure reason is a sphere so separate and self-contained that we cannot touch a part without affecting all the rest. We can do nothing without first determining the position of each part and its relation to the rest; for, as our judgment within this sphere cannot be corrected by anything without, the validity and use of every part depends on the relation in which it stands to all the rest within the domain of reason. As in the structure of an organized body, the end of each member can only be deduced from the full conception of the whole.* Cassirer pointed out in his 1945-article that this structural determinism (in a discussion of the work of Georges Cuvier) was a common-place in the late 18th and early 19th centuries.

as the process of definition will never end – or end with some undefined *units*[43]. If the word *defined* belongs to the level of logic, the definition is not particularly structural, as it is the metasigns that impose structural coherence on the object (Diderichsen 1966:21). Following Hjelmslev (1971) structural linguistics is

3.2.1ii un ensemble de *recherches* reposant sur une *hypothèse* selon laquelle il est scientifiquement légitime de décrire le langage comme étant *essentiellement*[44] une *entité autonome de dépendances internes*, ou, en un mot, une *structure* (pp. 28;109). (…) Le langage est la totalité constituée par la langue et la parole (p. 29). – [A structure is] un tout formé de phénomènes solidaires (p. 109) [so that] la manière d'être de chaque élément dépend de la structure de l'ensemble et des lois qui le régissent (p. 109). – And in Hjelmslev (1972b:111ff.) a structure is something that, in casu, a case system possesses[45].

Bynon and Brøndal do not consider the transcendence problem, supposing the empirical existence of the object of structural linguistics, while Hjelmslev advances no such existence postulates (1971:29; 1975:101). Hjelmslev adds the important notion of internal laws governing the totality's structure (pattern), and among those laws are historical laws controlling the existence of the totality as well as 'dispositions' which the system in question possesses in virtue of its 'specific structure' (1972b:109-110; cf. chapter 3.1. note 38). In anticipation of the discussion in chapter 4.1., it is interesting to note Hjelmslev's use of synonyms, *ensemble, totalité, structure, tout* as well as *système*, all of which are defined by *une entité autonome de dépendances internes*. However, Bynon, Brøndal and Hjelmslev do not address the question whether the system is an entity *sui generis, i.e.* a class as one, or is only a class as many, thereby existing in virtue of its parts (Hjelmslev 1963:92-93[46]).

In the synchronic literature neither the linguist's (Lyons 1982:6; 1977:30) nor the philosophical (Sartre 1966:89) definitions differ from 3.2.1, and the transcendence problem, the nature of the totality-concept, is by and large *terra incognita*. For instance, Lyons (1977:109-114) offers no comments on the transcendence question that have any bearing on the linguistic discussion proper. On the other hand, Sørensen (1958:17;32) is explicit about this problem: the system(aticity) of the object language is identical with the system(aticity) of the metalanguage. No everyday language has a system independent of or different from the system imposed by the synchronic linguist's theory, and only in so far as the metalanguage can be established as an integrated whole, *un système où tout se tient*, will the structural programme be implemented. The coherence of the sentences of the metalanguage will be brought about by a common set of well-defined terms (Sørensen, p. 31).

Hjelmslev's structural analysandum is a class or hierarchy and is as such preeminently suitable to being denoted subsumptively by the words of the everyday language (Hjelmslev

---

[43] Sørensen (1958:29-31) has demonstrated the impossibility of this enterprise.
[44] Nowhere does Hjelmslev define this ambiguous terem, *essentiellement*.
[45] Structure here means the particular configuration or pattern that the elements of a (sub)system exhibit.
[46] Unfortunately, Hjelmslev only discusses it in relation to syncretisms, and neither Gregersen (1991) nor Rasmussen (1992) considers this important issue.

3.2. Linguistic definitions of structure

1963:60). The extensionality of the scientific language within the nominalist tradition corresponds to the conception of any language process as being a synthesis of elements *déjà constitués*. Therefore this type of scientific language cannot cope with processes and phenomena which are to be described as constituents of developmental series 'in' which something new appears when a synthesis *qua* a relative totality is established. Description of an object is not only analysis of this object into gradually smaller parts, but also a synthesis in which the elements are seen both as components of a class (Hjelmslev 1975:def. V; cf. 2.5.1 below), and (against Glossematics) not as components of two different classes, but of a complex class, represented by 'A (of S1) >B (of S2)', which is 'more' than the sum of its components viewed from the perspectives of their respective synchronic states. In the diction of the previous section: the analytical mind cannot grasp the 'complex class' (the historical phenomenon) in one intuition and the synthetico-dialectical mind is not a mere extension of the former when this endeavours to understand the everyday language; the analytical mind does not 'surpass' itself, it defeats itself. The historical mind is not the bridge that enables the analytical mind to grasp the change of one state into another state as a process controlled by (the analytically understood) internal laws (of the predecessor state). In more conventional terms: *history is an integral part of the object, not an addition to description*[47]. Neither the spoken chain nor the developmental series (to be demonstrated) is merely a class of pre-established elements: no Middle English system could regard *helpes* 'helps' as an 'already developed' whole the first time it was created instead of the inherited *helpeth*. Similarly, the language-learning child that was exposed to both [hɔ:m] and [ha:m] in LateOE/EarlyME had no abductive means of subsuming both under a general rule, the phoneme /a:/ or lexical representation /ha:m/, to the exclusion of the 'new' form.

If the signs of a language are essentially static, a system *qua* a class of signs is essentially static. The synchronic sign assumes its static nature from two sources. Again Sørensen is quite explicit (1958:26): the synchronic properties of a sign have been attributed to it by the (synchronic) theory; a(n object) sign is static because the definiens of its metasign is static. Hence *there is nothing cogent about the static-synchronic conception of the everyday language*. The other source is the vague notion that man somehow needs something static – a fixed point, never a fixed line(?) – from which he takes his bearings. Thus successful communication presupposes supraindividual static units, and more specifically a sign is static because it is a social fact (Leach 1982:38).

It is a noteworthy feature that Labov's all-or-nothing statement – *a successful linguistic theory or practice* must be *social* (1978:xix; a position softened in his1994-work (pp. 303; 542)) – is not grounded on the nature of the sign, but rather, transcendently, on the instrumentalist view that language as a whole is a communicative means used by people living in sociologically definable groups: as such the autonomous language system reflects a given

---

47 Anttila (1989:402). – Despite Anttila's justified criticism of the dominant tradition's positivistic/naturalistic parameters, he does not draw the logical consequences of his stance: first of all, he works with the narrow, dualist synchronic concept of change: *Without synchronic variation, change would not have a launching pad* (p. 179); secondly, he does not generalize the change-concept (p. 402). Likewise, his definition of language: *a sign system that perpetuates itself through time* (p. 179), is also quite traditional, perpetuating the systems-concept.

unification sociale (Sartre 1966:89). Labov's a priori static stance obliges him to adopt the traditional view of change and synchrony: *the phonological system reacts to the mysterious process of sound change*[48] so that it becomes a problem how *linguistic change* is to be related to *the system's ability to transmit information* (1994:542-543). Labov does not see that the purported problem of how a language can preserve its meaning-carrying function when it undergoes (constant) change is a problem that stems from the adopted point of view of language[49]. Samuels (1972:28;64) also takes the fundamental unit of communication to be necessarily static – without arguing for his stance. The alternative is just as logical: the communicative or information-carrying viability of the everyday language presupposes change, an alternative that neither makes the *dépassement*-problem a problem nor excludes it from theoretical analysis (as *e.g.* in Hjelmslev 1972b:103); furthermore the alternative can be subsumed under the general conditions for existence.

Synchronically, structure/structural and system/systemic are to a certain extent used interchangeably, perhaps with structure as a systems-constituting property. And the distinction between the systems concept of an idealist-based ontological dualism (the system is a class as one) and the nominalist's extensionally motivated system (the system is imposed by the theory) is not, to repeat, an issue that mainstream synchronic theorizing finds relevant (cf. my comment on Anttila in note 47 above).

Does the systems concept of the dualism necessarily follow from the supposition that the fundamental unit of communication must be static (or a social fact)? If there is a solidary relation between staticness and communication, and I do not wholly disagree, we should establish what is static (and social) in communication and the everyday language and then relate it to the systems concept. The problem is that the communicative-sociological view only captures the cohering function of the everyday language, and so do the Prague School's teleologizing stance (Vachek 1972:13) and Anttila's return to a hermeneutical teleology (1989:402). Neither can cope with *the individualizing* and *dissociating* nature of language, two existence-modes that characterize human nature, too (Kant 1923; Hansen 1986). The concept of the individual is subsumed under that of the group, and in so far as the other concept is accepted, it is immediately backgrounded by either the unifying function of language (dialect) or the purported teleology of language as *rule-governed goal-directed and goal-intended* human action (Anttila 1989:399). Again we must say that the negative view of language development as only change (and variation) narrows the field of vision so that dissociation is regarded as 'destruction' (Sartre and the linguistic tradition) or as alienation (psychoanalysis; cf. Anttila 1989:399-400[50]).

---

48 It is the theory's task – still within a two-state framework – to *predict* how and what the reactions of the system will be.
49 Anttila (1989:404) reaches a conclusion diametrically opposite to Labov's views of change, change *being the very essence of meaning*. I can subscribe to this conclusion, but not to the way by which Anttila arrived at it. Anttila adopted the transcendent-dependent way, relying on hermeneutical analogies, instead of proceeding from language development itself in order – from this immanent point of departure – to arrive at an immanent theory of history. Anttila does not assume that the language historian may have something to tell other branches of scholarship. Like Lass (1997) he seeks help from the concepts of other sciences.
50 *changes pile up and alienate the participants and us from the original unity*.

3.2. Linguistic definitions of structure

If we do not accept the (problematical) systems-based synchronic view[51], we cannot ground a theory of history on the synchronic system, and we must therefore ask: where does the – empirically relevant – idea of language change come from? And where does the object – the historical process – of historical linguistics come from? With Anttila (above) I shall answer: from the everyday language itself, but disagreeing with Anttila[52], I shall argue – not by way of a teleological framework, but – from the immanent conception of language development that appears from this essay.

Nobody has seen the *langue* of any everyday language. Establishing the empirical relevance of the systems concept is the synchronic linguist's task. Similarly, nobody has ever seen a sound change or for that matter any other language change. Whether this fact is due to some principle, such as the Neogrammarian Paradigm (Bloomfield 1933:364) or is due to something more concrete, language change being too slow (gradual, infinitesimal (Strang 1972)) or too fast (sudden, abrupt (Hockett 1958:457)) is as such immaterial. In any case such positions imply that the historian is unable to capture or observe language change because of the empirical data and/or because of some kind of theoretical imperfection. To deduce the fact of change from such postulates as all things must or will change (cf. Anttila 1989:179) is both non-cogent and mediated, and perhaps circular. I shall argue as follows (cf. also previous paragraph):

As the empirically relevant concept of language change does not come to us from or is imposed on us by the concrete facts of change *qua* percepta, it is bound to come from the everyday language and from ourselves. The everyday language is an empirical phenomenon; all our knowledge of it begins with experience; but not all knowledge of the everyday language stems from experience. Some knowledge of language must therefore come from language itself. Knowledge of language change stems from the everyday language, the latter merging with language change: knowledge of language change cannot possibly come from our immediate experience of the spoken chain. Cognition and the cognized merge, just as perception and perceptum merge in sense perception until the merger is resolved by the intervention of the everyday language (Løgstrup 1978:9;111-114). The language in our case will be a theory of history, and as such we do not need Anttila's mediating proposition: *change is the telos of language.*

If we did not know that the everyday language was a changing phenomenon and were unaware of the mechanism of change, we could not speak, understand or use it in our creation of reality. Since the everyday language is an empirical phenomenon, language change is empirical; therefore it must somehow 'subsist' in the spoken substance – as a merger of

---

51 To underline: if the system is a class as one, the synchronic linguist must demonstrate how this system (entity) is so empirical as to be able to enter into *Kausalnexus*; if he cannot do this, it remains an ideal entity and has no role to play in language development. If the system is an extensionally defined entity (a *nomen* and therefore a class as many), a theory based on such a system is not cogent; thus not a totality, only a part – which now becomes a misnomer – will be imposed.
52 Unfortunately, Anttila makes do with such slogan-like phrases as *change is (...) use of language*, he does not implement them. In my diction we do not 'use' language, we 'exist by it and it by us', a notional turn that I think Anttila may not object to in the light of his own '*becoming is the very essence of existence*' in the cultural disciplines (1989:402).

dynamic and static (percepta) factors. Hence, if successful communication is solidary with the static sign, it is also solidary with its dynamic nature, (knowledge of) which comes from our temporally structured everyday language or mind. This merger-concept will now be elaborated.

In the connotation there is a merger of language (spoken) and speaker (speaking), which is resolved when Primus speaks: CrE([abc]) → [abc][53]. The material substance of the spoken chain will establish the necessary distance between Primus, his utterance and his interlocutor so that the utterance can be 'understood', enter into the hearer's connotation. Similarly, the historical theory (*qua* language) resolves the spoken chain merger of static and dynamic elements as it establishes a distance between its object and cognition of that object. Cognition and (the use of) the everyday language are coextensive[54]. Ideally, the historical theory is our historical sense or vice versa. This position makes the mechanical view of language change meaningful: the spoken utterance is more than mere raw material for change; it is change itself[55], the problem being – from a traditional epistemological stance – the resolution of the merger of cognition and cognized: understanding an utterance is also comprehension of change.

In the connotation, the speaker is the content functive of an expression that is a sign which in turn is the expression of a content functive *qua* the denotatum in question. At the theoretical level, the maker of the theory/method/operation (Hjelmslev 1975:101) must also be a content functive in relation to the theory/language, which becomes an expression *qua* a semiotic whose content is the object (language)[56]. – Note how such signs as *Kepler*'s *Laws* also signify the discoverer of certain empirical facts.

In *parole* the speaker (cognition) and denotatum (referred-to/cognized) merge with the utterance: $Cr^{con}E(Er^{den}C)Er^{ref}C$; a given utterance, be it scientific or otherwise, carries *la trace de l'homme*. The study of language is anthropocentric, hence historical, and the conclusion must be that synchronic linguistics as a form of thought is dependent on historical thought, and knowledge of (the nature of) the everyday language presupposes knowledge of what history is[57], neither what biology (Lass) is nor what hermeneutics is nor what 'philosophy' is[58].

---

53 From now onwards '[abc]' will represent an utterance together with the denotative function 'ErC'.
54 In Book 10 of his *Republic* (596a onwards) Plato is very close to identifying cognition (truth) with language seeing that his ideas or forms correspond to the extensional use of language signs; a language sign is the name of a group of particulars, and to each name there corresponds an idea. In 597a Plato establishes the connection: the word 'bed' denotes *the real object, the idea.* – Let's pursue the argument: the idea, the real object, is not made by any artificer; as such it is uncreated, but in the final analysis the idea is created by God, and as such it is not subject to change.
55 This conclusion is also not incompatible with Anttila, who is more restrictive, as he makes change and meaning correlate. The reason for this is obvious: Anttila does not work within a consistent sign-functional framework, where the Es and the Cs of the sign function are equilogical and his hermeneutical approach is primarily concerned with meaning.
56 Both complex sign functions have this structure CrE(ErC)ErC; to be elaborated below.
57 This conclusion has been couched in diction adapted from Collingwood (1964:177).
58 Again one must object to Anttila's attempt at linking historical knowledge up to the practices of other disciplines (1989:399; cf. Sartre in 3.1. note 41).

3.2. Linguistic definitions of structure

It is an irrefutable fact that the everyday language changes and that we know it changes. This knowledge comes to us neither inductively nor deductively nor abductively[59]. Differences extrapolated from comparison of two, in practice and theory, undefined states do not make change, a mediating factor is needed. The only empirical alternative is that such knowledge comes from the everyday language and therefore from its speakers. The everyday language is a changing phenomenon, in so far as it is used, otherwise it does not exist; man exists, in so far as he changes, is *en ergeiai*, and the historical existence of both is sign-functional (both E and C, not only C, *i.e.* meaning). Speaking (*parlant*) man is spoken (*parlé*). When us<u>ed</u> (spoken) the everyday language becomes the <u>active</u> (speaking) means by which we turn so-called hostile and objective nature into a life-preserving setting for our historical existence; we create totalities *qua* signs – not in Sartre's all-embracing sense – which lend their respective meanings to the parts that constitute them.

Historical **ex**istence, man's and the everyday language's, is a constant process in which the 'agents' discard of their respective pasts what is not needed in the present's development into the present. The everyday language is participatively creative: becoming a part of the present, it – unlike logic – assigns meaning to objective reality. **Ex**istence is historical selection, which in contrast to nature's is active not just re-active. This man-made selection process is structural, and it is from the domain of history that it has been transferred to nature (biology), where its biological relevance was corroborated by the findings of Darwin and Wallace, and then returned to the humanities as a metaphorical expression of language change so that (historical) linguistics had to labour under the value-loaded sense of 'natural selection' and 'survival of the fittest', to which we must add purposiveness[60]. The process of

---

59 Change is not guess-work. Anttila (1989:399) cannot have the best of two worlds: the basis of abduction is formal logic, and the application of formal logic (reductionism and positivism) to history is what Anttila argues against; only by an equivocation can he salvage his abductive element in historical explanation: formal logic (as well as mathematics) is a man-made creation through and through, but it is not tied to any situation, time- or place-wise. Not only beause it is the language of physical theorizing, it is also bound to carry anti-historical implications with it because of its very nature: it is reductive (essentialism), subsumptive, (arbitrary,) analytical, non-dynamic, static, object-hooked, not synthesizing and non-creative because the meanings of its relations do not convey meaning to the relates. – (Lévi-Strauss saw the dilemma (1962:328)).

60 Kant's teleological interpretation of history seems to be grounded on the supposition that man has a choice, to exist and not to exist. Kant (1923a) has it that if man *wants* to transcend his 'animal' existence, he has to create the proper setting (totality) for such transcendence. If man does not do this, his existence will be a contradiction in terms, because then he would not fulfil his 'basic purpose' (*Anordnung; Naturanlage*). Therefore man must create wholes (societies), which are bound to be structural and of which man becomes a part. Discard the telos-aspect, and we are left with value-less existence *qua* empirical necessity: man cannot but exist; he has no choice. Any action is the present's development into the present in accordance with what may be called Nietzschean conditions: the 'ist' that was does not become a 'nichts'. Man's choice appears in how to implement the process without being controlled completely by nature and nature-like laws: while Kant (1923a) saw the controlling factors (the teleology) of history as analogical extensions of the laws of nature, there is a better parallel between nature and history: Kant (1923b:159) regards as the epitome of Nature *alle[s], was nach Gesetzen bestimmt existirt*, adding *Mit Recht ruft die Vernunft in aller Naturuntersuchung zuerst nach Theorie und nur später nach Zweckbestimmung. <u>Den Mangel der erstern kann keine Teleologie noch praktische Zweckmässigkeit ersetzen</u>*. (My emphasis). – Likewise in history, no teleology can substitute for a **a theory of historical development.**

natural selection is structural, hence 'historical'⁶¹. This Darwin-Wallace lesson was the natural confirmation of what the historical linguist already knew then: categories and concepts *qua* words change. Whether we call this innate knowledge is as such immaterial; but language-wielding man seems to know 'instinctly' how to **ex**ist, and the infant baby works hard to implement, if not perfect this knowledge as fast as possible.

This change concept – historical selection or **ex**istence – was not available to the founding fathers of our Western Civilization, and it is impossible to incorporate it into an immanent or transcendent dualism whose dynamics are katadynamic or into the concept of the generic process or generic history – or into a theory conception whose basis is 'limits and exclusion'⁶².

The notion of the category as an expression of *the* stable stems from its man-made definition, which fixes the category, while non-arbitrary stability is only found in *parole*, the spoken substance being unchanging and unchangeable *qua* a perceptum. The static system is not established inductively – Paul saw this and it has ever since been a commonplace. Where else does the notion of the static system come from⁶³?

The historical point of view ensures the empirical status of our investigations; it deals with what exists and happens – *was geschieht*. The obvious conclusion to be drawn from the fact that nobody has ever seen a language change is that there is no such empirico-perceptual fact. There is nothing to be seen, but something to be comprehended. In a way Saussure (1972:105) was right: the *dépassement* problem becomes a 'linguistic' problem, and therefore also one of understanding, and without implying or admitting any explicit phenomenological/hermeneutical heritage or affiliation, I shall run the risk of quoting Gadamer (1975:450), because his slogan-like sentence sums up the position that I am advocating from a language-historical stance: *Sein, das verstanden werden kann, ist Sprache*, a position that is curiously congenial to Collingwood's (*loc.cit.*), to Løgstrup (1978:112), who refers to Gadamer, and not least to the old dictum 'Like knows like' (Jones 1969:209). The *Sein* we understand is our constant development into the present; this historical process involving change (and stability, non-change) is conditioned by and involves the everyday language: in existence we demonstrate understanding (cognition). This point of view owes its *raison d'être* to the sign function: *Sein, das verstanden werden kann* becomes the referred-to –

---

61 (...) *contemporary science is indebted to Darwin and Wallace for the principle of mutability of species* (Hamilton 1967:5). – Darwin is *the paradigm of the explanatory but non-predictive scientist* (Scriven 1959:477).

62 I do not want to press this analogy too far, but synchrony with its state-concept has a certain affinity and co-occurrence with (the rise of) nationalism (Frisch 1962:409; Appleby *et al.* 1994:65; 92; Berlin 1976: 145 (and p. 164 where Kant's opposition to Herder is voiced); cf. Collingwood's idea of *epoch-making events* as that which delimits a historical phenomenon (1973:53).

63 To repeat: to make the everyday language static by definition is no relevant solution, and Hjelmslev (1973:60; cf. 1972b:86) is here being inductive in order to establish the basic meaning of the category. A static, extensional systemic explanation of *parole qua* manifestation of meaning-carrying categories does not imply empirical relevance. The category has *eine Grundbedeutung* (Hjelmslev 1928; *signification fondamentale = La valeur [qui] est le minimum différentiel de signification* (Hjelmslev 1972b:84; 86) and the system consists of a finite number of elements (categories) defined oppositively through their intensional definition. This structuralist position leads straight into essentialism (*i.e.* linguisticism), which does not yield an exhaustive description of the nature of *parole* facts.

3.2. Linguistic definitions of structure

created or formed by man – or the content (= C) of a sign, which is the *Sprache* (of the quote), *i.e.* $Er^{den}C$. The sign-functional form of the Gadamer quote looks like this: $Cr^{ref}E(Er^{den}C)$. The everyday language makes historical phenomena amenable to empirical investigation because it is both a part of them and a part of their motivating factors (cause). The problem is that what forms historical phenomena, the spoken chain, is so fleeting.

With the child's growing sense of history grows her sense of reality, through which her sense of history, grows. The two senses are necessary complements in the creation of the child's rational make-up; and just as she acquires her native language, so she acquires the two senses by trial and error, by developing her language, making it precise in respect of her growing capability to make empirically relevant syntheses and analyses; her growing awareness of herself as 'becoming different' in time (Bernsen 1978:295-296) is accompanied by her language developing in time; the language she uses to refer to herself at different points in time is not the same. Her self-awareness (in traditional terms: self-identity) is not a function of her having the same language at her disposal: an empirically relevant notion of self-identity presupposes a structural process in which at least two differences contract a relationship of external endogeneity and katageneity; this process of change is to be comprehended through our historical sense – expounded by a historical theory – which we find embodied in the everyday language.

This dynamic view of self-identity has often been summed up in aphorism-like statements: *Language has to change to stay the same* (Anttila 1989:405); the only thing that is static is the fact of change; change occurs in accordance with an unchanging law (Christiansen 1997:12;38); the essence of history is the changes in the idea of unchangeability (Christiansen, p. 18) so that we may end with this oxymoron: identity is loss of former identity. It is such statements that the historian must make plausible beyond their catch-phrase stamp.

Language, the sense of reality and the sense of history are so intimately tied up with each other that if one breaks down, the others break down, too. The child experiments, with impunity and without running any physical risks, with her language, not so with (her sense of) physical reality. Experiments in her daily life as to the possibility of climbing stairs, walking and running etc., or other attempts at not only defeating but also coming to grips with the possibilities (the laws) of nature, often hurt. So her forays into reality had better be in the form of language acquisition. Each utterance by her is not only a linguistic experiment, but also an experiment into reality, into existence. Finally, her attempts at comprehending historical phenomena are not separable from her language acquisition, through which she develops her historical sense. She learns as well as unlearns. She unlearns that *e.g.* 'moon' and 'lamp' are contents of the same sign in modern English (Bolinger 1980:141). If she can (un)learn this, why shouldn't she be able to learn a sound that her parents tell her to pronounce in a special (the 'correct') manner? To the child who is beginning to develop her sense of history, the development is understood in terms of a growing sense of the meaning of the word *yesterday, gestern, hier*: from the moment she is capable of distinguishing between now and then, all that has happened, happened yesterday. Her growing sense of history turns the contents of her amorphous yesterday into an ordered series, placed at the appropriate distance of her present. She acquires one of the codes of existence, *chronology*. Thus the placement of 'things' in a temporally ordered sequence is the product of, not just a precondition for, our sense of history, and this ordering assigns structural meaning to the 'things' so that they transcend the synchronic stage of being 'historical accidents'. The child's grasp of (linguistic) reality is not the sole product of what her parents tell her: through her language she contributes to it herself.

History is an inseparable and inalienable part of and setting for our daily and scientific lives – so why this not only structuralist, but also 'synchronistic' attempt at abolishing history and removing from the everyday language – and from structuralism – what cannot be excluded, in Sartre's words, *la trace de l'homme*? And why the sociological attempt at reserving a huge lump of historical reality for itself – perhaps all of linguistics (Labov 1978:xix)[64]? All of this boils down to the double-question: why is it scientifically justifiable and necessary to regard the everyday language as a static-synchronic or a social fact (Lyons in Labov 1978:xvi) *and not* as a historical fact?

Into each everyday language a historical existence has been built (Danish *en førtegning* (Diderichsen 1966:21))[65], hence the (child's) need for transcendence. Even the logician admits that *our ordinary language is full of theories* (Popper 1974:26: Cf. Samuels 1972:1-2; and at another level Coseriu's *Wiederherstellung*). But this part of the everyday language is not only a pile of junk or relics, it is a sense of history and reality. The inherited, imposing language is not the infant baby's; therefore she is free to change it, creating her own historical existence in order to come to grips with (her) reality. The everyday language is individual in the sense that the child transcends the inherited, exploits from it what she needs in order to create a new synthesis so that she ends up being in complete command of her historical grammar, which does not lose its collective orientation in the process. This leads to a question correlative to the abovementioned: where does the notion of the everyday language being a supraindividual entity immune to *dépassement*, come from?

The notion of the supraindividual finds its motivation in our Western Civilization's predominant penchant for the idealizing effect of Linguisticism, the conceptual creation of **a whole** with the corollary that individuals, 'things', are baptized *parts*, entities that acquire their *raison d'être* in the totality. The various totalities that we have been exposed to – since Thales[66] – originate in the everyday language, an origin that the meaning of *supraindividual* testifies to: (Plato) the supraindividual exists beyond the world of individuals or in the individuals (Aristotle)[67]. The supraindividual is a totality such as the Danish language, the Danish nation, a dialect, a group, a class, a pattern, a social fact, a structure (cf. Lévi-Strauss), which we all – the so-called contingent, atomistic entities – participate in or are parts of. However, none of these collection-terms transcend the language that has them as signs, hence the

---

64 How come that history itself – if it is history – has created this misnomer: *contemporary history*, in modern Danish *samtidshistorie*!

65 It cannot be translated into English, perhaps 'pre-picture'; the prefix *for-* corresponds to *Vor-* in German *Vorurteil* and *pre-* in *prejudice* 'præ-judicium', and it carries no negative overtones such as *distortion* = Danish *fortegning* 'distorted picture'. Diderichsen's term, however, is not quite appropriate: the inherited language will of course carry part of the history of the language group in question. Again we must take note of how deceptive the everyday language is: the child does not change anything: she learns a language different from her parents, whose language is continued 'unchanged', unaffected by the fact that their child is learning a language.

66 See Christiansen 1997 for an excellent overview of the history of metaphysics.

67 What is the meaning of this: *La langue* is placed in the individual speaker *qua un dépositaire* (Saussure 1974:38)!

3.2. Linguistic definitions of structure

parts become denotata. The alternative, which is of no help, either, seems to be that they are established inductively or as a Cartesian essence[68].

An argument in favour of the synchronic, supraindividual system is Hjelmslev's epistemological defence with Kantian overtones (Hjelmslev 1972a:34-35; Jones 1975:9-10): 'even if one denies the reality of the system, the system being an abstraction without reality-features and not an objective fact, the construction of a system is an epistemological necessity, an inevitable working-hypothesis, if we want to explain language. The object language's nature must correspond to the system constructed by the theoretician: because only on this assumption can we conduct an empirical comparison of more languages or describe a language.' Hjelmslev continues: 'finally, it is possible ('thinkable') that only if we accept this working hypothesis, will it be possible to really explain all language changes, in particular the fundamental transformations affecting the system.' – The argument contains the following levels:

1) the system is an epistemological requirement (cf. chapter 1.1.3 above), because knowledge is systematic (Hjelmslev 1972b:86-87), hence it constitutes a system.
2) From 1) it follows that a language theory must be systematic, hence constitutes a system.
3) From 1) and 2) it follows that the everyday language must be of a specific nature, it must be systematic, hence constitutes a (specific) system.

Hjelmslev commits a number of mistakes: hardly anybody denies that knowledge is systematic; but there is no immediate leap from systematicity to the reified entity a system. Secondly, Hjelmslev has not established a transition to the postulate that language is static, a static structure, a synchronic system. Thirdly, Hjelmslev accepts the universality of change, hence: the theory *qua* a system (of level 2) constructed in accordance with level (1) is changeable: S1 changes to S2; both S1 and S2 have been constructed by the theoretician; thus the changes do not come from S1, but from the theory's calculus of possible (contingent) systems. The change concept is katadynamic, not an empirical process 'between' S1 and S2, each of which has now become two historical accidents.

Hjelmslev's argument (1972a:11ff.) sets out from an analysis of the relationship between the state and change, and lands by way of the notion of the systematic (pp. 13-14) in a contradictory contrast between change and system; the system makes the state static, and he concludes with this rhetorical question (p. 22), which does not connect with the epistemological systems-requirement: 'to what extent can a language state be regarded as a system?' The answer is of course: our object must be regarded as a system if we want to have knowledge of it. But as Hjelmslev cannot come up with a cogent argument in favour of the logical or empirical priority of the state-concept, he cannot establish a necessary connection between system and the state = the static: change as a systematic process is an equilogical hypothesis; and synchrony and the system enter into no necessary connection, either, because Hjelmslev's initial argument above lacks a proposition: let the propostion 'the system is synchronic (or static)' be our conclusion or what is to be demonstrated, then we can set up this argument

---

68 The supraindividual language is also static because *it serves to establish and maintain socially prescribed patterns of behaviour* (Lyons *loc.cit.*). Why should it do that? 'Prescribed' by whom? Not even Lévi-Strauss's simple kinship-structure above can be said to have permanence.

3.2.2 Knowledge of X is static, synchronic   (another epistemological requirement)
  <u>The system is necessary (for us to obtain knowledge of X)</u>   (= 1) above).
  The system is static, synchronic

But where does the major premiss come from, what does it mean, and how are its predicates defined? From a Platonic haven[69]. Hence Hjelmslev's *metachrony*, the discipline that explains the evolution of systems remains a postulate. The synchronic linguist must prove that it is epistemologically impossible to regard change as systematic, or that history and theory are antinomical; he must prove that the period, the concept of the period, cannot be conceived scientifically, unless the period with its historical phenomenon is analyzed into a sum of 'points' in each of which all the relevant entities exist simultaneously.

That Hjelmslev devalues the notion of 'epistemological necessity' by equating it with a working hypothesis is not surprising, after all his theoretical outlook is that of the nominalist, not Kantian; the system, constructed by the theoretician, is not an innate structure in man's mind and by 1943, the year that saw *Prolegomena*, Hjelmslev has abandoned any correlation between the simple Glossematic dualistic hypothesis and its epistemological foundation in favour of a strictly nominalistic conception of his object: the theory is arealistic and arbitrary, enunciating no existence-postulates, it cannot be falsified by data, it being accountable only to the Empirical Principle.

In the preceding pages I have tried to come to grips with the notion of the supraindividual totality, the synchronic and static system of the *langue-parole* against a structural horizon. The synchronic system rests on a limit-setting postulate that cannot exclude the empirical facts of change. Each synchronic analysis seems to end in history – the moment we ask questions that go a little beyond noncogent description or classification of positive (if not positivistic) facts. Let the origin of such systematization come from a working-hypothesis, the results of such systematizing process have not yet been aggregated into a *tout-se-tient* whole *qua* a *sui-generis* system. And the crowning aporia is this: despite the fact that any state *est devenú*, we must only regard it as *étant*, because there is no causality between a given change and the establishment of a system (p. 227). The simple question is: how do we know that *l'état est devenu*? And why are we not allowed to treat it as a historical phenomenon, represented above by 'A(of S1) > B (of S2)'?

Here we must recall Hermann Paul's words: the totality-concept in modern linguistics is a return to a Realism in which the content of a sign is made an objective fact. In the history of linguistics, the totality concept has been used strategically to make an absolute gap between history and synchrony, to justify the received critique of history for being atomistic[70].

A totality is a language sign. A language sign denotes objects, processes, mental and physical. As such a single act of scientific denotation constitutes a totality in which the parts *qua* denotata receive meaning from the sign: what is denoted by the grammatical term *a case* ac-

---

69 Hjelmslev (1928:227) equates the language state and the system axiomatically; equally axiomatically, the system is equipped with the property 'to be synchronic' (p. 49): *Le système est éminemment synchronique*, and even a before- or after-system can only be explained *synchroniquement*.

70 Historical linguistics consists in considering the systemic elements in isolation (*pris séparément*) (1972b:110). That historical linguistics may have been so practiced does not entail that historical investigation must always proceed along those lines.

3.3. The sociological analogy. Durkheim 253

quires the content 'to be a case' (Hjelmslev 1973:60-61; Sørensen1958:24-33). But the sign *a case* denotes no entity different from its denoted (subsumed) particular denotata, such as the nominative case, the dative case, etc. It cannot reach into a 'real object' in a transcendent world nor into an entity that inheres (hides itself) in the interior of a positively given fact. Collection signs such as *the Danish language* have no denotata that are not denotata of other signs, and since more than one substance hierarchy can be encatalyzed to a semiotic hierarchy (Hjelmslev 1975:92-93), it follows that if the semiotic hierarchy in question is denoted by *the English language,* then the sign only refers to parts, not a totality. Hjelmslev may have realized this: his *Language Test* is no independent test so as to be applied to a totality denoted by *a language*, but is divided into a *Manifestation Test* and *Purport Test*, whose objects are denoted by the ambiguous terms *hierarchies* and *derivates*. This interpretation is substantiated by Hjelmslev's diagram of *Manifestation* (pp. 12-13; def. 28), where **only derivates** can be manifestants (= substance-functives; def. 31) or manifesta (= form functives; def. 32) **and only functives** are objects that *have function to other objects* (def. 13): only as a functive can a function (a class/totality) enter into a manifestation. If we are to go by this, it follows that a totality – systems or states – is not a class as one and does not have a history.

The important upshot of this is that the synchronic programme's criticism of the historian's data – for being atomistic, hence not empirically relevant – backfires: totality remains a metaphysical concept, a *nomen* to be interpreted nominalistically or Realistically, or a totality is a relative totality, something that can be analyzed into parts or synthesized into another relative totality. It is not systems (classes) or states (stages) *qua sui generis* totalities that are subject to change; they cannot enter into *Kausalnexus*.

3.3. **The sociological analogy: the 'disappearance' of *le sujet* and *la trace de l'homme* from the social fact. What is *un fait social*? An analysis of Durkheim's views, which end in history. Linguistic parallels. Extensional classification is not explanatorily adequate. The social fact is interpreted sign-functionally**

The sociological approach deals with the sociologizing, non-individualizing, unifying, 'uniforming', function of the everyday language[71]: man's inborn tendency to individualize, dissociate himself from other people, to transcend the inherited *qua* a limit-setting and imposing 'norm' is not considered. Variation is systematic, not free variation (Lyons in Labov 1972:xvi), let alone innovative. No wonder that variation and change are disruptive and mysterious (Labov 1994:543) and seemingly incompatible with the functional view of language[72], and the logical generalization is Labov's: language is *a social fact, rather than the*

---

71 Meillet (1951:84) focuses on an instrument-based definition of man: *L'Homme use d'outils et il possède un langage articulé, ces deux traits par lesquels il s'oppose à tous les animaux (…) résultent d'une même supériorité intellectuelle.* Needless to say, Kant's historical view of the sociologizing tasks of man is far more pertinent and informative than Meillet's. Note that Labov does not offer any general view of man into which he could incorporate his synchronic view of change (1994:16;25) or on which he could base his view of the man-made process (trajectory) of change (pp. 300-301).

72 *The function of language is for the speaker (or writer) to communicate meaning to the listener (or reader)* (Labov 1994:548; and cf. Labov's rejection of *variation* being functional (p. 568)).

result of individual choice (Labov, p. 598). This, however, will only explain half the story of language, and following Labov's division of the subject matter into *dialect geography, sociolinguistics, phonetics, and historical linguistics* (p. 25) one could say that this half story is the one told by sociolinguistics. But what is a social fact? Neither the sociolinguist (Labov 1994) nor the theoretician (Meillet 1951:70-83; 1975:1-18;44-60[73]) provides us with a definition, so let us see what may come straight from the horse's mouth.

According to Durkheim (1927;1974) a social fact exists in its own right, independently of the individuals on whom it imposes itself. Its basis is nonanthropocentric. Durkheim determines *un fait social* negatively in relation to the entities of the natural and psychological sciences; knowing what a natural object is, we know what a social fact is not; it is also not a psychological fact, because it does not exist in the individual's conscience or mind (1927:5;8); in a word, a truly supraindividual phenomenon. Despite this Durkheim sets out from a naturalistic premiss: *La première règle et la plus fondamentale est de considérer les faits sociaux comme des choses* (p. 20). After these introductory remarks I shall go on to discuss some relevant steps in Durkheim's argument, demonstrating the tenuous quality of it, and draw some linguistic parallels. – A social act performed by an individual person is the latter's compliance with

i) des devoirs qui sont définis, en dehors de moi et de mes actes, dans le droit et dans les moeurs (p. 6).

The sociological domain is dualistic; *le fait social* corresponds to the systemic entities of *la langue*, and the individual must obey the rules of a given sociofact, just as the speaker/ hearer will observe the rules (linguistic habits) of his or her group's language – in order to understand and be understood. In either case the individual manifests the *organisation définie* and *objectivité* of the social fact or the supraindividual system in question *incompletely and imperfectly* (p. 9 – Plato's bedmaker in the tenth book of the *Republic* is still in great demand).

The individual did not make the social convention (rule); a social fact exists *before* an individual[74]

(ii) (…) ce n'est pas moi qui les ai faits, mais je les ai reçus par l'éducation. *And* si [les faits sociaux] existaient avant [chaque individu], c'est qu'elles existent en dehors de lui (p. 6).

Social facts, which also include *Le système de signes dont je sers pour exprimer ma pensée*, are independent of the individual person because and only because they existed prior to him; they *fonctionnent indépendamment des usages que j'en fais* (p. 6) and have *cette remarquable propriété qu'elles existent en dehors des consciences individuelles* (pp. 6;8).

---

73 In the former work Meillet (p. 73) refers to Saussure's definition of *la langue*: a collection of linguistic habits.
74 Note this difference: in the process of sound change the social content does not rise until late in the process (Labov, pp. 300-301; 542).

3.3. The sociological analogy. Durkheim

Notwithstanding their external, supraindividual existence, sociofacts influence *la conscience individuelle* because they have been equipped (*doués*)

(iii)   d'une puissance impérative et coercitive en vertu de laquelle ils s'imposent à lui, qu'il le veuille ou non (p. 6).

The distance factor (or its objectivity) prevents the social fact from being an integral part of what it determines; but its simultaneous existence in a synchronic vertical relationship with the individual is conditional on – as Sartre remarked on the Lévi-Strauss's 'frozen' society – its origin, so that there is bound to be a historical link between *le moi* and the sociofact. Despite this the social fact has neither *l'individu pour substrat* nor a psychological one, only *la société* (p. 8), and Durkheim argues: the impact on *le moi* of the social fact *s'accuse dès que j'essaie de lutter contre eaux* (p. 9), and when I think I am not being influenced by it, it is only because my own sentiments and behaviour tally with it. A social fact is always *en ergeiai*.

Durkheim has not ruled out the possibility that all sociofacts have an individual origin (pp. 9;10), so that his generalization of the social fact's social origin[75] may only be a reflection of imperfect sociological methods, *i.e.* the synchronization of historical phenomena. In any case, the continued existence of the supraindividual sociofact (p. 11) is dependent on an individual's or individuals' compliance with it or willingness to learn it.

Objectivity, the thing-quality of the sociofact, is a precondition for another constitutive feature, *collectivity*, which Durkheim makes a precondition for the property of being *général*:

(iv)   (…) ce n'est pas leur généralité qui peut servir à caractériser les phénomènes sociologiques (p. 11).

The reason for this distinction is the traditional *manifestatio*-dualism. Social facts are different from *leurs incarnations individuelles* (p. 12). Not the individual, but his group *constitue* a sociofact. As we have seen in (historical) linguistics, Durkheim brings about the transcendence from the individual to the group linguistically by saying that the individual is to be *pris collectivement* (pp. 12;125-126; 128).

The manifestation by a class of individuals of a social fact is merely a function of its distribution and spatial diffusion (generality) *and* the individual's liberty to demonstrate a degree of individuality. An individual is never able to reproduce the social fact in its entirety; what he produces is *en partie un modèle collectif* (p. 14). A sociofact is therefore not dependent on it being manifested, a criterion that would imply dependence on being generalized: to be collective does not imply manifestation, but to be more or less *obligatory* (p. 14). A social fact exists in an

---

[75] Durkheim's lack of agents in connection with passive constructions does no allow us to see if he accepts Paul's and Goldmann's division of labour among a group of people producing a transindividual phenomenon.

(v) état dissocié (p. 12), (...) *and social facts may even* être sans être actuellement appliquées (p. 13); *it is* distinct de ses répercussions individuelles (p. 13), *and takes on* un corps, une forme sensible qui leur est propre *and constitutes* une réalité sui generis, très distincte des faits individuel[le]s qui la manifestent. L'habitude collective n'existe pas seulement à l'état d'immanence dans les actes successifs qu'elle détermine, mais, par un privilège dont nous ne trouvons pas d'exemple dans le règne biologique[76], elle s'exprime une fois pour toutes dans une formule qui se répète de bouche en bouche, qui se transmet par l'éducation, qui se fixe même par écrit (p. 12. – As for 'education' see p. 129).

Being concerned with methodology, Durkheim bypasses the question of origin. The dissociation of a social fact('s collectiveness) from its manifestations implies that the sociologist's method is not inductive – Paul and Hjelmslev agree *mutatis mutandis*. But the sociologist must have recourse to manifestations, if he wants to examine social facts *à l'état de pureté* (= *isolés de leurs manifestations individuelles* (p. 57)). He must then *dégager le fait social de tout alliage* (p. 12; cf. Sartre 1966:90 and the Descartes vignette of chapter 1.). The disengagement-operation is brought about by a method heavily reminiscent of Kant (1923:17-18), by statistics, which reveals *un certain état de l'âme collective* (p. 14). Sociology studies entities that are *organized, objective, collective, independent of manifestation*.

Durkheim has thus determined the social fact by a choice of words that also characterizes the objects of logic and synchronic linguistics: 1) immanence, 2) form, not substance (= not dependent on manifestation), 3) independence of the individual and the group, 4) existence in isolation, 5) imposing power, 6) transmission under the aegis of stasis, 7) generality and diffusion derived from collectiveness, 8) supraindividual and no individual substratum, 9) *la vie collective* is its substratum. A social fact is collective because it is

(vi) plus ou moins obligatoire. – *The collective* est un état du groupe, qui se répète chez les individus parce qu'il s'impose à eux. Il est dans chaque partie parce qu'il est dans le tout, loin qu'il soit dans le tout parce qu'il est dans les parties (p. 14).

The general distribution of a social fact is not caused by *imitation*; its diffusion in a society is independent of the individual. On the contrary, its generality is the product of its collectiveness and *coercitive* nature (p. 16 note 1). Durkheim's argument is circular: he presupposes what should be established: he needs to reject the social fact's diffusion as caused by the individual, and he does this by presupposing the collective nature of the social fact as well as its imposing power. And his final definition is this one

3.3.1 *Est fait social toute manière de faire, fixée ou non, susceptible d'exercer sur l'individu une contrainte extérieure;* ou bien encore, *qui est générale dans l'étendue d'une société donnée tout en ayant une existence propre, indépendante de ses manifestations individuelles* (p. 19).

---

76 An interesting remark in the light of both Lass's and Labov's reliance on biological models (Labov 1994:580-598), and Durkheim commented critically on this transcendent methodology in an 1898 article (1974:13ff.).

3.3. *The sociological analogy. Durkheim*

The question is: is the everyday language a social fact of the above order? Is it a constraint on our day-to-day lives[77]? In our attempt at imagining what it 'constrains' us from, we must do so in and through the everyday language. A heavy fine may restrain my tendency to exceed speed limits, but I can also break the Highway Code at my own discretion and with impunity. There is no such linguistic parallel.

However, a social fact has something in common with a language change; neither is a sensible phenomenon, both seem to impose themselves on their respective groups of people; neither can be 'broken' or opposed, and the paradoxical situation is that in language no synchronic rule so exists that it is not 'broken'[78]; apparently, it is bound, metaphorically speaking, to lose its strength and imposing power. No word does not change; not so with the social fact. Let us next consider Durkheim's argument in greater detail. The social fact is objective because it is

3.3.2 en dehors moi

a state of affairs presupposing existence

3.3.3 avant moi,

from which it follows that the sociofact, not unlike Kant's *Naturgesetze*, is so obligatory, coercive that it

3.3.4 determines the individual unidirectionally.

Durkheim presupposes both the concept of the horizontal and vertical syntheses of existence

3.3.5i T1   T2
         sociofact >   sociofact
         ↓   ↓
         ('others')   le moi (and others)

and if we want to understand the state of affairs at T2, we must resort to our dialectico-historical sense, not only to analytical reasoning. Furthermore, Durkheim needs the synchronic cut before T1 so as to exclude questions about the rise of the sociofact; immanence and objectivity have been established arbitrarily; the independence of the sociofact can only be es-

---

77 Coseriu (1974:26-37) has aptly criticized Saussure and Meillet for their uncritical attitude to and tacit acceptance of this superficially defined sociological jargon. My aim is different from Coseriu's, but the result may be the same: I want to show how the uncritical adoption by one science of concepts from another science leads to transcendent and metaphysical analyses.

78 Theoretical statements to the effect that some linguistic phenomenon cannot/must be the case invites their own refutation: (…) "over-whelm (…) is NEVER lost" (from the English language) or: *whelm* disappears – UNLESS *over-* precedes (…). Hoenigswald's two types of laws (1966:3). Cf. 3.3.8 below). But: *His black audience, searching for a human beneath all the hair and rhetoric, was underwhelmed.* (*Newsweek* (July 20) 1992:16).

tablished by way of a thought experiment: the people who created it exist no longer, but it is continued by other people so that it seems to lose its historical nature. The connotation that it originally contracted is dissolved, and other people – at T3+ – deliberately disregard the sign-functional nature of the sociofact, which points to its origin:

3.3.5ii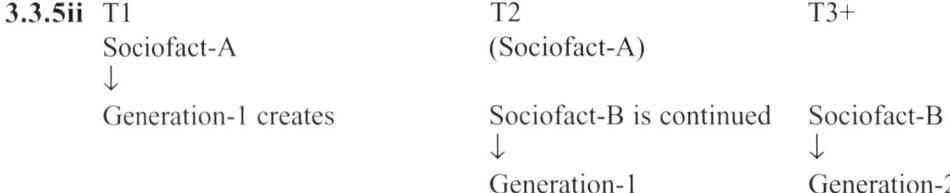

To labour my objection. Durkheim equivocates: the word *individual* is ambiguous. It may be correct that a given group of individuals did not create a sociofact; but it was created by other individuals. So Durkheim's equivocation consists in the different denotata of the word *individual*: now it refers to one group/individual, now to another, and in between we have the historical process of continuation without change[79]. Durkheim's dilemma is this: he must either idealize the sociofact and make it exist before the human race created the first society; or he must adopt a historical view of the rise of societies, as Kant did. Either the supraindividuality of the sociofact will be maintained at the cost of its empirical status, or man must assert his two basic, albeit conflicting faculties, in his **ex**istence: to individualize by being creative and dissociating himself from the community at hand, and to socialize by cooperating against a horizon of division of labour[80].

The constitution[81] of the social fact hinges on the simplification – through the notion of the arbitrary synchronic cut – of the historical phenomena, and its reduction to a stable synchronic fact consists of the following elements: Durkheim disregards the facts (i) that all sociofacts *had a before, when they did not exist, and will have an after, when they do not exist*; (ii) that a sociofact therefore has a man-made origin (is sign-functionally connotative); (iii) that man's nature is not that of an automaton (*l'homme parlé*), blindly obeying the commandments of the social code, but a complex of two opposing[82] characteristics; and (iv) that it is the collectiveness of the sociofact that 'disappears' the activity of the historical agent, turning man into an epiphenomenon.

---

[79] The originator *qua novateur* (p. 8) is well-known to Durkheim, but he is subsumed under the implicit characteristic of man as a purely receiving, passive, non-creative, law-abiding citizen, who does not resist a social law with impunity (pp. 129-130). In chapter 5 Durkheim is dealing with the (immanent) explanation of the social fact (pp. 117; 127); it is not to be explained in psychological terms (p. 128), only sociological ones (p. 126), but his field of vision is limited by his theoretical concepts, which rule out the possible relevance of the individual and historical factors, because *tout ce qui est obligatoire (…) a sa source en dehors de l'individu* (p. 129).

[80] Durkheim does mention other human faculties, but the individuel remains 'decentred' *vis-à-vis* the group (p. 130).

[81] If sociology is a historical discipline, then it is empirical; if it is not historical, its analyses are not exhaustive.

[82] Only from the point of view of our divalent logic and conventional A-or-nonA rationality, not from that of existence.

3.3. The sociological analogy. Durkheim

The empirical nature of a social fact is not in doubt; but it is participative and biuniformal (cf. chapter 3.2.7). Durkheim's constitution of it is not appropriate: we sense its 'coercive' reality as we try to resist it; when we do not sense the effect of a social fact overtly, he adduces an image from the natural sciences (p. 19 note 1). We do not feel the weight of air; despite this the air does weigh us down according to the pressure of atmospheres that prevails in a given place. Like a natural law, the social law is always enforced, owing to their respective types of necessity. And the imagery <u>both</u> creates this dilemma: from the empirical point of view, a social fact is either a natural law, which it is not, or it is an ideal entity, which it cannot be, <u>and</u> is deceptive to a fault: nobody is behind or enforces a natural law. A social fact is only *en ergeiai* in respect of man – *la masse parlante dans le temps* (Saussure 1972:112-113), and being a historical phenomenon, it is a constituent part of what it affects so that its participative and biuniformal 'energy' can be illustrated as follows in 3.3.6:

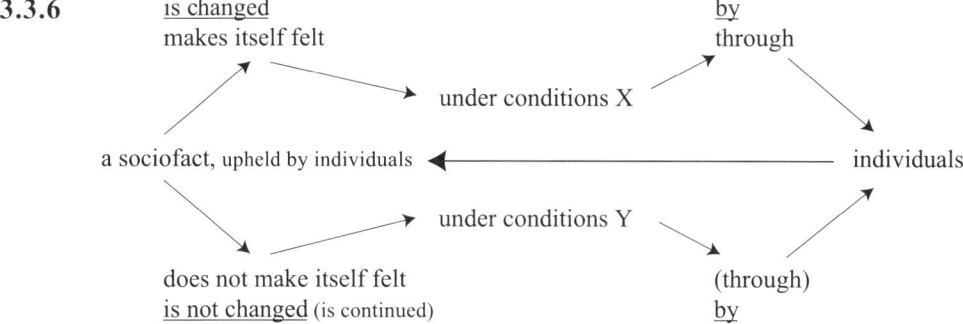

3.3.6

Unidirectional influence emanating from the autonomous social fact is incompatible with the reality of change. Unless we accept that the social fact is both determining and determined, (Durkheim's) sociology denies what it presupposes, history.

As neither the linguistic nor the sociological totality concept is well-defined in time and space – it may furthermore refer to a (language) state and to a part in it – I shall analyze the concept (*qua* a social fact) from the point of view of the linguistic part. Synchronic linguistics operates with a state concept that consists of immanent (systemic) units *qua* immaterial entities: nobody has actually seen the state or a unit; all the synchronic linguist has at his disposal is – in accordance with Durkheim – imperfect realizations of the parts of the state (Diderichsen 1966:21), these parts being rules which so enforce themselves on the speaker that the substance of the everyday language necessarily and obligatorily illustrates the coercive meaning of the rules. My illustrative data is taken from Late Old English and Early Middle English, it being the analogical extension of the nominal singular-plural pattern, $\emptyset :: es$: the two plural developments of OE *word* 'words': *word > word, word > wordes*, and the development of *stones > stones* (< OE *stanas*; the two sound changes from OE to ME, $y > i$ and $i > i$ will be added as illustrative material of the difference between sound change and analogy):

3.3.7   RULE 1: All nouns of type X must obey the $\emptyset :: -es$ rule
         RULE 2: All nouns of type Z must obey the $\emptyset :: \emptyset$ rule

The dynamic and biuniformal interpretation of the two rules looks like this:

3.3.8i

RULE 1 makes for	CHANGE and CONTINUATION	word > wordes stones > stones	Ø > -es -es > -es	y > i i > i
RULE 2 makes for	CONTINUATION and NON-CHANGE (Change	word > word stones > stones *stones > ston	Ø > Ø -es > -es *-es > Ø	y > y i > i *i > y)

Why does the second rule not make for change? Because, the synchronic grammarian must say, the rule (= a sign) under which the paradigm *word :: word* is subsumed exbraciates other specific nouns such as *ston :: stones*. Rule 1, in its specificity, makes for continuation, and the synchronic grammarian cannot explain the extension of the rule – unless by resorting to his synchronic cut, and then there will be nothing to explain, only data amenable to precise distributional description. The problem is twofold: Rule 1 makes *word :: word* disobey Rule 2, imposing itself on this type of words in defiance of both itself and Rule 2. In either case, the synchronic grammarian lands himself in an inconsistency. If change is law-breaking, the *dépassement*-problem is certainly a problem. But the synchronic dilemma is self-imposed, partly because of the empirically irrelevant synchronic cut, partly because of the unidirectional and binary nature of the concept of a rule *qua* a social fact (see 3.3.4). A rule as well as a language element is aitional (see below).

A rule's *puissance d'expansion* must be determined synchronically and in an empirically relevant manner, and so must the methodological step from either observed or deduced *distribution* to *rule*.

If a rule determines a specific group of words, there would be no problem if there were no (potentially) competing, *i.e. synonymous*, rules. The tacit ambition of the synchronic linguist is to be capable of explaining the extension of his rule in absolute contrast to the scope of any other rule. The moment he tries to do this, he shall be leaving the domain of descriptive[83]extensionality stepping on to the ground of intensionality: why does Rule 1 not subsume other phenomena under its scope? But it is a historical fact that the scope – number of denotata – of a synchronic rule will always change. So the question is: why did Rule 1 (in 3.3.8) expand its scope at the cost of Rule 2? Or: why did Rule 2 give up – part of[84] – its 'determining' (denotative) power? No extensional explanation can account for this synchronically; an exhaustive explanation will require the concept of a phenomenon's *Eigennatur*, a concept not foreign to historical linguistics; it can be detected in such interrelated factors as a form's 1) frequency, 2) phonetic substance, 3) distinctive quality, all of which factors may be seen in explanations of the survival of the English *-es*-ending[85]. Here the concept will be related to that of rule or 'cause'.

---

83 If we are dealing with metasigns, then 'denotative' replaces 'descriptive', and 'a rule' is replaced by 'the content of a metasign'.
84 I am being pedantic here: Modern English *deer :: deer*.

3.3. The sociological analogy. Durkheim

The type of the historical facts in 3.3.8 above demonstrates the need for a solution to the transition problem: a grammatical rule is **aitional**; it is enforced **but not only if** a number of extensionally specified conditions are fulfilled; a word is aitional, it being a language sign but not only if it is determined by a specific rule. Thus a word is not only the passive recipient (raw material for change) of external forces (synchronic rules): the word *word* was and had then become amenable to being influenced by a motivating factor, be it causal or a rule: productivity (creativity) and receptivity merge. The analogical extension of Rule 1 to neuter nouns in Middle English is not the whole story of Middle English nouns, because it does not explain its distribution, diffusion or the synchronic variation of the period, *i.e.* the continuation of the inherited paradigm. Conversely, the word *stones* had not become amenable to the influence of Rule 2:

3.3.8ii Rule 1   (Cr[Ø :: -es])   Rule 2   (Cr[Ø :: Ø])
   is extended   is not extended
   ston   *word*   *ston*   *word*
   remains   has become amenable   has not become remains   amenable

The aitional *Eigennatur* of a word is therefore expounded dynamically by: 'a word remains and becomes amenable to ...'; and without further ado Rule 2, like other signs, 'fades' away from the historical stage, it having nothing to 'determine' (3.3.4 and 3.3.7).

A word *qua* a social fact is not controlled completely by its rule; there is no uniformly determining 'public conscience' (*La conscience publique*, p.7) to check the linguistic behaviour of the speaker and hearer[86]. The rule's imposing and coercive power is relative. Being a fullfledged sign (ErC), the rule exists in accordance with its endogenous and exogenous nature, and far from being law-breaking, change is constitutive. Rule 1 above is certainly conducive to the continued existence of itself and the word *word*: changing *word*, it continues itself and this word **under the aegis of change**. Therefore it is not surprising that the synchronic (vertical) synthesis which a rule and its 'words' constitute is characterized by the properties of external endogeneity: [+attachment] and [+separation]. Their mutual attachment is neither obligatory nor necessary *sub specie temporis* (cf. Rule 2 above). The synchronic rule-word relationship is historical, not exclusively one that is motivated by the coherence of internal endogeneity.

The aitional nature of language elements explains their 'variation' or variability. Since it is inconceivable that Rule 2 *qua* a social fact would voluntarily relinquish part of its sphere of influence or rather would go against the epitome of its sole and whole *raison d'être*, the enforcement of its content, in relation to its rule the word has an *Eigennatur* that makes it amenable to the influence of other rules *qua* a part of its realization conditions. Let me pursue the argument a little more: the variability of OE/ME *word* (sing.) :: *word* (plur.) is due to its endogenous nature, which did not allow the word to be changed beyond recognition during the Middle English period; its development was controlled by the systematic opera-

---

85 But the survival of the fricative ending does not explain the demise of the equally frequent and distinctive nasal ending in the late Old and early Middle English periods.
86 Linguistic group pressure takes many forms: (social) polarization, hypercorrection, (prestige), imitation. Such pressures work both ways, being both rule-transcendent and rule-compliant.

tion of historical processes. Rule 2 contributed to these processes, but it was Rule 1 that eventually made the word continue (3.3.9i). This interplay illustrates the biuniformal nature of existence as a process of external endogeneity (3.3.9ii):

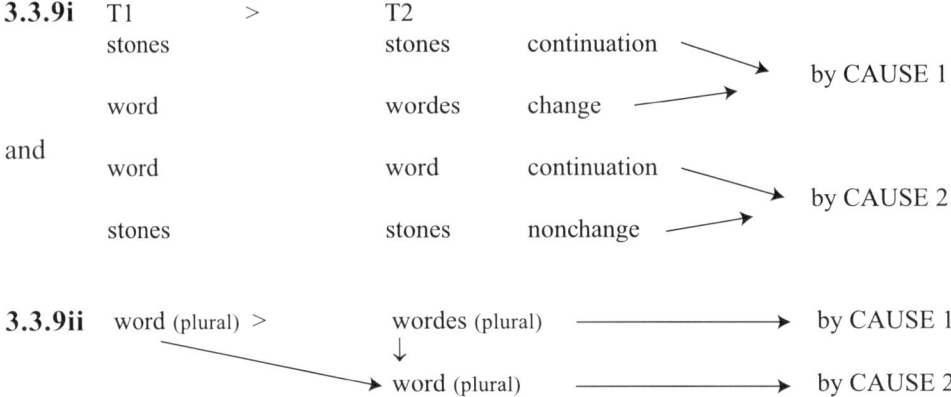

At T2 the historical phenomenon completes its meaning in *parole* with no law or rule functioning as a constraint; the new perceptual part (*wordes*) of the historical phenomenon receives its meaning from the structural relationship it contracts with the other, the inherited part *word*, which exists under the aegis of structural change[87]. I shall return to Durkheim's argument in favour of the non-individual nature of sociofacts, because it contains a type of logical fallacy that we came across in the corresponding linguistic discussion about the logical relationship between the concepts of change and stability in chapter 1. Durkheim's generalization-type in 3.3.10 is not cogent:

3.3.10 'I don't know X *qua* Y', hence 'X *qua* Y is not the case and X *qua* Z must be the case'.

The negative observational part of 3.3.10 refers to Durkheim's rejection of the sociofact having an individual substratum, while the positive conclusive part refers to his postulate that sociofacts have a supraindividual substratum. The presumed non-existence of 'X-is-Y' leads to the postulation of the existence of 'X-is-Z' by way of 'the negative conclusion: 'X-is-Y is impossible'. And from the perhaps correct observation that a given sociofact has not a specific *moi* as its (original) substratum, this fact is generalized into the non-individual substrum of sociofacts. As indicated above, the generalization hinges on an equivocation and a deliberate disregard for the historical facts, and the non-cogency of the argument is in fact due to its circularity:

---

87 *Excuse me, policemens, I know you busy.* James Baldwin *One Day when I was Lost.* P. 74. Corgi Books 1974. – In anticipation of my conlusion in chapter 6.3.B the structure in 3.3.9ii is summed up in dynamic terms: the synchronic meaning (of *wordes*) is its historical meaning (the meaning of *word*); the synchronic meaning is caused by the historical processes causing the form *word* to be continued against the horizon of change.

3.3. The sociological analogy. Durkheim

Durkheim argues from NonA to B, thereby presupposing two mutually irreducible universes of discourse: A :: NonA and B :: NonB; then he supposes the identity of NonB and A and that of B and NonA. Therefore the argument in 3.3.10 from

to
       NonA   (X is non-individual)

       B   (X is social)

creates a circle:

3.3.11

	Given	(1) the individual = A
we obtain by	(a) negation of (1)	(2) the non-individual = NonA
	(b) equation and unmediated substitution	(2a) the non-individual = the social = NonA
	(c) negation	(1a) the individual = the non-social = A.

Durkheim's conclusion in 3.3.10 asserts its premiss: a social fact is what it is because it is what it is:

    (2a) is equivalent to   (3) a (social) fact is non-individual, hence social

    (3) is equivalent to   (3a) a (social) fact is social, hence social.

Durkheim's argument presupposes the transition from the one to the other universe of discourse through the logical processes of negation and substitution; but his discourse of universe is empirical, therefore – like the universe of change and stability – contrary: A (= the individual) and B (= the social). In other words, the concepts of change/individual are not empirically identical with the concepts of non-stability/non-social.

The logical fallacies aside, Durkheim must be credited for the distinction between what is general and what is collective. A sociofact is supraindividual, hence collective; it is not the exclusive property of an individual – on condition that we deliberately ignore the historical origin of it. The sociofact is general because it is collective, becoming more or less generalized through the individual's agency (Durkheim, pp. 15-16). In linguistics the process runs the other way.

The step from variant to change is a generalization process in which a variant becomes general through spread, being gradually more and more imitated or used and thereby accepted by individuals[88]. Through generalization a variant eventually becomes different from what it used to be (from being accidental to being necessary), assuming the property of changehood and thereby the social properties of being collective, supraindividual and 'imposing'. How general(ized) must a variant be(come) for it to acquire changehood or systemic status?

---

88 See Charles F. Hockett's commendable exposition of the systematizing spread of a variant from its inception (*loc.cit.* – Cf. 3.4.1 below). – The concept of 'being accepted' is a synchronic indefinable; it can hardly be defined by 'continued existence despite loss of conditioning factors'.

A variant, say, an aspirated realization [ph] of /p/ in forceful style may be in general use in *e.g.* English, without being collective and necessary. It constitutes neither a change nor a systemic entity, */ph/, and for all we know [ph] may have existed in the English language during its historic existence with the abovementioned content; as such it must be said to have acquired at least some degree of collective-hood; we must again resort to the Commutation Test, this indestructible bastion of synchronic linguistics, in order to find an answer, which, furthermore, will throw light on the sociological problem.

A synchronic language element is social or systemic, if it satisfies the commutation test. As we saw above, the problem with this test is that it is not applicable to the dynamics of its object. The step from *[kyning] to */kyning/ in PreOE entailed no exchange of content elements. The variant *[ly:si] 'lice' had then the same commutative function as *[lu:si] in respect of, say, PreOE */læ:s/ 'pasture; less', and both */kyning/ and */ly:s/ merely took over the role in the commutative function that */kuning/ and */lu:s(i)/ used to have. Furthermore that preOE */u/ imposed itself on its speakers as *[y], let alone as */y/, makes no sense. On the contrary, the rise of variants testifies to the non-imposing, non-necessary power of the social fact, *in casu*, the phoneme. How can PreOE */u/ as a social fact be said to control a new variant such as *[y], its rise and spread? A possible answer: the role or function of *[y] is (more) similar to that of *[u] (than to the variants of any other phoneme): (some) words with *[u] and *[y] are in fact synonyms, just as in the modern language 'big' [big] and ['bi:g] are synonyms; but then the question: what is the social fact 'behind' these variants (cf. chapter 5.)? And in PreOE the rounded front sound could just as well be controlled by PreOE */i/.

Here is a case where Durkheim, the sociologist, could have learnt something from his changing everyday language and historical linguistics: the generalization of *[y] into /y/ does not presuppose the prior existence of the social fact */y/. PreOE */y/ could not have been created without the prior existence of its 'embodiments', and the point is that [y] becomes /y/ when [y] loses its conditioning factors and still *remains* in existence. The same goes for the social, more precisely, transsubjective fact, which becomes collective when its originators – the original group of *les moi*'s – perhaps no longer exist or are no longer needed for its continued existence; at this stage another group of individuals may have replaced the *original* one. With our loan from Gadamer (1972:14; – cf. *Prologue* above), Durkheim made his sociofact *eine historische Zufälligkeit*, as he lifted it out of its *geschichtliche Wirklichkeit*, in which it was bound to have a predecessor, from which it developed, and a successor, into which it developed, a historical reality that corresponds to the periodic nature of our linguistic examples: T1 /u/ > T2 /u/[u], [y] > T3 /y/ (and /u/)[89].

To sum up parallels between synchronic linguistics and sociology: (i) Durkheim's dualism (pp. 12-13), a social fact and its imperfect manifestations, is repeated in the *langue-parole* dualism; (ii) the sociofact is an immanent and individual-independent entity (p. 13); (iii) the social fact is *une résultante de la vie commune* and <u>un produit</u> des actions et réactions qui s'engagent entre les consciences individuelles (pp. 15; 117[90]), <u>which</u> becomes *eine den*

---

[89] Note that this explanation of the rise of /y/ does not explain the eventual systemshood of (Pre)OE /kyning/, where the conditioning factor is still present.
[90] As in linguistics, Durkheim does not deny the historical origin of his object.

3.3. The sociological analogy. Durkheim

Individuen übergeordnete objektive Wirklichkeit und Macht, die deren (: the individuals') *Verhalten regelte* (Arens 1969:467); (iv) *la trace de l'homme* and the creative agency of the individual (speaker/hearer) have been disappeared; (v) like the phoneme or grammatical category, the social fact is part of *un tout /qui/ n'est pas identique à la somme ses parties, il est quelque chose d'autre et dont les propriétés diffèrent de celles que présentent les parties dont il est composé* (p. 126); (vi) like the everyday language (*le langage*), *la société n'est pas un simple somme d'individus, mais le système formé par leur association représente une réalité spécifique qui a ses caractères propres* (p. 127).

Both Durkheim and the synchronic linguist postulate the empirical relevance of supraindividual entities. Both bypass the constitution of the social fact making do with summary statements like the ones quoted in 3.2. and above, the curious thing being that what makes a social fact a social fact is precisely what the sociologist disregards[91]. The above linguistic definitions do not imply a sociological basis, and in order to establish the link, we must turn to the father of modern structuralism, Ferdinand de Saussure, by way of A. Meillet's (1975:17-18) extreme version hereof: *ce sont les changements de structure de la société qui seuls peuvent modifier les conditions d'existence du langage*.

The final irony, however, is that in view of the parallels between the object of sociology and the everyday language it is curious why the direction of influence should proceed from the former discipline to the science of language: is it unlikely that the following **alterations** in Meillet's dictum, are irrelevant, a possibility that Labov (1994:24) does not even mention among his logical alternatives? That a language change can translate into a social change or a language structure into a social structure may at first glance seem a postulate with Realistic overtones, but following Labov it is not until at the later stages of the change process that a variant becomes a social variable and as such one of the fixtures of the social structure in question.

> Il faudra déterminer à quelle structure sociale (→ **linguistique**) répond une structure linguistique (→ **sociale**) donnée, et comment, d'une manière générale, les changements de structure sociale (→ **linguistique**) se traduisent par des changements de structure linguistique (→ **sociale**) (1975:17-18).

That *a social change* (**a language change**) may translate into *a language change* (**a social change**) is a corollary of the sign-functional nature of *man-made phenomena* (of **language phenomena**). Continuing this line of reasoning, which avoids the Scylla and Charybdis of either-or positions, I shall close this section with the following statements:

---

91 The dynamic existence of individuals: Durkheim is obliged to realize that the social fact *ne peut rien se produire de collectif si des consciences particulières ne sont pas données* And: interacting, *les âmes individuelles donnent naissance à un être (...) qui constitue une individualité psychique d'un genre nouveau* (p. 127). Neither Paul nor Goldmann would disagree, *pace* Durkheim's idealism. Durkheim does not draw theoretical consequences from the way a social fact is produced; it is a historical phenomenon, and Meillet (1975:17) makes this irrefutable fact hypothetical: (...) *s'il est vrai que la structure sociale est conditionnée par l'histoire, (...)*.

1) The social fact and the linguistic category (metaterm) exist through their man-made definitions, and the social fact determines (is imposed on) the members of a given social group (society) in virtue of its content, just as the (man-made) category denotes its denotata in virtue of its meaning and just as the words or utterances of the everyday language denote in virtue of their contents. A language sign may in some sense of the word be a social fact, but a social fact is certainly a language sign (Weber 1978:4), and in all three functions a human agent is necessarily involved and as such they contribute to man's historical existence.

2) The sociologist – like the linguist – generalizes the sociolizing nature of historical man to the exclusion of man's dissociating and individualizing faculty. The latter property permits man to seek and create differences in (the imposed) identity (at hand) so that variation and change have their own *raison d'être*, which does not make historical development (**ex**istence) mysterious, a rule-breaking or structure-destroying process in need of therapeutic revision, but which makes it a process of rational and systematic choice – freedom[92].

3) Closely connected with the notion of the dominance – priority – of the group *qua* supra-individual over the individual is the idealization of the sociologist's totality concept[93] as representing *une réalité spécifique* with properties different from those of the individual: this totality concept makes the sociological programme a metaphysical statement.

4) The interesting corollary of the above is that the social fact of sociology is to be compared to the linguist's metaterm, rather than to the phenomena of the everyday language. This can be corroborated in two ways: the everyday language provides some of the constituent factors of both the social fact and the metaterm (cf. chapter 4.). Secondly, the everyday language is – as Saussure emphasized – not a 'social contract', not a sociological phenomenon (cf. chapter 3.4.).

---

92 Descartes (*Med.*IV.57.27-58.25) did not make man's capability to choose 'freely', *voluntas* or *arbitrii libertas*, a cause of errors, *i.e.* irrational behaviour. This means that if we choose to apply our intellect appropriately, there is no cogent reason why we should not be able to create a theory of history – to regard history as inherently systematic.

93 It has become a class as one, nonresoluble into its individuals; see item (vi) on the previous page and *tout* in quote (vi) above.

3.4. A close-reading of Saussure's structural postulates about the immutability and mutability of the sign[94]. William Whitney's concepts of arbitrariness and convention. Saussure's arguments do not transcend the level of postulates, but Saussure affirms the historical nature of the everyday language and the incompleteness of the synchronic point of view

The possible influence on Saussure by the period's sociological debate is not my objective; nor is it my purpose to establish a special Durkheimean basis in Saussure's thinking (Harris 1987:224). About 1900 the scientific community had more or less explicitly adopted the then dominant notions – structure and function – of the general intellectual climate (see Topolski 1976[95]), although what happened in History came to play no decisive role beyond this particular field in these early decades of the previous century[96]. And there is no denying that it was a generally accepted attitude in the first half of the 20th century that the novelty of synchronic structuralism seemed to hold promises of scientific breakthroughs of a general nature, a belief that both Lévi-Strauss in his 1945-article and Cassirer's (1945:103-104)[97] sloganlike words testify to: the *human* sciences tried to bridge the epistemological gap between *les vérités nécessaires* and *les vérités contingentes*. – Glossematicians would not disagree (Hjelmslev 1963:127; 1972a:34).

To the properties of Durkheim's *fait social*, Saussure adds two well-known features: the language sign is characterized by *linearity* and *arbitrariness*. Above I argued for a third property, that of simultaneity, also fundamental to Sartre's criticism of structuralism. In the following we shall see how Saussure also presupposes this property, and as such corroborates my reinterpretation of structuralism as a predominantly historical point of view.

Saussure's greatest merit and contribution to the science of language is perhaps that he dared to ascribe the contradictory properties of 'immutability' and 'mutability' to the same entity (pp. 104; 24): the everyday language and the sign are both A and nonA[98]. The absurd – an entity is said to be both A and nonA – in our historical lives is something that goes against our expectations. And the fact that a change appears unexpected (para-doxy) to us

---

94 All page-references are to Saussure (1972), edited by Mauro Tullio.
95 Saussure's synchronic point of view was antedated by a decade by the Dane H.-G. Wiwel (1901).
96 It is a curious fact that the construction of theories of history was a prominent theme in history – both before and after – the days that saw the rise of the synchronic point of view in linguistics; questions of existence, explanation, the nature of historical phenomena occupied centre-position in the period's historical discussions (see Clausen 1963; and in particular pp. 95-149). *E.g.*, the complex relationship between the establishment of historical data and their interpretation (p. 97) has always been a decisive issue in history, an issue foreign to modern mainstream linguistics and of course nominalism. In my view Lass (1997:18) merely mentions some of the problems, and in his book there is no serious discussion of (what he means by) *the 'non-existence' of the past* and the (*epistemological, ontological*) *status of the objects of historical enquiry*. He does not envisage the possibility of an immanent historical theory, nor does he demonstrate its impossibility. – As for the notion of data-construction, see chapters 5 and 6.
97 One cannot help wondering at Cassirer's overflowing praise of linguistic structuralism, in view of the fact that in the very same article he shows how structural principles had been commonplace notions from the days of Goethe onwards (cf. chapter 3.2.).
98 I am not certain that the Editor's note on p. 108 is appropriate. There is no reason to find excuses for empirically relevant paradoxes, and Saussure's attempt at coming to grips with one of the purported paradoxes of historical existence is certainly commendable.

does not detract from its empirical relevance, so that perhaps *expected* change is more relevant than *unpredictable* change[99]. Furthermore, Saussure's introductory remarks (pp. 20-43) are less problematical than the chapters I am going to deal with: regarded as implementations of the former, the latter sections highlight the difficulties in combining the historical nature of the everyday language and its synchronic aspects[100].

A sign consists of a *signifiant* and a *signifié*, and the relation between them is arbitrary and so is the relation contracted by the sign and its denotatum. The *signifiant* appears as *librement choici* in relation to *l'idée qu'il représente*; but in respect of its linguistic community *the signifiant n'est pas libre*, because *il est imposé* (p. 104).

Who chose *le signifiant*? *La masse parlante* was not *consultée*. *La langue* did. That is impossible. Had *la langue* chosen its own signs, it must have existed prior to them[101]. The linguistic community cannot choose freely, either (p. 101; 104). So our first question, answered by Whitney (see below), must be: who or what imposed a given choice on a given *langue* and its group of speakers?

Saussure is right in emphasizing that the linguistic community is tied to (*liée à*) its language in a manner that prevents the language from being compared to *un contrat pur et simple* (pp. 104; 26). The group (*la collectivité*) is subjected to the sign; neither the language rule nor the sign function is a contract a group of people has consented to voluntarily (p. 104). On the contrary, a speech community has so *received* (p. 100) its language that *il n'est pas au pouvoir de l'individu de rien changer à un signe une fois établi dans un groupe linguistique* (p. 101). This type of argument is heavily reminiscent of Durkheim's (1927:128-129[102]).

Like Durkheim's sociofact, Saussure's language fact exists outside *notre volonté* (pp. 38; 104): *le signe linguistique échappe à notre volonté*). *La langue* is a social institution (p. 105), *reçu* (as a heritage) *dans une société* (p. 100).

La langue exists (at T1) *avant moi* and the society receiving it (at T2); it *is un héritage de l'époque précédente* (p. 105). Following Durkheim's diction, language exists in a state of dissociation from its speakers, and the generational transmission of a language turns out to be an *idealizing* agentless process in which the roles of the individual speaker and speech community are those of decentred, passive recipients: the individual transcends neither *in* nor *out* of the inherited everyday language, *déjà constituée*. Furthermore, the tacit assumption is that the language of T1 as well as Durkheim's social fact is inherited by the society of T2 against the horizon of non-change.

The process of imposition is described as follows: nobody has ever seen (*constaté*; cf. p 111) – we can only imagine (*concevoir*) – the process through which, *à un moment donné*,

---

[99] I am here referring to Samuels's (1972:2) division of change into those two categories as against the tradition's value-loaded conception of change as only *mysterious*.

[100] Despite the fact that *le langage est un fait social* (p. 21), *[à] chaque instant il implique à la fois un système établi* (by whom?) *et une évolution* (a thing? Yes, because) *à chaque moment, il est une institution actuelle et un produit du passé* (p. 24). Saussure's monolinear conception of history is clear, and the reification of the process is equally obvious: *une évolution = un produit*.

[101] Saussure is here being one of Paul's Realists. *Une langue* without signs is yet to be seen.

[102] Saussure's linguistic example of historical inevitability is here implemented – sociologically – by the necessary attachment of the individual to his or her nationality. Glossematically, this necessity appears in the connotation contracted by a particular nation and its language.

3.4. A close-reading of Saussure's structural postulates

the names (*les noms*) become distributed over the things (*les choses*), through *which un contrat serait passé entre les concepts et les images acoustiques* (p. 105), and this contractual relationship appears to the individual as both *un produit hérité* and *un idiome déjà constitué* (p. 105; cf. note 100 above), all of which leads up to this theoretical *fundamentum* of synchronic linguistics:

3.4.1 l'idée que les choses auraient pu se passer ainsi nous est suggérée par notre sentiment très vif de l'arbitraire du signe (p. 105).

3.4.1 is echoed in the structuralist's words (from Sartre 1966:88): *l'histoire est insaississable*. This unknowable transmisson of *Un état de langue donné* (…) *toujours le produit de facteurs historiques* (p. 105) from generation to generation is controlled by the arbitrary nature of the sign. No language historian really disagrees with Saussure over the incontestable fact that to the linguistic community *la langue* is *un produit hérité des générations précédentes* (p. 105). The curious thing is, however, that Saussure infers the immutability of the language sign from the fact of tradition: *Un état de langue donné* is always the 'product of historical factors' – whatever they are – *et ce sont ces facteurs qui expliquent pourqoi le signe est immuable*; and to be unchangeable means to resist *toute substitution arbitraire* (p. 105).

Saussure asks whether it is impossible to change the inherited language laws from one moment to the next? His discussion is extremely pertinent and the linguistic tradition has, I think, ignored it. Let me first recall Martinét's equilibrium concept (1964:94) constituted by the expressive needs (Saussure's *l'action libre de la société*) of the synchronic moment and the inborn inertia (Saussure's *la tradition imposée*) of that same moment as well as Samuels's extended version of it (1972:2). Both equilibria exist in a social context: *la langue* exists *dans son cadre social* (p. 105). Following Saussure, it holds true of any social institution – not including *la langue* – that there is a non-stable equilibrium between *la tradition imposée* and *l'action libre de la société*, and at one moment the forces of tradition will be more powerful than the present demands of the society, while the next moment experiences the converse relationship (p. 107). The equilibria of Martinét and Samuels constitute tensional fields in which the 'unhappy' balance between the two needs will cause change; their analogy is that of the sociologist's social fact. Not so with Saussure's social instituion called *la langue*. Here *le facteur historique de la transmission la* (*i.e. la langue*) *domine tout entière et exclut tout changement linguistique général et subit* (pp. 105-106). **The historical nature of language prevents change,** and our question 'What mechanism makes for stability?' has been answered, albeit paradoxically: *un etat de langue* is stable because it is a historical product. Despite himself Saussure is in a way right. A historical product is bound to have a static, stable or immutable element in it.

I agree with Saussure – and Durkheim – that a concept such as historical inevitability is needed; I have called it *empirical necessity* and determined it, not by the impossibility of change, but the non-contingent development of the present into the present[103]. I do not agree

---

103 To be explicit: in relation to human existence it makes no sense – is indeed absurd – to claim that there is an alternative to 'Man cannot but exist, live from the present to the present': we cannot choose not to let the present create the present. Durkheim's inevitability concept has also existential overtones, but is restricted to his particular purpose.

with Saussure as he limits his object (*objet réel*) of linguistics to the study of *la vie normale et régulière d'un idiome déjà constitué* (p. 105; cf.Sartre's criticism of Lévi-Strauss's similar use of 'preestablished'). As Durkheim did, Saussure arbitrarily excludes the historical factors, the agents of *constitué par* because language as *un héritage n'explique rien* (p. 105[104]); this is only correct, if the possibility of a historical theory has been ruled out a priori; what is more, he precludes the *raison d'être* of the immutability of the language state from the domain of synchronic linguistics. Thus Saussure, taking his point of departure from a two-state scenario, paves the way for his successors' emphasis on the single autonomous state: in terms of non-change the process L1>L2 does not prevent us from regarding L1 and L2 as two autonomous states (Hjelmslev 1928:227), but it makes no sense to do so, if L2 is the product of L1. – The problem is that Saussure – this also applies to Hjelmslev (*loc.cit.*) – has no theory to demonstrate how the process has repercussions in L2, and in this sense the fact that L2 comes from L1 explains 'nothing' – and we need not even ask such historical questions (Durkheim 1927:130).

In addition to the above metacondition for the immutability of the everyday language Saussure adds seven mutually incoherent factors making for stability (i – vii)[105]:

(i) Change (*changement linguistique général et subit*[106]) due to generational language transmission is impossible because in historical development one generation does not follow its predecessor in a straight and non-overlapping succession (p. 106). Generations exist in simultaneously overlapping and mutually interpenetrating S-curves and geology-like relationships.

(ii) The efforts needed by the child to acquire her native language exclude the possibility of *un changement général* (p. 106).

This argument is somewhat at odds with the next, which is reminiscent of sophistry:

(iii) The language users are normally unaware of their language, so how should they ever be able to change it?

Comments: Saussure presupposes a specific type of change; change is abrupt: we have either /a:/ or /ɔ:/. The change from /a:/ to /ɔ:/ presupposes absence of mutual cross-generational intelligibility, to which must be added the lack of psychological reality of the process represented by the A>B-formula. Had this process any psychological reality, then argument (i) would have to be abandoned, as the younger generation could use /ɔ:/, the older continue to use /a:/. In sum, generations do understand one another, hence no change is possible: (i) *if mutual intelligibility, then no change*.

---

104 Does language as *un fait social* explain anything?
105 It is interesting to note that Saussure criticized his predecessors for operating with incomplete arguments; they omitted the 'middle terms' or 'mediating propositions' (in Danish *mellemregninger*: Gregersen 1991 vol.1:149). But neither Gregersen nor *e.g.* Harris (1987) sees that Saussure must be blamed for doing exactly the same: his arguments here in chapter 2 of his *Cours* are logically flawed, incoherent and unsubstantiated.
106 It is to be noted how guarded Saussure's diction is in terms of the applied epithets.

3.4. A close-reading of Saussure's structural postulates 271

The child coping with the intricacies of language learning cannot also cope with change: (ii) *if language acquisition, then no change.*

The consequent of the third argument: *if unaware, then no change*, is a *non-sequitur*. The converse – change will follow from the language-users' unawareness – could be maintained with equal justification.

The consequences of Saussure's postulates are: (i) *if a language changes, then mutual intelligibility breaks down*; (ii) *if a language changes, then it is unlearnable by the infant*; and (iii*) if a language changes, then its speakers are aware of their language*. Hence: since nobody denies that all languages change, mutual intelligibility has always been preserved, and no infant has not become a native speaker of a language, the two first arguments are blatantly self-defeating, while the purportedly obvious falsity of the consequent of the third implication negates the antecedent, yielding an equally absurd result.

(iv)   The arbitrary nature of the sign makes it (and *la langue*) immune to change.

Saussure underpins the point in an oddly negative and indirect way. *La langue* cannot be discussed within a rational frame of reference: *(...) pour qu'une chose soit mise en question, il faut qu'elle repose sur une norme raisonable* (p. 106; Christensen 1962:11; Guthrie 1967:17-21). Since we cannot discuss the rationality – because there is none – behind a possible preference for *soeur* to *sister*, or vice versa, change does not rest on a rational basis; hence it cannot be discussed[107]; hence change is not possible, even though *la masse parlante* was *plus consciente qu'elle ne l'est* (p. 106). This argument is in conflict with the above remarks on the language user's alertness to his language; secondly, argument (iv) is equivalent to: *if a language changes, then it is not arbitrary.* And we must again say: language change is an irrefutable fact.

Saussure's problem is that he did not distinguish properly between the levels of logic and data: does the data, the synchronic *langue* and historical change, have no rational basis so that the linguist is unable to study them (Lass 1980:90; 143-144; 169-170; Labov 1994:10; Diderichsen 1966:21)? Or: has linguistics no theory within which to discuss *le langage*, because *(...) on ne sait comment dégager son unité* (p. 25)? In either case, the constitution of data is the name of the game, and we are in a way back to square one in that the former alternative is another statement of the claim that sensuous reality, in particular history is chaotic, while the second alternative raises the question: is it necessary to split up (analyze) the unity of language into its historical and synchronic components? Saussure elaborates the first alternative by an argument the intricacies of which I shall try to resolve by way of constructing an explicit argument in 3.4.2 below, his initial postulate, couched in the language of space[108], being that there is no reason why a speaker should prefer *soeur* to *sister*, hence the language in question does not change:

---

107 Paul Postal is oddly reminiscent of Saussure (see Samuels 1972:vii; Lass 1980:170; but cf. also Saussure, p. 110).
108 Saussure does not use the temporal development of Latin *soror* into modern French, but delimits his field of vision to data existing simultaneoulsy at a point in time.

3.4.2

(1)   If there is no reason for preferring *soeur* to *sister*,
then a language does not change.

         =

(2)   If a language changes,
then there is a reason why *soeur* is preferred to *sister*.

         &

(3)   If there is a reason why *soeur* is preferred to *sister*,
<u>then a language is not arbitrary,</u>

         hence

(4)   If a language changes, then it is not arbitrary.

         =

(5)   If a language is arbitrary, then it does not change.

Logically, (4) and (5) are synonymous propositions, and although conclusion (5) was what was to be demonstrated, it is falsified by (4), a proposition open to empirical verification, which (5) is not.

3.4.3

(1)   If there is no reason for preferring *soeur* to *sister*,
then a language does not change.

         &

(2)   If a language is a historical phenomenon,
<u>then there is no reason for preferring *soeur* to *sister*,</u>

         hence

(3)   If a language is a historical phenomenon,
then it does not change.

         =

(4)   If a language changes
then it is not a historical phenomenon

Conclusion (3) is Saussure's metacondition, while (4) is synonymous with it. (4) is patently absurd; if a language changes, then it is not transmitted by *le facteur historique* (pp. 105-106). The argument in 3.4.2(1) will be amplified by 3.4.4:

3.4.4

(1)   If *soeur* is a historical product,
then it does not change.

         =

(2)   If *soeur* is changed,
then it is no historical product

The evident paradoxes of arguments 3.4.3 and 3.4.4 are not salvaged by a consistent definition strategy, the involved historical concepts being undefined, while the concept of being arbitrary is – not defined by – only replaced by the synonym, *immotivé* (p. 101), which explains nothing.

3.4. A close-reading of Saussure's structural postulates 273

Only one supposition salvages Saussure's argument, and that is the essence-accidence parameter: if change relates to the accidental nature of the everyday language, it follows that, say, the English language has only one underlying system, still unbeknownst to the synchronic community, one system behind all its historically known manifestations. This (stable) system is (arbitrary and) historical and has therefore remained stable from ab. 6/700 until today[109]. The rescue operation results in absurdity: a system that has come into being is subject to ordinary existence conditions.

Saussure's problem lies in his yoking together of the properties, to be arbitrary and to be historical, without defining the latter one. He supposes the dependence of history on synchrony in order to infer the irrationality of the historical point of view: the arbitrary nature of *la langue* implies that its historical motivation has no rational basis; thus he also implies the irrationality in the time-based choice of either *soror* or *soeur*: there is no reason why speakers of Latin should have so chosen variants of Latin *soror* that we have French *soeur* today. Saussure did not see that the generalization of the fact that a choice was indeed made is the *historical inevitability (necessity)* that *there is no 'choice' to **ex**istence*; we cannot but choose to make the present create the present.

The notion of the arbitrariness of the sign is a commonplace in as well as one of the corner-stones of synchronic linguistics (p. 100). Whitney (1970:18-19) set great store by it in his 1875 book: it is a curious fact that synchronic linguistics grounds its theory on *a historical concept*.

Since Saussure's argument in defence of arbitrariness follows Whitney's, who also includes the concept of *convention*, I shall take my point of departure in the Whitney text. Furthermore, Whitney avoids any trace of cosmological reasoning, which Saussure does not (Saussure, pp. 100-101)[110].

Following[111] Whitney (1970:19), the sign is arbitrary *because any one of the thousand other words current among men or of the tens of thousands which might be fabricated, could have been equally learned and applied to this particular purpose*. Whitney's setting is generational transmission, we are not, I suppose, back in the days of the origin of language. To

---

109 When did Old English come into being?
110 Saussure's argument assumes a global stance: in respect of the class of all languages *qua* the class as one, the sign is arbitrary, but from the extensional point of view with the class of languages being a class as many, the case is not clear-cut; that *søster* is the preferred choice of Danes is certainly a motivated choice (cf. also Whitney's argument below). Anticipating the next note, I shall ask: how can we proceed non-transcendently from the motivated use of, say, Danish *søster*, to the word's unmotivated (arbitrary) nature, within the immanent state (cf. 3.3.10)? The answer is that arbitrariness presupposes at least two languages (space), and it presupposes what it should have proved, the impossibility of creating a historical theory (time). – As can be seen, synonymy is certainly a tacit assumption of Saussure's.
111 The quotes that I am analyzing all bear the mark of postulates; in order not to complicate the argument I do not ask this question of each postulate: how do Whitney and Saussure know this? The inference from the fact that there are cross-language synonyms, Danish *en hest*, English *a horse*, German *ein Pferd*, French *un cheval*, to the arbitrary nature of the sign is a *non-sequitur*. The argument also presupposes an adequate definition of synonymy. My point is that synchronic structuralism is self-defeating: as a sign acquires its meaning in and through the relationships that it contracts with other members of a given relational network, the abovementioned signs cannot be synonyms: their meanings are totally different through their different internal (immanent) relational dependencies. Therefore such sets of synonyms do not *non-transcendently* substantiate the arbitrary nature of the language sign.

the language learning child, there is no internal and necessary tie between word and idea, and any historical reason behind the tie is also non-existent to him. The sole and sufficient reason for the child's use of a particular sign is that it is used by those about him. Far from setting the ground for the immutatbility of signs, the argument highlights the historical nature of language.

A sign, according to Whitney, is conventional *because the reason for the use of this rather than any other lies solely in the fact that it is already used in the community to which the speaker belongs*[112]. The argument implies that there is no exclusively synchronic reason for a speech community's use of a particular (set of) of word(s), it is a both-and: the word was used prior to (time) and simultaneously with (time and space) its adoption by the language learning child. Whitney has it that there is no reason, *either in the nature of things in general, or in the nature of the individual speaker who uses it*, that *prescribes and determines it* (a given word[113]). The reason for the use of a given word is historical, and since no rationality based on the community *qua* a synchronic (*i.e.* social) fact can be advanced, synchronic (social) discussions about language are impossible, they do not rest (*reposer*) *sur une norme raisonnable*: the synchronic point of view must dissolve the paradox, that data has no rationality and the science studying this data must be rational – neither view rules out the possibility of a rational history. I conclude: the irrationality of data is a postulate; the possibility of a theory of history is still realistic. Whitney's and Saussure's position(s) will be summed up in 3.4.5i:

3.4.5i Of the members of a speech community it is true that *no reason* – inherent in either their individual natures or the nature of things – *determines* these people's use of X, a specific (set of) word(s), or the nature of the sign.

Hence the only reason why a speech community uses X, etc. is historical, and that this fact should imply the impossibility of change is absurd. The 3.4.5i postulate, telescoped into *No reason determines a person's use of X,* must appear as the conclusion of an argument; to this end a middle term such as *the synchronic use of X* is needed:

3.4.5ii A person's use of X expounds the synchronic use of X
  The synchronic use of X has no reason – (no rational basis)
  No reason determines a person's use of X.

From this we are to infer the irrationality of historical motivation:

3.4.5iii Historical motivation determines a person's use of X
  No reason determines a person's use of X
  No reason determines historical motivation[114].

---

112 Whitney presupposes the dualism's *same word*.
113 The sociologist's imposition-concept can easily be detected.
114 This assertion defeats the rationality of synchronic linguistics: 'Synchronic linguistics has no scientific foundation' (the major premiss) & 'Historical linguistics presupposes synchronic linguistics' (the minor premiss), ERGO 'Historical linguistics has no scientific foundation', Q.E.D.

3.4. A close-reading of Saussure's structural postulates

Logically, the study of historically motivated phenomena may not be 'reasonable'; they have no *raison d'être*, a conclusion not incompatible with the received view of history, but that historical processes cannot be translated into simple logical formulae is a fact that does not bother some historians (Clausen 1963:95), because the reason for this could – logically speaking – just as well be deficiencies in the logical apparatus; furthermore, the above conclusion is a postulate that Saussure is going to, indeed, has to renege on; and, to repeat, it does not disprove the possibility of a theory of history. The synchronic linguist's postulate that history is chaotic, hence no historical theory is possible, remains a value-loaded postulate whose whole argument follows that of 3.3.10[115].

Whitney and Paul agree: understanding language is a task based on historical principles. Synchronic explanation provides no rational basis for change.

Saussure's fifth argument in defence of the impossibility of change reveals his narrow view of history

(v) Historical development is the substitution of discrete units, and as such it carries no theoretical importance: *La multitude des signes nécessaires pour constituer n'importe quelle langue* makes it impossible to change language. – Cf. (vii).

Change *in toto* is impossible; but no historian would ever need to adopt such an extreme initial condition for a theory, and no historian disagrees with Saussure's observation.

Despite this, *in-toto* change is a corollary of the *tout-se-tient* principle of structuralism: if X in network S changes, then the other elements of S are bound to change and the totality S has changed[116]. In his sixth argument Saussure seems to have realized that argument (iv) backfires:

(vi) *Le caractère trop complexe du système* prevents change. The system – *Une langue constitue un système* – is not *complètement arbitraire*, but expounds *un raison relative* (p. 107). The system is so complex a mechanism that we can only grasp (*saisir*) it *par la réflexion* – (Parmenides and Plato would not disagree (Christiansen 1998:33;47)), and therefore *la masse parlante* is not competent to change the system and is on the whole ignorant (unaware) of it. Even specialists cannot interfere with *la langue*.

Saussure salvages synchronic linguistics as a scientific discipline, now that the system rests on *une norme raisonnable*. The consequence is, however, that he invalidates his corresponding point under (iv): there he claims that *la langue, système de signes arbitraires* (p. 106) lacks a rational basis, here he claims that *la langue qua* a system has a rational basis (p. 107), hence the system is not arbitrary despite the fact that it consists of arbitrary signs.

In Saussure's final argument, which contains two layers, his view of history is made explicit by the undefined notion of *une révolution*[117] (p. 107):

---

115 Cf.: "Because I do not know – have not been able to construct – a theory X of Y, a theory X of Y is impossible, and therefore a theory Z of Y is the case."
116 See Labov (1994:302) for a small-scale example. Chain shifting is in a way also an example hereof.
117 This corresponds well with Gregersen's investigation into Glossematic change: it is bound to be cataclysmic.

(vii)   change is *une révolution*. – Cf. (v).

Since change is a revolution, and since M is the case, change is impossible. What does this M represent? It is nothing but the negation of Saussure's change concept *une révolution*. In order to reject change Saussure becomes self-defeating; he is forced to logically negate his definition of change in (v) and (vii). The argument that Saussure presupposes is this (3.4.6):

3.4.6   language change is change *in toto* or a revolution
<u>*in-toto* or cataclysmic change is impossible</u>
Language change is impossible

Secondly, *la langue (...) est à chaque moment l'affaire de tout le monde* (p. 107), and this fact distinguishes it absolutely from all other social institutions, each of whose sphere of existence is limited in space and time[118]; *la langue, au contraire, chacun y participe à tout instant, et c'est pourqoi elle subit sans cesse l'influence de tous* (p. 107). Language is constantly under the influence of everybody, but, now the paradox, *Ce fait capital suffit à montrer l'impossibilité d'une révolution* (p. 107). The collective character of the everyday language prevents change. Saussure's argument is deceptively reasonable; firstly, it hinges on the empirical relevance of the idealized supraindividual entity called *la langue* or *le langage*. Not unlike Plato, he does not tell us how to understand the participation-relation[119]. Secondly, he creates premises that no historian has advocated and whose empirical relevance is zero:

It is not unreasonable to assume that it is impossible for all speakers of a given language to find one common rational reason for the change of, say, French *soeur* or OE *dom*, and therefore no change will occur; it is not unreasonable to assume that it is impossible for one single speaker to be able to persuade all his fellow speakers into adopting a change of his choosing – language change is impossible.

Saussure creates this paradox: man has the potential for changing his language, but precisely because of this faculty he does not change it, a type of argument that both Kant (1923:18) and Aristotle (Lear 1988:55-56) found absurd: a potentiality that is not *en ergeiai* is an empirical inconsistency. This will be tied up with the following:

Not only is the everyday language the business of *tout le monde*, it cannot change because of *La résistance de l'inertie collective à toute innovation linguistique*[120] (p. 107). Precisely because of everybody influencing their language, it cannot change.

In contrast to Durkheim and, in a way also to Saussure's initial position, the imposing power of language neither makes for stability nor prevents change. It is *la masse social* (p. 108) which makes for stability. Being inerte, it is *un facteur de conservation*, and language – in accordance with Durkheim's sociological view – is *une chose dont tous les individus se servent toute la journée* (p. 107). The above influence is conservative, hence the everyday language is stable.

---

118 Durkheim would not agree, cf. his metereological image.
119 Durkheim uses the non-committal copula to describe the relation between the whole and the part (see 3.3.vi).
120 This is another deceptively reasonable postulate – where does Saussure know this from?

3.4. A close-reading of Saussure's structural postulates

The logic of Saussure's argument is as follows: let's first recall that Durkheim's social fact did not find its embodiment (*faire corps avec*) in the behaviour of the individual persons of a community. Saussure is far more explicit: since *la langue fait corps avec la vie de la masse sociale,* it assumes the property of inertness <u>from</u> *la masse sociale*. The logic of the argument is fair enough, but it remains a postulate: how does Saussure know that man's fundamental characteristic is to be inert, passive, conservative, to resist change? How does he know – his stance is clearly Durkheimean – that *la langue* merges with the idealized group (as one), and not with each member of the group, so that the opposite properties 'to be inert' and 'to be active' could, as a theoretical starting point, be more or less evenly distributed over – and <u>in</u> – the members of *la masse parlante*? Secondly, is it necessary to maintain the presumed channel of influence, <u>from</u> the body social to language? Could it not be the other way round – with language being inert and thereby making *la masse parlante* inert (*parlée*)? True, the result would be the same, but not unconditionally so. The alternative postulate that *la masse parlante* is active (*parlant*) would yield the opposite outcome. With Saussure and Durkheim we see the passivization[121] of man, man is about to become *l'homme structural*, inert (spoken; *parlé*) and in so far as he is active (speaks, *parle*), he merely observes the rules of and is determined by the social facts *déjà constitués*. Like the social fact, language lies outside man's *volonté*. From such an 'anthropology' change and development, *dépassement* and existence cannot be inferred unless *qua* something mysterious, destroying and rule-breaking, something to be combatted.

To conclude Saussure's argument (p. 108): language is not *libre* because it is *un produit des forces sociales*, whose effectiveness presupposes time. *Si la langue a un caractère de fixité, ce n'est pas seulement parce qu'elle est attachée au poids de la collectivité, c'est aussi qu'elle est située dans le temps. (...). A tout instant la solidarité avec le passé met en échec la liberté de choisir. Nous disons* homme *et* chien *parce qu'avant nous on a dit* homme *et* chien (cf. Whitney). The past is the cause of the present, arbitrariness is the property of a phenomenon that is historical. Saussure's historical position emphasizes the process of continuation without change to the exclusion of continuation with change. The arbitrariness of the sign may make our choice free (*le choix est libre* (p. 108; 112)), but the argument is meaningless, as there is no free choice: that X could have been different, but is not different from what it is (was), is empirically irrelevant.

The sign's arbitrariness makes our choice free, but time makes it 'unfree' or necessary. This arbitrariness is grounded on the purported fact that the sign knows no other *law* than the law of tradition, which tells us to continue *homme* and *chien* because *avant nous on a dit* homme *et* chien (p. 108; cf. Durkheim's *avant-moi* argument and Whitney's notion of *already developed = déjà constitué*)[122]. The paradox is obvious: arbitrariness presupposes history (see Editor's note 150, p. 449); the historical nature of language prevents change; arbitrariness is the precondition for change: the dual nature of language is such that it makes for

---

121 Cf. also *La langue n'est pas une fonction du sujet parlant, elle est le produit que l'individu enregistre passivement* (p. 30), a postulate hard to reconcile with the fact that *la langue* is the product of *la parole* (p. 37; and note 128 below), *la parole* being human activity – *un acte individuel de volonté et d'intelligence* (p. 30).

122 Saussure does not see that the law of tradition also implies change: there was a time when people did not say *homme* and *chien*, but other forms of the respective Latin etyma; cf. below.

both change and non-change. In spite of this language change is impossible, and that makes the arbitrariness factor superfluous. This factor is a cosmological relic; we proceed from differences (among cross-language synonyms) to arbitrariness with no middle term (3.3.10) on two tacit assumptions: (i) (that two signs refer to) the same denotatum should entail choice of the same sign; (ii) no historical theory can explain the different choices. Saussure is implying that no theory of history is possible.

Saussure's view of identity leaves something to be desired in terms of precision: how does he know that *homme* and *chien* of T2 are the same as *homme* and *chien* of T1 or T3? *He is presupposing what he both wants to and should establish, the non-change of* chien *and* homme *through time* and *the historical theory behind the solidarity of the past with the present*, respectively.

The logic of Saussure's single arguments is tenuous, although the empirical relevance of each of them is not to be overlooked. The argument as a whole consists of mutually unconnected postulates. Saussure asserts the static nature of the language sign against the horizon of change. Neither the linearity, which here plays no role at all, nor the arbitrariness of the synchronic sign can be upheld without qualification.

Saussure does not succeed in amalgamating the historical with the social character of the everyday language. He insists on the former property as against the sociologist's attempt at defining his sociofact as anything but historical. Language as a sociofact is different from the sociofacts of sociology (p. 110), and the unique character of the everyday language is maintained: language as a symbol system is different from all other such systems (p. 111). What he owes to Durkheim could be the notion of the supraindividual constancy of sociofacts, but in contrast to the sociologist he anchors *langue*'s stability anthropocentrically, *dans la masse parlante*, and historically, in the past-present solidarity[123].

Saussure's treatment of the mutability concept supports the preceding critique. Nowhere in this section (pp. 108-113) does he forget the historical nature of his object, a nature without which *la réalité linguistique* is not complete (p.. 113). Duration, *la durée* (p. 113), is a precondition for the linguistic reality called *la langue*. Hence language grasped analytically or conceived only in *sa réalité sociale* (p. 112) is not empirically relevant. Similarly, *la langue qua* an immanent entity *en dehors de sa réalité sociale* refers to the individual aspect of language, *une chose irréelle* (p. 112). Saussure is quite explicit and the conclusion is inevitable: the study of language viewed as a sociofact only studies a linguistic subconstruct, a *membrum disjectum* that is *viable, non vivant* (p. 112), and the study of *la langue* without *la parole* is without empirical relevance (p. 37). Saussure's conception of *la langue* is incompatible with the immanent state-view, as the latter turns the everyday language into a subconstruct. Saussure's conclusion, that change necessarily follows from *la langue* existing *in* and *through* time ('dialectically'), is incompatible with his former affirmation of the impossibility of change.

[L]e temps altère toutes choses (p. 112), turning social facts into empirically relevant facts; it ensures *la continuité de la langue* (p. 108), which, in turn, suspends (*annule*, p. 113)

---

123 This illustrates how Saussure's views are narrowed by the social – say, Meillet's (chapter 3.3.) – interpretation of language, and to continue Meillet's diction: Saussure has language structure correspond to mental, not social structure.

3.4. A close-reading of Saussure's structural postulates

the freedom that the language user has to change his language because of its arbitrariness; and time changes the language signs as well as fixes them (p. 108). Time is a precondition for continuity (p. 108) and *le principe d'altération* rests on *le principe de continuité* (pp. 109; 113).

Change takes place against the horizon of *la matière ancienne* (pp. 108-109). Language material is always inherited by the succeeding generation. Therefore continuity or historical development cannot be represented by the monolinearity of the abruptness formula

3.4.9 /A/ > /B/

as the notion of the 'old material' implies both its own presentic aspect as well as its development into a new present. The concept of simultaneity follows from this endogenous process and the model in 3.4.10 is also an image of how the mind that tries to grasp language as 'une chose réelle' must be both discursive and analytical[124]

3.4.10 /A/ > /B/   (the combination of the archaeology- and the geology-metaphors).
$$\searrow$$
/A/

3.4.10 represents *la nécessité du changement*, but as Saussure puts it himself so very appropriately: *On nous reprochera peut-être de n'avoir pas été aussi explicite sur ce point que sur le principe de l'immutabilité* (p. 111), and he excuses himself by saying that he needs to find out to what extent a change-causing factor is necessary. Why is it easier to present a case for the immutability of language than for its mutability? The concept of immutability is elucidated in terms of change, to which fact and concept we all have privileged access, while that of mutability is comprehended through a static definition of continuity and a trivial 'time-changes-all-things' maxim. Saussure does not succeed in combining continuity with change and without change (pp. 108-109).

Saussure's general stance is squinting; although he asks both why a language changes, and why it is static, change is something to be combatted. A language is *radicalement impuissante à se défendre contre les facteurs qui déplacent d'instant en instant le rapport du signifié et du signifiant* (p. 110), language change being defined by *un déplacement du rapport entre le signifié et le signifiant* (p. 109). – Note that the notion of 'a revolution' does not crop up in this section.

Why should language defend itself against those factors? Development is not a fight and why should language not defend itself against the factors that make for stability? For there is no denying that we would be in real trouble if the everyday language did not change; and the squinting view is complete as this 'displacement', occurring against the horizon of continuity, is a consequence of the sign's arbitrariness, a static concept (p. 110). Saussure does not explain why language is to put up a fight against what happens necessarily and in accordance with the object's nature. Change, *évolution*, is *fatale* (p. 111), it is inevitable, and no

---

[124] Saussure distinguishes between continuation and change proper: *Les causes de la continuité sont a priori à la portée de l'observateur; il n'en est pas de même des causes d'altération à travers le temps* (p. 111).

language 'escapes' *cette loi universelle* (p. 112). This leads by way of the following statement anticipating the discussion in chapter 4 to my final assessment of Saussure:

> Saussure provides us with a viable definition of change; change is *relational*. But this does not conceal the fact that change is tricky; Saussure's profuse use of verb- and noun-synonyms testifies to that: language change is *changement*, even *général* and *subit* as well as *fatale, transmission, déplacement, évolution* (even *ininterrompue*), *le fleuve de la langue [coule] sans interruption* (p. 192), *substitution, altération, une révolution* and the verb *devenir*; the sign-functional relation can be displaced and *se relâcher* (p. 109). Instead of criticizing Saussure for being imprecise, opaque and 'metaphorical' – no agents apart from such abstracts as time, continuity and arbitrariness are specified – I prefer to interpret his diction as an expresssion of precision; Saussure's use of synonyms makes for precision, suggesting the rich, diversified and varied nature of historical existence. And because Saussure affirms the incompleteness of the synchronic point of view, what is needed is a historical theory as the rational foundation of the concepts/terms used.

To sum up[125]:
a) arbitrariness has no immediate link with the sign's immutability and mutability; both properties are solidary (p. 108). Language will change because it is continued, existing in time – *est vivant*[126]. *La langue* placed in a social, synchronico-communicative context is *une chose irréelle*;

b) time is a precondition for continuity, which is a precondition for change (p. 109), and since change is solidary with nonchange, time and continuity have paradoxical effects. Saussure cannot define the law of tradition, cannot find the basic causes of change, only factors that make for stability, factors that involve the change-concept;

c) Saussure's arbitrariness concept presupposes synonymy, and therefore polysemy, the very language universal which the *langue-parole* dualism either does away with or relegates to a *parole*-phenomenon. Secondly, both change and stability can be inferred from the concept;

d) in insisting on the social and historical nature of language, Saussure is bound to accept the priority of the latter property: language as a product of social forces is a subconstruct (cf. (a); pp. 112-113), while the historical forces make language 'real', constituting *la réalité linguistique*;

---

125 Whether this investigation reveals a specifically structural approach in Saussure's views is doubtful, and his metalingual deliberations (cf. 3.4.12 below) also turn out to be a general discussion of the relationship contracted by synchronic linguistics and historical linguistics (see Hansen 1982). So our three crucial concepts *structure, system, state* are virtually synonymous with *synchrony*, and we find no particular discussion of structuralism in Saussure.

3.4. A close-reading of Saussure's structural postulates

e) Saussure has no understanding for the *dépassement* problem, but grounds it on *notre sentiment très vif de l'arbitraire du signe*. He has no theory that explains language *comme un héritage de l'époque précédente* (p. 105), hence he cannot explain who or what 'constitutes' *la langue* at a given point in time so as to make it inheritable;

f) Saussure's 'anthropology' rests on simplifying postulates about the individual's *Eigennatur* and his relationship with the supraindividual in the form of either *la langue* (as a *déjà constituée* and historical product) or *la masse parlante*. The *langue-parole* relationship is also the product of idealizing postulates: he insists on *la langue* as a stable and stabilizing entity, and since neither the individual speaker[127], the group nor *la langue* itself makes for change, he must resort to time and continuity, and, as the ultimate irony, to *parole*: *c'est la parole qui fait évoluer la langue,* which is simultaneously *l'instrument et le produit* of the former (p. 37).

Saussure succeeded in highlighting the paradoxical, I would prefer, complex nature of the everyday language, but his work suffered the unhappy fate of being conceptually trimmed and logically streamlined by his followers and commentator-editors[128]. That Saussure's views were not an unqualified advancement on his predecessors is seen in the abovementioned analysis of Whitney's view of the language sign[129]; the latter's distinction between convention and arbitrariness is extremely appropriate because the former implies the *temporal* and *simultaneous* nature of language (signs)[130]. Saussure's continuity concept is undefined (cf. pp 111-112; note 124 above) and appears unrelated with the notion of the inevitability of the law of tradition; in a similar vein immutability is established on the basis of types of change that are impossible, while mutability is determined in trivial terms.

The two chapters that I have been dealing with in detail do appear contradictory, the one affirming the impossibility of change, the other the inevitable necessity of change. Instead of focussing on such inconsistencies and attributing the concluding view (see 3.4.12 below)

---

126 Saussure's Editor (pp. 448-449) does not establish a logical connection between the subject and predicate in: *la radicale historicité des signes les rend (...) radicalement arbitraires*. His argument repeats what Saussure did above: history is used to determine *la structure profonde* of the sign as static, ahistorical.

127 Cf. Saussure's Robinson-Crusoe thought experiment (p. 113).

128 Tullio de Mauro (the 1972-editor) tries to salvage the seeming inconsistencies in Saussure's argument and loose use of *signifiant, signifié, le signe* and *l'idée* (e.g. p. 108); Gregersen (1991) does not close-read Saussure's arguments with a view to their logic; Meillet generalizes the social aspect of language and Hjelmslev the immanence aspect (cf. Togeby 1965:6;194), and the translator of the German edition (Lommel 1967:23) makes *la parole* the instrument and product of *la langue*, a type of relationship that the post-Saussurean synchronic tradition would prefer to Saussure's historical one, *la langue* being the product of *la parole* (1972:37; and note 121 above). Coseriu (1974:25) steers a middle course, saying that all that is language is this *durch das Sprechen*, but *das Sprechen* only exists *durch die Sprache* (cf. chapter 1.5. above).

129 The commentator's defensive exposition of the Saussurean arbitrariness concept is not to Saussure's advantage.

130 To labour the point: the reason why a speaker uses, say, *homme* and *chien* is that those words are *already being used in the community to which the speaker belongs* (Whitney 1970:19). Saussure's view of history is monolinear (cf. 3.4.12 below), therefore he is unable to combine the simultaneity- and time-aspects in language transmission.

either to Saussure's debt to the 19th century, his understandable inability to cut himself completely loose from his Neogrammarian predecessor background, *or* to a theoretically regrettable *Nachlass*, I find Saussure's position perfectly consistent both with his overt attempt not to separate the complementary conditions under which the everyday language exists and with his attempt at viewing language as a whole in its empirical entirety. Historical linguistics is both 'archaeology' and 'geology' (3.4.10 above), and in accordance with the Critico-Philological Method Saussure was both critico-analytical in respect of his tradition and constructive synthesizing in respect of what became his legacy, *l'héritage de l'époque précédente*, to his successors:

3.4.12 La linguistique diachronique étudie, non plus les rapports entre termes coexistants d'un état de langue, mais entre termes successifs qui se substituent les uns aux autres dans le temps.
   En effet l'immobilité absolue n'existe pas (…); toutes les parties de la langue sont soumises au changement; à chaque période correspond une évolution plus ou moins considérable. Celle-ci peut varier de rapidité et d'intensité sans que le principe lui-même se trouve infirmé; le fleuve de la langue coule sans interruption; que son cours paisable ou torrentueux, c'est une considération secondaire (p. 193).

Not its social, but its historical reality makes *la langue vivante*; had Saussure followed his own observation (3.4.13), he would not have separated what cannot be separated in 3.4.12 – the temporal and the simultaneous (the endogenous nature of the everyday language):

3.4.13 A chaque instant [le langage] implique à la fois un système établi et une évolution; (…) il est une institution actuelle et un produit du passé (…); le rapport[131] qui unit ces deux choses est si étroit qu'on a peine à les separer (p. 24).

And one must conclude with Tullio de Mauro (Saussure 1972:472) that Saussure's view of language, reminiscent of Hermann Paul, was a

vision historicisante.

---

131 Undefined by Saussure, defined by internal and external endogeneity.

CHAPTER 4

On the structural totality concept and the logician's view of the everyday language. Polysemy and synonymy are language-constituting properties

---

> *The crucial point of contact between description and reality is to be sought in the utterance of a statement on the occasion of an experience which that statement utterance directly reports. The seeing of a green patch, and the simultaneous utterance 'Green patch now', constitutes the sort of composite event which, in its rare occurrence, gladdens the heart of the epistemologist.* Quine 1972:1

4.1. 7 tenets determining structuralism. Towards the anthropocentric basis of language. Historical and structural man. Analysis of the synchronic totality concept. Immanence and autonomy are acquired properties. The logic of the part-whole parameter does not justify synchronic linguistics' criticism of historical linguistics for being atomistic. The individual is a scientifically relevant concept: a class as many or a class as one

FROM ITS START as a scientific ism structuralism became closely associated with the transcendent use of logical methods. The extensionality of logic and its class concept were in perfect agreement with the structural point-in-time object, the latter eminently suitable for denotation (Quine 1972:80). Both this extensionality and the analyst's Archimedean vantage point (Rasmussen 1992:109), necessary in order to conquer the existential participative relationship, agree with the hierarchical autonomy view of language. The structural approach became identical with a basically logical approach to a given subject-matter through this line of reasoning: (1) structure is an ontological postulate and a scientific (epistemological) requirement: *La langue (...) est un tout en soi et un principe de classification* (Saussure 1972:25;34; Hjelmslev 1971:122; Rasmussen 1992:376); (2) structural classification like any other scientific method is necessarily founded on the principles of formal logic; hence (3) the latter mirrors the nature of the everyday language[1]. The historical consequences are well-known: history and structure *qua un système etabli* (Saussure 1972:24; Weinreich *et al.* 1968:98) became antinomical, and the notion of a(n immanent) theory of history a misnomer.

The below partly overlapping seven tenets are typical expansions of the definitions in 3.2.1; here I shall examine their logical geography before I turn to the nonformal part of the

---

1 Some historians did have serious reservations against this claim (Clausen 1963:95).

logician's basic view of the everyday language as having no *Erkenntnisswert* (chapter 4.2.). The result of the analysis is that the logic of synchronic structuralism does not corroborate the synchronic claim that the alternative of structural description is *une description nonscientifique* (Hjelmslev 1971:110[2]); and this amounts to an indirect rejection of the still current view that historical linguistics is dependent on synchronic linguistics for its scientific status:

1) a structural entity is a totality of parts, a system *qua* a hierarchy (Hjelmslev 1971: 110;27).
2) (structural) meaning is contextual (Hjelmslev 1971:109;122);
3) a structural totality is more or different than the sum of its parts (the *more-than*-principle);
4) a structural part depends on the other parts of the totality and the totality itself;
5) a structural entity is a system; an autonomous entity of internal dependencies, controlled by internal laws;
6) in a structure, a system, *tout se tient* and *tout est différence*;
7) the structural method is the registration of dependencies (relations) between 'dependents', not the explanation of the latter ones as *sui-generis* 'significant' phenomena (Hjelmslev 1971:122)[3].

My criticism of the tenets will focus on the concepts of totality, contextual meaning, the *more-than* principle as well as the concepts of immanence and autonomy, derived from these tenets. Tenets 6 and 7 and the concept of dependence will not be dealt with directly. The vagueness of the determination of the structural totality complicates the investigation: a structural totality appears both as a hierarchical and non-hierarchical entity, as is indicated in Figure 4.1.1 below. However, I shall be using the word as follows: a totality T is 1) the class of all objects and its two subclasses, the class of semiotics and the class of nonsemiotics; 2) a given everyday language or *langue*; 3) an analysandum *qua* a resoluble syncretism (merger)[4]. Whether a totality is more than a class as many, *i.e* a *sui-generis* entity with specific properties (a class as one) or just a sign, will continue to be a part of my discussion.

---

2 Note how Hjelmslev's terminology, if not point of view, changes: in 1972a it was the system that was the epistemological precondition for the study of language, here it is a structural principle implemented by both analysis and synthesis. Hjelmslev – this also applies to his 1943-views (in *Prolegomena*; Hjelmslev 1963) – commits the same logical error as Durkheim did (cf. 3.3.10 above): the binary relationship between science and chaos is not only an interpretation of science and reality, but it cannot be translated into one and only one universe of discourse: the universe 'science and nonscience' is not identical with the universe 'chaos and nonchaos', so that 'science' (= the Glossematic hypothesis) is synonymous with 'nonchaos' and 'chaos' with 'nonscience' (or subjective or poetic treatment). The falsification of the Glossematic Hypothesis or any other totalitarian theory is not *der Untergang* of the scientific status of the subject-matter at hand, perhaps the fall of the dominant reality-view *des Abendlandes*. – Nietzsche might not have disagreed.
3 See Rasmussen (1992:181) for a discussion of this; and Sørensen (1958:32; 30-31), who demonstrates the absurdity of the contextual criterion within a synchronic conception.
4 It is the historian's aim to demonstrate that a scientific object is also a synthesized one, so that knowledge of the period or the historical phenomenon, which contributes to man's **ex**istence, requires discursive reasoning.

4.1. 7 tenets determining structuralism

According to Saussurean structuralism synchronic functions A :: B – whose structural antithesis is the historical process A > B – constitute *la langue* of a given language state. Such synchronic functions have no history despite each of them being *[un] fait historique* (Saussure 1972:112; Hjelmslev 1961:58-59); and the paradox becomes even more curious as *la langue*, which has no history, not *la parole* is studied by the two antithetical disciplines synchronic and historical linguistics despite the fact that all *langue* elements have their respective origins in *parole*. A systemic element or synchronic opposition *was* a *parole* phenomenon that originated with the individual and which acquired *langue*hood through the dialectics of the historical process through which it acquired (gradual) acceptance by *la collectivité* (Saussure 1972:138-139; 193).

Saussure's argument ended in the anthropocentric foundation of *la langue* (in a connotation), and the view of man that is now beginning to materialize is that of **historical man**, whose *raison d'être* is his dual transcendence of the present into the present, the deconstruction of the inherited (system) and the construction of a new, presentic totality. In both processes *la parole* is an active factor.

After Saussure **structural man** involved a conception of reality that envisaged all entities, including man himself, as dangling on the relations of hierarchies, filling the gaps of, ideally, two-place functions: f(x, z) or ErC, in which the horizontal relation *qua* the applied analytical principle is the constant and the dependent variables, while the vertical relations somehow function as undefined substantiation of the analyst's claim[5] that the two relates are really the constituent parts of the totality (analysandum) in question. Structural man is a recipient, passive, inert *terminus* in a structural network, internally determinable by its parts. The paradox is that such a purported *sui-generis* entity consists of entities that are externally determinable through a synthesizing process (Hjelmslev 1971:110): each acquires its meaning from the horizontal relation, the *f* or *r* in the two functions above as well as the vertical relations in Figure 4.1.1.

4.1.1

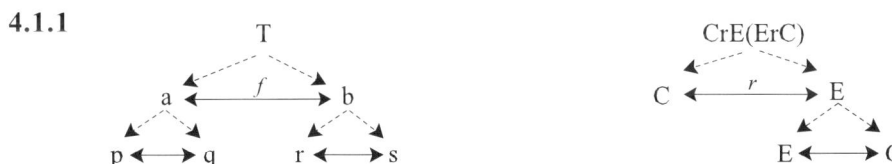

T is defined as the intersection of vertical lines a and b (cf. 1.5.2)
a is defined as the intersection of vertical lines p and q and T and of horizontal line b
s is defined as the intersection of vertical line b and of horizontal line r

No structural relate has a meaning of its own, an *Eigennatur*, and the structure T is *the immediate datum* or the *unanalyzed whole* (a syncretism) (Hjelmslev 1963:31;92-93), and with reference to *la langue* an entity *déjà constituée* (Saussure), *already developed* (Whitney).

---

5 In the Glossematic definition of *analysis* (Hjelmslev 1963:Def. 1, pp. 131; 28-29) the so-called vertical relations remain undefined; we are told that they are *uniform*. Only the horizontal relations expound dependencies, while it follows from the theory that a given analytical level is always unidirectionally dependent on the previous level(s).

T in 4.1.1 represents any phenomenon regarded as an autonomous totality, a *spectaculum*, that can be seen or comprehended in one clear and distinct act of the mind at a distance of anything else, *i.e.* an entity separable from anything else. This object view of reality, however, implies both separation and coherence, and therefore T as a structural relate is different from the other relates of the hierarchy as long as it represents the initial datum, entering into no synthesis (Hjelmslev 1971:110).

As a scientific object T is to be seen in virtue of its constituents. But these internal determinants do not guarantee its autonomy. What establishes the property of immanence (autonomy) is cognizing man, N.N. From this 'seeing'-relation T becomes the relate of an object-subject function, TrN.N., and structurally T acquires part of its meaning from the horizontal (connotation-like) relation that connects it with another relate, N.N. (Hjelmslev 1973:128).

For T to become autonomous it must detach itself from the N.N.-relate. This separation (analysis) is brought about by a linguistically motivated negation process: if T is T, then T is no other object; and 'to be no other object' implies the presence of at least one more object, T*. And the moment this other object transcends its purely linguistic existence, it takes the place of the first relate, T, which thereby becomes separate from N.N. We may then say that T and T* constitute a structural relationship, in which both relates are parts, f(T, T*).

Let T be the Glossematic class of all objects, then the property 'to be the class of all objects' is created by the premised fact that there is no other object; nonT represents no class of objects. Structurally the concept of the class of all objects is determined by the concept of the empty class, a class that has only linguistic existence; hence the class of all objects is determined by its only equiempirical phenomenon, the analyst N.N. As N.N. is a part of the class of all objects, the TrN.N. relationship turns out to be participative: T(N.N.) :: N.N.

Consequently, (1) if T is a totality *sui generis*, the structural principle cannot be upheld, except axiomatically, and immanence, autonomy and totality become correlative concepts; (2) if T is a structural entity, it acquires part of its meaning from a horizontal relation, which may be the anthropocentric connotation: the everyday language is what it is because of something else, and that something is the language user – which turns language into a part.

The interesting thing is that neither alternative rejects the concept of a phenomenon's Eigennatur. If (1) is asserted, the meaning of T stems from itself. If (2) is asserted, T must have an Eigennatur seeing that it has more meaning than the structurally acquired meaning of *being no other object T** (T is not nonT): a structural object must have more meaning than it acquires from its relates, otherwise it could not enter into a structural network, where meaning as in a give-and-take process is exchanged among the parts. Immanence, autonomy and totality are therefore acquired properties, and with the affirmation of the Eigennatur of a phenomenon, meaning is more than contextual meaning.

A theory, whose object is an autonomous structure, describes it as a class of parts. No analysis, however, is concerned directly with the totality[6], only with the entities *in* the totality. The class of all objects is not a member of itself. This is made explicit by the Glossematic procedure, which begins with the articulation of the class of functives, not with that of the text/process or the language/system of a semiotic (Hjelmslev 1975:deff. 38, 39; Ggb). Secondly, the metaterm and its definiens (def. 24), a semiotic, does not denote an absolute

---

6 *(...) the as yet unanalyzed text in its undivided and absolute integrity* (Hjelmslev 1963:12).

4.1. 7 tenets determining structuralism

totality, *e.g.* the same as the signs *the English language, the French language* or *the Danish language* purportedly signify in virtue of their respective contents.

Totality is a metaphysical notion, and with the empirical relevance of this concept gone – part of the standard criticism against historical linguistics for being atomistic vanishes.

The notion of parts deriving part of their meaning from their totality is only true from the extensional perspective: T is a sign assigning meaning to its denotata: the metasign *a semiotic* denotes each member of the subclass of semiotics, which subclass is no semiotic and whose members are not *sui-generis*-totalities. And from this it follows that the *more-than* tenet should be defined by 'different from', seeing that a totality, which is a sign, is certainly different from the 'class' of data that it delimits. I shall elaborate my criticism of tenets 2 and 3.

Together the tenets explode the basis of the structural point of view because of the equivalences created by the theory's definition system and principle of analysis (cf. 4.1.2 below). If the immanent totality is more than the sum of its parts, its nature cannot be established through its analysis into parts. And if it is not more than the sum of its parts, the parts do not derive part of their meaning from the totality. In the analysis of a totality, the different levels are characterized by determination so that a given successor level is dependent on its predecessor. Thus the former level is different from the latter one; but a relation of equivalence is implied in that analysis does not add anything new to the analysandum: the analysandum *qua* a totality can be replaced by its component parts (cf. below).

However, the *more-than* tenet is true in so far as it is understood that a structural totality is relative: both a sum of internally intersecting lines and itself intersected by external lines. Empirically, the totality is a part – because of its external relations: the totality concept is heuristic: *a given totality always implies something else, points to something else from which, structurally speaking, it derives part of its meaning.*

The structural totality is defined through a binary division into its constituents; of these we know that they are characterized by their horizontal relation, r; the two parts are then regarded as relative totalities and divided into their respective parts. The first analytical step establishes an equivalence which defines T by aRb, and the second step defines a by pRq, and b by rRs. The initial object T is known partially seeing that the statement 'T consists of aRb, which consists of pRq and rRs' cannot be finished. Hence there is some truth in the *more-than* tenet. The alternative is that the analysis stops with the so-called smallest parts which are no objects *qua* analysanda, *i.e.* relative totalities: the structural totality T consists of a class of elements that can be analyzed (= A) and a class of elements that cannot be analyzed (= nonA): the structural programme, explicitized as in Glossematic theory or in its undefined linguistic versions, is either cognitively inadequate, ending in ignorance, or theoretically inconsistent, ending in a paradox. This leaves the structural programme with three equally unacceptable options:

A totality is aitional: T but not only if its parts. A structural totality is only more than the sum of its parts if (1) synthetically, it becomes a part; if (2) analytically, the analysis is inconclusive; if (3) non-structurally, the structural part has an Eigennatur. The aitional nature of the structural part preserves its relative totality-part property, which means that the sharp contrast between analysis and synthesis cannot be upheld.

The English word *on condition that* derives its Englishness from the linguistic totality of which it is a part, the so-called English language. What is the English language? The English language, *at a certain stage of its chronological transformation*, <u>connotes</u> *a definite nation* (Hjelmslev 1961:58-59). Hjelmslev confuses the relates in the part-totality parameter. Firstly, a purported totality, the English language, is defined by entering into a synthesis; secondly, what connotes the English nation (a relative totality) is a sign *the English language*, which is a part of the purported totality *qua* a class of signs, that class of signs that *the English language* is meant to denote *qua* a class as one, a totality with specific properties different from its parts. Thirdly, Hjelmslev does not see that the English language is a constitutive part of the English nation (its connotatum), an entity that would not be what it is without its 'native' language.

Totality is a concept linguistically motivated, a word; and denoting a class of words, it becomes a part of itself: it participates in itself. Such macrototalities as a national language, a dialect, a state, a system etc. do not denote or do not exist over and above their parts; they are classes as many, and as such they do not enter into *Kausalnexus*, they being signs.

From where does the English language – granted there is such an empirical phenomenon – so acquire its Englishness that it can transmit this meaning to parts such as *on condition that*? – This discussion will be resumed at the end of this chapter.

As a purportedly immanent and autonomous entity it must acquire its meaning from the fact that it is not the French language, the Danish language, Swahili, etc. – all this in accordance with the structural principle of *tout est différence*: T is not nonT. But this expression of static self-identity conveys no knowledge, being a tautology. Where does French acquire its meaning from? From the fact that it is not Swahili, English, etc. Again we see how a structural totality is partially definable through analysis and oppositively through synthesis. As indicated this position is untenable, and its below specification ends in the aitional and participative nature of totalities:

4.1.2		English	$=_{df}$	Non-French, Non-Danish, Non-Swahili
		A	=	Non-B, Non-C, Non-D
=		A =	=	Non(B, C, D)
and		B	=	Non(A, C, D)
hence		A =	=	Non(Non(A, C, D), C, D)
=		A	=	Non(Non-A, Non-C, Non-D), C, D)
=		A	=	A, C, D, Non-C, Non-D

From its initial negative definition, it ends up being defined both positively and negatively, so that A is, structurally, more than A: A is aitional and participative: A but not only if A.

A structural entity is more than a part – or a totality, because its properties arise from both external and internal structural processes. As a part the English language acquires its meaning from the part(s) with which it contracts relationships: f(English, French, Danish). This implies that English and other native languages are constituent parts of a totality that is more or different than its constituents. The new totality transcends empirical reality seeing that typological classification does not establish the *type* as an empirical fact, the type denoted by – in English – *language* or *a native language* or *a semiotic*. Structural entities such as individual languages exist without having to appear from the analysis of a totality *qua* the class of all objects, articulated into the class of semiotics and the class of non-semiotics.

4.1. 7 tenets determining structuralism

They exist – as far as they exist as totalities – as different individuals each with an Eigennatur in compliance with the sixth tenet.

A totality is a mental short-cutting device, one of whose functions is the converse of the *pars-pro-toto* law of the human mind's secondary processes (Jacobsen 1976:20-21;35). Here the part takes the place of the whole, there the whole represents a part. It is not only grammar that has this least-effort function (Samuels 1972:49), each word has it. Concepts are words and a word is a totality of determinants ('sub-concepts') comprised in and by the word. The principle of definition bears me out:

4.1.3 definiendum = $_{df}$ definiens
concept = meaning (= a sentence)
T = a+b+c
'a father' = 'a male parent'.

The definiendum is a class as one and is *transformed into* a class as many by the analysis (Hjelmslev 1963:92-93). The definiens is the class of determinants defining the definiendum *non-exhaustively*[7]. Being concepts themselves, a, b, and c (in 4.1.3) are totalities turned into parts through synthesis, just as T, a word, has been turned into an immanent totality. Only as parts are the elements of the everyday language *en ergeiai*; the utterance is also *en ergeiai* as a part of the speech situation. This does not diminish the empirical status of the defined word as an individual. The words are parts of an utterance, which, in turn, contributes to the constitution of a speech situation; the degree of reality of the determinants of the definiens is the same as that of the definiendum; there is no empirical relevance in the introduction of the concept of 'logically different levels'. The distinction belongs to the level of logic (Sørensen 1958:35-41; 45-46;51). The definiendum is a short-hand expression for the defining sentence[8]. But as a relative totality it does have an Eigennatur; secondly, any such definiendum contains something undetermined, but determinable. This property distinguishes the word of the everyday language from the terms of the scientific language (Hansen 1985:87-89). And that is why the Glossematic parellelism between *a diachrony* (the scientific language) and *a metachrony* (the object language) cannot be upheld without qualification (Hjelmslev 1975:124-125). Because the two types of language have different etiologies, they will develop differently; but their existence conditions can be captured by the general theory of historical development, as both contribute to the present's development into the present.

The dialectics of the totality-part parameter may be summed up as follows: any phenomenon regarded as a totality is also a part, and in so far as a part can be analyzed (divided) it is also a totality. Therefore, of our two mental processes, point-in-time reasoning is neither logically nor empirically prior to discursive reasoning through time.

---

7 Sørensen's semantic analysis (1958:34-43) does not exhaust the meaning of a word, *e.g.* "a father" = "a male parent"; analysis is more than the Glossematic definition. Such man-made definitions acquire their relevance contextually, in respect of the theory in question.

8 As such it expounds the *pars-pro-toto* law, and it is not surprising that identity, both at the level of logic and in our mind, is derived from this law by way of the mental law saying that 'similarity is the same as identity' (Jacobsen *loc.cit.*). When the phonologist works with phonemes or the generativist with his own intuitive understanding of his native language, his procedure finds its justification in the psychic laws that control the secondary processes in our over-all mental make-up.

The relativization of the totality concept entails a departure from the metagenous conception of linguistics: the empirical relevance of the synchronic language type must be substantiated, and immanent totalities such as *the English language* have at best a heuristic value. This leaves the historian with a totality concept whose fundamental property is that of openness – amenable to influencing and being influenced, and the relativization of the everyday language places it in a multiplicity of phenomena where continuity and change reign with no all-embracing uniformity or singularity. The purported separateness of the everyday language is to be seen against the horizon of attachment and contact, a fact that is compatible with its openness and dual transcendence (see below).

The theoretical consequence is that no (typological) theory of the class of languages, a theory potentially true of all languages, is possible; there is no empirically relevant theory of language in abstracto. The universalist deals with the *disjecta membra* of the individual languages (*e.g.*, Hawkins 1979; 1983:119). The metalanguage describing its object language is only the *disjecta membra* of the latter, and *language in a metalanguage* is not the same word as we find in *e.g. the English language*. The conclusion in 4.1.2 will receive a dynamic interpretation immediately below and follows from the abovementioned remarks:

4.1.4		If T = a+b+c
	and	if T is more than the sum of a, b, and c (its definiens)
	and	if T is an object sui generis, a totality,
	then	T contracts a structural relationship with T*,
	& then	T and T* are parts (the definiens) of another totality X
	=	X = T+T*
	hence	T ≠ T
	or	T ≠ a+b+c

The empirical relevance of the *sui-generis* totality is clear; what T represents is relevant as a structural part with an Eigennatur; T is an individual, and so is T*. What X represents is also empirically relevant on condition that no inductive fallacy be committed. Taking the equiempirical relevance of both the definiendum and the definiens's constituents as our point of departure, we may suppose the empirical relevance of X and substantiate it, not structural-typologically, but structural-historically: T and T* may spring from the same mother language, X. In a sense a historical totality is more than its present constituents: it also has a past (Saussure 1972:113)[9].

---

9 Rasmussen (1992:372-380) tries to establish the empirical and theoretical relevance of typology over the historical point of view, but he merely reproduces *le maitre*'s circular argument: "*idet synkronisk betragtning går forud for diakronisk (her i termernes tilvante betydning), opstiller forskeren inventarerne for to (eller flere) sprogtilstande (hvorimellem der er kontinuation) ...* ".The last parenthesis gives the argument away: the synchronic point of view presupposes what is to be established, continuation (change). In spite of the fact that change cannot be sensed, only registered (pp. 376-377), the synchronic linguist knows immediately that and when there is continuation between the systems of two or more states. – My translation of the Danish sentence is: "as the synchronic point of view precedes the historical point of view, the linguist sets up the inventories of two (or more) language states (between which there is a function of continuation)"; or more precisely: " ... language states (between which we know immediately there is a function of continuation)".

The anthropocentric basis (cf. Paul 1966:36-37; Weinreich *et al*. 1968:104-105)[10] of the everyday language is participative, and it is *energeia* in so far as it is *en ergeiai*. This dynamic property arises when the everyday language becomes a part of the structural network of the connotation, not when it becomes a part of a synchronic totality – necessarily *sans la trace de l'homme*: N.N. :: Language(N.N.).

In sign-functional terms, a language element becomes *energeia* when it becomes the expression for a content and as such a part: $Cr^{con}E(Er^{den}C)$. The sign element ($Er^{den}C$) becomes the expression for a content when it is used by the individual speaker in a speech situation: **no sign is not a part (E or C) of a sign**, which agrees with: **no sign-relate (E or C) is not a sign**.

The argument in 4.1.4 implies an inductive movement which its dynamic interpretation avoids, but which the synchronic systems concept does not. Paul (1966:24; 11, 31) focusses on the empirical aspect of it: *(...) zwischen Abstraktionen gibt es überhaupt keinen Kausalnexus, sondern nur zwischen realen Objekten und Tatsachen*. Synchronic linguistics agrees with Paul's, the historian's views[11]. Methodologically, the system cannot be established inductively. That leaves us with the deductive alternative: the systems concept is a hypothesis and the system comes from the applied theory (*E.g.* Hjelmslev 1963:8-11; def. 28; Sørensen 1958). As such the system is an abstraction, and so is the correlation of efficient communication and (the stability of the units of) the system. As a totality, the system is only a class as many, not an entity with specific properties. Hence it has no history. But it is a curious fact that hardly any synchronic linguist denies that the everyday language has an anthropocentric basis and is a historical phenomenon, and in spite of this the everyday language is regarded as an immanent, objective totality: man has been decentred structurally[12]; linguistically, the denotative function has been cut off from its connotative function, *la langue* from its *parole*, the latter becoming individual, subjective, contingent (Saussure 1972:30): the supraindividual system was created as an *imposing* entity (Hjelmslev 1973:125-126), an entity whose continual failure to grasp the empirical process that turns a variant, from its inception in an idiolect, into a systemic unit (Paul 1966:21-22; 24; 33; 37), is striking.

Martinét's functionalism strikes an anthropocentric note (1964:94), but its psychological basis rests on a biassed view of man. In accordance with the structuralist position, he generalizes one human characteristic: *inertia* and receptivity, to the exclusion of its correlative characteristics, activeness and *engagement*: the synchronic linguist does accept that the state, the stable system, is the result of historical processes, but he disregards it when he constitu-

---

10 And neither Saussure nor Hjelmslev nor for that matter the modern generativist denies this fact.
11 In their representation of Paul, Weinreich *et al*. (1968:104-120) put the cart before the horse. They highlight similarities between Paul and modern synchronic linguistics after having translated Paul's terminology into synchronic terminology. In so far as there is a relevant parallelism, it is modern synchronic linguistics that follows Paul: Hermann Paul was a historian, not a synchronic linguist in disguise (Hansen 1982).
12 Samuels (1972:136) seems to accept the empirical relevance of the individual's idiolect, but abandons the stance levelling one individual interlocutor under his *habitual interlocutors* (p. 89), so that one obvious source of variation, the individual speaker-listener situation, is eliminated; and as the individual speaker *qua* an individual goes by the board, so does the possibility of constructing an idiolectal theory, and the possibility of creating an idiolectal basis for a theory of change (see Hockett 1958:457; Keller 1985).

tes his analysandum. I shall so generalize this activity that 'inertia' (the stable) is the spin-off of man's productive and creative existence.

Even the anti-idiolectal conclusion of Weinreich *et al.* (1968:186) is anthropocentric; their definitorial exclusion of the idiolectal origin of change through the distinction between the 'rise of a feature of speech variation' (inherent variation in speech (p. 187)) and the 'beginning of change' does not override the fact of the latter's idiolectal origin. As above, they imply the theoretical relevance of the synchronic cut, which invalidates their claim that a spreading 'linguistic feature' already in its idiolectal inception does not carry a 'certain social significance' – through its speaker's social status. Their conclusion may follow from either type of origin: the spread of a feature throughout a 'specific subgroup of the speech community' may either reinforce its original socio-idiolectal 'value' or alter it (p. 187).

If synchronic linguistics does not heed Paul's warning, it becomes the Medieval Realist's discipline. The way to avoid this fallacy is to take the distinction between a class (a totality) as many and a class as one seriously:

The empirical relevance of both the class as one and the class as many is not in doubt. But the empirical relevance of neither Durkheim's nor the structuralist's totality has been substantiated: when, in an analytico-deductive procedure, does a given subclass assume empirical relevance as a class as one? Synchronically, the initial datum, Language X, is regarded *a priori* as a class as one. I shall illustrate my point here with a Glossematic example:

In the Universal Component (Hjelmslev 1975:3-16) the articulation of the class of objects is linguistically motivated, while the articulation of the class of functives assumes empirical relevance. But Hjelmslev does not explain why the Universal Component is to be partitioned into those two classes – objects and functives. It would seem far more natural if Hjelmslev had continued the analysis of classes of languages and texts (def.s 38 and 39) into their respective component parts, namely the class of functives, instead of establishing them by definition as two equilogical subclasses between which *There is no relationship of premission* (p. 8). The theory, however, provides the answer to the question of empirical relevance: a class as one and a class as many are both empirically relevant if manifestation and purport are present, *i.e.* if manifesting hierarchies (language substances) can be encatalyzed to a semiotic form *qua* a totality (Hjelmslev 1975:92-93; N17); such an encatalyzation process does not result in a totality manifestans and a part manifestans. Thus the two crucial totalities denoted by the metaterms *a text* and *a language* do not transcend their theoretical level and both remain signs denoting *manifesting* entities (parts) in virtue of their respective definienses[13]. Therefore this denotation yields only minimal knowledge of its denotata (manifestantes), in so far as the purpose of the denoting definiendum (metaterm) is only to establish differences in accordance with the principle of *tout est différence*, and the *tout-se-tient* principle will follow from the coherence which the theory exhibits internally and which two or more denotata exhibit when denoted by the same metaterm. With the latter principle we are back to Figure 4.1.2 and argument 4.1.4, and I shall demonstrate how the *tout-se-tient* principle is incompatible with extensionality.

---

13 *A text* denotes a syntagmatic ('sequence´) of elements characterized by 'both-and', while *a language* denotes a paradigmatic (´class´) of elements characterized by 'either-or', in traditional diction, an utterance and a grammatical category, respectively.

Phenomena may cohere either extensionally as members (denotata) of a class (a denoting sign) or intensionally through a common property.

Extensionally, the class of languages (semiotics) will have a property L that does not characterize the class of nonsemiotics. Thus for the universe of the class of all objects to constitute a coherent class, it must not be characterized by the properties L and NonL. Intensionally, the common property or properties of the class of all objects will not be the L-NonL property that distinguishes the class of semiotics from the class of nonsemiotics. Extensionally, it is impossible to establish a totality that exists over and above the so-called individual parts, because then we should be asserting an absurdity: P is both L and NonL. If the class of objects is not an object itself, then we may infer that the class of languages is not a language itself. We have only individual parts of the class. Modern English will be distinguished from Modern French by the property E, which French does not possess. For these two languages to belong to the same type, T cannot be established by the universe of E and NonE, but must be characterized by a property that both languages possess. To establish this property is the synchronic typologist's task, but it is his concomitant task to explain how French and English will acquire (have acquired) their distinguishing properties. There is only one answer: French and English are both similar and different because they have developed from the same origin. French and Swahili are different because they have developed from different origins. The totality-part point of view will have to make do with a katadynamic explanation: our three languages are similar because they are members of the class of languages, they being different because of historical accidents. Typologically (see chapter 4.4. below) the individual language states become variants of the type. But what is this type – in certain linguistic quarters, universal language?

In the light of the harsh criticism that synchronic linguists have levelled at historical linguistics over the years for not dealing with totalities – history being unscientific for being atomistic and too preoccupied with the philological detail (see Mitchell 1992:94) – the degree of attention that synchronic linguists have directed at the totality concept is surprising. They have been unable to establish the empirical relevance of their basic datum, the denotatum of such words as *the English language*, *the French language*, let alone *an English language state*. Is their datum to be defined extensionally or intensionally? What is the relationship between the word *the English language* and the purported states of English, Old English (Late- and Early-), Middle English (Early- and Late-) etc. It is not an extensional common denominator, expounding a coherent universe of A and NonA, nor an intensionally defined entity, consisting of all the properties that the various states share.

What the synchronic linguist deals with is the same as the language historian explains, the philological detail, a word, a sound, a sentence. The English grammarian describes language phenomena that are only relative totalities, **parts**, which can be denoted by the sign *the English language*, a sign that denotes no totality with specific properties. Denoting all the words, sounds and sentences that we call English, the sign is certainly extremely polysemous. The disjecta membra that a grammar book decribes are not transformed into an entity different from the sum of parts and with a life of its own.

Only that exists – has empirical relevance and can enter into *Kausalnexus* – which is selectable by a manifestans (the Glossematic formulation) or which completes itself in *parole*.

4.2.
The totality concept is ambiguous and polysemous. A critique of the logician's view of language, language being necessarily polysemous (ambiguous) and synonymous (precise). These two language universals are not equilogical. The *Erkenntniswert* of the everyday language. The everyday language transcends into reality in two ways

While it is essential to structural analysis, the part-totality parameter and totality itself assume no independent position in modern symbolic logic over and above the concept of a set or class; and a set is defined intuitively in mathematical logic as *any collection into a whole M of definite and separate objects m of our intuition or thought* (Delong 1971:71); Quine (1971:1-6) following this nonhierarchical definition underlines the distinction between the class as one and the class as many. But the significant point in both conceptions is that logical **totality** is a mental construct and therefore linguistically motivated, in short **a sign**. The concept's empirical relevance becomes a decisive issue in our search for *Kausalnexus* in order to justify the synchronic point of view's totality-based criticism of historical linguistics. Synchronically, the resolution of certain syncretisms preserves the relevance of both points of view; and so does the analysis of the written/spoken 'text' into parts[14]. From here, however, there is no immediate transition to the over-all structural view (3.2.1) in the sense that the class denoted by *the English language,* as a class as one, has the same empirical relevance as *the English language*, as a class consisting of many parts.

The conclusion in chapter 4.1. was that the synchronic point of view is atomistic at the level of data, and coherence – the *tout-se-tient* principle – is brought about by the extensionality of the metalanguage. The properties of a linguistic substance stems from the applied theoretical form. And the implication of this is that the synchronic point of view takes for granted what it should establish, the autonomy of its object, and from this a priori assumption the priority of the analytico-deductive method follows – as well as the other basic structural principle, *tout est différence*. While the latter principle may exclude the human factor, the other is bound to involve it and not to make it a *quantité négligeable*, as the logician and synchronic grammarian do.

The logician's point of departure is his conception of the everyday language as an anthropocentric and historical phenomenon, and therefore as an inefficient communicative means. Human languages are ambiguous and do not serve as a scientific means – unable to hold 'truth' – a truely Platonic stance. They have been *evolved over a long period of history to serve practical purposes of facility of communication* (Church 1952:2; 1-9). From here another postulate is inferred: *In a well-constructed language of course every name should have just one sense* (p. 7)[15], in absolute contrast to the univocity of formal languages, the words of the everyday language are homophonous, at best polysemous[16], in any case ambiguous and not precise.

---

14 The synchronic linguist ignores the fact that both analytical processes are matched by the resolution of mergers: ME /i/ < OE /y/ and /i/; and the split process, *e.g.* PreOE /u/ > /y/ and /u/. The split process will be so generalized that any sound (element of the L-continuum) is a syncretism resoluble when it develops into the present; see chapters 5 and 6.

15 Curiously, a stricture that crops up in grammaticalization theory with no motivation.

16 I do not distinguish between polysemy and homonymy. I shall only be using the former, which refers to the content-side of the sign, while the latter considers the expression-side.

4.2. The totality concept is ambigous and polysemous

In the present century mainstream logicians and many linguists have regarded human language as the greatest obstacle – apart from subjective man himself – to man's acquisition of knowledge. Objectivity and immanent historical analysis have been impossible. Therefore all man could hope for was to be able to perfect his analytical instruments, and in the course of this process radicalists hoped to do away with the human factor – in linguistics, the content-side, too. This ideal is embodied in the modern meaning of the words *objective*, *symbolic* and *formal*: 'anything that has not been tainted by subjectivity (and by 'already-developed' meaning)'. An objective and formal statement is inherently a true statement about reality. Or less presumptiously: a true statement must be objective and formal.

As Saussure pointed out, man cannot really change his language, so what was left for logicians – and linguists – to do was to create a perfect language. Symbolic logic – the (meta)language of symbolic logic itself and mathematics, of science and history (Hempel) and linguistics, but not that of our everyday lives, or of the logic (of exsistence) that we all as nonlogicians master to a fault – has in no way lifted the veil human languages purportedly cast over reality, be it synchronic or historical, macroscopic or microscopic. The synchronic totality concept and the methods of symbolic logic invariably end up in a nominalist position and therefore remove historiography one step from reality, and that with a theory which entails a reality of a premised nature, in order to make the latter amenable to scientific analysis.

The objective of science is transcendence into reality *qua* an already constituted entity; the historical narrative is transcendence into a preexisting reality (*res gestæ*), but also transcendence *qua* **ex**istence – creation of reality, past and present. In the logician's, linguist's and historian's conception of the everyday language it is understood that man cannot 'access' reality without the activity of language. Man's transcendence into a section of reality is privileged, and this section is the historian's: the part of reality that man creates is of course amenable to being transcended-into by man, while the natural scientist – and other sciences that have objects for their data – may not have such privileged access.

Creating the present out of the present, man so creates his historical reality with his language that both become constituents of this section of reality. Thus even the amenability (Glossematic appropriateness) of the nominalist's theory has its origin in this privileged access to reality, and therefore in man's historical existence; the nominalists[17] could not come to an agreement about the significant properties of their common object, *in casu* the everyday language, without their respective everyday languages and therefore without the dialectics which constitute the in-time reasoning that resulted in this agreement.

The logician and the linguist agree that the everyday language is ambiguous. The linguist sees the ambiguity as springing primarily from the historical development of so-called extralinguistic reality – as such the nature of language is stability and univocity, while intralinguistic development suggests that the nature of language is changeability and ambiguity; change is imposed and accidental (Saussure 1972:30), and change is essential and not im-

---

17 *From certain experiences, which must necessarily be limited even though they should be as varied as possible, the linguistic theoretician sets up a calculation of all the conceivable possibilities within certain frames. These frames he constructs arbitrarily: he discovers certain properties present in all those objects that people agree to call languages, in order then to generalize those properties and establish them by definition* (Hjelmslev 1963:17-18).

posed, respectively. The logician makes do with the abovementioned type of characteristic[18] and leaves the domain of human language, while the language historian is interested in coming to grips with the role of language in human existence and with how the very same language manages to remain at least first among peers when it comes to communication and any other intersubjective activity, and to constantly maintain its collective orientation while changing constantly.

Following the logician and the historian the everyday language is a changing, historical phenomenon, it is polysemous and synonymous. These properties also follow from the synchronic *langue-parole* dualism: more than one substance, thereby implying synonymy, in fact *all purports* (Hjelmslev 1975:13-14) manifest the same form; polysemy is a precondition for this synonymy[19].

What does the predicate *ambiguous qua* a language universal mean? It means two things: the everyday language *qua* a system/*la langue* is not precise; precision is a property of a formal language; secondly, the spoken chain is a disjunction[20]; a propositional function is univocal.

The moot point is how to interpret synonymy *qua* a language universal and whether the everyday language may be precise so that it is paradoxical in respect of the properties, *precision* and *ambiguity* (non-precision: L is both A and NonA). The logician generalizes the latter property, inventing another language. In the course of this process he forgets that his own everyday language is the metalanguage of his purportedly precise and ambiguity-proof set of formulae[21]. Otherwise he has merely invented a private language or an *effective* method by which, under certain conditions, to generate tautological, contingent, or inconsistent propositions. A simple experiment will reveal the logician's dependence on ambiguous language. He may claim that the truth-functional definitions of his well-formed formulae will suffice; anybody seeing the truth table of a function will *immediately* grasp the meaning of the latter. However, the logician cannot explain nonarbitrarily why the truth-functional definition of, say, the logical conditional must be *TTFT* and not *TFTT*. Neither can he explain why asserting the consequent is no rule of inference of the system[22]. With no everyday language at his disposal, the inventor of the truth-functional system presupposes his would-be fellow logician's a priori knowledge of the proviso: what has not been stated explicitly is not possible. (The alternative is that the logical system is innate knowledge or self-evident (Haack 1974:29)). The logician presupposes the 3.3.10 inference rule so that

---

18 The linguist may easily counter: what is the *practical* purpose of continuing the verbal ending in modern English, *-es*, or of the replacement by *-inge* of *–inde* in Middle English?
19 Only logical formulae are pure synonyms: 'p & q' = 'q & p'; or 'p ⊃ q' = '-q ⊃ -p', while *great* and *big* in modern English are synonymous because both are polysemous (cf. 6.3.B).
20 The irony is that if sentence X is ambiguous – and we know it, then we have resolved the syncretism immediately.
21 The logician's argument is self-defeating: if his supposition, that human language is ambiguous, is true, *i.e.* precise, then his reason for constructing a precise language is superfluous, his everyday language being precise. And if it is precise, it is ambiguous; and he may or may not construct a precise language. – If his supposition is false, then the everyday language is not ambiguous, but precise. – (Logical hair-splitting may sometimes be an appropriate attack on common places).
22 Let 'If it rains, then the streets are wet' be our conditional: then it cannot be the case that it both rains and the streets are <u>not</u> wet; but it is the case that it both doesn't rain and the streets are wet. Secondly, we cannot by asserting the consequent fact that the streets are wet, infer that it also rains: '(If p, then q) & q' ⊃ p' is a contingent formula.

4.2. The totality concept is ambigous and polysemous

4.2.1 what is absent is not the case = inference from *absence* to *cannot be the case*

whose historical version is[23]

4.2.2 inference from *now-and-then absence* to *always absence*.

As the logician has only one metalanguage at his disposal, the only viable conclusion is that if logic is a theoretical panacea, and since the everyday language is logically indispensable, the latter one is second to none.

The next step in my argument sets out from the unbiassed hypothesis that the everyday language is both ambiguous and precise, which will be rejected if a univocal word is found, and the question to be answered is

Why is the everyday language both ambiguous and precise?

Why, following the psychologist (Jacobsen 1976:11-23; cf. Løgstrup 1976:36; Hansen 1985:89-92), does the everyday language let us down, and why, following Wittgenstein (1963:§§98-99) is it in order as it is?

My answer will aim to interpret and thereby provide a reason for the ambiguity and precision of language so as to establish a necessity behind the language historian's causal factors of ambiguity (the mechanicists) and precision (*Deutlichkeit*; the functionalists). Polysemy makes for ambiguity, synonymy for precision, both universals expounding the dual nature of the language sign, ErC.

The logician[24] sees synonymy not as a language property equiempirical with polysemy, but as a result of polysemy. Each word is anti-structurally viewed as an autonomous entity.

In the L-continuum, I shall suppose, each word exists in *a synonymy field*. Each word in such a field is therefore necessarily polysemous (*tout est différence*), and each polysemous word exists in a synonymy field (*tout se tient*). The logician's ambiguity concept should be seen against the horizon of precision and the structural nature of signs. Synonymy is the indispensable relation between the definiendum and its definiens as well as between the levels that arise from the deductive movement of analytical reasoning (see 4.1.3 above).

What is really at stake is the structural nature of language, which the logician has not come to terms with. To repeat: the fact that a word enters into a synonymy field makes for precision, neither ambiguity nor redundancy. Since there is no such thing as a set of *pure synonyms* and *this fact is not likely to be accidental* (Samuels 1972:65), synonymy is not mere

---

23 Modern theorizing conceals the rule under the name of hypothesis. A hypothesis is based on 'what is now and has been the case will always be the case'. Why else should the synchronic linguist establish by definition what he and some other linguists have agreed on!

24 Sørensen (1958:39-40) follows suit; his semantic definition system is not sufficiently exhaustive for historical analysis; his system makes the word *a father* a homonym. How can Sørensen decide that the sign *a father* is identical with and only with *a male parent*? The definiens does not permit the semanticist to strike the sign *a father* from the vocabulary of English. It makes no sense to say that the sign *a father* with the meaning 'a male parent' has been struck from the vocabulary of English, but not the sign *a father* with the senses 'the male person who invented or started something' or 'God' (cf. 6.3.B). Synonymy has no place in synchronic grammar books (see *e.g.* Bache and Davidsen-Nielsen 1997).

redundancy. The everyday language is precise, and if precision does not inhere in the polysemous nature of language, it must be placed with the synonymy property – seeing that the system cannot solve all meaning-problems. What lies behind the correlation between synonymy and superfluous redundancy is the commutation test's intractable notion of *intellectual meanings* expounding *significant* distinctions (Hjelmslev 1963:60-65; see *The Prologue*). In the final analysis both the synchronic linguist and the logician presuppose this version of the dualism: denotative (central, cognitive) meaning vs. connotative (accidental, secondary, associative; attitudinal, subjective, social) meaning[25].

Polysemous words contribute to the non-ambiguous nature of the spoken chain; calling the complex sign *flying planes can be dangerous*, ambiguous may not be wrong, but the view does not reject the structural interpretation and facts that such signs acquire their respective meanings from the other entities of the involved synonymy field: *flying* patterns with other adjectives or verbs, and that the involved signs complete their meaning in *parole* as structural intersections. This well-known example in Anglo-American linguistic circles, however, is no more ambiguous than the equally meaningful sign *the bank*, which is ambiguous both as to the article and the noun-form[26]. The latter alternative leaves us with the two problems of establishing (i) the respective contributions by the speech situation and synonymy fields to the precision of the spoken chain, and (ii) conditions for its ambiguity.

What may already appear from the above, synonymy and polysemy are equilogical *langue* properties (see chapter 1.6.1) in accordance with the equilogical nature of the sign function's relates, E and C. Thus 4.2.3 is a preliminary statement of the role of the two properties:

4.2.3 The precision-making property of the everyday language is a synthesizing, *structural* property. It is a *langue*-property as it originates in the synonymy field of the L-continuum, a spoken chain property through the intersectional nature of the latter[27].

The ambiguity-making property of the everyday language is a synthesizing, *non-structural* property. It is a *langue*-property as it originates in the (word's) polysemy field of the L-continuum, a spoken chain property through the intersectional nature of the latter. – Cf. 1.5.2.

I shall now demonstrate that the two properties are not strictly parallel, despite the structural parallelism that the simple sign function seems to suggest, ErC (cf. 4.2.3).

---

25 In this section I am using the words *denotation* and *connotation* and their respective derivatives in the sense of *central, primary* and *secondary, accidental*, respectively.
26 In generativist jargon: both complex signs 'can be assigned more than one structural description', but only on condition that we remove the linguistic context, *e.g.* suprasegmentals, that makes such signs language signs.
27 The use of, say, *great* is 'intersected' by the other synonyms of this field or: '… occurs against the horizon of …'.

4.2. The totality concept is ambigous and polysemous 299

4.2.4   SYNONYMY   POLYSEMY

One meaning   One form
having more forms   having more meanings

Consider the following thought experiment:

4.2.4i   The *structural* nature of *synonymy* is obvious. Something that can be seen and heard enters into a fairly concrete structural network of forms influencing each other, one form acquiring its meaning through or in relation to the meanings of other forms. From the point of view of their respective *meanings* Form F1 is not that which Form F2 is, and vice versa.
F1 /great/ acquires its meaning M through meaning M of F2 /big/, where the meanings M are different from (the denotative, cognitive) meaning M, **magno-** *(4.2.5).*

4.2.5   SYNONYMY

The same cannot be said about polysemy *mutatis mutandis*:

4.2.4ii   The *structural* nature of *polysemy* is obvious. Something that can be 'grasped' enters into a fairly 'abstract' structural network of meanings influencing each other, one meaning acquiring its form through or in relation to the forms of other meanings. From the point of view of their respective *forms* Meaning M1 is not that which Meaning M2 is, and vice versa.
M1 "blessed" acquires its form F /blessed/ through form F /pious/ of M2 "pious" where the Forms F are different from (the denotative, cognitive) Form F, **sely** *(4.2.6).*

The italicized parts of the two statements are sufficient to suggest the nonparallelism (cf. Samuels (1972:66), where the Middle English data has been borrowed from).

4.2.6   POLYSEMY

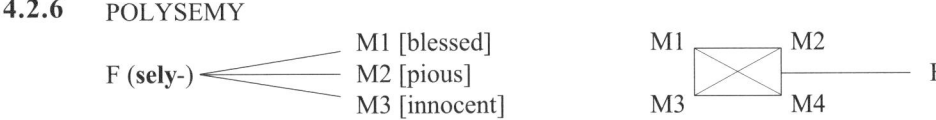

Logically the structural network of a synonymy field is redundant because of the idealized entity called the denotative meaning; it is assumed that the denotative meaning **magno-** can be equipped with a form different from any specific language form. The logician's – and linguist's – problem is that the moment he asserts the idealized entity, he is also asserting the participative relation in that the chosen form of the denotative meaning **magno-** will be a form of the synonymy field he is trying to suspend. Furthermore, because of the denotative meaning **magno-** the forms of the synonymy field are polysemous (4.2.7), just as the polysemy experiment involves synonymy.

4.2.7

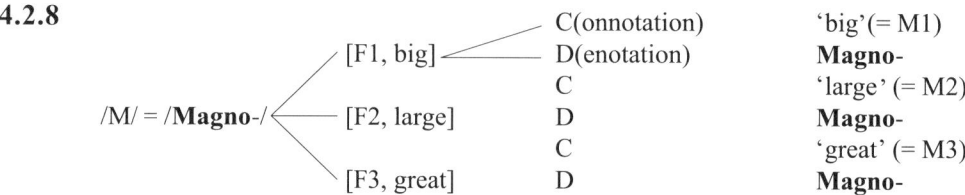

The formal background of the logical view is the dualism with its variant (*parole*) and systemic (*langue*) levels so that we have two types of network, the synonymy network being reducible to this formula:

4.2.8

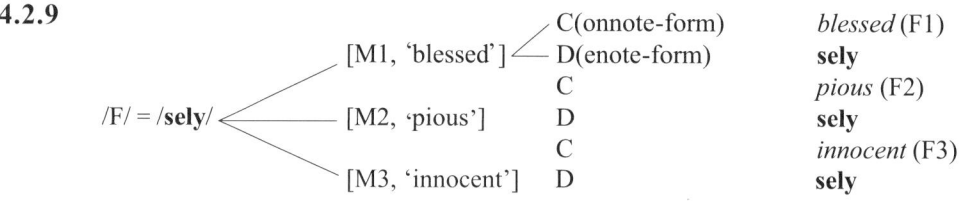

In Figure 4.2.8 /M/ represents the denotative meaning that the group of synonyms (of more than one language) shares or participates in connotatively, and it is somehow distinguishable from the structurally acquired connotations, the distinguishing marks of the forms in question. Similarly, the polysemy field of a word may be represented as in 4.2.9:

4.2.9

In Figure 4.2.9 /F/ does not represent something denotative in the sense that the denote-form *somehow* appears every time a meaning and its (structurally acquired) form appears. But again if **Magno-** turns our three forms into a unit *qua* a structural synonymy field, then the form *sely* turns our three variant meanings into a unit *qua* a structural polysemy field. For the sake of symmetry, we may say so, but in practice and theory, there is a significant difference between the structural cohesion of a synonymy field and that of a polysemy field; the controlling factor is an idealized entity in the first case, while the controlling factor into the second is an empirically relevant time- and space-specific entity. The latter enters into *Kausalnexus*, the former is an *Abstraktion*.

The horizon of the synonymy field is static. **Magno-** is an absolute concept with no history. But when it comes to polysemy, there is nothing absolute about a given form. No correspondingly stable and permanent form controls the meanings of a polysemy field. The Middle English form *sely* is no universal; it even has a meaning of its own and a history. The polysemy field contains equilogical and equiempirical language signs, while the universal of

4.2. The totality concept is ambigous and polysemous

the synonymy field has no such form of its own[28]. That forms should be identical in respect of one meaning expounds a dated dualist metaphysics, while the fact that meanings are identical in respect of one form represents an empirically relevant scenario, on condition that once we assert the polysemy of a form, then we shall also establish a synonymy field.

What the two fields have in common[29] is not ideal entities, but a historical theory[30] that establishes *Kausalnexus*; our sense of history is the dynamic factor that brings cohesion to both fields, and the synchronico-historical conflict may now be voiced as follows:

Is the rationality of the synonymy field *really* immediate and based on static principles? Is the rationality of the polysemy field *really* dependent on those static principles? The reduction of the synonymy field to the domain of logico-empirical redundancy is mediated by the metaphysics of the Dualism: the relational identification of a number of forms under one 'concept-meaning' (an *eidos*) in spite of obvious formal differences is not empirically relevant *vis-à-vis* the corresponding relational identification of a number of meanings under one 'concept-form' in spite of obvious semantic differences. While the polysemous word is an undeniable fact, the former view, based as it is on the analogy of polysemy, is either false or ends in idealization.

This means that *the roots of the Dualism are to be found in the polysemous nature of the everyday language. The notion of finding identity in differences is only empirically relevant from the point of view of polysemy. By analogy, the notion of dualist entities has been transferred to the synonymy field where it creates idealized entities, thereby making substance accidental; this notion was then imposed (back) on the historical domain, with the well-known results of the empirical science of language being idealized and history made, at best, dependent. In a word, the notion that we speak the same language in spite of obvious idiolectal or* parole-*based differences is conditional on the empirical fact of polysemy.* The everyday language *qua* a totality must be viewed in analogy with the polysemous word so that this totality becomes as empirical-concrete as the word – the utterance, which we hear. That is impossible, hence such a notion as the English language remains an idealized entity whose motivation is purely linguistic. This conclusion corroborates what was said above about the totality concept.

The semantic network of polysemy does not make substance accidental, but preserves it as essential. **Each of the meanings of a polysemy field is identical in relation to their common form and different in relation to their respective forms**, thereby establishing a synonymy field. The same cannot be said about synonymy *mutatis mutandis*: **Each of the forms of a synonymy field is identical in relation to their common meaning and different in relation to their respective meanings**, thereby establishing a polysemy field. This

---

28 This analysis demonstrates how and why the synchronic reduction of history was brought about by the synonymous function (cf. the Functional School of Horn and Lehnert), while the Neogrammarian view was based on polysemy, a language universal that does not invite immediate transcendence into the Dualism's ideal haven.

29 There is a curiously distinct difference between the reduction of forms (expressions) to one systemic element (phoneme) *without a meaning*, and the far less successful reduction of meanings (contents) to one systemic element *without a form*. A semantic primitive is always a sign – unless we are staunch Platonists.

30 Hopper and Traugott's 1993-analysis of the dynamics of the grammaticalization of modern English *to be going to* signalling 'immediate future' expounds a processual theory.

difference provides a reason for the dynamic priority of polysemy over synonymy in language evolution (cf. The Neogrammarians and Samuels 1972). Polysemy arises in the spoken chain when the everyday language completes its meaning, and the two notions 'differences in identity' and 'identity in differences' are therefore two aspects of the same dynamic process: the empirical search for identity in (the) differences (of sensuous reality) leads to the establishment of synonymy *on the basis of* polysemy, with the latter making the notion of 'differences in identity' empirically relevant.

The synchronico-logical dualism arises from a conflation of the two fields in 4.2.8 and 4.2.9 into the figure in 4.2.10:

4.2.10
/F/ ⟵ [F1] with D1 + Ca
      [F2] with D1 + Cb   + a *parole*-based, idiolectal meaning.
      [F3] with D1 + Cc

Here /F/ represents one word, characterized by its cognitive/basic/denotative meaning, and which may be manifested in innumerable variants [Fx]; each manifestation implies both the denotative meaning and a distinguishing, structurally acquired style-/code-/register-connotation as well as an attitudinal connotation produced by the concrete speech situation. Criticizing the everyday language, the logician emphasizes the connotative aspect of a word, disregarding the denotative, while the linguist pays particular attention to the possibility of reducing the formal variants to one, thereby making the sign-functional (denotative) relation between one *expression* and one *content* the fundamental linguistic unit. But: the linguist's simple sign function $Er^{den}C$ is not immediately empirically relevant (cf. Hockett 1958:440-442); the logician's view of the sign function is inconsistent.

In normal everyday usage we are never in doubt as to the denotative power of a form, be it a word or a sentence. If we are, the trouble-shooter is *What do you mean by ... ?*. The connotative meanings may create problems, but again, being what they are, connotative and insignificant, they may never create significant problems in respect of the denotative meaning. This leaves the logician with an alternative: he must either maintain that the denotative meaning of a word is never deciphered; hence the word remains ambiguous – which is absurd, or he must work with this type of polysemy, thereby confusing polysemy and what may look like homonymy; *a bank* or *a bachelor*:

4.2.11
F1 ⟵ D1
      D2
      D3

Homonymy cannot be so generalized as to be the basic characteristic of language. Thus the logician has no justification for his claim that the everyday language is imprecise or opaque.

The polysemy of lexical entries in a dictionary is a far cry from the precision of the spoken chain; on the other hand, the empirical relevance of the above trouble-shooter and misunderstandings in our historical lives are not in doubt. My point is not to explain away polysemy-related ambiguity, but only to clear the stage for an unbiassed investigation into polysemy and synonymy, ambiguity and precision, as language universals. Under what con-

4.2. The totality concept is ambigous and polysemous

ditions are such sentences as those below both precise and ambiguous? The sufficient and necessary condition is not the synchronic system 'behind' them. In the Anglo-Saxon Chronicle (Year 1140, p. 268) we read that

> te king sculde been lauerd and king wile he liuede. And æfter his dæi ware Henri king. And he helde him for fader and he him for sune.

As a structural phenomenon the Middle English sentence is precise; but this precision does not come from the systemic meaning (rule) of the personal pronoun, which in fact constitutes the reason for the ambiguity! The next ME sentence cannot be deciphered grammatically, but is completely unambiguous:

> þis and te othre foruuardes þet hi makeden suoren to halden þe king and te eorl and te biscop & te eorles and rice men alle (*ib.*). – *Cf.* Crist self and his apostolas us tæhton ægþer to healdenne þa ealdan gastlice and þa niwan soþlice mid weorcum. (Ælfric's *Preface to Genesis*).

Another type of ambiguity that comes from the system is seen in these nonambiguous clauses

> Nu þincþ me, leof, þæt þæt weorc is swiþe pleolic me oþþe ænigum men to underbeginnenne[31]. // ... þæt seo boc is swiþe deop gastlice to understandenne. (Ælfric *op.cit.*).

If the two verbs (infinitives) are classified systemically as transitive verbs, then *þæt weorc // seo boc* are the direct objects; if the two noun groups are subjects, then the verbs are not transitive. Note that concord is no trouble shooter; in this opening statement of Year 755 in the Anglo-Saxon Chronicle, only all the 'extra- and intralinguistic lines' that intersect constitute its meaning

> Her Cynewulf benam Sigebryht his rices ond Westseaxna wiotan for unryhtum dædum.

The same goes for King Ælfred's (*Preface*[32]):

> þa hie Creacas geliornodon, þa wendon hie hie on heora agen geþiode ealle, ond eac ealle oþre bec. Ond eft Lædenware swæ same, siþþan hie hie geliornodon, hie hie wendon ealla þurh wise wealhstodas on hiora agen geþiode.

The sentences complete their meanings in *parole*, and their precision inheres in them. This does not explain our initial question above, so let's resume the discussion from 4.2.3:

---

[31] Danish has no concord to indicate a possible systemic solution: "At gå i gang med dette arbejde er meget farligt for mig (...)" = "Dette arbejde er meget farligt for mig (...) at gå i gang med".
[32] See also the perplexing use of pronouns in the Parker Chronicle's entry for the year 755.

4.2.12 What is the *raison d'être* of the ambiguity and precision of language?

4.2.13 A spoken chain is always precise in respect of its *originary* conditions[33].

The empirical relevance of the irreducible speaker-hearer relationship follows from both 4.2.13 and the absurdity of defining the concept of communication in terms of two identical persons. The spoken chain is meaning completion and *parole* makes communicative capital of the speech situation; it is extralinguistically and intralinguistically motivated. A spoken chain is *un fait structural* par excellence. It contracts relationships with its speaker, hearer(s), extralinguistic and intralinguistic contexts; it is a part of a man-made structural network and itself, a structural network. The statement in 4.2.13 will be expanded by 4.2.14:

4.2.14 The meaning and form of a spoken chain are the product of their endogenous, exogenous and katagenous realization conditions and their extralinguistic motivation.

And the figure in 4.2.15 will represent a preliminary sign-functional interpretation of the historical movement of the present into the present. The historical function of the everyday language is to form and create the present (= EL). The process involves three sign functions:

4.2.15 $L \longrightarrow EL$ (formed) $(NN)\underline{C}r^{con}\underline{E}(ErCx)r^{ref}\underline{C}(= EL)\mathbf{E}$

$\searrow$ L $\searrow$ $(ErCx)\mathbf{C}$

4.2.15 indicates that a given utterance becomes – be it ever so fleeting – a constituent of the present *qua* the latter's forming content (= **C**), a contributing cause and a part of the present. Otherwise the extralinguistic situation, which becomes the content *qua* the denotatum (= $\underline{C}$) of the spoken chain (= $\underline{E}$) would not become the present *qua* the expression of the historical phenomenon.

An utterance is precise in respect of its speaker, while in the ears of the listener it is ambiguous, because the latter cannot reproduce all the realization conditions of an utterance. In relation to both speaker and listener, the utterance is static (Paul 1966:28). The spoken chain is *necessary* because its possible alternatives are not infinite. When it comes to history, the contingency principle is absurd: it is inconsistent to claim that man may just as well not create a new present. The necessity, however, is unbound by the utterance's becoming ambiguous in the eyes of the listener; the utterance becomes polysemous, acquiring the connotative meanings of the speaker and the listener(s).

The necessity of linguistic ambiguity and precision is twofold. Firstly, it follows from the empirically irreducible difference between the speaker and the hearer, so that 4.2.16 will expound the polysemy function of the everyday language:

---

33 The word *originary* will be explained below.

4.2. *The totality concept is ambigous and polysemous* 305

4.2.16 Because the speaker and the hearer are irreducibly different, a spoken chain will always be 'different' or vary in respect of the speaker and the hearer.

To the speaker his spoken chain is precise, hence it is not precise in relation to the hearer (4.2.16). It is precise because there is no alternative to it and it is part of man's existence[34]. Secondly, the two properties are rooted in the transcendence complex: how to acquire access to reality, how to bridge the gap between theory (man, language) and data (reality) in order to establish knowledge, and, to repeat, how to **ex**ist, to constantly develop into the present.

Transcendence into reality in order to establish knowledge is an essential function of language, and our everyday languages are the indispensable means *by which* we live our lives, scientific and everyday ones, *by means of which* we come to grips with experience, comprehend reality, and orientate ourselves in the empirical world, *by means of which* we conduct our behaviour in relation and in *respect* of other people; the everyday language is an inalienable part and instrument in the creation and continuation of the man-made world (Bernsen 1978:295; 300).

The everyday language is always with us; when we forget it, reality will remind us immediately that if we want to cope with, indeed survive in nature for any length of time, we must necessarily rely on language. An everyday language is more than dispensable memory, dated theories and experiences: our sense of reality and of history grow with and are dependent on our command of an everyday language. Knowing one's language is also knowledge of reality; no language does not carry existence postulates with it, of which the everyday language's contribution to and participation in man's historical existence is the clearest expression. Bernsen's – and Kant's – principle of the primacy of experience is correlated with the everyday language, and is not to be reduced to having a purely instrumental starter-function in man's everyday life.

According to the logician an everyday language may be adequate for the execution of man's daily chores, and is anything but perfect when it comes to scientific transcendence into reality. I argued above that that is a postulate, showing that the everyday language is precise, a type of precision which hinges on the very fact that our existence is conditional on the everyday language. If it was not precise in the most perfect sense of the word, the survival and continued existence of the human race would be impossible to understand in a nonreligious myth; if the everyday language were imprecise, reality would be imprecise, too[35]. Amidst objective nature the everyday language is the minority 'gene' that enables man to survive, and the logician should ask *Why is that so?* now that human language is such a bad epistemological instrument. Why has man been able to develop such intellectual masterpieces as symbolic logic, mathematics, (non-)Euclidean geometries, or why has man not adopted symbolic logic as his native language?

---

34 'To be imprecise' or 'to be ambigous' is no meaningful property to be assigned to man's historical existence.
35 It makes no sense to let reality be predicated of such concepts as *imprecise, opaque* or *ambiguous*. Reality is not ambiguous, the everyday language is part of reality, hence the everyday language is not ambiguous. It is absurd to claim that modern English *wicked* (Quine's example, *loc.cit.*) is not true of or does not denote *each wicked person individually,* and instead includes non-wicked persons. Therefore it is meaningless to describe language change in similar negatively value-loaded terms (see *The Prologue*).

An answer lies in the answer to the questions: of what is a (scientific) statement S true, and in respect of what is S precise? If S is to be true of its denotatum, the circle is obvious. For us to declare a statement true, we must a priori know its denotatum precisely: *true knowledge of X presupposes knowledge of X*. If we know that S is ambiguous or imprecise in respect of its denotatum, we may just dissolve the disjunctive statement or formulate a precise one[36].

From his nominalist stance Hjelmslev (1941:103-104) cannot agree with Bernsen's 'primacy of experience', saying that there is nothing, no data, <u>before</u> or <u>after</u> the application of the nominalist's theory, which – creating its own data – provides an exhaustive description of reality. The dynamic version of the view(s) runs as follows: **there is no historical existence, no experience, before or after the 'application' of the everyday language**. The present moment exists neither before nor after itself. What exists before the application of the everyday language is the present moment, created by it, and what exists after its application is the past – the present turned into the past: the moment we 'apply' our language (or experience), the present develops into the present.

How can the individual relate to what *was* before him? How can the individual grasp the (historical) nature of what is contemporaneous with him? Through his historical sense. To elaborate: a statement S is true, because I know the denotatum of S such that the nature of the denotatum 'corresponds to' the meaning of S. The only conclusion (*pace* extreme nominalism) to be drawn from this is that scientific statements presuppose some less mediate transcendence into reality, and this kind of transcendence can only be brought about by the everyday language (Hjelmslev (1963:17-18) must agree; see note 17 above). At this crucial intersection the historian and the nominalist agree: **the everyday language is a provider of knowledge of reality**; not all our knowledge (*Erkenntnis*) *entspringt* from *Erfahrung*. However, where the historian elaborates and indeed turns the position into the basis of his further work (theory), generalizing this *Erkenntniswert*, the nominalist leaves it and constructs his theory on few arbitrarily selected premises, without generalizing this necessarily discursive start of any scientific analysis. The arbitrariness of the nominalist theory is incompatible with the necessity of historical description. However, the nominalist's theory also presupposes the everyday language after its application in the sense that the latter is its metalanguage and is needed when the theory is to be 'posed' among data (cf. below).

Transcendence is dual: into and away from – reality; the transcendence-into movement is synthesizing, the transcendence-away from movement is analytic. The energy of the everyday language involves transcendence away from reality, and this notion is perhaps particularly adequate in the eyes of logicians and formalists, seeing that ambiguity/falsehood may be derived from it.

We try to ward off hostile reality by means of the everyday language. When we feel a need to withdraw from the world – man's dissociating and individualizing faculty – we retreat into the everyday language, a movement epitomized by the monologue or prayer. Transcendence-

---

36 It is easy to see how the *Erkenntniswert* of the everyday language asserts itself. But it is this initial fact that the nominalist cuts off from his theory.

4.2. The totality concept is ambigous and polysemous 307

away-from implies there is no a priori gap – between language and reality – to be bridged; on the contrary, there is also a bridge to be burnt[37].

The everyday language is able to accommodate these two complementary faculties or needs in man, enables him to transcend into and away from reality. How does it accomplish this? Transcendence-away-from occurs when we say that the everyday language implies no existence postulate. We need not appeal to Eliot's *human kind Cannot bear very much reality* in order to see the movement's relevance. So since it is man's inalienable possession, the everyday language has been evolved to fulfil the two faculties, needs or socio-psychological features, which Kant called man's unsociable sociability (Hansen 1986:105).

A man-made formal language emphasizes the movement-away-from in that it must be made to transcend into reality, thereby transcending its own inherent limitations[38]. The logician's postulate, tacitly adopted by formalists, that the everyday language epitomizes transcendence away from reality through its imprecision and opaqueness, puts the facts upside down: his purportedly univocal formal language is a retreat from reality.

In our Western Civilization man's conception of himself and his language (logos; ratio) hinges on the horn's of a self-inflicted dilemma: *homo loquax* is a 'poet', who *poiei*, or an 'author', who *auget*: a creating or augmenting person. Our cultural trend-setters opted for the latter, redefining creativity, innovation and originality in terms of theory construction, finding the method (the right *hodos*) by means of which true knowledge could be re-cognized and retrieved, not created. This is not *poiesis*, only the establishment of that which is already the case, 'constituted' or 'developed': a proces of augmenting. Consequently, 'original', 'unique', 'innovative', 'creative' became the poet's properties unless he was performing in his mimetic mode: man's energy was expended on approximations of varying reality-degrees. The S-is-P conclusion of Aristotle's syllogism became the image of the process of cognition with 'S is P' already present (hidden) in the premisses. The conclusion, as such redundant, need not be explicitized. Historiography does not establish knowledge as its narratives do not uncover what is present and lies unchangingly hidden; the historian creates, enunciating existence postulates whose reality-degree is rather low, unstable with no permanence; basically he repeats what only has shadow-existence.

A nominalist theory, a means of transcendence into reality, is in fact a transcendence away from empirical reality. It carries no existence postulates; it is its own object, and if, venturing into reality, it catches any fish[39], such a stroke of luck will only testify to its appropriateness, not its theoretical status, whose fate lies solely at the hands of the Empirical Principle.

---

37 Curiously enough, this movement is also implied by Hjelmslev's opening characterization of language in *Prolegomena*: language is *the ultimate, indispensable sustainer of the human individual, his refuge in hours of loneliness, when the mind wrestles with existence and the conflict is resolved in the monologue of the poet and the thinker* (p.3). This language universal is not generalized by Glossematic theorizing. – Falsification is also a retreat from reality, or rather acknowledgement of this retreat.

38 The extreme nominalism of Glossematics makes Hjelmslev forgo deliberations of the transcendence problem temporarily, but this limitation in the over-all field of vision is arbitrary in view of the abovementioned starter-problem; and the bridge-building function of 'one substance hierarchy selecting a formal hierarchy' is certainly a weak statement of the transcendence problem.

39 Note how the nominalist does not leave any degree of activity to the theory, it being selected by the substance.

And leaving Plato aside, we may ask (1) whether Aristotle creating the first formal language, or Thales reducing everything to 'water', retreated farthest away from reality, a question that deserves some reflection and not off-hand dismissal; and (2) whether William of Occam's Revolt, dichotomizing reality into matters linguistic and non-linguistic reality, is a generally viable solution to the transcendence problem; and whether a historically accurate (*geschichtliche*) comprehension of that Revolt[40] does not presuppose that it is not turned into a *historische Zufälligkeit*?

As indicated above, the nominalist cannot do without the both-and nature of the everyday language. The initial 'Agreement' and the first definition of his theory, premised by all succeeding definitions, necessarily rely on the theoretician's language; indeed a very unhappy situation! The indefinables of the first definition are only precise in relation to its progenitor. Therefore the nominalist has a reckoning – with the everyday language – after or before the application of the theory. Furthermore, in the construction of his theory, the nominalist and the structuralist go hand in hand, eliminating the agent: who or what *performs* the universal operation of Glossematics (Hjelmslev 1975:def. 1)? Who *describes* (= *analyzes*) a given object (Def. 3)? It is obvious that the connotative function cannot be relegated to a position 'low down' (= late) in the procedural Glossematic hierarchy, when the perspective of the theory is widened. As such the analytico-deductive point of departure is necessarily anthropocentric, historical and a synthesis[41]: It is an *I* who analyzes the class of objects, availing himself of the everyday language, a member of the necessarily non-empty class of all objects.

The speaker-related precision of an utterance does not imply that the utterance is not *open* to the listener's understanding. No utterance or written statement is the speaker's/writer's exclusive property (see Coseriu 1974:64-65). To this extent both are *collective*:

4.2.17 An entity becomes collective if its existence is characterizable by distance *qua* both attachment and separation, in relation to both a speaker and a hearer.

The features that determine this basic speech situation (in 4.2.17) are not surprisingly not different from those that specify the historical process of external endogeneity:

4.2.18 SPEAKER → [UTTERANCE] ← HEARER
        [+attachment]{is}   {becomes}[+attachment]
        [+separation]{becomes}   {is}[+separation]

An utterance is 'attached' to its speaker and becomes so to the hearer; it becomes 'separated' from its speaker and is 'separate' from the hearer. The utterance assumes the listener as its new connotatum.

---

40 Occam's Revolt was empirically relevant in the light of its historical context; nominalism's relevance today cannot be justified only by reference to its original premises. Occam's was a defence of empiricism, and I have argued that modern nominalism as it appears in synchronic linguistics has more than a superficial coating of Realism.

41 Bernsen (1978:295) puts it like this: *(...) it is not a purely contingent question if in reality there are persons in cognitive situations, with a history and bodies in space and time.*

4.2. The totality concept is ambigous and polysemous

The scenario in 4.2.18 involves polysemy; one utterance with two meanings, the speaker's and the hearer's. The utterance is static against its palimpsest horizon, it is fleeting and cannot change. Its polysemy is a precondition for the dynamic collectiveness of the utterance as well as for the speech situation as a whole. The meaning of the utterance is not preestablished, but made precise. The elements (*langue*-elements) of the L-continuum are underdetermined by meaning and experience; that is why the everyday language is not all of reality. The L-continuum's elements imply that more meaning or experience will appear: they are not sign(-element)s in their own right, but signs awaiting the role of a part of a sign function: $E(Er^{den}C)r \ldots$.

It is a contradiction in terms to uphold both the possibility of fixing a word/utterance by a definiens/preestablished meaning and the notion that the everyday language is *energeia*, not *ergon*. And with a slight alteration of Roland Barthes's view of myth-making[42], the definiens turns the sign into a myth, a theoretical myth; in both cases it is the task of the listener or speaker-turned-listener to try to break the stand-still or, with a loan from Gilbert Ryle (1970:17) *explode the myth*.

There is some truth in the claim that modern structural linguistics (Cassirer (1945:116-117), referring to Viggo Brøndal) has vindicated the *energeia-ergon* parameter of Wilhelm von Humboldt. This, however, is only correct if the opposite of *ergon* is not static, abstract *energeia*, but concrete, actual activity and reality in the word's Aristotelian sense (*Met.* 1048a; Hansen 1985:75-78): the everyday language is *energeia* only in so far as it is *en ergeiai*; it is not only the product of a process, but also the process out of which it itself exists. It is *en ergeiai*, when spoken or when the speaker transcends in or out of reality on the basis of reality. The Wittgenstein metaphor (1963:§132) may be so elaborated that *das Leerlaufen der Sprache* creates confusion, not so *wenn sie arbeitet*. And with another, an archaeological loan, the answer to 'What is the everyday language?' is 'It is in process of becoming' (Renfrew 1989:211-249). This may tally with Wittgenstein (1972:4-5) saying that the life of a sign is its use, but when he adopts the synchronico-static assumption that *The sign (sentence) gets its significance from the system of signs, from the language to which it belongs* so that *understanding a sentence means understanding a language*, his notion of *use* does not agree with the historian's. Wittgenstein's dilemma is this: either the system of signs contains all meanings, and we are back to the *déjà constitué* point of view, or extralinguistic reality is static. An utterance is dynamic, lives, is *en ergeiai*, when it completes its meaning in *parole*, becoming a constitutive part of reality. When it does this, the changing everyday language makes reality amenable to our transcendence into and away from it, and knowledge of reality can be achieved on the basis of the everyday language: in so far as we understand the everyday language, we also understand reality, not all of it. Understanding a language is more than knowledge of an 'already-developed' set of rules.

An utterance is precise in relation to its speaker, and because the hearer cannot reproduce all its realization conditions, it is also ambiguous (polysemy). Therefore the utterance is not the common denominator that permits the transformation of an existential situation into a static, frozen form, be it the immanent state or the knowledge of an ideal hearer/speaker.

---

42 The distinguishing feature of the myth is *de transformer un sens en forme* (Barthes 1970:217). The Naturalistic Reduction is a special type of myth-making: *de fonder une intention historique en nature* (p. 229).

An utterance is precise in relation to its own – I hesitate to use the word – contingency, because the historical theory – the speaker's knowledge of his language – gives the speaker *a* choice[43] – not an unlimited number. This synonymy factor inheres in the L-continuum, and will be given its theoretical foundation in chapters 5 and 6.

A speech situation is completed against the horizon of both polysemy and synonymy: the utterance ($Er^{den}C$) enters into two connotations (synonymy), and these connotations exist against the horizon of polysemy: $Er^{den}C$ r ^{con} C(= Primus) and $Er^{den}C$ r ^{con} C(= Secundus).

4.3. A comparison between the historian, the nominalist and the 'poet'. The historian is poised between the poet-maker and author-augmentor. An outline of a historical theory in the light of polysemy and synonymy; the historian is the hearer, historical phenomena the speaker (the sign): historiographical transcendence is sign-functional. The historical phenomenon's form is not captured by the formed-unformed parameter

Between the *poet-maker* and the *author-augmentor*'s products we find the historian's, whose ultimate arbiter is historical accuracy: the distinction between what really happened and what did not happen is absolute (Passmore 1987:73-74); the observation seems trivial, but assumes theoretical relevance because it is the historian's Razor banning potentialities and ideal entities from his narrative. Modern historiography, according to Passmore, is witnessing a struggle between Modernism's view of history and history's evident demand for *clarity* about the past and *evidence* (Martin 1993; Kosso 1993). Linguistics, on the other hand, has not really attained to such a degree of self-consciousness as to pose the question: is what the linguist is saying is the case really the case? Linguistics has more often than not taken the question – and the implicit affirmative answer – for granted, and the retreat into the haven of nominalism and idealism has been thorough[44].

Historical accuracy is closer to reality than a self-validating nominalist theory. Like the poet, the historian's body of primary data, the past, is not '(t)here'. If found, historical phenomena are structural constituents of the present, *i.e.* of continued reality. Fictional writing is also constituents of reality, but our sense of reality is unable to read into a novel, a poem, a play, an equiempirical past out of which the persons, places, actions (the plot) have developed. The writing of fiction like a nominalist theory and the making of an urn involves no questions of empirical veracity or accuracy, and the meaning of a theory or poem does not entail existence (postulates) beyond the degree of existence an urn has. A nominalist theory like fiction has an absolute beginning; a novel or an urn is not falsifiable, a historical narrative or a theory is; the historical narrative has no absolute beginning.

The historian is a maker in his attempt at creating the past, but his *poiesis* is with no poetic licence as he does not asssume *complete control over his material* and he does not *make it mean what he means*. The historian's *first loyalty* is not to *his own conception of the truth but to the facts* (Kitto 1966:291-292). The nominalist is loyal to his scientific principles, the

---

43 This choice is the resolution of the syncretism through the present's development into the present.
44 In his attack on the dominant paradigm of the Chomskians, Antilla may be said to imply the question and an answer in the negative (1979, 1992).

4.3. A comparison between the historian, the nominalist and the 'poet'

Platonic craftsman to the Idea. The historian is an augmentor in that he contributes to the sum total of our knowledge of reality, but with no licence to create redundancy, potential or ideal entities. If he did not do his job, his data would not come into being – unlike the natural scientist and synchronic linguist, who both retrieve what is already the case; this cognition is *re*cognition with the object of the theory existing – or not existing – irrespectively of the analyst's work.

This may be given a sign-functional interpretation. To the synchronic nominalist the theory is either a denotator or a potential denotator (Sørensen 1958:12-18). The poet's work does not denote and is not a potential denotator. The historical narrative denotes and is not a potential denotator, but in a special sense. Løgstrup (1976:57-60) makes the distinction between words that 'have' and words that do 'not have' their denotata, roughly a distinction between ostensive and demonstrative/interpretative words. Words for the historical phenomena do not have their denotata[45], they are closer to (the meaning of) the utterance, being interpretations as their meaning arises contextually, in their linguistic context.

The poet's work <u>does not denote, does not have its/a denotatum</u>; it is neither true nor false – nor meaningless. It does not create its denotatum or entail existence postulates.

The nominalist's theory (its metaterms) <u>denotes</u> and <u>denotes potentially</u>; it has its denotatum (ostensive) and it does not have its denotatum; its sentences are either true or false. It creates its denotatum as a scientific object and does not create its denotatum as a positive given; it does not entail existence postulates.

The historian's narrative <u>denotes</u> and does <u>not denote</u>; as such it is either true/accurate or false/inaccurate, respectively. It does not have its denotatum and creates its denotatum; it entails existence postulates.

The Realist's text does not denote, does not have its denotatum; implies truth and falsity; and entails existence postulates and it creates its denotata.

As can be seen the historian has a Realistic aspect: knowledge of its subject-matter comes from its (utterance-like) narrative and it entails existence postulates. And the decisive difference between the historian's narrative and the nominalist's sentences is that – in Kitto's words (1966:79) – there is *only one standard of Truth*, and this truth merges indistinguishably with a given existence postulate. The difference between the historian and the poet was formulated above by Kitto, a formulation that establishes an affinity between the nominalist and the poet: there is no poet that does not set out from reality – the poet's work exists against the horizon of empirical reality, which also applies to the nominalist; once this has been said, they also share complete control over their material, making it mean what they want it to mean.

The nominalist and the historian also differ in relation to contingency (potential denotation): that the present and the past are solidary by means of the law of tradition (Saussure) means that the process of the present so creates the present that this empirically necessary

---

[45] Løgstrup refers to the phenomenal level, which the language philosopher does not leave (pp. 172-175). I shall adapt the distinction to my purpose, which involves both a distinction between the narrative and the historical events (cf. above) and a more expansive definition of a historical phenomenon. – Løgstrup did not see that the nominalist's metaterms also denote in virtue of the structural interaction of the words of an utterance-like type of sign, namely in virtue of their definientes.

directionality establishes the solidarity – the *solid* connection – between the present and created past, to which there is no alternative and between which nothing else can occur. My argument leads up to this unfalsifiable statement: *the historian knows both that something – not anything – is the case (the present) and that something – not anything – really happened; the historian does not know when a sentence stating something about what really happened (about the past of the present) is empirically relevant in an absolute sense, because the part of the everyday language that contributed to a given historical event has vanished*[46]. This is a theoretically more appropriate restatement of Labov's *Historical Paradox* (1994:21)[47], which boils down – not to a paradox – but to a regrettable and defeatist statement: *to the extent that the past was different from the present, there is no way of knowing how different it was*.

Labov's Historical Paradox is also a – desperate? – cry for truth. The nominalist has a decision problem: which statement is true (Hjelmslev 1971:110-111; Rasmussen 1992:180-181)? The historian has a decision problem: which narrative is correct? The nominalist has his Empirical Principle[48], the historian his theory[49]. I shall illustrate my point with reference to Thomas Aquinas (*Summa*, 1a.1,10), whose words will be interpreted dynamically: *the correct or accurate statement (narrative) is not to be found because historical phenomena are signs: ipsæ res significatæ per voces* (words) *etiam significant aliquid*; this denotative power of things is due to the meanings of the signs that are constituent parts of them *qua* historical phenomena. These (*res*) will therefore be both structurally precise and ambiguous, and 'equivocation' does not necessarily spring from the words (*voces*): *(...) sensus* (interpretations of a text) *(...) non multiplicantur propter hoc quod una vox multa significet, sed quia ipsæ res significatæ per voces aliarum rerum possunt esse signa*[50]. The historical phenomenon is a sign, pointing to its origin (predecessor). The nominalist, on the other hand, simplifies things implying a theoretical true-or-false alternative: *(...) dans le cas où il y a plus d'une analyse possible, le réaliste reste libre de croire qua l'équivoque est inhérent à la nature de l'objet soumis à l'analyse* (Hjelmslev *loc. cit.*).

A sign and a historical phenomenon are precise in relation to their respective originators. The nominalist's definiendum contains indeterminate elements (Hjelmslev 1963:131); the ambiguity inheres in the definiendum, not only in the definiens. The nominalist's system of definitions is precise through convention. Thus the Realist, the nominalist and the historian

---

46 And because the meaning part of the historical phenomenon is not the essence of the phenomenon.
47 Labov's Uniformitarianism is too narrow a view of the historian's task: *to explain the differences between the past and the present*. Secondly, Labov does not justify his claim, which seems to rest on the ideal assumption that the historian's task should be to recreate all of the past, an ideal that is existentially impossible and therefore empirically irrelevant.
48 Some historians would refer to the Covering-Law model, others to abductive reasoning.
49 This will be elaborated below, but I shall suggest this much: because of the absence of contingency in history – there being no potential denotators – the theory will guarantee the accuracy of the narrative.
50 In Ælfric's *Preface* we read: *[Jacob] hæfde feower wif*, which 'voces' denote the fact that the patriarch Jacob had four wives. However, this fact is itself a sign of something, a something that it is imperative for Ælfric to explain to his 'dysigum men'; to the lay people in Ælfric's days such Old Testament situations were extremely ambiguous (equivocal). To complete the picture: the events in the Old Testament were – in addition to what they originally were – signs of what was to come: *seo ealde æ wæs getacnung toweardra þinga*.

4.3. A comparison between the historian, the nominalist and the 'poet'

have the equivocal nature of data *qua* signs in common, with the nominalist displacing the equivocal to being a property of the arbitrary theory (Hjelmslev 1963:18;61), and the close relationship between synonymy and polysemy has been corroborated: more statements about the so-called same object constitute a synonymy field based on polysemy in respect of the multisemous nature of both the object and the synonymous statements[51]. The cause-effect explanation of historical phenomena – *e.g.* in the form of a multi-conditional parameter (Samuels/Jespersen) – also testifies to my conclusion: finding *the* cause of a historical phenomenon is impossible, and by way of a short logical excursus which relates to the question of cause in historical analysis I shall proceed with my sign-functional interpretation of historical development.

Logically, a true statement is implied by any statement, *i.e.* a false and a true statement, this being the ultimate in ambiguity – and synonymy: the true statement is ambiguous/polysemous in relation to its implicantes, which are synonymous in relation to their common implicatum[52], and their structural precision lies in the fact that they – A and NonA – exhaust a given universe of discourse. I shall illustrate this as follows:

Great Britain declared war on Nazi Germany in 1939 is a true/accurate statement whose denotatum[53] (*pace* Løgstrup) had for its cause the fact that Nazi Germany invaded Poland (A) or the fact that Nazi Germany did not invade Poland (NonA). Logically correct, historically absurd. The truth of the historical statement does not hinge on antecedent propositional possibilities, but on the occurrence of the historical event in question. Historical accuracy is a function of historical coherence, not logical possibilities. Today the meaning of *Great Britain declared war on Nazi Germany in 1939* is understandable[54] when placed in a temporally well-ordered structural whole *qua* a historical narrative consisting of, among other sentences, *Germany invaded Poland in 1939*.

The denotatum of *Great Britain declared war on Nazi Germany in 1939* becomes an expression (E) of its origin, both as its cause and as its content (C), since the historical cause is a constituent of its effect (to be illustrated below). The sign-functional interdependence contracted by the sign's content and expression mirrors the cause-effect interdependence of historical solidarity; and the historical theory will then set up contents *qua* 'causes'. A curious consequence of this – sign-functional or cause-effect solidarity – is that the accuracy of

---

51 The host of synchronic theories (descriptions) of so-called modern English that the past 100 years have seen bears my position out at a practical level.

52 Hjelmslev's offhand dismissal of a discussion of truth is unacceptable in view of Lewis's two Startling Theorems (Hjelmslev 1973:128; curiously, Rasmussen (1992:180) does not discuss this either; cf. Hansen 1983a:459).

53 In Bache and Davidsen-Nielsen (1997:191) the denotatum is called *a situation*, this being the meaning *that sentences express*. Following these grammarians, a historian's narrative 'informs' his readers of an 'instance of there being declared war', an instance 'in which there are two participants: Great Britain, which does something by which Nazi Gemany is affected'. As such the meaning of (this type of) sentence is made up of three *components*, components that somehow exist before the historical situation itself, *Jack fixing the old motorbike* (Bache and Davidsen-Nielsen) and 'GB declaring war on Germany in 1939'. I shall demonstrate that such componential analysis is not sufficient to expound the meaning of 'situations', because it splits *das geschichtliche Wirklichkeit* into *historische* – the grammarians would say: *synchronische* – *Zufälligkeiten*.

54 As such it is not different from the synchronic grammarians' sample sentence: *Jack fixed the old motorbike*.

the historical narrative may be doubted in the short run but in the long run – under the aegis of the period – it is impossible for the historian to create a consistently false account of a comparatively long historical period leading up to the present: he cannot make the elements into a structural whole whose coherence is demonstrated theoretically[55].

The formula A>B is the customary way of saying that B has developed from A. Since B comes from A, A is somehow a part of B such that our historical phenomenon is a structural synthesis as well as a sign (= ErC). Hence what the B of the formula represents consists of an expression part E and a content part C so distributed as to make B a sign (E(ErC)) of its origin A* (= C): the historical phenomenon (= AB) connotes its origin in virtue of its content-cause (= A**) with the origin A* continued under the aegis of change as A**:

4.3.1   A*   >   B   CrE(ErC)   OE *word* (plur.) //-*as*   >   ME *wordes* //-*es*
              ↘ A**   ↘ ME *word* //-*as*

The model in 4.3.1, to be expanded below, is a simplification of the sign-functional nature of the historical phenomenon; but it suggests that the historian does not stop at the manifested meaning (of ME *wordes* // -*es*), but moves on to the structural meaning.

The historian's transcendence problem consists in the gap between his knowledge (at T2) that something did happen at T1 and the meaning of the statement about the T1 event he is going to produce at T2. The transcendence mirrors the activity of the everyday language when it is *en ergeiai*, closing the gap involved in the present's development into the present. Knowing the present does not imply precise knowledge of the origin of the present. As the past (a part of it) is continued into the present, the historian knows that not anything could have been the origin of the present, and because the historical event cannot be recreated, we are bound to conclude that in the historical narrative truth and existential postulates merge[56].

That the accuracy of the historical narrative is not established through correspondence between theory and data, which is gone or has changed, means that we must be able to sift the historically accurate statements from the inaccurate ones; the historian needs the help of a theory to close the gaps that his evidence presents him with so that an inaccurate statement does not take the place of an accurate one in a historical series. Thus the historical theory will take care of the logician's or formalist's critique of the everyday language for being ambiguous. And I shall now move on to the problem of synonymy, as a cause of ambiguity and redundancy.

---

55 The corollary is that the longer the time span of the historian's period is, the more likely it will be that his account is accurate. This may support Toynbee's call for a world history to replace the scientific monograph's relatively immanent historical event (see Kosso (1993:13) for a discussion hereof).

56 In Bache and Davidsen-Nielsen (p. 192) there is a seemingly similar, but curious merger of truth and existence: referring to what they call the *primary actional distinction (...) between* dynamic *and* stative, they say that *A dynamic situation* happens *or* takes place *while a state* exists *or is* true *of someone or something*. If it is the case that John did fix the old motorbike, then the sentence *John fixed the old motorbike* is true of John – not of the motorbike? – just as the sentence *Ottawa is the capital of Canada* is true of the 'stative situation that Ottawa is the capital of Canada.

4.3. A comparison between the historian, the nominalist and the 'poet' 315

The problem presents itself immediately we know that something happened, *e.g.* that Trotsky was X-ed in Mexico in 1940. Synonymy expounded by a plurality of narratives of the same event does not express historiographical imprecision or redundancy; the fundamental pragmatics of the speech situation are mirrored in the historians' narratives. An example:

    Was Trotsky

                    murdered
                    killed
                    assassinated

Or   did he

                    die

                                        in Mexico in 1940?

Similarly, what did really happen when

    OE *a:, -as*

                    developed into
                    evolved into
                    became
                    changed into
                    was replaced by

                                      ME *ɔ:, -es*

Or when

    ME *ɔ:, -es*

                    developed from
                    evolved from
                    came from
                    sprang from
                    (was) changed from
                    was derived from
                    replaced

                                      OE *a:, -as*?

In 3.4.1 this type of scenario was interpreted as one of descriptive precision, while the historical event may seem ambiguous. The idealist solution is to interpret the scenarios dualistically as variants on the same 'theme', where the variants expound the common, denotive or cognitive meanings of the respective processes (verbs).

The historical event is collective, but no two persons/historians (*qua* hearers) can (re)produce two exactly identical events – a commonplace in synchronic linguistics. Thus the fact of synonymy and the synonymy field of the everyday language will come in handy for a plurality of narratives; secondly, a given set of synonyms may testify not necessarily to the ambiguity of the event, but to the complexity (meaning<u>ful</u>ness) of it. The nominalist's requirement[57] that only one statement (or word or narrative) be applied to denote a phenomenon presupposes that denotation is only a process of naming, but as was argued above denotation is also interpretation in the sense that the full meaning of historical phenomena can only be described through a complex sign whose meaning is created contextually. Let's return to Trotsky.

The fact 'Trotsky, dead' could not even as an epistemologist's godsend have occurred unless something conditioned and caused it, so that it is bound to exist as a part of the structure of a historical series. As a historical phenomenon the event is a sign of its origin (predecessor) and its structure will illustrate how the historical statement *Trotsky is dead* is understood contextually.

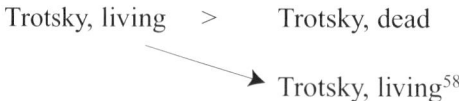

We know that what happened to Trotsky was caused by another event; somebody Xed him, which is a part of the situation at T2, 'Trotsky, dead', as well as of the historical series in question. We need a theory to unify the individual statements (expounding our understanding), as the mechanism of change – exogeneity – is not identical with the theory. The theory describes the coherence between the idio-events, 'Trotsky, alive', 'Somebody does something to Trotsky' and 'Trotsky, dead'. The continuation of the content of 'Trotsky is alive' at T1 becomes a structural content-factor in our understanding of the situation at T2. Structurally, the '*Eigen*-existence' (endogeneity) of Trotsky is intersected by an external line (exogeneity) – somebody did something to him, just as the internal history of Great Britain was intersected by the Nazis' invasion of Poland in 1939, thereby contributing to Britain's exogenously motivated development into the present, expounded by Britain's declaration of war on Nazi Germany (see below). As to the causes of the T2 situation, 'Trotsky, dead', they cannot be inferred unambiguously from the present, and the statement *Trotsky died in Mexico in 1940* must be explained by a number of other synonymous statements so that the verb *to die* will embody these meanings in its polysemy field: it assumes its meaning contextually, as it does 'not have' its denotatum. – The historical process will now be interpreted sign-functionally.

The empirical necessity of the present's development into the present is an existential condition, and since idio-events embody this necessity, they are scientific analysanda *per se*: the origin becomes a co-constituent of its effect and this solidarity is sign-functionally due to the sign function's expression-content solidarity.

The historian is the *hearer* in relation to his data; the alternative is unacceptable: the historian *qua* speaker would imply that he recreates the past. AS in a normal speech-situation:

---

57 which implies the structural dictum of searching only for differential minima.
58 Consequently, the synchronic grammarians overlook the historical *component* in their 'situations': the sentence *John fixed the motorbike* does not make synchronic sense unless the hearer knows the history of the situation, the situations out of which 'John fixing the motorbike at T3' has developed: – the situation that the motorbike had been out of order at T2 had developed from the situation that the motorbike was in order at T1. The 'past' is the horizon against which the present is understood.

4.3. A comparison between the historian, the nominalist and the 'poet'

*what the hearer understands is the product of the solidarity of the spoken chain and its context, in which solidarity – constituted by the intersecting lines of the speech situation – the hearer becomes the content of the **internal** manifestation of a connotative function*, SO in the historian's writing of a narrative: *a historian's narrative is a representation of what he has 'heard' (understood), the representation being the **external** product – the expression (E) – of the connotative solidarity constituted by the historian* qua *the content (C) of the expression that itself is a complex sign denoting – in virtue of its content – the meaning of the historical events in question*.

This means that the narrative, being an expression, contracts a solidary function with a sign function's content *qua* 'what actually happened', the historical event which we know occurred. The narrative's denotative power is different from that of fictional writing: neither 'has' their denotata, but the historical narrative carries existence postulates with it through its sign-functional solidarities.

Historical phenomena are man-made and the everyday language is a forming part of them. As the destruction of *La Bastille* was a disaster to some, it was a blessing to others, so in the eyes of various people Trotsky was murdered and assassinated. The historical phenomenon is collective in virtue of its material substance, but is also the product of individuals, as such it is an expression of its maker(s) *qua* its content, and with this we are back to Lucien Goldmann's sign-functional nature of *tout fait humain* (1966:152) as well as the transindividual nature of the historical phenomenon created through 'cooperation' and 'division of labour' (Hermann Paul):

4.3.2 Product *by* N.N. → **E** *r* **C**

The function in 4.3.2 is a simplified extension of the Glossematic connotation (cf. 4.1.1; 4.3.4). The product (E) formed by N.N. is part of a complex sign in whose content is a sign function:

4.3.3 E (= the material substance or material of the fact or event)
  r
  C(ErC) (= the everyday language forms the material in question).

4.3.4 E(ErC)r^{con}C (= N.N.) (= the connotative content is added to 4.3.3).

The connotator, N.N.'s utterance, in 4.3.4 becomes the content of the historical phenomenon, and since there is no permanence in the connection between the historical agent and his product, the connotatum disappears (cf. 4.2.18)[59]; to put it metaphorically, 'history speaks to the historian' (4.3.5):

---

59 The formula **E**(ErC(ErC))rC conflates 4.3.3 and 4.3.4, with the last Content being N.N.

4.3.5 $E(ErC)^{con^1}rC$ (= N.N.1)   [+separation]; (ErC) is left as the content of the spoken chain, itself a sign of its progenitor.

E
r
C(ErC)

(= the material substance of the spoken chain)

[+separation] (ErC) is left through the fleeting nature of the spoken substance.

$E(ErC)^{con^2}rC$ (= N.N.2)   [+attachment] (ErC) becomes the denotative sign in the hearer's internal connotation.

The historical product being a sign becomes the expression in a complex sign with the historian as its connotatum – in accordance with the normal speech situation. In contrast to the hearer (N.N.2), the historian 'speaks out' what he 'hears', thereby producing another connotative sign, the narrative, which becomes the expression for the historian *qua* its content: the narrative is always *somebody*'s narrative:

4.3.6   E (= the material substance or material of the fact or event)
r
C(ErC) (= the forming language) $E(ErC)^{con}rC$ (= historian)

The two underlined sign functions in 4.3.6 will become a complex sign, and they exhibit the compatibility requirement: the historical fact is amenable to being described – understood – in virtue of its language part, its meaning. Secondly, the solidarity between the content and expression of a sign function is a precondition for the historical accuracy that the historical theory expounds. Thirdly, for the representation in 4.3.6 to become a proper sign function the latter's *r* must be added as in: C(ErC) r E(ErC). Fourthly, the historian is neither an augmentor nor a poet; the function that the two signs in 4.3.6 contracts is a (Glossematic) correlation; they do not contract a *both-and* relationship, but an *either-or* one (Hjelmslev 1963:deff. 26-29, and 67-68): when the historical fact exists, the narrative does not, and vice versa. It follows – sign-functionally – that the correlation is a bilateral presupposition (Hjelmslev 1970:136-137)[60].

As man's existence is a forming process, the historian's narrative is a forming process, too. The amenability-criterion rejects the katadynamic unformed-into-formed (manifestation-) process. Like the hearer and the sculptor, the historian does not form something previously unformed nor is his narrative an arbitrary rendition of amorphous historical facts. In history the theory of a narrative assumes the place and function of the nominalist's self-validating scientific principles.

Any present phenomenon, >B, is the successor of its present, >A>, and it **exists** (**out of** this >A>) simultaneously, in a structural (solidary) relationship (vertical synthesis), with the changed continuation of its predecessor, >A.

---

60 This sign-functional consequence may seem paradoxical; it means that historiographical analysis is an integral part of man's historical existence: we cannot imagine a society without historians.

4.3. A comparison between the historian, the nominalist and the 'poet'   319

What we denote by the word *a marble statue* is a structural man-made historical phenomenon, which is also a sign. The marble statue expounds *the synchronic formed substance-unformed substance* parameter against a dynamic and temporal horizon. Before the marble block was made into a statue, it was also formed. Formed matter is another horizon against which the logical parameter of formed-unformed matter is to be seen. Secondly, when the utterance *That's a marble statue* denotes, underlying our understanding of what we see, the statue in question, is a sign function that turns the denotatum into a sign of the unformed-formed parameter, which, in turn points to the origin of the statue. Our historical sense enables us to understand the simultaneous relationship: unformed matter :: formed matter, as mirroring the historico-structural process which deconstructs a given totality: 'unformed matter < formed matter' (cf. Samuels 1972:11; 114). Our historical sense interprets the process as follows: the concept of 'the unformed' is the product of katageneity; as such the unformed 'comes from' what was formed. A historical phenomenon is the inseparable merger of formed matter and the concept of *relative* non-formed matter:

4.3.7   formed matter (>A>) >   formed matter (>B)

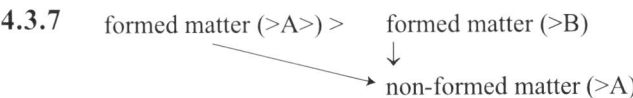

non-formed matter (>A)

The figure in 4.3.7 is so interpreted that the registration of a perceptum as a marble statue is mediated by the resolution of the perceptual merger through our historical sense; the formedness of the statue structurally endows the material out of which the statue has been made with the property of not having the same form as the marble statue. The original marble block is unformed *qua* not having the same form in relation to the statue's form. The property of being unformed is an acquired property in that no language sign denotes unformed amorphous entities[61].

The everyday language will immediately form the purportedly unformed marble block the moment this object is perceived as a marble block through, say, *There's a marble block.* The concept of the unformed (NonA) is grasped through the concept of formedness (A)[62].

Historically, description of the origin of – what is represented by – >B involves description of >A as an entity that participated in the origin of >B. Description of >A> is impossible; it is not 'there' anymore; as a present it has of necessity developed into the present. Since the historian, searching for the origin of the present, describes part of the present, >A, the sign function 'C(ErC)rE(ErC)' will be established automatically, in which >A as relatively non-formed matter is formed by the narrative into formed matter resembling >A> as closely as possible – which we know did occur. The historian's narrative of T2, >A'>, can never recreate the content of >A>[63].

---

61 Hjelmslev's *Gedankenexperiment* (1963:52; 49-50) presupposes the empirically irrelevant concept of unformed substance. Hjelmslev may have felt that because he takes great pains to emphasize that the unformed substance has *no possible existence* as a *sui-generis* entity. The processes Hjelmslev is describing are all processes involving the change of one formed substance into another one. In spite of this Hjelmslev creates *the unformed purport*, which is *extractable from* …[in casu] *linguistic chains*.

62 Logical contradiction is a linguistically created and participative relationship: A :: NonA(A).

63 There are limits to Collingwood-like recreation of the past (cf. 6.3 below).

The synchronic linguist's presupposition is that the object under investigation *may* be regarded as a self-sufficient entity, B (*étant*) of >B, despite the fact that B *est devenu*[64]. As merely 'being' the synchronic object does not point – through >A – to its origin. Sign-functionally, the relation, r, is removed: a historical narrative without this r would have been severed from its content and have become fiction. The synchronic state cannot be *pensé* in isolation from its origin, our knowledge of language, as of any other historical phenomenon, is dialectical, not only, as Hjelmslev (1928: 225) insists, analytical. And the curious paradox is that by insisting on B – as *la première réalité en matière linguistique* – the description of B carries no existence postulates with it, because B does not exist!

The synchronic point of view salvages the extensionality of the synchronic theory in that the theory 'has' its object – in so far as it exists. The historian's counterargument is as follows: an extensional theory cannot cope with the simultaneity of the katagenous synthesis of the historical phenomenon, through which the latter points to its origin: the historical narrative does 'not have' its object, because its meaning will appear contextually, in the dialectical interplay between other narratives: a narrative of the French Revolution is only meaningful as part of a larger narrative of the history of France[65].

To conclude: firstly, the notion of the everyday language being ambiguous (the synchronic linguist) and imprecise (the logician) comes from a superficial analysis of the two universals, polysemy and synonymy, and of the everyday language's existential function.

The distinctions of a synonymy field make for precision in the service of the speaker, and the polysemous word/utterance completes its meaning in the spoken chain. Trotsky was killed, murdered etc. because of the collective stratum of any historical phenomenon, which becomes a sign-functional constituent in individual connotations. As no sign has a primary, denotative or cognitive content, so the historical phenomenon has no basic or core meaning: an essence; and the fleeting nature of the meaning component of the historical phenomenon makes room for its interpretation by another 'hearer'. The conditions for the accuracy of such interpretations were mentioned above. In conclusion I shall give three examples of the dialectical foundation of historical phenomena:

1. The formula A>B exhibits this structure: we know that ME *person* developed into *parson* and *person*, and then contracted a synchronic-katagenous relationship ME *person* :: *parson*; and as the latter form is a historical product, >B, the former is, too, >A:

4.3.8   >A   ::   >B   (the synchronico-simultaneous function at T2 (*e.g.* ME *person* :: *parson*))
   >A> > >A   (the present is continued into the present under the aegis of change, T1 > T2)
   >A> > >B   (the present's development into the present; to be amplified in chapter 6).

The minor premiss indicates that, in our case, ME *person* (of the synchronic function in the major premiss) existed prior to T2. The elements of the major premiss point to something else, namely their common origin (the Jones Hypothesis). This something is a present that has become a past. And the target of the historian's narrative being the past as a present, as a successor, the historian cannot possibly catch this empirical moment.

---

64 The synchronic object (Hjelmslev 1928:227) is in Saussure's diction only *viable*, not *vivant*.
65 Being a constituent part of the present, the past cannot be removed from the present without destroying the present: the synchronic project is self-defeating in terms of empirical relevance.

4.3. A comparison between the historian, the nominalist and the 'poet' 321

2. The other type of historical event will be analyzed through the principles that have been established for language development, those of the developmental series, endogeneity, exogeneity[66] and katageneity. The historical scenario consists in the two events, the invasion by Nazi Germany of Poland in 1939 and the subsequent declaration of war by Great Britian on Germany, *qua* the respective contents of the two historical statements: *Nazi Germany invaded Poland in 1939* and *Great Britain declared war on Nazi Germany*.

The historian deals with two idioevents or two developmental series, of which one intersects the other:

4.3.9

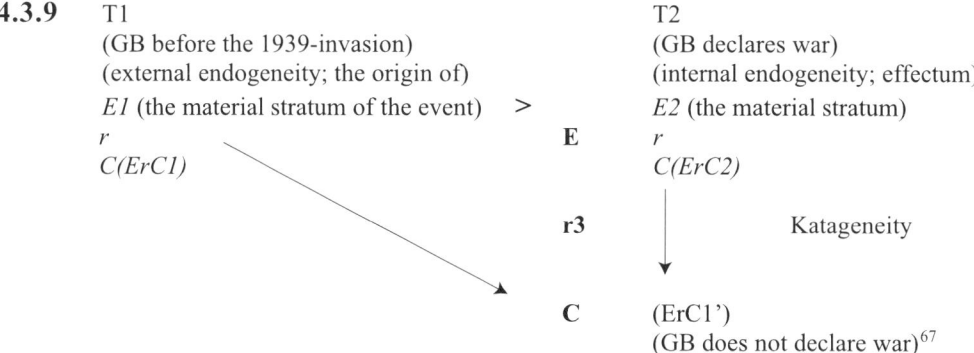

What becomes a predecessor event (E1) contributes to the creation of a new present in terms of external and internal endogeneity and is itself continued under the aegis of change (katageneity). The new present (E2) is to be understood against a situation of war having not been declared on Germany, and this is what is left of or continued of the former present (E1) into the new present. Germany's invasion of Poland is a content cause of E2 ('extrinsic exogeneity').

The sign (**Er3C**) is a complex sign whose **E** is a sign ($E2rC(ErC2)$) and whose **C** is a sign (ErC1'): **E** ($E2rC(ErC2)$) **r3 C** (ErC1'), and it points to its origin – GB before the 1939-invasion – in virtue of its content, an origin which has been continued and changed katagenously. In describing what was the case at T2 the historian has a privileged access to the historical phenomenon because of the meaning compatibility between event and narrative; the narrative becomes the expression of the contents of the historical phenomenon; and since anything can be deduced from NonA (= 'GB does not declare war'), restrictions on this contingency is yet to be stated. To underline: the historian describes how the original successor has become a predecessor.

3. In the language example the predecessor form does not disappear as the event of T1 above. It enters into a style-based synonymy relationship with the new form, while the old form remains available as a changed form; our sample data is the development of the OE plural ending *–as* into ME *–es*.

---

66 This is no proper exogenous relationship; as with loan words or extrasystemic motivation, we need a type of motivation that defines 'extrinsic exogeneity'. Like linguistics, history needs the extra-intrasystemic parameter.
67 The *raison d'être* of the contradictory formulation at T2 will appear from the semantic system constructed in chapters 5. and 6., and partly in 4.3.10 below.

4.3.10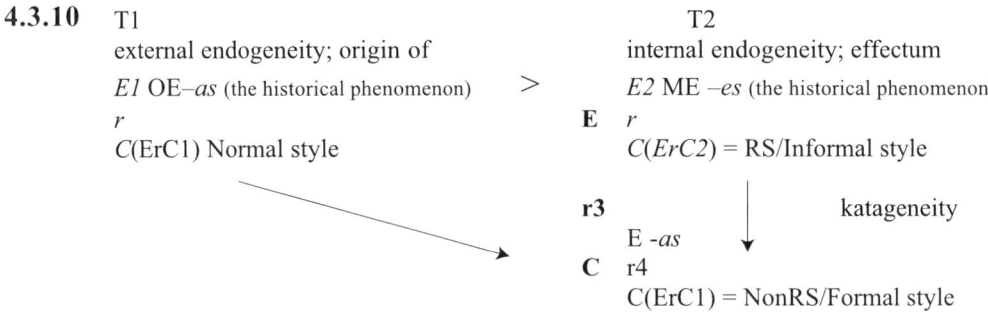

The figure in 4.3.10 says that ME –*es* is the inflectional successor of OE –*as*, which is continued with changed content in respect of its original content. It continues to exist in a structural relationship with its successor, as a choice available to the speakers then. The concept of change is a semantico-relational concept, as the process C(ErC1) > C(ErC1) represents a difference relation: C(ErC1) ≠ C(ErC1'). The development of –*as* of T1 into –*es* of T2 is not one of change, but of replacement; as such the English language *varies* from T1 to T2.

There is an alternative to the above historical interpretation of synchronic facts, namely that of abduction, and I shall briefly examine the Trotsky-example above abductively (see Hopper and Traugott (1993:39-40), who are here referring to Andersen 1973): 1st step) the 'observed result' – note the diction – is 'Trotsky, dead'; 2nd step) a general law is invoked to the effect that all men are mortal; the 3rd step) the guessing step: Trotsky was a human being! This may be abductively correct, but if we replace the initial observation with 'Trotsky, assassinated', what kind of empirically relevant conclusion, be it ever so uncertain, can be inferred? For instance that Trotsky was a counter-revolutionary; or that Stalin ordered the killing of Trotsky?

It follows from the sign-functional relationship that the elements of the everyday language consist of an expression and a content, and that only such binary elements can have a history; change will be defined as a semantic process, and as such something to be understood, not perceived. This position is so strengthened as to be embodied in this universal, which asserts the anthropocentric basis of historical processes[68]: *no historical phenomenon has no meaning; no meaning is not man-made.*

---

[68] This conclusion demonstrates the inappropriateness of using analogies or parallels to explain language change that come from the domains of natural science.

4.4. The universality of synonymy and polysemy and their *raison d'être*. Structure and polysemy are language-unifying phenomena. The language-learning child is creative and transcends into reality with her language. Analysis and synthesis are both empirically relevant. The everyday language has *Erkenntniswert*, is not an expendable cognitive means

Neither the historian's, the nominalist's nor the language historian's data are immediately given – apart from the historian's monument and the linguist's historically transmitted initial data. The historian's data are **polysemous** (or polyfunctional) and not something to be denoted ostensively, which **synonymous** narratives testify to (cf. the Trotsky scenario); and since the nominalist's datum may be analyzed into components and (synthesized) into classes (Hjelmslev 1975:7; 1963:31), knowledge of this polysemous data is couched in synonymous statements; and of two equally simple descriptions, the one that involves the simplest procedure is the correct one (1963:18; 122): the nominalist removes these data ambiguities arbitrarily: by accepting procedural levels or by a principle, respectively.

Synonymy and polysemy arise in the speech situation with the connotations of the spoken chain [abc] (plus Ccon = Primus, and Ccon = Secundus) and are as such two aspects of the same coin. **Synonymy** is involved in the everyday language's **ex**istence, and expounds meaningful distinctions inherent in the polysemous datum; it is presupposed by the *langue-parole* dualism, by syncretisms, and the synchronic linguist's definition system.

Polysemy is a property of each sign of the everyday language and appears explicitly in the Glossematic *implication* (Rasmussen 1992:211-215; Hjelmslev 1963:91-93)[69]. Hjelmslev's example highlights the interrelatedness of polysemy and synonymy and corroborates Paul's by now often quoted words, that the moment we ask certain relevant questions, we shall step on to the historical domain. In the implication *p/b* unvoiced *p* implies voiced *b*, if the former merges in a *b*-sound when appearing before *b*; /p/ + /b/ is manifested by [bb]. This mechanical variation creates synonymous variation seeing that [p] = [b] in relation to /p/, while the spoken chain becomes ambiguous: how is [bb] deciphered – interpreted or understood – *immanently*? Synchronically, it can only be done if we already know what to be looking for – 'un fait déjà constitué' or a rule to the effect that 'all [b] followed by [b] are syncretized /p/'. The historian's counter-argument to this synchronic circle is that the synchronic [bb] syncretism can only be resolved non-circularly if we know the history of the language in question, namely that the predecessor [bb] *was* [pb]. The synchronic explanation presupposes the historical split+merger-process, and if the synchronic linguist appeals to his synchronic cut, emphasizing the *all* of his rule, there is no reason why the two *b*-sounds should not be re-duced to /b/. Siertsma (1965:183-187) puts it differently: to say that (com)mutation of /b/ and /p/ has been suspended in modern Danish *top* is meaningless as the two phonemes do not contract a function in postvocalic position: -V__#.

The *raison d'être* of polysemy and synonymy will now be discussed with reference to basic processes in the language-learning child's acquisition of her native language. I shall regard this 'struggle' as existential, which means that in learning her native language the child transcends into and away from reality, in order to achieve a realistic grasp of the latter, and

---

69 An implication is the manifestation of a syncretism.

my starting point is the psychological observation that it takes time for little children to identify the same thing or person 'as the same' because in the early stages of her mental development the child cannot keep the emotional value she attaches to an object stable and fixed so that varying emotional values in respect of the same object will create different objects (Jacobsen 1976:18-19).

The language-learning child is faced with two (embedding) problems: how to make (her) *reality conform to her language* (reality is embedded in the linguistic matrix), and how to make her *language conform to reality* (language is embedded in the socio-historical matrix (Meillet)). Her synthesizing and analyzing faculties are put to good use in these processes, and this before any questions of truth/empirical accuracy or falsity appear[70].

Polysemy and synonymy are two means to the same end; they arise from the child's forays into reality in order to come to grips with it and her own existence, and both are used to salvage the child's preferred perception of a particular part of reality: the child creates a language-tied conception of reality, in spite of reality's obvious differences; conversely, she must obey the physical, social and psychological requirements, facts and conventions that she cannot control or form, in spite of what her language tells her.

Reality may impose on her a division that is in conflict with her linguistic proficiency, and similarly her language may impose on reality syntheses that are in conflict with reality. Furthermore, the child has to develop her historical sense; at this early stage in her mental development (Jacobsen 1976:27), the child's mental make up appears to lack the temporal sense that enables her to grasp the process of 'a formed phenomenon having been formed into another phenomenon'; secondly, totality as a hierarchical class is a late development, and the infant baby is unable to distinguish between or analyze a given empirical situation into herself and her surroundings. The 'I :: non-I' or the 'I :: you/it'-relation is not a priori at her disposal (Cassirer 1953:272). During this development her language never fails her.

However, she will *become* able to *distinguish* – and *identify*. Only relatively late in her development does she realize that, say, her mother[71] is one and only one person, as the latter may have been both not so nice to her and nice to her; one and the same person or object assumes two different identities in the eyes of the child. In other words, the language-acquiring child <u>distinguishes wrongly</u>: she has not grasped the synonymy-function, as she creates two entities on the basis of the polysemous (polyfunctional) nature of, in our case, her mother. Put positively: the child creates a synonymy field, as she separates the good mother from the evil mother; and what is more her language will accommodate her socio-psychological needs, creating the preferred perception of reality. <u>Therefore she makes her reality conform to her knowledge and command of her everyday language</u>. The child transcends away from (so-called objective) reality in order to transcend into reality (in a crude term, to survive).

The child <u>identifies wrongly</u> – as the *lamp–moon* example in chapter 3.2. showed. She makes her language polysemous. This polysemy-making is needed when the child is going

---

70 Aristotle (*Inter.* 16a) saw this: 'for combination and division are essential before you can have falsity and truth' *peri gar synthesin kai diairesin esti to pseudos kai to aleethes*. This position is continued in his view of the word as both a totality and a part (16a20-22). – Like the savage mind, the mental activity of the language-learning child is both analytic and synthetic (Lévi-Strauss 1962:290), characterized by creation of internal coherence and constant expansion (p. 287).

71 The example is Erling Jacobsen's.

4.4. The universality of synonymy and polysemy and their raison d'être

to use a word correctly, *e.g. a mother*, so that the child will not go on distinguishing wrongly. Similarly, the created synonymy field is a precondition for the child's eventual ability to distinguish correctly. In spite of identities (synonymous properties) the two words *a lamp* and *the moon* are different in English. <u>The child makes her reality conform to her knowledge of her language</u>. Again, she transcends into reality by transcending away from it – perhaps for reasons of expediency and economy, respectively.

The child gradually creates a language that will enable her to make the proper distinctions (separation) and identifications (attachment). If her linguistico-mental development was merely a coming to grips with how to differentiate (analyze), the everyday language would certainly oblige her, it being the perfect match for precise observation of differences through its changeability, with the result that the empirical world would seem to be populated only by differences[72]; her language would become univocal. Her wrong identifications would then be incomprehensible. Conversely, if synthesizing was the sole process the child was to learn to master, her wrong divisions would be incomprehensible.

However, the child also makes correct identifications, (eventually) identifying her different 'mothers'. Her mother is one: this identification process is not one of generalization or subsumption, a whole created on the basis of more (manifesting) entities, but the process creates a **synthesis**, a historical phenomenon, existing *through time*, which can then be denoted by the so-called same sign. A sign's polysemous nature is amenable to the equivocal-polyfunctional existence of 'extralinguistic' phenomena: the child's grasp of reality is not a product of the analytical part of her mind, so that understanding reality is not an analytico-deductive movement from classes to components of classes – a fact that abductionists will not disagree with.

A factor in the child's eventual ability to distinguish between the two words *a lamp* and *the moon*, thereby resolving the created syncretism in *a lamp*, is her sense of history: one meaning of a form will gradually 'develop out of' another meaning. Both the moon and a lamp are 'functions' that exist simultaneously, and the child learning *English* is unable to identify the two (now wrongly) through time – the way she came to – correctly – identify her mother's contrasting functions. Instead she makes linguistic capital of the abovementioned ability to distinguish, now correctly, by means of two words.

Reasoning in time – dialectical reasoning, synthesizing – is necessary in resolving the child's wrong identifications and wrong divisions. <u>Reality was not conformable to her language or the child could not maintain her knowledge of her native language in the light of reality</u>. To this extent reality has an imposing influence on the everyday language, with the corollary that a given speaker or group of speakers are unable to continue the synchronic state against the influence of its circumambient reality: reality does not remain conformable to a given language state (*un fait déjà constitué*)[73].

It is now apparent why homophones, (*to*) *let*, (*to*) *lie*, (*to*) *bear* in modern English may

---

[72] A type of argument that Paul Kiparsky also availed himself of in his attempt at rationalizing 'The Great Vowel Shift'.

[73] Ironically, this is also due to the fact that the everyday language changes: the everyday language – because it changes – does not remain conformable to reality. The only conclusion seems to be: reality and language see to it that they remain mutually conformable through their dynamic interaction.

create problems: no child's developing sense of history and reality is able to make the respective denotata cohere by means of historical links or appear functionally related over a period of time; on the other hand, the distinctions in the empirical world are not able to impose themselves on the English language. Saussure was right in claiming that the formula A>B has no psychological reality; the child does not understand the homophonous contents of the verb *let* through reference to the sound change that shortened some ME *e:* into *e*. But this only obtains for the expression side of a language; from the non-psychological reality of some A>B we cannot infer the absence of man's historical sense or of the psychological reality of empirical processes. The rise of homophones is but one example of how the everyday language imposes itself on reality, or of transcendence away from reality[74].

In sum, polysemy is the historically based identification of different meanings (Cs) in respect of one form (E). Synonymy is the identification of different forms (Es) in respect of one meaning *qua* an extralinguistic 'function'. The latter process follows from the child's inability to identify meanings or functions in respect of only one form, through her knowledge of her language. The child capitalizes on the above ability to create distinctions in (her) reality. The difference between the pairs *great – big* and *a lamp – the moon* is one of degree[75]. The processes behind the rise of both pairs have the same objective, that of precision; the <u>identification of functions</u> is a process that makes for precision within a continuum, with the identification – as a type – of the child's mother at the top resulting in a minimal potential for ambiguity (and subsequent synonymy), while other identifications (*e.g.* of the functions of a lamp) result in the greater potential for ambiguity (and subsequent synonymy), the more difficult it is for the child to identify functions.

The <u>differentiation of functions</u> is also a process of precision-making within a continuum, with the linguistic differentiation of a lamp from the moon as a process with a relatively high potential for ambiguity, while the distinction between the functions of *big* and *great* will be less prone to ambiguity.

Why does the child synthesize certain synonyms into one word, discarding a word (or construction)? Why does the child analyze 'polysemes' into two words, creating two words? Why does the child appear not to do anything to certain polysemes (*let*) – creating more words – or to certain synonyms (*great*, *big*) – discarding words?

These are questions that have been constantly asked – sometimes tacitly – in the literature. In my view they have not been asked as equilogical and equiempirical parts of a whole, and have therefore been answered separately – and as such each of them may suggest a certain *mystique linguistique*[76], reinforced by the dominant metaphysics and mode of reasoning

---

74 The former process seems to be more dominant than the opposite one; but this is probably due to the choice of examples; when we come to the words denoting the historical phenomena, in the Løgstrup sense, the influence of language on reality is presumably far greater.

75 We are now back to the previous discussion about the inability of synchronic linguistics to distinguish nonarbitrarily between insignificant and significant contents, between (in)significant differences. The commutation test has no criterion for the essential distinction between the respective contents of the word pairs *big – great* and *a lamp – the moon*: in what cogent sense is the latter content difference cognitive/intellectual whereas the meaning difference is non-intellectual/non-cognitive in the former?

76 "Why do we burden our brain with a pile of synonyms?" – "Why do we put up with polysemy?" – questions whose general horizon is negative, because we assume that it need not be like that.

4.4. The universality of synonymy and polysemy and their raison d'être

in our Western Civilization. One School – the Mechanical – highlights polysemy to the exclusion of synonymy, another School – the Functional – synonymy to the exclusion of polysemy, the synchronic conception relegates both universals to the status of linguistic epiphenomena.

As for the makeshift, temporarily created synonymy field, the child does not create different histories to become contents of sign functions with the synonyms in question, because she finds no socio-historical and linguistic support for such histories – her eventual grasp of reality is not conformable to the continuation of such synonyms. The child resolves her makeshift polysemes because the resultant (two) words involved have different histories *qua* contents supported by the speech community. Despite the impossibility of identifying the histories of the meanings involved in a homophone, the child does not find linguistic and 'extralinguistic' support for the resolution into more forms of the different histories that constitute the meanings of the homophone.

In the preceding discussion I have upheld an explanatory parallelism between identity and difference, polysemy and synonymy as well as contents and functions, and two things seem clear: synonymous forms are discarded in relation to the eventual rise of historically based polysemes (the child's mother is a historical phenomenon). In this process the extralinguistic functions play a central, imposing role. Where two functions cannot be identified, synonymous forms are learnt; this process embraces such differences of degree as our two pairs of words above exemplify, so that the greater the degree of identity, the more likely it is that a form will be discarded. As regards homophony, the imposing role of language is such that it makes two non-identifiable functions or histories merge in one form. The continuum in 4.4.1 summarizes the inalienable interaction between language and extralinguistic reality:

4.4.1

Eventual imposition by		Phenomena	Functional similarity,	(Non)resoluble
reality	language	*qua* syncretisms	degrees of …	syncretisms
****	*	*a mother*	f1 = f2	Nonresoluble
**	***	*big/great*	f1 ≈ f2	*Resoluble
***	**	*a lamp/the moon*	f1 ≠ f2	**Resoluble
*	****	*(to) let*	f1 >< f2	***Resoluble

To the language learning child her mother's contrasting ways of behaviour appear to her with emotionally incompatible values, thereby corresponding to the contents – f1 >< f2 – of the homophone, a syncretism easily resoluble into a synonymy field. Her need to distinguish matches the ease with which her language allows her to do so. Conversely, reality will impose itself on her language; there will always be a large element of 'similarity' in her mother's behaviour (f1 = f2). On the other hand, the dissimilarity in the contents of the homophone makes it easy to resolve, to which resolution the respective synonymy fields of the everyday language will come in handy – *e.g. prevent* and *permit*. But eventually, the everyday language will impose itself on the child, continuing the homophone.

The similarity of the moon's and a lamp's functions conditions the language merger, but the different functions – f1 ≠ f2 – will eventually come out so different as to make the child's merger seem a makeshift one. Her need to distinguish may not seem decisive, but reality and language seem here to lend mutual support to the resolution.

The similarity[77] of the contents – f1 ≈ f2 – of *big* and *great* conditions a merger, and there may be no great need for a resolution, and the contribution of the everyday language to the resolution of the (potential) polysemy must be pretty great.

The empirical world does not invite its inhabitants to make unmediated generalizations; subsumption has no psychological reality to support its empirical relevance. The child does not create her reality, linguistic and extralinguistic, through the subsumptive generalization and reduction of differences into identity, even though some of her identifying processes are reminiscent of the PreSocratic Monists. Her generalizations and identifications are psychologically real, grounded as they are on her developing sense of history and reality.

The present outline of a theory of polysemy and synonymy against the horizon of language acquisition in which the child's creative, historico-structural, faculties have been taken seriously, is to be seen against the static conception that we have seen from Whitney and Saussure to the present: the arbitrariness of the sign enables the child to do whatever she wants to do with her language, but she cannot do so owing to the imposing power of social convention and the law of tradition. This alternative implies that, when learning her native language, the child becomes the passive recipient of bits of information (from her parents) to the effect that, say *a lamp* does not denote the moon – the content of *the moon* is part of the content of the other word. Where does this leave the child's creativity? Corrective information or instruction is not the decisive factor in the child's transcendence into reality, in the growth of her historical sense and her existence.

To understand the polysemy field of a word requires the speaker's historical sense; different, even contrasting contents may be identical (in relation to a form) because there are historical links between them; and there is nothing to disprove the possibility that the language user is capable of reading historical links into a given polysemy field so that one content has developed out of the other, and the greater the meaning span, the greater our efforts to bridge the gap, but the easier it is for the polysemous form to preserve its self-identity[78].

If the language sign were univocal, it would not be able to continue its existence under the aegis of change, because it would disappear when it became synonymous with another sign or polysemous.

The same – *mutatis mutandis* – does not obtain for the synonymy field, be it of the *lamp-moon* type or the *big-great* type; no expression, say, *great*, develops out of another one, *large*, and different expressions do not cohere relationally through historical links between them. It is significant that the split types *person/parson, also/as* in English do not constitute synonymy fields, except for a relatively short period after their creation (*pace* the indefinite article *a/an* and shortenings like *phone/telephone*; but cf. *bus/omnibus*). The psychological reality of synonymy is not in doubt: two forms are synonymous in virtue of their respective polysemies, a fact open to immediate observation; but the child's historical sense is not involved here as in the case of the learning of the polysemous word.

---

77. I am assuming a certain time gap in the child's acquisition of all the nice distinctions in a synonymy field. This also applies to the homophone, all of whose contents may be learnt at different stages of the language-learning process.
78. This historical sense is seen from the expression's point of view in our ability to understand archaisms.

4.4. The universality of synonymy and polysemy and their raison d'être

A synonymy field constitutes a vertical synthesis in the L-continuum with whose elements the word selected into *parole* enters into a structural interplay on the basis of polysemies. The rise or the loss of a synonym has overt structural consequences for the field in question; both processes occur in respect of the individual word, which either loses a content, leading to differentiation, or adds a content (Samuels 1972:65-67)[79]. We also saw how resolution of syncretisms by the language-learning child involved synonymy: the syncretism in *let* was easy to resolve because of the existence of other signs, such as, *prevent* and *permit*; the *lamp*-syncretism was also easy to resolve in the same way. But the polysemy of the *mother*-type can hardly be resolved into more signs[80], while the fourth type, the possible syncretisms in either *big* or *great*, was envisaged as resolved through time when the child learns the meanings of the words contextually.

Saussure was right stating that no one can change her or his language *qua* a supraindividual entity. My point is: why should a child or for that matter anybody else do that? We can make reality conform to the everyday language[81]. Saussure's starting-point was biassed: the child's objective is neither to change nor to fix her language but primarily to acquire a realistic conception of reality which enables her to exist as a historical person. In the below 4 summaries of the scenario in 4.4.1 I shall underline the dialectical element – a temporal principle – in both the identity-making and differentiating function of the everyday language. My aim is also to generalize these processes, ideally spanning the language-learning child's efforts to exist and the scientist's efforts to 'exist scientifically':

1) *The expansion of a form through extralinguistic factors (acquired knowledge of reality): the everyday language is embedded in a historico-social matrix:*

The polyfunctional self-identity of the child's mother could not be synthetized unless through time or the present's constant development into the present with the linguistic result that a (synonymous) form[82] is discarded and a meaning is added, because the child discards a form and adds the meaning of this form to another (synonymous) form. – The abolition of a synonymy field through the creation of a polysemy field: extralinguistic distinctions have been syncretized.

A historical narrative or a theory of X expands its scope so as to include the content of another (synonymous) one. Originally, the two narratives/theories were not synonymous[83].

2) *The restriction of a form through linguistic factors (acquired linguistic knowledge): reality is embedded in the everyday language (in a linguistic matrix).*

---

79 The loss of a form may also occur, *e.g.* the ME -*eþ* and –*ind(e)* endings change the contents of ME –*es* and –*ing(e)*, respectively.
80 This does not mean that the word has no synonyms: *ma, mum*; *a female parent* (the semanticist's definiens!).
81 This is the Realist's dynamic version of the arealist's static theory conception: as a form the everyday language contributes to the constant creation of man's historical existence.
82 Whatever makeshift forms the child may have created.
83 The proto-type of this reduction is transcendent theories, theories that aim to re-duce a discipline to another one.

The polysemous self-identity of a sign could not be resolved unless through time with the linguistic result that a meaning is discarded and a form added, because the child discards a meaning and adds this meaning to another (synonymous) form. The creation of a synonymy field through the abolition of a polysemy field: a(n extra)linguistic distinction has been continued.

A historical narrative or a theory of X is added so as to produce more knowledge of X. The narrative or theory grows more detailed or the analysis in question proceeds further[84].

3) *The restriction of a form through extralinguistic factors (acquired knowledge of reality): the everyday language is embedded in a historico-social matrix.*

The polysemous self-identity of a language form (*lamp*) could not be resolved unless through time with the linguistic result that a meaning is discarded and a form is added, because the child discards a meaning and adds this meaning to another (synonymous) form. The abolition of a polysemy field through the creation of a synonymy field: a linguistic distinction is continued.

A historical narrative or theory Y is added so as to produce more knowledge of reality. The scope of a theory is so restricted as to create a new object (scientific) with its own theory[85].

4) *The expansion of a form through linguistic factors (acquired knowledge of language): reality is embedded in the everyday language.*

The polysemous self-identity of a language form (*to let*) could not be synthesized unless through time with the linguistic result that a form is discarded and a meaning is added, because the child discards a form and adds a meaning: the creation of a polysemy field through the abolition of a potential synonymy field[86].

(1) and (3) are dualities: in the former the child cannot maintain – create – through her language two separate extralinguistic entities (an extralinguistic 'split'); in the latter she can-

---

84 On the *big-great* analogy this is an instance of what may be called the growth of knowledge outside a falsificationist conception; another (a new) point of view may be applied to the datum in question. – The continuation of the synonymous indefinite articles *a/an* is an example of the dominant influence of intralinguistic factors, whereas that of *person/parson* may have had strong extralinguistic motivation with the continuation of *also/as* (and *so*) being motivated by both intra- and extralinguistic factors.

85 Linguistics abounds with such theories, synchronic linguistics, sociolinguistics, psycholinguistics, universal linguistics, general linguistics, contrastive linguistics, not to mention particular ones: systemic grammar, functional grammar, generative grammar, transformational-generative grammar, binding-and-government grammars, to which must be added Luick and Jespersen's versions of 'the same' as well as John Anderson's and Paul Kiparsky's.

86 The Church's monopoly on (knowledge of) reality in the Middle Ages could exemplify this homonymy, expounded by the Bible's status as reality's be-all-and-end-all. The synchronic conception or synchronization of language development is also an instance hereof.-Thales's *All is water* is a further example. – Homonymy may be defined as follows: the creation of a polysemy field through the abolition of what is not a linguistic synonymy field. The above shortening of ME /e:/ made two OE words such as *lettan* 'to hinder' and *le:tan* 'to permit' merge in LateME.

4.4. The universality of synonymy and polysemy and their raison d'être

not maintain – create – through her language the (seeming) unity of reality (an extralinguistic 'syncretism'). (2) and (4) are dualities: in (4) the child cannot maintain two separate extralinguistic entities (or functions) because of her language (a linguistic syncretism), and in (2) she cannot maintain the unity of (her) reality because of her language (a linguistic split).

The *raison d'être* of polysemy and synonymy is grounded on the relates of the sign function. The role of the form (E) of a polysemy field is to preserve and contain a span of meanings (Cs), thereby ensuring the continuation of the sign's self-identity and individuality against other forms; to this end the structural tenet *tout est différence* is meaningful, and the polysemy field expounds the notion of differences in identity.

Similarly, the structural context-principle, which we see in a synonymy field, expounds the notion of identity in differences (*tout se tient*). Both the structure of a synonymy field and the form of a polysemy field have a unifying function. The former preserves and controls the mutual coherence – against the horizon of separateness – of a number of forms, thereby balancing the individuality of the single signs. In (1) above the imposing unity of reality asserted itself over and above the differentiating potential of the everyday language, and it was matched by the unifying potential of it (polysemy). In (2) the unity of reality could not assert itself over and above the imposing differentiating potential of the everyday language, but the structural principle behind the synonymy field provides the everyday language with unity. In (3) the imposing multiplicity of reality asserted itself over and above the unifying potential of the everyday language, and it was matched by the differentiating potential of it. In (4) the multiplicity of reality could not assert itself over and above the imposing nature of the everyday language, but the structural principle controlling the spoken chain provides the everyday language with multiplicity.

Structure and polysemy, the linguistic correlates of the notions of identity in differences and differences in identity, are language-constitutive properties through their respective totalizing and individualizing functions. They are two non-parallel constitutive aspects of the same phenomenon.

The child transcends away from reality. She cannot identify the unpleasant, unkind aspects of a kind and pleasant object – her mother. Such differentiation is a retreat from reality which quite obviously is an attempt by the child at salvaging a more realistic, the preferred view, of her mother. By differentiating – dividing, analyzing – her mother she safeguards the good mother from the 'evil' one[87]. What seems to be a transcendence away from reality is but a devious way of transcending into reality. By means of the synonymous poten-

---

87 This is the psychologico-realistic basis of the classical Dualism. The Dualism is a view of reality that splits reality up into a preferred (essential) and a not so desirable, an expendable (accidental) part. Plato made reality conform to his theory of reality or simply <u>to his language</u>, in which the basis of his ontological dualism inhered. Similarly, the wife in Løgstrup's story (1976), tells a lie, because she knows the truth – her husband's hiding-place. Thereby she may save her husband from the secret police. Her telling this lie is an obvious transcendence away from reality/the truth, but is *really* a devious way of transcending into reality, thereby creating the reality she prefers. – Hjelmslev's working hypothesis (1972a:34) is also an expression of a preferred reality, couched in the postulate of it being an epistemological precondition for the scientific study of language. But in the case of Plato and Hjelmslev (and Thales) reality fought back, so that the preferred unity (polysemy) turned out to be a homonymy (4 above) that had to be resolved into a 'synonymy' of type 3 above.

tial of the everyday language the child separates what she will later on unify, the contrasting properties or behaviour of her mother. The philosophical version is more sophisticated: Plato creates an empirical scenario in which the so-called real world is seen against the horizon of the ideal one in order that the latter may assert its autonomy and the former appear in its biassed-preferred garb: "This sensuous chaos cannot be reality!" Plato did not arrive at an empirically relevant synthesis, which the child did, but drew the only logical conclusion of this unbridgeable gap separating his ideal from the empirical world. Hence his transcendence through participation[88] must remain obscure, a less than whole-hearted compliance with or anticipation of the expected, but to him irrelevant question of how to make the two worlds cohere. While the child created denotators, Plato created a potential denotator, which 'did not have its denotatum'[89]. However, the unity of reality did impose itself on Plato's ontological dualism, ontological dualisms becoming dated metaphysics.

All transcendence is transcendence into reality – we cannot but **ex**ist. Transcendence away from is a process equiempirical with, but logically different from it, and 4.4.2 sums up the (child's) four types of transcendence above:

4.4.2   transcendence away from(transcendence into) :: transcendence into[90].

In (1) and (3) the child transcends away from extralinguistic reality, in (2) and (4) from linguistic reality; but as the everyday language is constantly *en ergeiai*, both 'realities' will assert themselves. And even extreme formalism, the extremest nominalist and staunchest idealist, exists under the *energy* of the everyday language, it being – with all its purported weaknesses – the ultimate metalanguage (Church 1965:1). So if the nominalist's theory is a (preliminary) transcendence away from reality, it is but a devious way of transcending into its datum, waiting to become selected by a substance, but this selection cannot occur without the involvement of what it tries to do away with, the everyday language. As a consequence of 4.4.2 – if we want to keep the parallelism between the processes behind the language-learning child's acquisition of her language and the scientific process, we must also say that

4.4.3   falsity(truth) :: truth

i.e. a false statement exists at the mercy of truth. If somebody says: "That's not a cow ruminating over there – it's a horse", the false statement exists against the horizon of truth. Historical accuracy may be less than an absolute concept, but it is understood against the horizon of truth merging with the fact that something did happen: we know that something did occur in the present, but not only if a specific man-made epistemological, theoretical or methodological criterion is satisfied, and the only precondition for this knowledge is existence as it is: our Selbstbewusstsein tells us that we exist as formed entities, and our sense of historical development tells us that as a formed entity we are bound to constantly **ex**ist out

---

88 Empirical world(ideal world) :: ideal world.
89 See Gellner (1969:1-3; 58) and Cassirer (1953:271).
90 When the poet (Keats, *Ode to a Nigthingale*) *leaves the world unseen* in order to *fade (...) away into the forest dim*, his transcendence is but a statement on reality.

of another formed entity (cf. 4.3.7 above). In a word, the nature of man is the implementation of some kind of 'innate' principle of continuation, change[91] and changeability, and this principle is our historical sense, which is inalienable from the everyday language. As such the everyday language is able to convey more than dated theories, distorted views of reality, but a real premonition of what reality is. Even Hjelmslev is obliged to accept this:

Glossematics is a nonmetaphysical theory (Hjelmslev 1973:104). A metaphysical theory contains undefined terms. The Glossematic theory contains undefined terms (pp. 105-106). The impossibility of creating a nonmetaphysical theory is the indispensable foundation of theory-construction, implying the everyday language. Hjelmslev's elegant attempt at solving the inconsistency remains elegant and is no solution: the theoretically undefined terms will be defined as data for the theory; furthermore some definienda have two definienses (note the ambiguity of data), because certain well-defined definienda may also be defined by the theory – semiology's analysis of the constituents of the definienses in question[92]. To sum up: *a scientific theory consists of undefined and defined terms, the former being words of the everyday language; assuming that a scientific theory yields knowledge of (a section of) reality, the everyday language yields knowledge of reality*. Q.E.D.

That the historian does note write his narrative on the basis of formally well-defined sentences, on the basis of a hypothesis about there being a system behind any process, is no argument against the scientific status of his work. *Das Erkenntniswert* of the everyday language lies in the fact that it has privileged access to reality – and reality to it – because of its **ex**istential function.

---

91 This is an Aristotelian stance (Lear 1988:57-58).
92 In other words, the theory creates a synonymy field expounding the polysemy of the definiendum.

CHAPTER 5

Towards the theory and its static application. The *break* concepts are introduced. The interidiolectal network and the theory. The developmental series

> *For he, Adeimantus, whose mind is fixed upon true being, has surely no time to look down upon the affairs of the earth (...); his eye is ever directed towards things fixed and immutable, which he sees (...) all in order moving to reason.* Plato *Republic* 500c

INTRODUCTION

IN 1914 LUICK (1964:7) wrote that it is *gebieterisch* that the study of language transcends the stage of description where only *mehrere Entwicklungsstadien* and their *Verbindungslinien* are described; the new stage consists in giving *einen Einblick in den Werdegang einer Sprache*. The crucial term here is *der Werdegang*, within which there will be some room for the state concept (*ein Sprachzustand*, p. 5). The problem with Luick's *Historische Grammatik* is that in it we do not find definitions of what he means by *Werdegang* and *Sprachzustand qua eine Entwicklungstufe*. We can only infer from his explanations what this – in Danish – *bliven-til* 'becoming' (*devenir*) may be. I have called it – in English – **existence**, and somewhat redundantly *historical existence* in contrast to the existence of objects of nature. In this part of my essay I shall set up a theory of sound change which can cope with the simplest type of sound change that we have, the isolative type we see in *e.g.* OE (Southumbrian) *a:* > ME *ɔ:* and OE (Northumbrian) *a:* > ME *a:*, changes that our tradition has not dealt with to any serious extent and, as regards continuation, which have not received any theoretical treatment for the simple reason that history is viewed as differentiation processes only (cf. Labov's Historical Pardox). The theory will establish *Verbindungslinien* between both types[1]. Conditioned sound change was defined in chapters 1 and 2. The state concept has some empirical relevance in that it implies the notion of a break, a notion presupposed by our tradition's 'synchronic cut': when the synchronic linguist makes his cut – a process that we often – always? – look in vain for in synchronic grammar books – he creates a break between two states – stages of the same language. This break is not a gap of the black-hole type, but must possess some empirical relevance. And in order not to become bogged down in Zenonian thought-experiments about the irreality of movement, I have introduced two relational concepts, **a break-before** and **a break-after**, so that the dynamics

---

1 Luick does not explain continuations, at all; and his *Verbindungslinien* remain undefined.

5.1. *Last words on the synchronic systems concept*

of existence can be expounded by the dynamic concepts (processes) **to break into, not to break into, to break away from, not to break away from**: these concepts also enter into the determination of e**xistence: the present's constant development into the present, thereby creating the past** *qua* **the present-turned-into-the-past-by-the-present**.

Before this we need to tie up some loose ends about the empirical irrelevance of the systems concept as a synchronic, supraindividual entity, and in particular how it is constituted.

5.1. Last words on the synchronic systems concept: it is an *explanandum*. The concept is tentatively replaced by collective orientation. The constitution of the synchronic system; origin is common origin, structure is history. Synonymy and polysemy are further generalized

Following the philosopher (Wittgenstein 1963:§§146,527,531) the system is an empirical phenomenon that enables us to understand the everyday language; following the cognitive-oriented linguist and semanticist (Bierwisch in Pinborg 1976:254-256) the system is constituted by the rules that enable us to understand a sentence: a system is an entity or some rules. Following the synchronic linguist (Sørensen 1958) the system is the metasigns that denote the signs of the everyday language. Following Hjelmslev (1975:8-16) the system is introduced by definition (deff. 38, 39) through the articulation of the class of objects (Gga) independently of the class of functives (Ggb), between which two component classes (of GgA) there is a purely arbitrary relation: and in *Prolegomena* the system was a part of a hypothesis. In atomistic grammar books like Bache and Davidsen-Nielsen (1997), Kuiper and Allan (1996[2]), the system plays no role at all[3], and in *The Prologue* the very number of (phonological) systems defeat the empirical relevance of both the systems concept and this type of formalism.

The confusion and profusion of candidates oblige the historian to ask the question: is the synchronic system a man-made construct – and is the everyday language therefore a monoplane phenomenon (Sørensen 1958) or a dual phenomenon (cf. Hjelmslev above) – or is the synchronic system an empirical phenomenon, which mainstream linguistics from Saussure onwards suppose, so that theoretical linguistics is a return to ontological dualisms (cf. Samuels above)? I shall approach the problem from an angle that considers both views.

The function of the synchronic system 'behind' the everyday language is to *rationalize* the class of utterances. A spoken chain is meaningful if and only if it complies with the rules determining the structure of the system. This rationalizing function is based on the supraindividuality of the system (Meillet 1951:72-73; 1975:16-17). The rationality of the individual speaker's utterance is guaranteed by the system: rational behaviour in and through time is the systems-conditioned product of the individual: the system like a mathematical calculus

---

2 "(...) specific topics are as far as possible dealt with *separately* and *exhaustively*." (p. 1; my emphasis). This is a structural contradiction in terms. A phenomenon lifted out of its context is removed from the mutual interaction between it and other phenomena and as such cannot be described exhaustively. Thus the authors must suppose the falsity of the *tout-se-tient* principle.

3 The 'systems' in suchlike grammar books are the brain-children of their respective authors. This is also characteristic of historical works, *e.g.* Lass (1987) and Jones (1989).

determines the allowed, allowable and disallowed utterances; but since the system is the necessary condition for neither the spoken chain nor variation and change, it is only one of several preconditions.

The latter two facts have unpleasant consequences ignored by the synchronic tradition: firstly, any rule infringement entails a disallowed utterance; hence language change is rule-breaking and change becomes incomprehensible – and mysterious because any language undergoing change does preserve its meaning-carrying ability intact. The escape clause is that change is manifestations of the system's allowable potential, but then change is no longer mysterious, is no longer rule-breaking. This means, in turn, that East Anglian – as well as many American dialects (Hughes and Trudgill 1987:16-17) – which does not have the present tense ending, -es, has exploited the 'allowable' alternative by the system: *she walks* : **she walk* > *she walk*. In standard Southern British English the latter form is therefore also an allowable form, and is not forbidden by this system: what happens to the synchronic grammarian's rule, according to which the third person singular of the present tense requires the *–es* ending in standard English? The concept of rule-breaking entails a shift in the *burden of comprehension* (Gellner 1969:155), away from the 'already established' systemic entities to the spoken substance of the communication itself, which is (made) intelligible by the hearer's (and the speaker's) dynamic knowledge of his or her own language; this knowledge includes what will be called **attitudinal** systems (cf. above 0.5).

The existence of the spoken chain is therefore characterized by multi-conditioning; since a given system is one of more *parole* conditions, other *parole*-determining factors/systems may be envisaged besides the factor that leads to an absurdity, namely the factor logically represented by nonA (non-systemic): how can *la langue* and *la parole* constitute a whole *le langage* (Saussure 1972:24-25) if *la langue* has all the properties (A) that *la parole* does not have (nonA), and *vice versa* (pp. 37-38)[4]?

The system is the *raison d'être* of the spoken chain. No synchronic theory explicitizes the rationality of the system. What guarantees the a priori rationality of the system — *sub specie temporis* the Ptolemaic world picture was not rational – or was it? Saussure touched briefly on the question of language rationality. Hjelmslev pushed the problem one step away from the everyday language, trying to establish criteria for the rationality (the scientific status) of the study of language with the corollary that data is rational if it can be denoted by a man-made theory[5].

Our tradition has it that historical rationality is no independent apriori, whereas the system is the reason for its own rationality; the causal principle of the dualism is **participative**: the system is (1) the cause of itself; (2) a precondition for the spoken chain; (3) a precondition for the historical development of the everyday language. As a cause, the system is **bi-uniformal**: while producing *parole* manifestations under the aegis of constant variation, it

---

4 It is not only language change that creates paradoxes and inconsistencies; the synchronic conception also has its share once the logical geography of it is investigated; furthermore, synchronic linguistics has this alternative: a product (the spoken chain) is rational because it comes from a rational system or it is rational because it creates a rational system; the latter alternative is not incompatible with Saussure (1972:37).

5 In many linguistic treatises prediction seems to have replaced the Glossematic Empirical Principle, see Hopper and Traugott (1993:63) with further references. In either case scientific adequacy has replaced empirical relevance.

5.1. Last words on the synchronic systems concept

continues and upholds itself; and conversely, while changing itself (A>B), the system produces unvarying *parole* manifestations: /A/ [a], [b] > [b] /B/. However, the existence of no everyday language leads to a preexisting supraindividual system; the rationality of historical development is not explained by the postulation of what is to be established; secondly, explaining a part, *la parole*, of a phenomenon (*le langage*) by means of the other part, *la langue*, (see Hjelmslev 1973:120) of that phenomenon is tantamount to committing the Phenomenal Errror (Topolski 1976:140), or in the positivistic tradition of conflating cognition – the explanatory power of the system – and data – the (communicative) function of the system (pp. 133-134)[6]. In sum, synchronic linguistics has a data constitution problem.

When we speak, we *move from* the system, making choices (Coseriu 1973:95). The very nature of the implementation of the dualism is that a choice between at least two non-exclusive possibilities – synonymous variants – will always be the starting point; one of these variants will be preferred. Systemic change is *movement into* the system and implies a choice, no matter whether the process is envisaged in the Functional terms of redundancy and synonymy or in the Mechanical terms of ambiguity and polysemy. Synchronically and statically, the speaker's choice is between 'already developed'- preferably uncreated – supraindividual (preexisting) possibilities, the rules and the systemic structure (Peeters 1995:62) of which determine his choice, whereas the dynamic point of view has it that a speaker so chooses that he is at liberty to produce a different utterance from the one that he actually produces. This liberty is – my argument – neither arbitrary nor contingent, but restricted by the speaker's **sense of history**, expounded by the historical theory behind the involved developmental series: the speaker considers both the facts of the past and the fact that the present utterance will contribute to the continued existence of (the L-continuum of) his language. If a synchronic, preestablished system is involved in the process, it is the synchronic linguist's task to prove its role[7]. Let's therefore use the system, which the synchronic point of view cuts from its anthropocentric moorings, in the explanation of the history of a system (cf. 0.3.2, 0.3.3). The system of King Ælfred's West Saxon dialect is different from the system of the West Saxon language spoken two hundred years later. This supraindividual entity must have made certain choices in Ælfred's days, when it manifested words such as *ham*, *stan*. The system manifested during the 11th and 12th centuries chose /ɔ:/ as the new systemic (collective) choice, *hom, ston*. Why and how did the respective systems do this ? How did the latter one choose between /ɔ:/, /a:/, /a/, /ɑ:/?

From the vantage point of the so-called completed change, the preferred variant emerges virtually by default (Samuels 1972:34-39): a systemic element appears as if through an elimination race when competing variants are no longer continued. – The early system continued /a:/. This process occurs in the language of the individual and at the level of the group. The

---

6   As was indicated in the *Prologue*, grammaticalization-theoreticians labour under the same dilemma (Hopper and Traugott 1993:xv-xvi).

7   Current attacks in the linguistics literature levelled at the synchronic conception are not based on immanent internal criticism of the logic of synchrony; that certain synchronic theories are historically inadequate does not 'falsify' 'synchrony'. So Anttila's criticism (1989) and the criticism from in particular grammaticalization-theoretician's (Heine *et al.* 1991; Hopper and Traugott 1993:xvi) do not aim at a logical refutation of synchrony, but are attempts at – by demonstrating the strength of their respective theories – rejecting synchrony.

continuation of a(n individual's or dialect's) system is brought about by the dropping of synonymous variants, while seen from the inception of a change its continuation is not impeded by the creation of synonymous forms. The systems problem is no pseudo-problem. It is an indubitable fact that the dialectal distribution of changes indicates that change involves a collective choice: a majority of speakers south of the Humber did eventually choose /ɔ:/ in the 11th and 12th centuries, while the equally empirical possibilities did not become collective. Statements to the effect 1) that *one system chose X, another system Z, for systemic, supra-individual reasons*, 2) that *the respective systems north and south of the Humber so imposed themselves on their speakers that these two groups of speakers had to choose one specific sound to replace (continue) the inherited one*, or 3) that *the respective choices subserve some telos* remain hypostatizing postulates. The Jakobsonian requirement that a sound change be interpreted in terms of the phonological system is putting the cart before the horse[8].

What characterizes both types of development is that neither (macro)dialect loses its collective orientation at the interpersonal level, while at the level of the group something happened. At the interpersonal level the language of two people in constant contact never loses its collective orientation, this also applies to people's language at the level of the group; but the language of groups of people not in constant contact will eventually lose its common collective orientation in the processes that eventually create two different orientations (cf. Sartre's *dépassement*). The final stage is the rise of a new language. How does the synchronic linguist explain this: *the system behaves differently at the level of immediate interpersonal contact and at the level of the group seen against the horizon of separation*?

No matter how we view the systems concept, one problem crops up constantly: how to constitute the system of a language, a dialect? In the remainder of this chapter I shall analyze some scenarios that are presupposed by the synchronic grammarian. In the 5.1.1 scenario two words or sounds, x and y, appear in a distributional relationship defined by a linguistic context {a, b, c ...} and by an anthropocentric matrix of speakers {1, 2, 3, ...}. The two forms fill slots in a section of a spoken utterance so as to make up a synonymy field and for simplicity I shall make do with phonetic examples[9].

5.1.1i

	a	b	c	d	e	f	g	h	i
	[pin	pen	pæn	pan	pɔn	pon	pun	phu:n	phy:n]
1	x	x	x	x	x	x	x	y	y
2	x	x	x	x	x	x	y	y	y
3	x	x	x	x	x	y	y	y	y
4	x	x	x	x	y	y	y	y	y
5	x	x	x	y	y	y	y	y	y
6	x	x	y	y	y	y	y	y	y
	[pin	pen	phæn	phan	phɔn	phon	phun	phu:n	phy:n]
	T1	T2	T3	T4	T5	T6	T7	T8	T9

---

8 Already Paul (1968:18) rejected the teleologcial approach: *die Absicht der Mitteilung ist zwar (...) vorhanden, aber nicht die Absicht etwas Bleibendes festzusetzen, und das Individuum wird sich seiner schöpferischen Tätigkeit nicht bewusst.*

9 The forms could be [a great :: big country], [a green telephone :: phone]; [ph- :: p-eil] 'pale'. As regards phonetic variants; we say either that [p-] and [ph-] share the distributional field that /p/ controls or that two synonyms split up a semantic field *qua* a stylistic field [pheil] :: [peil] /ʔeil/.

5.1. Last words on the synchronic systems concept

The synchronic linguist will interpret this data as follows:

A1   Conditions: no telescoping of the points in time or of the spatial spread
         At T1 all speakers use x{a}: no interidiolectal variation
         At T3 speakers 1-5 use x{c} and speaker 6 y{c}: interidiolectal variation
         At T9 all speakers use y{i}.

These facts can be translated into causal statements:

A2   Why x at T1? Because of contexts {a} and {1-6}
         Why y at T9? Because of contexts {i} and {1-6}
         Why x at T3? Because of contexts {c} and {1-5}
         Why y at T3? Because of contexts {c} and {6}.

At T3 [ph] and [p] are combinatory variants in respect of anthropocentric contexts and free variants in respect of the linguistic context {c}. They are not reducible to systemic elements because their mutual exchange does not create a ?significant, intellectual, denotative distinction. The synchronic statement cannot say anything about the temporal significance of the appearance of [ph-]. Description merges with (causal) explanation.

B   Conditions: telescoping of time-points and of linguistic contexts; contingency is supposed:
         Of Primus it is true – for any point of the period – that x{a,b, c, d, e, f, g} and y{h, i}. Conditioned intraidiolectal variation.
         ...
         Of Sextus – it is true – for any point of the period – that x{a,b} and y{c, d, e, f, g h, i}. Conditioned intraidiolectal variation.

The temporal dimension has been made redundant by the contingency principle, and in relation to the speaker [ph] and [p] are combinatory variants and to the speakers they cannot be reduced to the same systemic element. The /X/ [ph] and [p] model presupposes the elimination of *le sujet*.

C   Conditions: telescoping of time-points and linguistic contexts and the speaker contexts into a group permits us to set up this 'system':
         Of the group {1-6} it is true – for any point of the period – that x{a, b}, y{h, i}, x,y{c, d, e, f, g}. Conditioned idiolect-based variation with x and y both as combinatory and free variants.

Still, the synchronic linguist has no aggregating method through which to create a dualism: the simple model '/..?../ [ph] and [p]' is an arbitrary conflation of three descriptive statements: '/..?../ [p] + /..?../ [ph], [p] + /.../ [ph]'. The synchronic linguist can only generalize the variation to all speakers, and then disregard contexts {a, b, h, i} or treat them under some kind of lexical-diffusion heading – or make the synchronic cut after T2. The ideal as in figure 5.1.1ii remains an ideal

5.1.1ii $\{a \ldots i\}$
              $\downarrow$
$\{1-6\} \rightarrow$ /z/
                        $\uparrow$
                        $\{T1 - T6\}$

Our brief analysis sends us back to square one: 1) no rule can describe the conditions for the manifestations of /z/ immanently; 2) the implementation of /z/ is also not obvious; 3) no systemic element can be established nonarbitrarily[10]. Let's therefore look at another type of scenario as in 5.1.1.iii

5.1.1iii

	a	b	c	d	e	f	g	h	i
	[pin	pin	pin	pin	pin	pin	pin	phin	phin]
1	x	x	x	x	x	x	x	y	y
2	x	x	x	x	x	x	y	y	y
3	x	x	x	x	x	y	y	y	y
4	x	x	x	x	y	y	y	y	y
5	x	x	x	y	y	y	y	y	y
6	x	x	y	y	y	y	y	y	y
	[pin	pin	phin	phin	phin	phin	phin	phin	phin]
	T1	T2	T3	T4	T5	T6	T7	T8	T9

A descriptive statement will conflate T1 and T2 into one point characterized by [p], and T8 and T9 into another point characterized by [ph], and in between there is a period characterized by two variants. Are we in the presence of three or one state of the same language? Is the middle state a transition state? And if the data is conflated into '/p-in/ [ph-in] and [p-in]', then the systemic element expresses a historical fact.

This type of data cannot be described in the language of 'stasis', only by means of dynamic concepts; and the synchronic point of view has not solved its data-constitution problem expounded by the complementarity between the *methodological* priority of the text or spoken chain and the supposed *theoretical* and *empirico-ontological* priority of the system.

The language of time interprets the data in the 5.1.1iii scenario dynamically by means of the historical concepts of *irreversibility* and *directionality*. These imply the following four equilogical possibilitites:

---

10 The Glossematic generalization principle does not permit us to reduce this equivocal result to a univocal one (Hjelmslev 1963:69): the problem is that the 5.1.1 scenario yields three options, so that each option may be regarded as the univocal solution; 1) [p] alone, 2) [ph] alone, 3) [ph], [p]; and more significantly the principle kills the dynamics of the language: Hjelmslev's *sine* example governing no ablative in Latin (pp. 94-95) is equivocal: it could be a new independent (adverbial) usage of the form *sine* and not necessarily deficient textual transmission, where an ablative may be supplied.

5.1. Last words on the synchronic systems concept

5.1.1iv
a) irreversible directionality   The historical series
b) reversible directionality   The non-chronological series
c) irreversible non-directionality   Synchronic functions
d) reversible non-directionality   Absence of order

The a)-row means: if A develops into B, then B cannot develop into A
The b)-row means: if A develops into B, then B comes from A
The c)-row means: if A develops into B, then B comes from A
The d)-row means: if A develops into B, then B develops into A.

The difference between a) and b) is that in a) the point of view is fixed, representing the present's development into the present, while b) allows the point of view of the successor and that of the predecessor, but only at the level of theory.

The difference between b) and c) is a dynamic-static distinction; the former is not redundant, saying that 'A develops into B' and 'B comes from A', the static one is, saying that 'A precedes B' and 'B succeeds A'. In synchronic linguistics it does not matter whether the adjective *green* is said <u>not to be a necessary condition for</u> *a telephone* in *a green telephone* or *a telephone* is said to be <u>a necessary condition for</u> *green*[11]. Let the autonomy of synchronic linguistics be expounded by the c)-row, then the autonomy of historical linguistics is expounded by the a)-row; either alternative is irreducible to the other.

The dynamics of the data will now be analyzed against the horizon of irreversible directionality in order to establish a definition of the notion of **common origin**. The scenario in 5.1.2 consists of a 4-speaker network and two forms in four different contexts at T10. The development from T10 into T11 reveals that y is receding and x the innovative form, whereas the development from T9 to T10 exhibits a process with x receding and y innovative:

5.1.2   L2   T10   1   x,y{axa, axb, bxa, byb}
                             2   x,y{axa, axb, bya, byb}
                             3   x,y{axa, ayb, bya, byb}
                             4   y{aya, ayb, bya, byb}
             >
        L3   T11   1   x{axa, axb, bxa, bxb}
                             2   x,y{axa, axb, bxa, byb}
                             3   x,y{axa, axb, bya, byb}
                             4   y,x{axa, ayb, bya, byb}
or   >
        L1   T9   1   x{axa, axb, bxa, bxb}
                           2   x,y{axa, axb, bxa, byb}
                           3   x,y{axa, axb, bya, byb}
                           4   x,y{axa, ayb, bya, byb}

---

[11] I do not want to labour this important point, because that would imply a logical excursus: but "P is the sufficient condition for Q" is logically equivalent to "Q is a necessary condition for P". – That the four possibilities in 5.1.1iv may constitute a logical conditional: reversibility ⊃ directionality, is immaterial to the present discussion, but will be dealt with in the essay's logical sequel.

According to the two-state model, the two independent developments are equally possible; but as developmental series, the process is impossible – from T9 to T10 to T11. *Two empirically possible developments, T9 > T10 and T10 > T11, which imply logically that the state at T9 is the predecessor of the state at T11, constitute an empirically impossible process with two indistinguishable states.* The two-state model implies the empirically irrelevant possibility of historical reversibility and cannot decide which one of two states is the predecessor, which one the successor. The synchronic state L2 is ambiguous in respect of both its predecessor L1 and successor L3; it may imply a possible future state L3 whose possible realization is improbable because of L1, the origin of L2. The theory must capture this irreversibility. Consider the T10 state in 5.1.3

5.1.3 T10 1 x,y{axa, axb, bxa, byb}
              2 x,y{axa, axb, bya, byb}
              3 x,y{axa, ayb, bya, byb}
              4 y{aya, ayb, bya, byb}

Given idiolect 4, its predecessor may be a continuation 'y{aya, ayb, bya, byb}' or a state with receding x and spreading y: y{aya, ayb, bya, bxb}. – I shall disregard continuations. Given idiolect 3, its predecessor may be a state with x spreading or x receding. The predecessors of idiolects 1 and 2 may be a state with spreading x or x receding. What is the predecessor state of the state at T10 like?

5.1.4 T9 1 x,y{axa, axb, <u>bya</u>, byb} > T10   x,y{axa, axb, <u>bxa</u>, byb}
              2 x,y{axa, <u>ayb</u>, bya, byb}   x,y{axa, <u>axb</u>, bya, byb}
              3 x,y{axa, <u>axb</u>, bya, byb}   x,y{axa, <u>ayb</u>, bya, byb}
              4 y{aya, ayb, bya, byb}   y{aya, ayb, bya, byb}

Per se no idiolect is in conflict with orderly, irreversible gradual development. Let's proceed from the T8-state, on which the development of the T9-state into the T10-state puts certain strings. (The same holds true of the successor state at T11 *mutatis mutandis*).

5.1.5i > T8   T9   T10
         1 x,y{axa, <u>ayb</u>, bya, byb} >   x,y{axa, <u>axb</u>, <u>bya</u>, byb} >   x,y{axa, axb, <u>bxa</u>, byb}
         2 x,y{<u>aya</u>, ayb, bya, byb}   x,y{<u>axa</u>, <u>ayb</u>, bya, byb}   x,y{axa, <u>axb</u>, bya, byb}
         3 x,y{axa, axb, <u>bxa</u>, byb}   x,y{axa, <u>axb</u>, <u>bya</u>, byb}   x,y{axa, <u>ayb</u>, bya, byb}
         4 y{aya, ayb, bya, byb}   y{aya, ayb, bya, byb}   y{aya, ayb, bya, byb}

I shall demonstrate that origin ensures orderly development within the group as well as in the idiolect, which means that the predecessor(s) of T8 will also be relevant:

            > T7   T8   T9
         1 x,y{<u>aya</u>, ayb, bya, byb} >   x,y{<u>axa</u>, <u>ayb</u>, bya, byb} >   x,y{axa, <u>axb</u>, bya, byb} >
         2 x,y{aya, ayb, bya, byb}   x,y{<u>aya</u>, ayb, bya, byb}   x,y{<u>axa</u>, ayb, bya, byb}
         3 x,y{axa, axb, bxa, <u>bxb</u>}   x,y{axa, axb, <u>bxa</u>, <u>byb</u>}   x,y{axa, axb, <u>bya</u>, byb}
         4 y{aya, ayb, bya, byb}   y{aya, ayb, bya, byb}   y{aya, ayb, bya, byb}

5.1. Last words on the synchronic systems concept

At the idiolectal level a speaker is not unlikely to say *phone* in a number of contexts at T7 or to say /ɔː/ for Old English /aː/ in *ham, ban, stan, gan*. It is also not unlikely that a number of other speakers will use *telephone* or /aː/ in the above words. But that they – as a group – should start (continue) a developmental process as outlined in 5.1.5 is unlikely. No common origin can be established for the group of idiolects at T7, which may be continued as in 5.1.5. Hence synchronic interpretations of the state at T10 in 5.1.3 will *per se* not be empirically relevant: **a synchronic network always represents an irresoluble syncretism in respect of its dynamics**. This is due to the historico-structural coherence of the group, within which the 'same' element may not be both receding and spreading. No matter how great the tendency might have been in the late Old English and early Middle English periods to reverse the development, so that the common origin of our sample material could not be established, the collective orientation of the language was maintained. The developmental series of Tertius did not survive. The below scenario in 5.1.5ii illustrates what is meant by the concept of common origin; idiolect 4 will not be used in the argument apart from corroboration of the conclusions drawn for the other idiolects. – The typographical lay-out is meant to facilitate the reading of the steps in the argument.

5.1.5ii

	T7		T8		T9		T10		
1	aya		**axa**	hom	axa		axa		(x spreading)
	ayb		**ayb**	ban	axb		axb		
	bya		**bya**	stan	bya		bxa		
	byb		**byb**	gan	byb		byb		
2*	aya	=	<u>aya</u>	ham*	**axa**	hom	axa		(x spreading)
	ayb	=	<u>ayb</u>	ban	**ayb**	ban	axb		
	bya	=	<u>bya</u>	stan	**bya**	stan	bya		
	byb	=	<u>byb</u>	gan	**byb**	gan	byb		
3*	axa	*hom*	axa	≠	<u>axa</u>	hom*	**axa**	hom	(x receding)
	axb	*bon*	axb	≠	<u>axb</u>	bon	**ayb**	ban	
	bxa	*ston*	bxa	≠	<u>bya</u>	stan	**bya**	stan	
	bxb	*gon*	byb	≠	<u>byb</u>	gan	**byb**	gan	
4	aya		aya		aya		aya		
	ayb		ayb		ayb		ayb		
	bya		bya		bya		bya		
	byb		byb		byb		byb		

Certain assumptions must be stated: bold-faced data has been selected according to the Jones Principle: certain similarities between speakers of different ages, speaking the same dialect cannot have been produced by accident; from this it follows that idiolects 1 of T7, 2 of T7-8 and 4 of T7-10 are not immediately relevant data, and will be rejected on this assumption: no change is complete so that no network is completely homogeneous. And to underline the empirical irrelevance of a network expounded by such idiolects, I shall use the Middle English verbal forms (plural, indicative), (OE -*aþ*) > -*eþ* > -*en* and the relevant forms of the personal pronoun (singular, feminine) *heo* as further illustrative material:

5.1.5iii

	T7		T8		T9		T10	
1	-eþ	heo	**-en**	**hjo**	-e	scæ	Ø	she
2*	-eþ	heo =	**-eþ**	**heo***	**-en**	**hjo**	-e	scæ
3	?	?	Ø	she ≠	-e	scæ	**-en**	**hjo**
4	-eþ	heo	-eþ	heo	-eþ	heo	-eþ	heo

There are two candidates for the common-origin nomination, idiolect 3 at T9, with x receding, and idiolect 2* at T8, with x spreading: *the common origin, be it of a form or of a state, exists prior to at least two of the respective forms or states in question*[12]. Only idiolect 2 at T8 satisfies this criterion. Hence from the point of view of development, the split process is fundamental and so is the synonymy-structure relationship rising against the horizon of polysemy. Thirdly, the causal principle is participative and biuniformal and is expounded by the processes of (external) endogeneity and katageneity: what changes something – the idiolect of Primus from T7 to T8 – makes something else continue – the idiolect of Secundus from T7 to T8.

The fundamental difference between the advanced definition of genetic relations and such traditional ones as Meillet (1966:15, referring to grammatical similarities) as well as: *deux langues sont dites parentes quand elles résultent l'une et l'autre de deux évolutions différentes d'une même langue parlée antérieurment* (p. 16) is that this one does not appear from a historical theory, only from the purported immanence of the *langue* relate of the *langue-parole* dualism, the former from a historico-structural theory.

Historical development is a *structural* process, not atomistic; the everyday language is not a private language despite idiolects. The life of the everyday language unambiguously exhibits a collective orientation controlled by *irreversible directionality* and *common origin*[13]. Synchronically (at T9 or T10), the four idiolects above do not tell against the possibility of the network's collective orientation. But absence of a common origin for the four idiolects limits the empirical relevance of the network as a whole.

---

12 It would be deliberately paradoxical to say 'The origin also exists prior to itself'.
13 These two intralinguistic factors are disregarded in the literature when it comes to what ensures 'mutual understanding'; the 'system' is no guarantee, and *die Neigung, zu sprechen wie der andere* (Coseriu 1974:65) cannot be generalized either (cf. social polarization and the child changing her parents' languages). The very existence of different languages indicates that a communicatively successful implementation of the relation between – in Coseriu's words – *dem Etwas-Sagen* and *dem Es-jemandem-Sagen* does not reduce to a simple question of the intentional factors that constitute and control the dialogue of a given speech situation.

5.1. Last words on the synchronic systems concept

The concept of the common origin is structural; it combines both such traditional historical properties as *to be(come) a predecessor of a successor*, and traditional synchronic properties such as *the simultaneous existence of a successor and its origin* – under the aegis of change. This will be illustrated here through an investigation of the concept's logical geography. – I am still supposing the principle that the processes on the above interidiolectal (diatopic) axis may be *equated with* the corresponding process of change on the idiolectal axis of time (Samuels 1972:11; 114)[14]. To that end we shall consider a two-state scenario with OE /a:/'s development into ME as illustrative material: the development of L1 of T9 into L2 of T10 is characterized by spreading x (= /ɔ:/) and receding y (= /a:/)

5.1.6 T9
1 x,y{axa, axb, <u>bya</u>, byb} > T10 x,y{axa, axb, <u>bxa</u>, byb}
2 x,y{axa, <u>ayb</u>, bya, byb}   x,y{axa, <u>axb</u>, bya, byb}
3 x,y{axa, axb, bya, byb}   x,y{axa, axb, bya, byb}
4 y{aya, ayb, bya, byb}   y{aya, ayb, bya, byb}

In Primus's idiolect, having developed from OE *ham, ban, stan, gan*, the following replacements of context classes have occurred:

5.1.7 'x{a-a, a-b} > x{a-a, a-b, <u>b-a</u>}   /hɔ:m, bɔ:n/ > /hɔ:m, bɔ:n, stɔ:n/'
implies
'y{<u>b-a</u>, b-b}> y{b-b}'   /sta:n, ga:n/ > /ga:n/'.

The model in 5.1.7 does not exhibit the correlate of the split process, the choice concept. The elements of the choice process are called *the possible alternative(s)* and *the positive alternative*. The model will be reformulated in 5.1.10 by way of 5.1.8 and 5.1.9, which introduce both the *choice* factor and the *simultaneity* concept, hence the *synonymy*- and *structure*-concepts:

5.1.8 x{a-a, a-b, <u>b-a</u>}   (*contains*)   the *possible* alternative, +[stɔ:n]
x{a-a, a-b}   (*lacks*)   the <u>positive</u> alternative, -[sta:n]

5.1.8 implies 5.1.9

5.1.9 y{b-b}   (*lacks*)   the *possible* alternative, -[stɔ:n]
y{<u>b-a</u>, b-b}   (*contains*)   the <u>positive</u> alternative, +[sta:n]

Both 5.1.8 and 5.1.9 are needed because the two <u>positive</u> context classes do not contract meaningful structural relationships (katageneity) with their respective *possible* context classes. The origin of the possible alternative of 5.1.8 must be explicitized; the <u>positive</u> class which is, biuniformally, the origin of both *possible* classes becomes the <u>positive</u> alternative of the *possible* alternative which becomes the present moment. The syntagm 'y{b-a}', not 'x{b-a}', is the origin, having prior existence.

---

14 The scenarios in 5.1.5 demonstrate that Samuels's principle cannot be used without placing conditions on the synchronic material; the T10-scenario in 5.1.5ii is empirically irrelevant.

5.1.10 y{<u>b-a</u>, b-b} > x{a-a, a-b, <u>b-a</u>}   the possible alternative, +[stɔːn]
      y{<u>b-a</u>, b-b}   the positive alternative, +[staːn]

At T10 the context class x{a-a, a-b} cannot become the origin because the moment the context class x{a-a, a-b, b-a} comes into existence, x{a-a, a-b} does not exist; similarly, y{b-b} as a one-member context class is also not empirically relevant at T10.

The model in 5.1.10 partially represents the choice involved in historical development: *if the possible alternative had not been produced, then a positive alternative could have been produced*, in this case in the form of y{b-a}. The synchronic tradition has it: *the freedom of the present moment suspends the necessity of the predecessor*, which will be reformulated structurally by: *the freedom and necessity of the present moment so interact as to produce orderly existence qua the development of the present into the present*. And this freedom prevents us from immediately establishing the past of the present moment, *i.e.* within the two-state model's confines; the historical relationship is aitional: the here-and-now intersection of the present moment, but not only if a specific predecessor; and the past, but not only if a specific present.

The possible and positive alternatives are parts of a historical, structural phenomenon, in which either part influences the other. The generality of this position includes the interidiolectal networks. The idiolects above constitute a 'macro'- historical phenomenon, each influencing the other through the simultaneity of katagenous syntheses, the structure that keeps a synonymy field together. The impossibility of idiolect 3 implies that the concept of synonymy is structural with a historical basis. And I shall set up the following correlate of the universal that no everyday language does not change: *no network of idiolects, each of which is the respective native language of each of the individuals of the group of native speakers, has no common origin*. The empirically relevant concept of the **common origin** is not the sufficient condition for, but **the necessary condition for the mutual understanding between – or collective orientation of** – a language group's **native idiolects**.

Concluding this section, I shall set up some of the structural processes that represent the seemingly simple historical process in 5.1.11, OE /staːn/ becoming EarlyME /stɔːn/:

5.1.11 /staːn/ >   /stɔːn/   Possible alternative of {Primus}
          ↓
          /staːn/   Positive alternative of {Primus}
                    Possible and positive alternative of {Secundus}.

The model in 5.1.11 also shows that /stɔːn/ is a new form, carrying the property of being 'marked', while the positive alternative assumes the content of being 'unmarked'; but 5.1.11 is part of a larger structure. In relation to the individual speaker the spread of a sound or a form makes the *old* context class 'x{a-a, a-b}' non-existing. The same goes for the *new* context class *qua* a possible alternative 'y{b-b}', seeing that its positive alternative is continued:

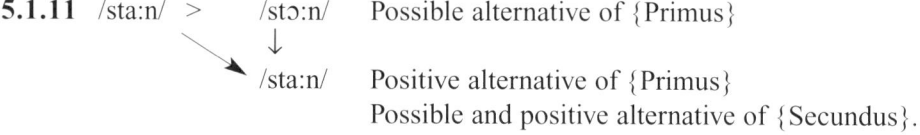

5.1.12 x{a-a, a-b} >   x{a-a, a-b, b-a} new context class
                        ↓
                        x{a-a, a-b}   old context class (non-existing).

5.1. Last words on the synchronic systems concept

```
y{b-a, b-b}    >      y{b-b}                    new context class (non-existing)
                        ↓
                      y{b-a, b-b}                old context class
```

This leads to the dynamic structure of the context class in 5.1.13:

5.1.13 x;y{a-a, a-b; b-a, b-b} > x{a-a, a-b, **b-a**}   marked/new
                                       ↓
                              y{**b-a**, b-b}   unmarked/old.

The marked-unmarked parameter is the historico-structural product of endogeneity and katageneity through the creation of a new (one-member) context class (Primus of T8 in 5.1.5ii) followed by the expansion of a context class (Primus of T9); both processes occur against the **common-origin** horizon. The marked form is the possible alternative and the positive alternative becomes the unmarked form. But as the marked form is the present here-and-now intersection and this property cannot be generalized to all present-here-and-now intersections, or all historical development, it will be replaced by more appropriate sign-functional properties which take into consideration the polysemy of non-marked forms.

The possible alternative of the vertical synthesis of katageneity is not only a constituent of a historico-structural phenomenon, it is also an empirical possibility with a function of its own acquired structurally. Thus *temporal linearity and spatial simultaneity have been united*. The rise of a new phenomenon does not entail the disappearance of another phenomenon, but that something else occurs. On the contrary, the rise of a new phenomenon presupposes the continuation and subsequent simultaneous presence of its origin. And the cause of a given intersection, becoming the positive alternative, participates in the realization conditions for the historical phenomenon in question.

History and structure are two sides of the same coin: the concept of origin is structural, causing contextual meaning to arise. Synonymy – constituted minimally by the possible and positive alternatives – is the simultaneity horizon, controlled by common origin, against which the historical process of development from the present into the present occurs. The precision of the possible alternative (markedness) occurs against the horizon of polysemy seeing that the positive alternative (unmarkedness) assumes the contextual meanings that the possible alternative does not carry – this will be modified in chapter 6.1 and 6.2. The positive alternative is polysemous, and the polysemy property of the everyday language has now been given a non-connotative basis. As the marked-unmarked parameter suggests, historical development is sign-functional.

The synonymy concept has been interpreted in accordance with polysemy: as a content may be said to develop from another content in a polysemy field, so an expression may be said to develop from another expression, creating a synonymy field. This means that *the raison d'être* of the type of synchronic synonymy field analyzed in chapter 4 will be controlled by the dynamics of the respective polysemy fields – again *pace* the indefinite article in English *a/an*.

The concept of common origin combines the notions of identity in differences and differences in identity and has been given an empirically relevant interpretation. The *Stammbaum* theory does not contain the simultaneity property of the split process (see *e.g.* Nielsen

(1998:57;35); Beekes (1995:31)), while William Jones's alternative – as to the existence of the common source of Greek, Latin and Sanskrit – *which, perhaps, no longer exists* is correct given an empirically relevant conception of historical development. This in contrast to Meillet's *Stammbaum*-based view of the genetic relationship, where the origin no longer exists (*parlée antérieurment*).

5.2. All language change is equiempirical. A theory of change must not commit the Phenomenal Error. The line-point imagery of synchrony is conditioned. The concepts of the series and the element are not parallel; both are theoretically relevant. An outline of data. Definitions of the developmental series of isolative sound 'change': *inertial* and *ertial* developments; *breaks*. Ertial transitions reveal the inadequacy of the two-state model

In the history of linguistics theories about language change have mostly been constructed on the basis of large-scale processes which characterize the history of the dialects of the Indo-European language family; and these changes are more often than not regarded as *raw-material* for the demonstration of the scope or strength of a given theory[15]. To mention some of these laws (see Collinge 1985) that are of particular interest to the Germanic branch of the family: the First Germanic Consonant Shift (The Rask-Grimm Law) and the Second Germanic Consonant Shift, both of which Prokosch *exploited* in his periodic conception of sound change[16], and which together with the Great Vowel Shift are *used* to exemplify Push- and Drag-chain mechanisms; Verner's Law and PreOE Breaking and *I*-mutation *exemplify* conditioned change *qua* exceptions to changes that operate without exceptions (isolative change). The type we see in LateOE and Middle English loss of grammatical endings is *used* to demonstrate cyclic swings in grammar, the transition from predominantly analytic/synthetic states to predominantly synthetic/analytic states or grammaticalization processes. Grammaticalization studies set great store by seeming parallel developments in different languages: modern French *je vais le voir* / English *I am going to marry Mary*, constructions to be compared with similar constructions in certain African languages (Hopper and Traugott

---

15 Changes have been lifted out of their chronological series and empirical contexts in order to illustrate a given theory as in Chomsky and Halle (1968:252), or to prove the Neogrammarian Regularity Hypothesis or otherwise. The same criticism must be levelled at Stiles's dualism-interpretation of Luick and Labov's treatment of chain shifts (1994); Labov, furthermore, makes his analyses controlled by the Uniformitarian Principle. This type of theorizing is subsumptive and rests on the – in my eyes – fallacious belief in the reification of historical phenomena, in the final analysis, on a belief in the necessity of transferring subsumptive-positivistic principles from the natural sciences to history.

16 Prokosch's Fundamental Principle was in a way a circular shift stretched out, a phonetic interpretation of Grimm's circle view of the Germanic Consonant Shifts. His shift was a full circle in that the process had a start, unvoiced plosive, T, to which the process returned and had as such completed its pre-established or 'natural' course. The process was characterized by strengthening and weakening (lenition) so that the Principle was in fact a theory of consonantal change – it being *a consistent system of changes* taking place *during the early history of the Germanic languages* (Prokosch 1966:50-51). The theory was also the speaker's in that it was based on the two articulatory gestures: release or check of breath. The problem with Prokosch's theory is that in the positivistic tradition he confuses data with theory.

1993:84[17]). The LateME and early 15th century voicing of unvoiced fricatives is both interpreted as a gap-filling process and *used* to strengthen the phonetic explanation of Verner's Law. But most typically, all those purportedly independent changes – each with a start and an end – are used in the standing debate about mechanical and functional/systemic theories[18].

So-called 'minor' changes such as the Old/Middle English raising of /a:/ to /ɔ:/ have not roused particular interest, let alone the unrounding of OE/ME *y*-sounds to Kentish *e* and Middle English *i*, and in accordance with the modern synchronization of history historical descriptions that we see in *e.g.* Campbell, Luick and Dobson, are virtually unscientific, because they apply the proper code of history, that of chronology; I have demonstrated that chronology need not end in 'mere' registration of atomistic facts, but is structural when couched in the language of time, not in that of space[19].

In sum, the reason for this predominant type of methodological approach is not only the synchronization of history, but also the identification of historical theory with causal explanation, the (positivistic) merger of theory and data, and the view of historical development as being discontinuous and disruption of a stable state: change is made to constitute a delimitable process, eminently suitable for subsumptive treatment. A change may also be in progress (Weinreich *et al.* 1968:184; Labov 1994), while 'to be in progress' is not a property of the continuation of the stable synchronic system of a state. A system is not in progress, not in process of becoming (cf. below).

A theory of change based on only one type of change is bound to commit the Phenomenal Error (Hansen and Nielsen 1986:1-3). The distinction between circular and non-circular shifts in phonology is clear. Isolative non-circular (minor) shifts are merely registered, albeit as laws, whereas circular shifts or chain shifts are subjected to (theoretical) explanations

---

17 That such common developments should have a cognitive basis (see chapter 6.3.), as is assumed by grammaticalizationists, is still to be proved, because we are not told why similar processes do not occur in all languages: German *gehen*, Danish *at gå*, or Faroese *ganga* (Lockwood 1977) exhibits no such grammaticalization tendencies.
18 Bynon (1977) is an example of how various theories are assessed against the 'same' material.
19 One must therefore agree with Hopper and Traugott (1992:10), who lament the inadequate state of the available theories for historical analysis: *"Unfortunately our available linguistic vocabulary or "metalanguage" for expressing the relationship between earlier and later linguistic phenomena is poor."* The question is: why do they not *attempt to change* this deplorable state of affairs, but *follow custom*, using the spatial language of abduction (pp. 34, 39). The two writers refer in particular to *renewal* (Meillet's *renouvellement*) and *replacement*. The latter concept will be given a dynamic definition below and renewal makes no sense within the constant-change notion of a proper dynamic theory. Had Hopper and Traugott followed Paul's warning and had liberated themselves from the 'tyranny' of the synchronic 'custom', they might have separated the historical wheat from the synchronic chaff as in the next footnote. In Harris (1994:33-34) we see a mix of synchronic and historical metalanguage: in his discussion of compensatory lengthening (*qua* 'quantity stability') Harris takes for granted what he wants to prove, that the [ai]-diphthong in words like *night, fight* emerges from lengthening in Middle English in connection with the loss of the unvoiced velar fricative [x]. How does Harris establish non-discursively that the words in Middle English did not have a long vowel /i:/ and that their modern spelling was not caused by some spelling 'reform' in the early modern English period? In addition to this, it is interesting to note how the phonologist, being concerned with the pronunciation of English, relies on the written medium. Had the phonologist used his data consistently, he would have had to prove how he could dissolve the polysemy of the spoken language's [mait] into *might* and *mite*, and [rait] into *right* and *rite*.

(Lass 1976:52). The distinction is a *Nachlass* of Realism couched in nominalistic diction and is dependent on the omniscient historian's non-chronological vantage point: a change has an existence of its own that spans a couple of hundred years – despite the fact that change is viewed as discontinuous and disruptive.

All change is equilogical and equiempirical; hence all sound change is either circular or noncircular; both terms are redundant: change is what it is. The Great Vowel Shift possesses no more properties than the abovementioned isolative shift: *both (?)types of change must be explained by the same type of principle*. The notions of large-scale change or long-term change as we see in grammaticalization could be subsumed under the questions: *Why sooner?* and *Why more change?* in particular periods of a language, questions concerning the speed and degree of change (Samuels 1972:88) so that the Realistic Fallacy will not be committed. However, the unitary nature of long-term change has a reckoning with the static systems view of change, a change being 'disruption of a state into a new state', or, to put it in dualistic diction, 'the step from variant to change' or in grammaticalization terms, 'when a lexical item comes to serve a grammatical function' (Hopper and Traugott, pp. xv, 2[20]).

The synchronic point of view implies that circular shifts must be divided into their component parts (states) so that the development of ME /i:/ into the modern language's /ai/ and ME /ai/ into /ei/ are to be explained as a finite sum of state transitions: the synchronic two-state model is incompatible with the purported unity of large-scale change. Our sample development[21] will bring out the problems:

5.2.1   OE   >   ME   ME   >   N(ew) E(nglish)
        State 1   State 2 State 2   State 3
        /a:/ >   /ɔ:/   /ɔ:/ >   /o:/

It is heuristically appropriate to emphasize the transition concept with a view to avoiding entelechy-interpretations of long term change. We still have to see the telos that sums up the English /i: > ai/-development, either as an individual development or as a part of an entity called *The Great Vowel Shift*. – No consistent telos either accounts for the disparate developments of OE /a:/ into ME /a:/ and /ɔ:/, and ME /a:/ into NE /æ:/. No change is more exceptional than other changes: no change is major in respect of other – minor – changes; no change

---

20 In English *going to* is polysemous in *I am going to marry Bill*, not so in *I am going to college* or *I am going to go to college*. The historical question is: when did *going to* (= *gonna*) become grammaticalized, when did this particular change happen? Samuels (1972:11-12) summed the problem up as regards the expression plane: what is the *precise diachronic link between the two [phonetic variants]*? Hopper and Traugott, viewing the issue from the content side, have an odd way of synchronizing the change: *The change occurs only in a very local context* (p. 2), such as the *marry*-context above. They seem to imply that modern English /go/ has two main functions; it is a lexical item and then /go/ can somehow be changed into its auxiliary meaning, *i.e. In the absence of an overt directional phrase, futurity can become salient* (p. 3): the *change* (p. 3) from lexeme-hood to futurity as in *I am going to marry Bill* is somehow brought about by displacement of (by *inference*) the implicit future fulfilment of the telos of the whole proposition. In other words, they imply that the diachronic link *qua* a cognitive process is to be sought in the modern language so that one lexeme /go/ undergoes *fresh and repeated* (Samuels) devaluation *qua* semantic change by each new user or generation.

21 From OE/ ha:m/ to ... modern English /həʊm/ 'home'.

5.2. All language change is equiempirical

is less causally motivated than other change. No change has no cause; isolative change and conditioned change, our Neogrammarian heritage, are equiempirical, and if either has a cause and is amenable to theoretical treatment, then the other is so, too (Samuels 1972:25-27).

The notion of the isolated and autonomous fact, the philological detail, is the synchronic linguist's and structuralist's punching bag, and the reaction of both to what came to be known as the Neogrammarian Paradigm can only be characterized as a seemingly never-ending series of rounds of shadow boxing. It is absurd to claim that the abovementioned great chronicles by Luick, Campbell and Dobson and others are not scientific.

The synchronico-structural rejection of the period concept hinges on the arbitrary supposition that a line is so defined by a class of points that its existence is contingent on that of the point. This definition is not immediate: it is mediated by distance *qua* both separation and attachment and succession *qua* reversible directionality: a point always exists in absolute separation from other members of its class, but when combined into a line, each point is made to cohere inseparably with at least one other point; the line is reversible: reversibility does not come from the point. The line-point imagery of synchronic linguistics does not account for the irreversible directionality of either the spoken chain's palimpsest successivity or the developmental series. The line-point imagery implies **reversible nondirectionality** (chaos).

Irreversible nondirectionality with its concomitant concept of non-successive simultaneity is the contribution of the human mind (the theory in question) to the historical phenomenon, while **irreversible directionality** and **simultaneous successivity** are rooted in our human **ex**istence. The paradigmatic-syntagmatic parameter constitutes no sufficient means of interpreting the process that makes a syntagm arise out of the combination of paradigmatic elements. The English language is hardly in need of keeping **klim* as a spare word besides *milk*, my point being that the attempt at bypassing the synchronic fact of the spoken chain's irreversible directionality, mirroring the irreversible directionality of the historical process, through the theoretical acceptance of allowable, potential but unused words (syllables) and disallowed syntagms, is empirically unjustifiable. Saussure (1972:103) and with him the synchronic tradition did not determine linearity. The system reduces the irreversibility of the spoken chain to nondirectional functions (cf. above), thereby suspending a fundamental property of the everyday language, *i.e.* the merger of *parole*- and historical directionality. Following my interpretation, synchronic linguistics tries to combine irreversible nondirectionality (the system) and irreversible directionality (the spoken substance), through the imposition of the property of contingency on the elements of the spoken chain, whereby the rule concept finds its place, a rule being an accidental constraint on a historically given synchronic class of possible sequences: synchronic grammar has become universal grammar.

In the following I shall incorporate the concepts of irreversible directionality and simultaneity into a historical theory, and I shall so circumscribe my field of vision as to exclude loss- and addition-processes[22]. Secondly, I shall – continue to – argue for this thesis:

---

[22] The disappearance or appearance of a sound is a general property of all sounds, controlled by neither the *Eigennatur* of a sound nor a telos; any L-elements may disappear, while the sentences of the spoken chain do not disappear or develop; they are not L-elements. The loss (or addition) of X in 'CVC-X-C' is not the product of the history of X: a schwa sound is no more prone to disappear than to become a non-high centralized *i*- or *u*-sound; cf. the recent change in modern English /-ɪ #/ in *e.g. city* [-ɪ > -i]. Loss and addition are mechanically imposed on the stuff of language, and belong to exogenous development, which was dealt with in chapters 1 and 2.

The synchronic two-state model is inadequate, historically and empirically, and so is the concept of immanent synchrony. Synchronic linguistics does not define such crucial concepts as 'the synchronic cut, a break-before and a break-after' (the breaks around the state) let alone the dynamic concepts 'to break away from, not to break away from, to break into, not to break into'. Synchronic linguistics does not explain how the everyday language preserves its collective orientation, without having recourse to systems postulates. The two-state model conceals the similarity between continuation with change and continuation with no change and has in fact no theory for the latter type of process. Synchronic linguistics hopes to be able to 'systematize' itself out of the purportedly intractable facts of sensuous reality.

Logically possible types of developmental series will be enumerated in 5.2.3 below, some of which may be more empirically relevant than others, and I shall only be dealing with one type exhaustively. The analytic weftage that will lead up to the schema in 5.2.3 and the figure in 5.2.4 is these two theoretically equirelevant questions: *What makes for stability?* And *What makes for change?* Our tradition has it – or implies – that

> What makes for **stability** is factors that **maintain stability**, and
> what makes for **change** is factors that **disrupt stability**.

Change and stability are determined in terms of stability. The alternative point of view is as logically viable:

> What makes for **change** is factors that **maintain change**, and
> what makes for **stability** is factors that **disrupt change**.

Accordingly, a biased point of departure seems to be the determination of *change* and *stability* by the respective factors which

> *disrupt stability and maintain change* and *disrupt change and maintain stability.*

The prevalent idea that stability somehow dominates change is deeply rooted in the everyday language and therefore in the human mind (cf. chapter 1). E.g., we say that (1) *A thing changes / A thing is modified //* Danish *En ting ændres / En ting ændrer sig //* German *Ein Ding verändert sich //* French *Une chose change / Une chose se modifier* and (2) *A thing is stable*. The former type of sentence implies that the thing was stable and that its stability has been disrupted or not maintained – or, dualistically, that something remains stable through change[23]. The latter type implies that the thing – *e.g. une langue* – was stable in its immediate past and will remain so into the near future. In any case, the sentence does not immediately imply the concept of change, let alone a dualism.

If *A thing is stable* is to imply change, some kind of emphasis or contrast must be added to it: *A thing is stable now // This thing is stable.* Change is implied when the present moment is marked. However, the concept of the marked present is not generalizable (cf. chap-

---

23 In the content of this simple (in)transitive verb, *to change*, together with *to become* (see chapter 6.3.) lies the dominant metaphysics of our Western Civilization.

ter 5.1. above). The generalization of the *now* into an *always* makes the markedness of the synchronic point redundant, but it also denies the possibility of change. A viable synchronico-static conception of the everyday language must incorporate the universality of change. But the dilemma of our tradition is this: *the object of stability is less something which has become what it is than continuation of what it was*, while the sentence under which our sample utterance *A thing changes* is produced is not that the thing changes from being in a state of change, let alone in process of changing, to something stable, hence *the object of change is less something which is continuation of what it was than something which has become what it is*, with the dualistic version: *the object of change is less something which has become what it is than something which is what it was*.

I have repeatedly said that my starting point is the possible equitheoretical nature of change and stability, and that is still valid. But the general inclination of my essay is that a possible asymmetrical relationship will appear to be in favour of the concept of change so that this will be an argument in favour of the untenability of the synchronic stance as the foundation of the life of the everyday language. To sum up the traditional point of view:

1) In *A thing changes* the supposition is **from** stability **to** change; change presupposes stability.
2) In *A thing is stable* the supposition is **from** stability **to** stability; stability does not presuppose change – in the short run.
3) In *A thing changes* the supposition is **from** stability **through** change **to** stability. Change, presupposing stability, is a dynamic relation betwen two immanent stable states and entails the concept *to be in a state of transition*.

The supposition of 3) incorporates that of 1) and accounts somewhat paradoxically for the generalization of the short-term view so that the periods of stability are always end goals of their respective changes. A process has a start and an end. Furthermore, 1) and 3) imply the abovementioned concepts, 'to break away from' and 'to break into': X of State 1 breaks away from State 1 and into State 2 – under the aegis of continuation; the gap between the two states occurs against the horizon of coherence.

A modest conclusion must be that despite the object-hooked bias of the everyday language no argument can remove the fact of change and development from the empirical world.

It follows from 2) that a state has duration; this continuation in time is a property of the stability of the synchronic state; had the state no duration, we would have constant change. While a sentence is being uttered, the stable system 'behind' it must be enforced and so durable that my interlocutor can use this supraindividual entity to understand what I said and I cannot change it while speaking.

The duration of change emerges not only from the concept of change of change, but form the very change-stability relationship. Change has a beginning and an end; a given state begins to change – 'changes away from' – and it stops changing – 'changes into'[24]. The notion

---

[24] The logic of this state-process-state model is more than tenuous: State 1 begins to change, and it(?) – or as State 2 it(?) – stops changing having become State 2. Does the process have a beginning and an end? How does this beginning and end connect with State 1 and State 2, respectively? If the states are the beginning and the end, it is the process that has the states. It is easy to see how the everyday language forms a given

of long-term change – chain shifts and grammaticalization – implies the duration of the process. I shall argue that change and stability have different logics: change is brought about by *disrupting* factors; a state remains stable owing to factors that *maintain* it.

Disruption of change means both disruption into stability and disruption into change; two (new) sets of motivating factors, one making for stability, the other for continued change, respectively, replace the factors that made for change (cf. D and E in 5.2.3 below).

Continuation of change means that the maintaining factors of a given change process remain table, just as the maintaining factors that cause a given state to remain stable may be continued (cf. B and A in 5.2.3 below).

Disruption of stability means disruption into change as one set of motivating factors is replaced by another set (cf. C in 5.2.3). The sixth possibility (F in 5.2.3) is the moot point. Is it conceivable that a state may continue its stability, if the original set of motivating factors is replaced by another one? Can two sets of motivating factors that both make for stability replace each other? I shall argue that this is impossible[25], but the possibility has some empirical relevance (see chapter 6.). The schema in 5.2.3 sums up the logical possibilities that end in the asymmetry of change and stability:

5.2.3	Change *Inertial* Development	+Continuation of motivating factors (A and B)	Stability *Inertial* Development	Stability > Stability or Change > Change
	Change *Ertial* Development	−Continuation of motivating factors (C and D)	Stability *Ertial* Development	Stability > Change or Change > Stability
	Change *Ertial* Development	−Continuation of motivating factors (E and F)	Stability *Ertial* Development	Change > Change or ?Stability > Stability

The asymmetry of change and stability will be interpreted as follows: <u>change</u> implies itself in two ways inertially and ertially, while stability only implies itself inertially:

Def. 5.1[26] Inertial development is continuation of a phenomenon with its motivating factors constant.

Stability is continued (**A** in 5.2.3 and 5.2.4), if the motivating factors do not cause change.

  – dualistic – metaphysics: State 1 appears as State 2 after a process of change, but then State 1 must also be the 'appearance' of its predecessor state, etc. Pending the synchronic linguist's answer, the historian may conclude that the two-state-model is empirically irrelevant.

25 Two sets of factors may produce the same effect: /y/ may appear from rounding of /i/ or fronting of /u/; this is not the issue here.

26 A more precise definition is this: '(…) the present's development into the present with motivatings factors constant.́'

5.2. All language change is equiempirical

The same type of change is continued (**B**), if the motivating factors cause change[27].

Def. 5.2[28] Ertial development is continuation of a phenomenon with its motivating factors not constant.

Ertial development captures the E-, D-, and C-processes in 5.2.3 and 5.2.4, and the definitions make the traditional concepts – minor and major isolative change, circular and non-circular change – redundant[29]; the two types of change capture continuation with and without change.

The figure and schema in 5.2.4 represent data – developmental series – that is determined by the advanced notion of motivating factors; the data is somewhat idealized but does not represent empirically irrelevant phenomena (cf. *e.g.* Dobson's trajectories of the long vowels in the *Great Vowel Shift*). The figure, to be read from left to right, is a crude idealization of articulatory movement in the oral cavity, and does not represent the asymmetry of the latter.

The advanced theory will not decide which of the types of the developmental series below may be more empirically relevant than others. What is at stake here is the independence of the trajectories of chain shifts: '… > A > B > C > D > …', or whether such a developmental series is a combination of other series: '… > A > A > A > A > B > B > B > C > C > C > C > C > …'.

Finally, the F-possibility of 5.2.3 does not really correspond to **F** of 5.2.4, if we may say that it represents non-articulatorily based identity of forms – an issue that will be resumed in chapter 6.

5.2.4 ARTICULATORY MOVEMENT IN TERMS OF RELATIVE TONGUE POSITION (HEIGHT) UNDER THE AEGIS OF DURATION

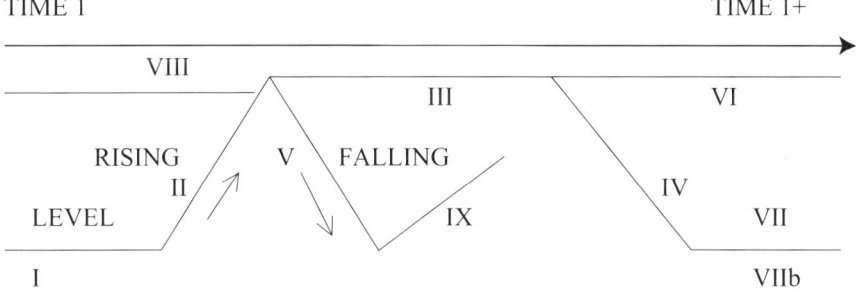

---

27 Inertial development is an empirically relevant conception of a teleological drive. The telos is defined dynamically in relation to something language-internal, not language-external – the continuation of motivating factors. As can be seen, our tradition has a biassed view of telos: the A>A-type of process may also have a telos! Or: is A>A not historical development?

28 In this context the term *ertial* is a cover term; it will be defined in chapter 6.1. and 6.2. and Appendix 1.

29 As regards chain shifts proponents of the empirical relevance of such change needs to explain the conditions under which they occur and do not occur, and not least how they fit into the various static synchronic models.

INERTIAL DEVELOPMENT
The members of the calculus: motivating factors are constant

A	B
I, III, VI, VII, VIIb, VIII	II, IV, V, IX
Stability > Stability	Change > Change
Continuation with no change	Continuation with change

ERTIAL DEVELOPMENT
Combinations of A and B: motivating factors are not constant

C	D
I > II, III > IV, VIII > II	II > III; IV > VII, IV > VIIb, II > VIII
Stability > Change	Change > Stability
Continuation with change	Continuation with change

E	F
II > V, V > IX	III > VI
Change > Change	Stability > Stability
Continuation with change	Continuation with 'change'

Comments

I shall here look into the logical geography of the linguistic commonplace that the same sound cannot be reproduced. If the assumption were correct, the everyday language would be characterized by constant fluctuation and variation. The reason why the axiom must at least be redefined is that it entails historical inevitability: a sound that has been produced cannot possibly be produced again. Historical prediction and retrodiction would then consist in the historian knowing what cannot be the case; secondly, the more sounds a speaker produces, the realization-probability for all other possible sounds will be the greater. I think that the absurdity of the latter consequence is evident.

Let's look at the truism from another angle. Synchronically, the probability of a given sound's being uttered at T1 will be the same as that of all other possible sounds. At T2 the same applies, on condition that we are speaking about the same class of possible sounds. Hence a given sound may be reproduced by the same speaker. That is the only viable interpretation of the slowness and gradualness of sound change, **level** inertial change. The (easily recognizable) survivals of the IE stem * *mus-* in various dialects are incomprehensible unless repetition of the same sound be accepted. Furthermore continuation without change: OE (= State 1) /e:/ > ME (= State 2) /e:/, would not be possible if any definition of linguistic repetition was disallowed *a priori*.

What is inertial development? The **A**-type is characterized by **level internal & external endogenous motivation**, so that the product of the individual's speech remains the same *qua* a structural phenomenon. The same goes for the **B**-type, **nonlevel internal & external endogenous motivation**. What is involved here is the overt interplay of the product's *Eigennatur* and its realization factors so that the empirically relevant concept of 'the same sound' can be established. Following Samuels (1972), Horn (1936) and Horn and Lehnert (1954:35-68) and

5.2. All language change is equiempirical

Luick (1930)[30], I shall say that subglottal and supraglottal as well as glottal factors strike a delicate balance in the sense that 'more' of one factor entails 'less' of another and *vice versa*. Examples: the physiological effect of relatively high subglottal pressure tends to entail devoicing as well as, in the case of vowels, rasing and/or fronting, and aspiration in the case of consonants; rasing of the back of the tongue naturally entails the production of a back vowel with (some degree of) lip rounding. Emphatic speech tends to produce longer vowels than less emphatic speech. The implementation of the principle is here seen in isolation from exogenous influence, including *Ruhelage* or basis of articulation (Samuels 1972:18-25; 41-43); and it agrees well with the dynamic principle: 'for something to occur, something else must have happened or happen' and the structural commonplace that a phenomenon is not the mere sum of its constituent parts. The acoustico-auditory effect of a sound may be the same despite different realization factors: the over-all level of a sound's prominence is constant[31].

This way of preserving constancy in speech production contributes to the fact that the everyday language always keeps its collective orientation. The listener's contribution is not only a monitoring function (Samuels, *loc.cit.*), it is also an active factor (Coseriu 1973:60-61)[32] in the gradualness of change, owing to his structural sense of history expounded by the interplay of external endogeneity and katageneity.

The figure in 5.2.4 operates with three types of inertial development, level, rising and falling. They will be implemented positively with sounds and structurally in respect of their individual histories. This means that the spoken (phonetic) substance of the everyday language will be the basis of the theory and will therefore be more than raw-material for the latter.

Ertial development is more complex and is here regarded – in accordance with the calculus view, as combinations of different inertial developments, characterized by **breaks**. The types will appear below. – The various developmental series as combinations of level and nonlevel series are the following ones

1) level + rising   C: stability > change
2) level + falling
3) rising + level   D: change > stability
4) falling +level
5) rising + falling   E: change > change
6) falling + rising
7) level + level   F: ?stability > stability // A + A

---

30 See also Parmenter 1933; Peterson 1961; Lira 1976; Back 1979. – This identity principle is also related to Labov's near-mergers, where measurable articulatory differences are not perceived as such, but as identities, by the hearer.

31 If we see the principle in the light of a simple type of sound change, it may not sound paradoxical: any sound may stem from at least two different causes: /e/ from lowering of /i/ or rasing of /æ/. – A simplified equation runs like this: the sound /i/ requires more pulmonary force ('stress') than /e/, but the latter is more sonorous than the former: [+ more pulmonary force → less sonority] = [+ less pulmonary force → more sonority] in respect of prominence. I am using the latter term in a sense slightly different from its received sense in mainstream Anglo-American phonetics. By 'prominence' I mean the over-all acoustico-auditory effect of a sound, produced by sonority, pitch, length, loudness, expiratory stress'. See Jones (1976:134-152 and in particular note 12, p. 59).

32 or as Paul pointed out: speech production relies on both *eine Arbeitsteilung* and *Arbeitsvereinigung*.

As indicated each inertial developmental series is a member of a calculus (of the A- and B-series) and the below series are sequential combinations (*parole*-manifestations). In these combinations the structural principle will immediately assert itself, seeing that the first or last elements of a series will be influenced by its context; this is a process unforeseen by the combinatory rules of the calculus. – The calculus-determined gaps[33] have been indicated by extra spacing; and the structurally determined series have been marked by 'NB':

C stability > change
1) **level + rising**
I>II   a3 > a4 > a5 > a6   >   α > æ > ɛ > e > ɩ   (From C 5 to C4 to-wards C 1).
or
            æ2 > æ3 > æ4 > æ5   >   α > æ1 > ɛ > e >   (From C3-4 to C4 to-wards C 2).

level + falling
III>IV   u3 > u4 > u5 > u6   >   ʋ > ǫ > o > ǫ̣ > ɔ   (From C 8 to C8-7 to-wards C6-5).
VIII>II   ɛ2 > ɛ3 > ɛ4 > ɛ5   >   e > ɛ1 > æ > α > a   (From C3 to C2 to-wards C5).

D change > stability
3) **rising + level**
II>III NB   a > α > æ > ɛ   >   e1 > e2 > e3 > e4   (From C5 to C3 to-wards C2).

II>VIII   a > α > æ > ɛ1 > e   >   ɛ2 > ɛ3 > ɛ4 > ɛ5   (From C5 to C2 to-wards C3).
4) **falling > level**
IV>VII NB e > ɛ > æ > α   >   a1 > a2 > a3 > a4   (From C2 to C4 to-wards C5).

IV>VIIb   e > ɛ > æ1 > α   >   æ2 > æ3 > æ4 > æ5   (From C2 to C4 to-wards C3).

E change > change
rising > falling
II>V NB   a > α > æ1 > ɛ1   >   e > ɛ2 > æ2 > α2 > a2   (From C5 to C3 to C2 to-wards C5).

6) **falling > rising**
V>IX NB   e > ɛ1 > æ1 > α   >   æ2 > ɛ2 > ẹ1 > e/ɩ   (From C2 to C4 to C3 to-wards C2-1).

or NB   e > ɛ > æ > α   >   a > ɔ > ǫ̣ > o > ǫ   (From C2 to C4 to C5 to-wards C7).

F ?stability > stability (by way of change)
7) **level > level**
III>VI   a3 > a4 > a5 > a6   >   α1 > α2 > α3 > α4   (From C5 to C4).
            α1 > α2 > α3 > α4   >   a3 > a4 > a5 > a6   (From C4 to C5).

            ɛ1 > ɛ2 > ɛ3 > ɛ4   >   e1 > e2 > e3 > e4
            e1 > e2 > e3 > e4   >   ɛ1 > ɛ2 > ɛ3 > ɛ4

---

[33] The letter 'a' is used as a C5-vowel, lower than Cardinal 4, α.

5.2. All language change is equiempirical

Only from the omniscient historian's Archimedean vantage point can the above cuts be made; and I shall demonstrate that if the synchronic linguist makes these cuts, he will not produce a historically relevant narrative; secondly, it will be demonstrated how the compound series has an *Eigennatur* so that it cannot be reduced to a sum of its elements. Furthermore, the transitions around the 'cuts' are not similar. These three facts go to demonstrate the theoretical irrelevance of the two-state model.

In Group 1 there seems to be a clear-cut transition so that the last element of the predecessor-series (= P) and the first element of the successor-series (= S) are univocal. The cut or transition will be specified as follows: the break after P will be defined below and so will the break before S

Group 1 P > S
III>VI   a3 > a4 > a5 >   a6 > α1 > α2 > α3 > α4   level > level
IV>VIIB e > ε > æ1 >   α   > æ2 > æ3 > æ4 > æ5   falling > level
II>VIII a > α > æ > ε1 > e   > ε2 > ε3 > ε4 > ε5   rising > level
I>II   æ2 > æ3 > æ4 >   æ5 > α > æ1 > ε > e > ɪ   level > rising
VIII>II ε2 > ε3 > ε4 >   ε5 > e > ε1 > æ > α > a   level > falling

All **predecessors** are characterized as follows:
e.g., a6   P does not break away from (the series of) its own predecessor   -First
              P does not break into (the series of) its successor   +Last
              P breaks into (the series of) its own predecessor   -First
              P breaks away from (the series of) its successor   +Last

All **successors** are characterized as follows
α1   S does not break away from (the series of) its own successor   -Last
      S does not break into (the series of) its predecessor   +First
      S breaks into (the series of) its own successor   -Last
      S breaks away from (the series of) its predecessor   +First

This defines a synchronic cut as an absolute break: **-P [-First, +Last] & S- [-Last, +First]** in contravention of Metacondition 2 (see chapter 1.5.3).

In this group there is no reason to distinguish between the individual element as either a predecessor or a successor and the series. In the next groups *element* and *series* cannot be used interchangeably:

Group 2 P/S > S
I>II   a3 > a4 > a5 >   a6 > α > æ > ε > e > ɪ   level > rising
III>IV u3 > u4 > u5 >   u6 > ʊ > o̞ > o > o̝ > ɔ   level > falling

II>III a > α > æ > ε > e1 > e2 > e3 > e4   rising > level
IV>VII e > ε > æ > α > a1 > a2 > a3 > a4   falling > level

Here we see how the calculus fails us: what are the last and first elements of two series become equivocal when the series are combined. The elements marked by '**P/S**' are the last elements of their series in the calculus, but become possible first elements in what is manifested as the successor series. The ambiguity of data is due to the applied calculus point of view. And the following definitorial analysis of the synchronic gap is conditional on this stance.

All **predecessors** are characterized as follows:

a6	P does not break away from (the series of) its own predecessor	-First
	P breaks into (the series of) its successor	-Last
	P breaks into (the series of) its own predecessor	-First
	P does not break away from (the series of) its successor	-Last

All **successors** are characterized as follows

α	S does not break away from (the series of) its own successor	-Last
	S breaks into its predecessor	-First
	S does not break into the series of its predecessor	+First
	S breaks into (the series of) its successor	-Last
	S does not break away from its predecessor	-First
	S breaks away from the series of its predecessor	+First.

The break is defined by: **-P [-First, -Last] & S- [-Last, -First, +First]**.

The four predecessors (**P/S**) of Group 2 so intersect the whole developmental series that no unequivocal break occurs after them, while the four successors so intersect their respective series that a break before them is established in relation to the series ([+First]) of which they are constitutive elements and to the predecessor series of which they are not constitutive elements. They indicate that a <u>series</u> has come to its end and a new one begins. The gap, however, is not absolute as in Group 1; it is mediated by the successors' immediate affinity with their respective predecessors, and by the fact that the predecessor element (P/S) becomes a constituent of both series. As such the difference between two such transitions as 'L1 /e1/ > L2 /e2/' and 'L2 /e2/ > L3 /e3/' (of II>III) cannot be captured by the two-state model.

The four developmental series of Group 2 are reversible. The II>III-series could be reversed into a level-rising series:

$$e4 > e3 > \underline{e2 \ge e1} > \varepsilon > æ > \alpha > a$$
II>III $a > \alpha > æ > \varepsilon > \underline{e1 \ge e2} > e3 > e4$

But the definition system brings out the structural-material differences that arise in the overall context: while /e1/ keeps its definition as above, remaining a P/S-element, the fact that /e2/ appears in a structurally different context changes its properties; in the original developmental series it has the above successor-properties, but as a predecessor of /e1/ in the reversed series it belongs to the predecessor-series, and as such it does not break into [+Last] and breaks away from [+Last] the series of its successor. We can again see the inadequacy of the two-state model; it does not capture the structural differences between the transitions

5.2. *All language change is equiempirical* 361

'/e2/ > /e1/', and '/e1/ > /e2/', differences due to the chronology. – The third group is also equivocal:

Group 3			P/S	>	S	
II>V	a >	α > æ1 > ε1 >	e	>	ε2 > æ2 > α2 > a2	rising > falling
V>IX		e > ε1 > æ1 >	α	>	æ2 > ε2 > e1 > ẹ/ι	falling > rising
		e > ε > æ >	α	>	a > ɔ > ǫ > o > ọ	

The definition of the breaks looks like this:
The **predecessor** elements:
e   P does not break away from (the series of) its own predecessor -First
         P breaks into (the series of) its successor -Last
         P breaks into (the series of) its predecessor -First
         P does not break away from (the series of) its successor -Last

The **successor** elements:
ε2   S does not break away from (the series of) its own successor -Last
         S breaks into its predecessor -First
         S does not break into the series of its predecessor +First
         S breaks into (the series of) its successor -Last
         S does not break away from its predecessor -First
         S breaks away from the series of its predecessor +First

The break is defined by: **-P [-First, -Last] & S- [-Last, -First, +First]**.

    The preceding analysis has demonstrated (1) how different the predecessor and the successor of a two-state transition are and (2) that there are more different transitions. The synchronic point of view levels all combinations, series and elements under one type. The synchronic cut of the omniscient historian's two-state model constitutes data in an empirically irrelevant manner. The developmental series of Group I appear as conjunctions of two well-defined subseries; the subseries of the other two groups exhibit some ambiguity in this respect. For the sake of relative completeness I shall define the transitions of three elements in an inertial series such as ... a4 > a5 > a6 > ... .

    Here a4 and a5 will receive the same specifications; neither breaks away from (the series of) its predecessor [-First]; both break into (the series of) its successor [-Last] and into (the series of) its predecessor [-First]; and neither breaks away from (the series of) its successor [-Last]; and if the successor a6 has the same successor as above, its specification will be [+Last, -First, +Last, -First] as above.

    Not only has the inadequacy of the synchronic two-state model been demonstrated, but the relevance of the structural principle (of context) and that of the series and the element as two different phenomena have been shown. The latter two points will be elaborated in the following.

    The present creates the past; the present's movement into the present entails a structural history in which the meaning of the past and the meaning of the present arise through and in structural processes. The meaning of the past is never identical with the meaning of the present that became this past. Thus change *qua* a replacement (cf. Traugott and Hopper 1993:10, 33; Samuels 1972; Hoenigswald 1966) which never replicates the replacé has been given a rational basis.

The serial and structural nature of the developmental series is invitable because of the empirical necessity of existence (see chapter 6.3.): a successor creates a predecessor and develops into the present.

1)	a4 >	a5 >	**a6** >			a6 as the successor
	-F, -L	-F, -L	**-F,**			
	-F, -L	-F, -L	**-F**			
2)	a4 >	a5 >	**a6** >	a7 >		+ level inertial element[34]
	-F, -L	-F, -L	**-F, -L**	-F		
	-F, -L	-F, -L	**-F -L**	-F		
3)	a4 >	a5 >	**a6** >	a7 >	a8	+ level inertial series
	-F, -L	-F, -L	**-F, -L**	-F, -L	-F	
	-F, -L	-F, -L	**-F -L**	-F, -L	-F	
4)	a4 >	a5 >	**a6** >	α >		+ level ertial element
	-F, -L	-F, -L	**-F, +L**	+F,		
	-F, -L	-F, -L	**-F, +L**	+F,		
5)	a4 >	a5 >	**a6** >	α1 >	α2	+ level ertial series
	-F, -L	-F, -L	**-F, +L**	+F, -L	-F	
	-F, -L	-F, -L	**-F, +L**	+F, -L	-F	
6)	a4 >	a5 >	**a6** >	ɔ >		+ level ertial element
	-F, -L	-F, -L	**-F, +L**	+F,		
	-F, -L	-F, -L	**-F, +L**	+F,		
7)	a4 >	a5 >	**a6** >	ɔ >	o̜	+ rising ertial series
	-F, -L	-F, -L	**-F, -L**	+F, +F, -L	-F	
	-F, -L	-F, -L	**-F, -L**	+F, +F, -L	-F	

As the specifications indicate, a6 can only be decribed partially when it is the here-and-now intersection (1), and its historical nature depends on the material nature of its successor. The former fact illustrates the relevance of the series, and we can infer from the series that the minimum series is a four-element one: the period concept has now been given a rational basis independent(ly) of the point concept. A given intersection, a synchronic point existing at T1, cannot be analyzed adequately as a structural phenomenon, synchronically influenced by and influencing its context: each member of the two-state model is dependent on what

---

34 The full specification of a6 indicates that a series will be created by its respective successors.

precedes it and succeeds it. Not only does the successor determine the nature of its predecessor, but the latter is also determined by the successor's successor: *the two-state model yields no exhaustive analysis of its two constituting (synchronic) phenomena, each of which is dependent on the historical conditions under which they exist.*

To sum up: a dynamic conception of the everyday language as a historico-structural phenomenon has been demonstrated under a general theory (sentence) of **ex**istence; there is an empirically relevant distinction between the element and its series; a developmental series consists of at least four structurally related elements; the synchronic linguist's synchronic cut can be given a rational basis; but it is not to be generalized under one definition; no successor of a transition – the synchronic linguist's datum – appears as specified immanently by only [+First] and/or [+Last] features; and finally a definition system is to be set up that captures the temporal period of the developmental series and this in a language that is consistently dynamic and sign-functional[35].

The state concept of synchronic linguistics interferes with the historical process and data; neither can the combined developmental series be reduced to a sum of combinations of the series in 5.2.4 *sub* A and B nor can the series be reduced to being a sum of elements. *That historical phenomena are members of an autonomous synchronic class delimited by synchronic cuts is a metaphysical position*; the postulation of such an object, to use Hjelmslev's echo of Hermann Paul, is *a metaphysical hypothesis which linguistic science will have to be freed of* (1963:23). The so-called *object under examination*, be it a Glossematic analysandum or a structuralist system or a generativist module (cf. Traugott and Hopper 1993:32-34[36]) cannot logically be an immanent totality. Immanence implies no external relations, only internal ones, which, however, are external in relation to the parts of the immanent object. The immanent object of the analytical mind invariably leads to a thing-in-itself concept, be it a totality or an unknowable (Cassirer 1953:305). The analytical mind must suspend its own fundamental principle – in Cassirer's words – *we never know things as they are for themselves, but only in their mutual relations* and even the datum's permanence is a *relation of permanence* besides its relation of change (p. 305). Any element has a transobjective sig-

---

35 The exposition so far has considered the expression, not the content of the sign function.

36 Hopper and Traugott's theoretical (synchronic) basis limits their vision of language development: change as merely rule-change – criticized as early as 1972 by Samuels – presupposes that we know what 'a rule' is: a synchronic rule carrying universal/general validity is yet to be set up; and in connection with what was – they – said above about the need for a metalanguage they commit this 'analytical' fallacy: *Language does not exist separate from its speakers. It is not an organism with a life of its own* – I agree with this, but when they continue: *(...) language is characterized by an abstract set of rules independent of language users (...)* (p. 33), they continue the astructural-analytical tradition of separating what also coheres: the synthetic corollary of their position is that no speakers exist separate from (their) language and no speaker is an island unto himself. They eliminate *le sujet,* analyzing an entity into autonomous parts. Secondly, although they accept a simultaneity (p. 36) aspect in their theory, and in spite of the fact that they are critical of the generativist model of change (in Andersen's conception; see above), their over-all conception is a static two-state model with longer or shorter periods of 'developmental gestation' between the start and end of the process (pp. 34-38). Thirdly, it is surprising that they do not not consider Samuels's far more comprehensive theory of grammaticalization, seeing that the over-all validity of their conclusions is not only contingent on a specific synchronic state-model, but also dependent on the assumption that sound change plays no role in grammaticalization.

nificance that the analytical mind cannot grasp by *applying* a synthetical approach, when this is needed (Lévi-Strauss). It follows from the generalization of the sign function that the historical phenomenon is 'transcendent to itself' in that it can never 'mean itself'. And, if I may borrow Cassirer's succinct phrase (p. 306; cf. p. 299): we grasp *the category of thing through the category of relation*. The logic of the *langue* concept is not as simple as its autonomy conception supposes (see Meillet 1966:18): *la langue* presupposes *la société*, and *les sociétés humaines* presuppose *le langage*, hence *la langue* presupposes *le langage*, and as the latter consists of the conjunction of *la langue* and *la parole* (Saussure 1972:112), *la langue* presupposes itself and *la parole*; the purported point-in-time quality of *la langue* (or the state) both presupposes the extension-in-time quality of *la parole* and is transcendent to itself: and we are back to Saussure's historicism: *pour être dans la réalité* (*vivante*) the everyday language must be envisaged as a historical phenomenon, which 'never means itself'.

5.3. The definition system continued; the historian's focus; chronological series and nonchronological relationships: resumption of the trancendence problem. The notion of self-difference *vis-à-vis* self-identity. The concept of a cut is provisionally defined by the concepts *a break around* and *no break around*, i.e. around the past of the present's here-and-now intersection. State-transitions are polysemous

The developmental series A>B>C>D represents a chronology that may be seen from two angles: as a chronological series, characterized by the empirical relation 'to develop into', which, in turn, is defined by the irreversible directionality of **ex**istence. As a nonchronological relationship, it is reversibly directional and characterized by two relations which are the contributions of the human mind (to reality): 'to develop into' and 'to come from', the predecessor's and the successor's respective movements when the latter movements have been lifted out of the present's constant development into the present, *i.e.* their existential context[37]. Both relations are transphenomenal, derivable from the development-into relation of **ex**istence and belong to the theoretical level[38].

The historian's *focus* is the present phenomenon that has become the past of the present that it developed into – a phenomenon that has completed its historical nature becoming both a successor and a predecessor. C and B in the above series are such foci.

The dynamics of the chronological series cannot be grasped analytically, because existence is a synthetical-structural process in which A develops into B, becoming the past of B, which develops into C, becoming the past of C, which develops into D, becoming the past of D, which … . This empirical process will be translated into the level of theory as the basic historical point of view: something (the present) develops into (creates) something else (the present). I shall say that C and D contract similarity and difference relationships, C and B being both different from and similar to D and C, respectively.

---

37 The arguments of the syllogisms in 5.3.2 below illustrate my point.
38 The synchronic tradition does not recognize the polysemy of the notion of 'to develop into'.

5.3. The definition system continued

In a nonchronological relationship the predecessor or successor contract similarity- and difference-relationships with their respective successors and predecessors.

The historian cannot predict from his so-called synchronic datum (= B) either what the next successor is going to be or what B has come from. Man's historical sense is not strong enough to elicit the specific nature of a successor *qua* the present-to-be and its predecessor *qua* the present-that-was. But we do know that B has a predecessor and has or will have a successor (see note 39): *a given successor S is historically related with a successor S and a predecessor P, but not only if S develops into S' or comes from P'*. This sentence excludes entelechy-, telos-based or eschatological explanations, as it focusses on the inevitable process itself[39], not on the course (trajectory) of development. It also makes room, as it were, for the historical theory's contributions in terms of establishing nonarbitrary and noncontingent developmental series: historical accuracy.

In synchronic linguistics it is more often the case than not that no overt attempt is made to distinguish between the levels of theory and data (*pace* Glossematics). Coseriu set great store by the fact that 'synchrony' belonged to the level of theory or description, not to that of data; but he did not pursue the question: why do we need the concept if it has no empirical relevance, having no (class of) denotata? The synchronic state belongs to the level of theory. Grammaticalization proponents are *initially* aware that the word *grammaticalization* refers both to data and to the study of this data, and emphasize that their data makes the synchronic concept inadequate[40]. The synchronic linguist assumes that he can grasp his object with no help from his historical sense. The synchronic description *is* how the synchronic state (or system) would appear to us if we were able to empirically grasp it in one instantaneous act (of the analytical part) of the mind. This assumption conforms to the reification of the everyday language, and as such it implies the notion of the synchronic cut: the synchronic conception reads a foreground into its initial (surface-)picture of data (cf. Cassirer 1953:291), thereby automatically creating a background that is discarded, as lying 'outside' (before or after) the 'foreground motive'. The next step is easy to see: the 'motive' is cut free of its external relations, it loses its relational, foreground-properties, becoming a synchronic *sui-generis* object – by the undefined cut.

However, we need the concept of a theoretical cut in order to match our grasp of **exist**ence. There is no absolute gap between two elements in a developmental series; there is no gap in our historical lives; but there is a difference between the present and the past. We shall say that the cut is produced by a cutting element, which, accordingly, has the property of creating varying degrees of coherence. Theoretically, a cutting element is any one element of a given series; empirically there is only one: the here-and-now intersection of the present: the element that constitutes the present moment in conjunction with what has been continued of the previous present into the new present (see 1.5.12-1.5.16 for *parole*'s cutting elements).

---

39 It is impossible to imagine the present as a present that does not create the present: from this existential inevitability – Nitzsche's *vivo*, not the Løgstrup-Augustine *moriar* – we may say that a successor always creates a successor (cf. 5.2.)

40 The basic problem with this position – with which the historian cannot disagree – is that *e.g.* in neither Heine *et al.* (1991) nor Hopper and Traugott (1993) do we see an independent analysis of the synchrony concept(,) which they criticize, and it is not clear what they mean by the word *synchrony*.

The concept of a cut will now be defined with reference to some of the developmental series set up in chapter 5.2. The analysis will show how data is more than raw material for a theory, how it has an *energeia* of its own. – I shall begin with the III>VI series of Group 1.

Here /a6/ so cuts the series that it becomes an element of the preceding series, while /α1/ so cuts the (preceding) series that it does not become an element of it. The successor of /α1/, /α2/ so cuts the series that a new series is about to be established, in that it makes /α1/ cohere with it.

In Group 2 the 'energy' of /a6/ of I>II and its successor /α/ is the same as above, while the successor of /α/, /æ/ makes /a6/ a member of the same series as /α/. Even from this semi-dynamic stance it is possible to see the operation of structural processes: both /a6/ and /α1/ constitute the same transition as in Group 1, but their respective successors change this picture, establishing different gaps[41]. However, the logic of a cut(ting element) is more complex than that.

The present relates to its past in two ways: it breaks away from and breaks into what becomes its past. It does not break into the future, something that does not exist! – an ideal construct. As the present cannot have the property of breaking away from its past, the fact that it also breaks into the latter is a corollary of the former process – against the horizon of continuation and under the aegis of change. The present element does not consist of (structurally acquired) [+Last] properties, but of dynamic [+First] properties; hence there cannot be a break after the present: * '... > C > D]'. **Ex**istence creates a difference with the past which is translated into an immediate break before the present, and therefore also a break after what became the past of this present: '... > C] > [D'. And as the next corollary of **ex**istence, the past C must still have its break before it: * '... > [C] > [D', so that the immediate past of the present will appear as an autonomous, delimitable entity. The synchronic paradox will now become clear: the seeming immanence or autonomy of an entity follows from the historical process of existence; the cuts round an entity are consequences of this process, so that the constitution of the synchronic element – state or system – is a historical fact which has been deprived of its dialectical-synthetic properties in order that it may become an analytically conceivable phenomenon.

5.3.1

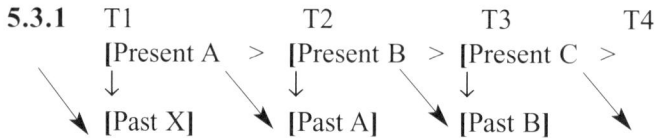

	T1	T2	T3	T4
	[Present A] >	[Present B] >	[Present C] >	

Existence from T1 to T2 to T3.
(The present's development into the present under the aegis of change)

The 'existential' break before Present C is co-existential with Present B's development into Present C, and this development implies the break after Past B, which carries with it its existential break before it, etc. The interesting thing here is that the autonomous phenomenon [Past B] is different from what it was as Present B, a fact that accords well with Labov's Historical Paradox: we do not know how much the present and the past or for that matter Past

---

[41] This also applies to the series of Group 3.

5.3. The definition system continued

A and Past B differ. Therefore the Law of Transitivity does not apply to historical phenomena: the following syllogisms commit the fallacy of the Four Terms, seeing that the 'B' is the successor in A developing into B and the predecessor in B developing into C[42]:

5.3.2 A develops into B   A is the past of B   B is the present of A
       <u>B develops into C</u>   <u>B is the past of C</u>   <u>C is the present of B</u>
       A develops into C   A is the past of C   C is the present of A.

Secondly, the conclusions establish nonchronological relationships, while these appear as chronological series in the premises. **A nonchronological relationship is therefore a resoluble syncretism, resoluble into chronological series**, and the historian's task is to resolve these syncretisms into such series on the analogy of his (innate) knowledge of how the present develops into the present – his historical sense.

The form of a break is not as simple as 5.3.1 suggests; in chapter 5.2. it was demonstrated how the period in the shape of different developmental series created different breaks; these breaks will now be specified more fully, when the definition system of the period has been set up. Two sets of definitions will be established of which the former highlights the facts of discontinuity, and the second set will expound the empirical nature of the breaks in the developmental series.

The arrows in Figure 5.3.1 represent the following processes: endogenously ('>'), the present is determined by its past, and the present determines its past katagenously ($\downarrow$), while '$\searrow$' represents the part of the process of external endogeneity that continues the 'past' into the present under the aegis of change. The nonchronological relationships in 5.3.2 will be summed up in the figure in 5.3.3, and are translations of the knowledge of our historical sense into the level of descriptions[43]:

5.3.3   [past A] > [past B] > [present C]   Cf. 5.3.1
     $\Rightarrow$

       Predecessor relation *is the past of*   Successor relation *is the present of*
       [A] is the past of [B] and [C   [C is the present of [B] and [A]

Whether there is a break around a given element is a question of definition; this was implied above by the introduction of the two concepts 'a break around' and its corollary 'no break around'. These concepts will be defined below.

---

42 Again one must agree with Hopper and Traugott (1993:39) in their criticism of syllogistic reasoning; but they do not see that the substitution of abductive methods for the former type of methodology is a Scylla-Charybdis choice.
43 The conclusion of the first syllogism translates the structure of the empirical process into the level of description; this is the reason why processes that we represent by 'A>B' – always at the level of logic – will be regarded as syncretisms: the applied historical theory sets the limits to the resolution. The historian tries to eliminate the gaps between the past and the present (see chapter 6.3(1)-(3)).

Two new concepts will now be introduced: an *irreversibly uniform* relation and a *reversibly uniform* relation[44]. The former characterizes the chronology of A, B and C: A is the past of B and B is the past of C; the latter characterizes the nonchronology of the three phenomena: A is the past of B and C, or C is the present of B and A. In a nonchronologcial relationship there is a fixed vantage point, A (the past of ...) or C (the present of ...), and seen from these two points the respective classes of successors and predecessors would have been turned into amorphous masses, but for the sign-functionally ordered nature of the present's development into the present and for the historical theory at the level of theory.

The historian also looks for similarities – not only differences. So the structure of non-chronological relationships and chronological series looks like this: the class of successors/predecessors is similar in respect of the predecessor relation/successor relation that also makes them mutually different: B and C contract a predecessor relation with their common predecessor A which makes C and B different at the same time. And: a predecessor A and a successor C are similar in respect of the historian's *focus*, B, which makes them different. Similarity is brought about by the fact that two elements contract a relationship with one entity which makes them different – because of the dynamics of existence. This structure combines the relational-formal and substantial-material nature of the historical phenomenon.

In inertial development there are no breaks around the elements of the developmental series, and our first data will be the level and nonlevel series of Group 1 (III, I, VIII and IV, II); let /a6/ be the present's successor, then /a5/ is the focus, and the relations will be illustrated through the translation of the chronological series into nonchronological relationships:

5.3.4 $a4 > a5 > a6 \Rightarrow$ <u>Predecessor</u>: $\underline{a4} = a5 = a6$; (= a4 is similar to a5 and similar to a6)
         <u>Successor</u>: $\underline{a6} = a5 = a4$

The predecessor and succesor of /a5/ constitute **externally symmetrical** and **internally non-contrasting** relationships – this is seen from the point of view of substance and against the horizon of similarity and difference. – This dual supposition is posed in all nonchronological relationships. In 5.3.5 the predecessor and successor of /æ1/ (in IV>VIIb) constitute **externally symmetrical** and **internally contrasting** relationships:

5.3.5 $\varepsilon > æ1 > a \Rightarrow$ <u>Predecessor</u>: $\underline{\varepsilon} = æ1 = a - and - \underline{\varepsilon} \neq æ1 \neq a$;
                 (= ε is similar to æ1 and similar to a; ε is different from æ1 and different from a)
                 <u>Successor</u> : $\underline{a} = æ1 = \varepsilon - and - \underline{a} \neq æ1 \neq \varepsilon$

An **internally non-contrasting** relationship (5.3.4) is defined as follows: *a given element – (P and S) – is relationally similar to other <u>not mutually different</u> elements in relation to a specific property in respect of which it is not different from those elements.* – See 6.3.6(2a/b).

An **internally contrasting** relationship (5.3.5 above and 5.3.7/9 below) is defined as follows: *a given element – (P and/or S) – is relationally similar to other* either (5.3.5) <u>mutual-</u>

---

[44] The concept of uniformity is introduced in order to separate out irreversible contact-motivated processes such ME *hie > thei* 'they', which are not motivated by external endogeneity.

5.3. The definition system continued 369

ly different or (5.3.7/9) *not mutually different elements in relation to a specific property in respect of which it is different from those elements – or different from at least one of those elements. – Ad* 5.3.5 see 6.3.6(1a/b).

An **externally symmetrical** relationship (5.3.4/5) is defined as follows: *the predecessor and successor (in a nonchronological relationship) contract similar internal – noncontrasting or contrasting – relations with their respective successors and predecessors. –* NB 5.3.8 and 5.3.10.

It will be demonstrated how the system of definitions[45] will eventually (see 6.3.6) combine the relational-dialectical nature of development (external endogeneity) with the material (articulatory) nature of the historical phenomenon (internal endogeneity).

Relational analysis is Janus-faced: the similarity relationships and difference relationships in 5.3.5 apply in respect of the material property *lower than* and *higher than*: the successor /a/ is 'lower' than either of its predecessors (/æ/, /ɛ/) and the predecessor /ɛ/ is 'higher' than either of its successors. – The first part-definition of the break-concept may now be set up:

Def. 5.3.6 There is **no break around** an intersection in a developmental series, if its predecessor and successor constitute either of these two nonchronological relationships: (i) symmetrical non-contrasting similarity- and difference relationships; (ii) symmetrical contrasting similarity- and difference relationships[46].

Corollary: *if there is no break around an intersection, the intersection is an element in an inertial developmental series.*

The following analysis applies to noninertial developmental series and leads up to a definition of 'a break around'; our foci will now be /a6/ of III>VI; (cf. /æ5/ of I>II and /ɛ5/ of VIII>II, i.e. predecessors in the transitions of Group 1).

5.3.7 $a5 > a6 > \alpha1 \Rightarrow$ Predecessor: $a5 = a6 \neq \alpha1$ (= a5 is similar to a6 and different from α1)
Successor: $\alpha1 = a6 = a5 - and - \alpha1 \neq a6 \neq a5$

The predecessor and successor of /a6/ constitute **externally nonsymmetrical** relationships such that the predecessor /a5/ constitutes an **internally noncontrasting** relationship and the successor /α1/ an **internally contrasting** one. The predecessor is similar to /a6/ and different from /α1/ in respect of the same material property. The successor is different from and similar to its predecessors in respect of the same material property, it being 'fronter' and/or 'higher' than the others.

---

[45] The sign-functional analysis of the developmental series in chapter 6 will make these definitions cohere as well as introduce the difference-factor in the first definition above.
[46] In this and the following definitions *symmetrical* refers to external relationships and *internal* to (non)contrasting relationships.

An **internally noncontrasting** relationship (5.3.7/8/9/10) is defined as follows: *a given element – (P and/or S) – is relationally similar to at least one of two other <u>mutually different</u> elements in relation to a specific property in respect of which it is not different from both of those elements.*

An **externally non-symmetrical** relationship is defined as follows: *the predecessor and successor (in a non-chronological relationship) do not contract similar internal – noncontrasting or contrasting – relations with their respective successors and predecessors.*

With a nonlevel predecessor series (IV>VIIb and II>VIII) the other breaks of Group 1 look like this – with /a/ and /e/ as predecessors:

5.3.8 æ1 > α > æ2 ⇒ Predecessor: æ1 ≠ α = æ2
                    Successor: æ2 ≠ α = æ1

The respective predecessors of /α/ and /e/ constitute **externally symmetrical** and **internally noncontrasting** relationships; the seeming similarity betwen 5.3.8 and 5.3.4 will be dealt with below.

The last transitions of Group 1 will be those or rather the type in III>VI, IV>VIIb and II>VIII with /α1/, /æ2/ and /ε2/ as foci. – The corresponding transitions in I>II and VIII>II (5.3.10) are similar to 5.3.8.

5.3.9 a6 > α1 > α2 ⇒ Predecessor: a6 = α1= α2 – *and* – a6 ≠ α1 ≠ α2
                     Successor: α2 = α1 ≠ a6

5.3.10 æ5 > α > æ1 ⇒ Predecessor: æ5 ≠ α = æ1
                     Successor: æ1 ≠ α = æ5

The definition of **internally contrasting** relationships (the predecessors) in non-chronological relationships was anticipated above.

This analysis, which leads to definition 5.3.6 above and definition 5.3.11 below, corroborates the previous analysis of the gaps in the developmental series in Group 1: the previous 'absolute gap' is translated into the following statements based on the break-concepts: there are **breaks around** both the predecessor and the successor of the five series in Group 1:

III>VI	a6 > α1
IV>VIIb	α > æ2
II>VII	e > ε2
I>II	æ5 > α
VIII>II	ε5 > e

5.3. The definition system continued

Def. 5.3.11 There is a **break around** an element, if its predecessor and successor constitute one of these nonchronological relationships: (i) nonsymmetrical noncontrasting (= P) and contrasting (= S) relationships; (ii) symmetrical noncontrasting relationships NB; (iii) nonsymmetrical contrasting (= P) and noncontrasting (= S) relationships[47].

The definitions oblige me to resolve (a) an ambiguity and (b) an inconsistency: (a) *the no-break-around* concept implies there is no break either before or after the element in question; the *break-around* concept implies that there is at least one break in the vicinity of the element in question. The definitions based on a chronological series of three elements do not indicate where the break occurs, before or after, say, /a6/; and the same goes for its successor /α1/. – A static solution may add an *ad hoc* definition to the effect that 'if breaks occur around both of two successive predecessors, then there occur a break after the predecessor and a break before its successor'.

The inconsistency refers to definitions 5.3.11(ii) and 5.3.6(i), as such to the level of theory, and may be resolved with reference to the chronological series itself: *if a chronological series exhibits a difference-relation, then there is a break around the historian's focus if the latter enters into internally noncontrasting relationships*[48].

The three-element period that has been the background of the above exposition has certain weaknesses which will be removed when we expand the historian's datum to being a period of four elements; with this proviso the definition system has been so adequate as to distinguish between the three subgroups of the developmental series of Group 1, and the below summary is meant to show that the reversibility that the series seem to possess is dispelled as regards the falling/rising-level series *vis-à-vis* the level-falling/rising series: 'to develop into' is not the mirror-concept of 'to come from'.

Group 1   P > S
III>VI $a3 > a4 > a5 >$ $\mathbf{a6} > \alpha 1 > \alpha 2 > \alpha 3 > \alpha 4$ level > level

The transitional break of this reversible series (cf. above)[49] will be defined as follows:

Def. 5.3.12 $a3, a4, a5 =_{df}$ 5.3.6(i)
             $a6 =_{df}$ 5.3.11(i)
             $\alpha 1 =_{df}$ 5.3.11(iii)
             $\alpha 2, \alpha 3, \alpha 4 =_{df}$ 5.3.6(i)

---

47 This definition will be changed in chapter 5.6., where it will be demonstrated that the break-around concept is not an immediate one; the break before/after a historical phenomenon is mediated by the concept of *a potential break*.
48 This inconsistency will be dissolved in chapter 6.3.6, where the material definitions will be combined with devaluation definitions, the concept of devaluation being defined in the following sections.
49 I shall not tax my reader's patience: α1 and a6 will exchange specifications when α1 becomes the predecessor of a6.

IV>VIIB	e > ɛ > æ1 >	α	>	**æ2** >	æ3 > æ4 > æ5	falling > level
II>VIII	a > α > æ > ɛ1 >	e	>	**ɛ2** >	ɛ3 > ɛ4 > ɛ5	rising > level

The breaks in these two series, which are not reversible, will be defined as follows:

Def. 5.3.13   …, æ1/ɛ1 =$_{df}$ 5.3.6(ii)
α/e =$_{df}$ 5.3.11(ii)
æ2/ɛ2 =$_{df}$ 5.3.11(iii)
æ3/ɛ3, … =$_{df}$ 5.3.6(i)

I>II	æ2 > æ3 > æ4 >	**æ5** >	α > æ1 > ɛ > e > ι	level > rising
VIII>II	ɛ2 > ɛ3 > ɛ4 >	**ɛ5** >	e > ɛ1 > æ > α > a	level > falling

Def. 5.3.14   …, æ4/ɛ4 =$_{df}$ 5.3.6(i)
æ5/ɛ5 =df   5.3.11(i)
α/e =$_{df}$ 5.3.11(ii)
æ1/ɛ1, … =$_{df}$ 5.3.6(ii)

5.4.

The material definitions of the other developmental series. Further justification of the period *qua* an analysandum in its own right. Different types of 'breaks around'. Serial relationships are not symmetrical. Gradualness and multiple conditioning *qua* factors that 'bridge' gaps solve a problem

The series of Groups 2 and 3 will now be analyzed in accordance with the same Archimedean-like principles as those of Group 1, with the point of view still the material one:

Group 2

I>II	a3 > a4 > a5 >	**a6** >	α > æ > ɛ > e > ι	level > rising
III>IV	u3 > u4 > u5 >	**u6** >	ʊ > ǫ > o > ǭ > ɔ	level > falling

will be defined by   a4/a5//u4/u5   5.3.6(i)
a6//u6   5.3.11(i)
α//ʊ   5.3.6(ii)
æ//ǫ, …   5.3.6(ii).

II>III	a > α > æ > ɛ	> **e1** >	**e2** >	e3 > e4	rising > level
IV>VII	e > ɛ > æ > α	> **a1** >	**a2** >	a3 > a4	falling > level

will be defined by   æ/ɛ // æ/α   5.3.6(ii)
e1//a1   5.3.11(iii)
e2//a2/   5.3.6(i)
e3//a3, …   5.3.6(i)

5.4. The material definition of the other developmental series

To repeat: the two series level-falling/rising and falling/rising-level are not mirror developments of each other; the intersection around which a break occurs changes its material nature according to that of the elements preceding and following it. – In Group 3, the V>IX2 series of Group 3 will be dealt with separately:

Group 3

II>V	a > α > æ1 > ε1 >	e >	ε2 > æ2 > α2 > a2		rising > falling
V>IX	e > ε1 > æ1 >	α >	æ2 > ε2 > e1 > e/ι		falling > rising
V>IX2	e > ε > æ >	α >	a > ɔ > ǫ > o > ọ		

will be defined as follows:

        æ1/ε1//ε1/æ1   5.3.6(ii)
        e//α   5.3.11(ii)
        ε2//æ2   5.3.6(ii)
        æ2//ε2, ...   5.3.6(ii)

These definitions are sufficient to differentiate each group from the other two as well as the subgroups within the main groups. But the definitions entail three questions: are the two (three) types of break empirically relevant? Are the breaks in Groups 2 and 3 similar? Does it make sense to claim that in Groups 2 and 3 there is a break around the predecessor of the transition?

In Group 1 the gap was clear, defined by a transition around whose predecessor and successor there were breaks. By definition the developmental series may be regarded unequivocally as conjunctions of two inertial developmental series. Other combinations of inertial (sub)series and their consequent definitions create no clear gaps. The structural nature of the elements of a transition alter the pre-existing nature of the elements in question; and the gaps in Groups 2 and 3 suggest that the notion of a cut is intimately tied to the notion of the cutting element's *Eigennatur*.

My definition system has vindicated the period as a theoretically and empirically relevant concept. Consider the following three-element series, A>B>C series:

1)   a6 > α1 > α2
2)   ε > e1 > e2

Both α2 and e2 (= C) create a break around their respective predecessors (= B); but the breaks differ depending on their respective predecessors:

1a)   a5 > a6 > α1 > α2
2a)   æ > ε > e1 > e2

Knowledge of the historian's focus, B, presupposes knowledge of its predecessor A and the successor C: the fact that /a5/ is the predecessor of /a6/ makes /α2/ create breaks around both B and A, while the fact that /æ/ is the predecessor of /ε/ makes /e2/ create only a break around B.

To conclude: **the period has been vindicated as an empirical entity that necessarily consists of 4 'states'.** How long such a period is in so-called real time is not the issue. Secondly, **the principle of structure as a dynamic-historical one is not in doubt**.

The problem mentioned above is due to the phonetic properties of Cardinal-4 and Cardinal-5 vowels, and in general to the representation of the articulatory basis of sounds. The traditional quadrilateral is a gross idealization of the oral cavity – from glottis to the front teeth. I have treated the sounds represented by α and a as if they contracted the same relationship as, say, u and o, or e and ε.

V>IX2 $e > \varepsilon > æ > \alpha > a > ɔ > \underset{.}{o} > o > \underset{.}{o}$ a falling>rising series

Following the ellipsis diagram in Jones (1962:36; cf. Ladefoged 1975:13; Labov 1994:259[50]), Cardinal 5 [a] is lower than Cardinal 4 [α], and may as such belong to the predecessor series, with /α/ specified by definition 5.3.6ii, as a member of a nonlevel inertial series. The break will therefore be established around C5 [a]:

5.4.1i $æ > \alpha > a$ ⇒ Predecessor: $æ = \alpha = a$ – *and* – $æ \neq \alpha \neq a$
                                Successor: $a = \alpha = æ$ – *and* – $a \neq \alpha \neq æ$

5.4.1ii $\alpha > a > ɔ$ ⇒ Predecessor: $\alpha = a = ɔ$ – and – $\alpha \neq a \neq ɔ$
                                Successor: $ɔ \neq a \neq \alpha$

The point is that C5 represents an absolute low in respect of tongue-height, and will as such be lower than both /α/ and /ɔ/. Therefore 5.4.1i is a straightforward inertial relationship based on 'higher than' and 'lower than'. In 5.4.1ii /α/ may be higher than both its successors; but there is no gradualness involved, defined by 'higher than'.

With Cardinal 5 as the articulatory nadir /ɔ/ is higher than /a/ and lower than /α/, a possibility that breaks the gradual orderliness of the definition system. Finally, /α/ and /ɔ/ may be of the same height, a possibility that may not be physically viable, if there is a fixed correlation between tongue-height and 'roundness', in the sense that the roundness of back vowels requires a certain degree of raising of the back of the tongue.

The notions of gradualness and multiple conditioning will solve the problem. The theoretical importance of gradualness follows from existence, where abruptness or suddennes has no relevance: we see no gaps in existence or revolutions that do not create coherence with the past. The Law of Tradition is grasped against the horizon of coherence and attachment. The parameter exists at the level of theory: the basic principle of gradualness is the abovementioned constant hierarchy-like resolution of the 'syncratic' nonchronological relationships: 'B<A', into 'A > ... > ... > B', where the historian tries to establish as many trans-

---

[50] The feature tree of generative grammar (Kenstowicz 1994:146) is both too abstract and not elementary enough; furthermore, there is no consensus about how the details of such a tree are to be *worked out* (p. 146). In Harris (1994:114-115) it is virtually impossible to extract a well-defined English sound system.

5.4. The material definition of the other developmental series

itions as possible to create gradually greater degrees of coherence. The relevance of 'degrees of coherence' follows from the different types of gap that our developmental series have established; this implies a notion of 'bridging factors', which will be developed in chapter 6.1 and 6.2., and used here in connection with multiple conditioning. In the above case we shall need another conditioning feature, namely [more/less retracted]: in the series /α > a > ɔ/ the predecessor is less retracted than either of its successors, and the successor is more retracted than its predecessors:

5.4.2 α > a > ɔ ⇒ Predecessor: α = a = ɔ – *and* – α ≠ a ≠ ɔ
                    Successor: ɔ = a = α- *and* – ɔ ≠ a ≠ α

The transition is defined by being externally symmetrical and internally contrasting. The principle of multiple conditioning asserts itself where asymmetrical physical conditions for speech production – the oral cavity (Labov 1994:256-257) – impose themselves, so that the determination of gradualness is a participative function with abruptness – (the higher/ lower parameter being inapplicable) – understood against the horizon of gradualness (more/less retracted)[51].

Labov (1994:258) sets up two gradual trajectories from the low *a*-area, C5, moving up either in the front region or in the back region; he does not see a similar connection in the low area, regarding C1 [i] as the stopping place for all rising sounds.

The preceding analysis results in these Archimedean definitions of the types of break that we have been dealing with:

5.4.3 A SUMMARY OF THE SERIES' ARCHIMEDEAN-LIKE DEFINITIONS

Group 1
III>VI   5.3.6i + **5.3.11i** + **5.3.11iii** + 5.3.6i
          LEVEL>LEVEL

II>VIII   5.3.7ii + **5.3.11ii** + **5.3.11iii** + 5.3.6i
IV>VIIB   RISING>LEVEL
          FALLING>LEVEL

I>II   5.3.7i + **5.3.11i** + **5.3.11ii** + 5.3.6ii
VIII>II   LEVEL>RISING
          LEVEL>FALLING

---

51 I have chosen height as the elemental feature because retraction/fronting is in general not as easily recognized *by the perceptive apparatus* (Labov 1994:238). Furthermore retraction seems to be a concomitant characteristic of lowering (laxing), both in the front and back region (see *e.g.* Labov's ellipsis p. 259; cf. p. 236). It is to be noted that the non-peripheral – laxing correlation only applies to the front vowels (including Cardinal 5), while retraction by lowering of the back vowels – from C /u/ – entails 'more peripherality'.

Groups 2 and 3

I>II2	5.3.6i + **5.3.11i** + 5.3.6ii + 5.3.6ii
III>IV	LEVEL>RISING
	LEVEL>FALLING
II>III	5.3.6ii + 5.3.6ii + **5.3.11iii** + 5.3.6i
IV>VII	RISING>LEVEL
	FALLING>LEVEL
II>V	5.3.6ii + 5.3.6ii + **5.3.11ii** + 5.3.6ii
V>IX	RISING>FALLING
	FALLING>RISING

In these definitions the original difference between Group 1 and Groups 2 and 3 has been formalized and the phonetic conditions, stated in def.s 5.3.6 and 5.3.11, for the breaks were the applied point of view; the problem with this static application of the definitions is that the position of the break is to be inferred from the data. Next I shall set up a number of **transition** definitions from a more period-oriented point of view, but still against the general horizon of the two-state model, with the latter implying the logical possibilities in 5.4.4[52]:

5.4.4i 1) a break after   X] [Z   a break before   def. 5.4.10-11-12
          2) no break after   X [Z   a break before   def. 5.4.7-8-9
          3) a break after   X] Z   no break before   def 5.4.13-14-15
          4) no break after   X Z   no break before   def. 5.4.5-6

But the principle of immanence or autonomy implies that there is a break around the entity in question, be it an element, a state or a system: [X] or [Z], so that the analysandum has been cut off from its entire historico-structural context:

5.4.4ii 1) a break after and before   [X]   (Synchronic) scientific datum
           2) no break after and a break before   [X   No scientific datum
           3) a break after and no break before   X]   No scientific datum
           4) no break after and no break before   X   No scientific datum

A period of at least four intersections is a precondition for the establishment of a structurally relevant interpretation of the break concept. The first group, where the the transition X>Z is defined by no break after X and no break before Z consists of the inertial developments in series III>VI, I>II, VIII>II of Group 1 and I>II, III>IV of Group 2. – What X and Z represent can be seen from the appendix to this chapter.

---

52 These possibilities established against the horizon of logical contradiction will be modified in chapter 5.6.4 below.

5.4. The material definition of the other developmental series

Def. 5.4.5 With X and Z as the historian's data, there are no break after X and no break before Z, if the respective predecessors and successors of X and Z constitute symmetrical non-contrasting similarity-difference relationships.
→ Transition definition: **5.3.6i + 5.3.6i**.

The second group consists of the inertial developments in II>III, IV>VII of Group 2, IV>VIIB, II>VIII of Group 1 and II>V, V>IX of Group 3:

Def. 5.4.6 With X and Z as the historian's data, there are no break after X and no break before Z, if the respective predecessors and successors of X and Z constitute symmetrical contrasting similarity-difference relationships.
→ Transition definition: **5.3.6ii + 5.3.6ii**.

Inertial developments have now been defined in relation to the basic phonetic differences between their respective transitions.

The next transition is characterized by there being no break around X1 and a break around Z1, which <u>implies</u> a break before Z1; The transition is found in III>VI, I>II, VIII>II; I>II, III>IV:

Def. 5.4.7 With X1 (= Z) and Z1 as the historian's data, there are no break after X1 and a break before Z1, if the predecessor and successor of X1 constitute symmetrical noncontrasting similarity-difference relationships, and the predecessor and successor of Z1 constitute nonsymmetrical noncontrasting (= P) and contrasting (= S) similarity-difference relationships.
→ Transition definition: **5.3.6i + 5.3.11i**.

X1 and Z1 will now refer to II>III, IV>VII, yielding the following definition:

Def. 5.4.8 With X1 (= Z) and Z1 as the historian's data, there are no break after X1 and a break before Z1, if the predecessor and successor of X1 constitute symmetrical contrasting similarity-difference relationships, and the predecessor and successor of Z1 constitute nonsymmetrical contrasting (= P) and noncontrasting (= S) similarity-difference relationships.
→ Transition definition: **5.3.6ii + 5.3.11iii**.

Referring to IV>VIIB, II>VIII; II>V, V>IX, intersections X1 and Z1 will yield this transition-type:

Def. 5.4.9 With X1 (= Z) and Z1 as the historian's data, there are no break after X1 and a break before Z1, if the predecessor and successor of X1 constitute symmetrical contrasting similarity-difference relationships, and the predecessor and successor of Z1 constitute nonsymmetrical noncontrasting similarity-difference relationships.
→ Transition definition: **5.3.6ii + 5.3.11ii**.

378   *Chapter 5. Towards the theory and its static application*

The next transition type, X2>Z2, is the ones in Group 1, where there are breaks around both elements; from this a break after X2 and a break before Z2 are implied:

Def. 5.4.10 With X2 and Z2 as the historian's data, there are a break after X2 and a break before Z2, if the predecessor and successor of X2 constitute nonsymmetrical noncontrasting (= P) and contrasting (= S) similarity-difference relationships, and the predecessor and successor of Z2 constitute nonsymmetrical contrasting (= P) and noncontrasting (= S) similarity-difference relationships.
→ Transition definition: **5.3.11i + 5.3.11iii**.

Def. 5.4.10 formalizes the transition in III>VI, while definition 5.4.11 captures the transitions in IV>VIIb, II>VIII:

Def. 5.4.11 With X2 and Z2 as the historian's data, there are a break after X2 and a break before Z2, if the predecessor and successor of X2 constitute symmetrical noncontrasting similarity-difference relationships, and the predecessor and successor of Z2 constitute nonsymmetrical contrasting (= P) and noncontrasting (= S) similarity-difference relationships.
→ Transition definition: **5.3.11ii + 5.3.11iii**.

Definition 5.4.12 formalizes the transitions in I>II, VIII>II:

Def. 5.4.12 With X2 and Z2 as the historian's data, there are a break after X2 and a break before Z2, if the predecessor and successor of X2 constitute nonsymmetrical noncontrasting (= P) and contrasting (= S) similarity-difference relationships, and the predecessor and successor of Z2 constitute nonsymmetrical noncontrasting similarity-difference relationships.
→ Transition definition: **5.3.11i + 5.3.11ii**.

The final three definitions refer to the transitional ambiguity of the transitions in Groups 2 and 3: there is a break around X2, which implies a break after X2, and there is no break around Z2, which implies no break before Z2:

Def. 5.4.13 With X2 and Z2 as the historian's data, there are a break after X2 and no break before Z2, if the predecessor and successor of X2 constitute nonsymmetrical noncontrasting (= P) and contrasting (= S) similarity-difference relationships, and the predecessor and successor of Z2 constitute symmetrical contrasting similarity-difference relationships.
→ Transition definition: **5.3.11i + 5.3.6ii**.

5.4. The material definition of the other developmental series

This definition refers to transitions I>II, III>IV of Group 2, while definition 5.4.14 refers to II>III, IV>VII:

Def. 5.4.14 With X2 and Z2 as the historian's data, there are a break after X2 and no break before Z2, if the predecessor and successor of X2 constitute nonsymmetrical contrasting (= P) and noncontrasting (= S) similarity-difference relationships, and the predecessor and successor of Z2 constitute symmetrical noncontrasting similarity-difference relationships.
→ Transition definition: **5.3.11iii + 5.3.6i**.

Definition 5.4.15 refers to II>V, V>IX of Group 3:

Def. 5.4.15 With X2 and Z2 as the historian's data, there are a break after X2 and no break before Z2, if the predecessor and successor of X2 constitute symmetrical noncontrasting similarity-difference relationships, and the predecessor and successor of Z2 constitute symmetrical contrasting similarity-difference relationships.
→ Transition definition: **5.3.11ii + 5.3.6ii**.

The static application[53] of the material definitions in 5.3.6 and 5.3.11 entails a dilemma. In the developmental series of Groups 2 and 3, the element around which a break occurs is tied off as an immanent entity with a break after and before it (cf. 5.4.4ii); the same applies to the corresponding elements in Group 1. As such the synchronic linguist has an instrument, namely the applied historical, periodic definitions, by which his immanent state can be constituted; but as we can see, it cannot be generalized to the elements of all transitions.

The basic material of this preliminary exposition has been two-state developments, expanded by the intersections required by the theory; the predecessors of the class of X and the successors of the class of Z have been supposed as 'unknowns' in respect of coherence; and we saw how what is historically unknown becomes knowable under the aegis of historical development and a historical theory. Despite the fact that the applied point of view freezes data into a still life, the outcome of the analysis does demonstrate how historical data has a structural foundation, as an element changes its properties in accordance with its chronological context. This indicates that the two material definitions have not been restricted *a priori* as regards their scope, they can be applied statically and dynamically.

These problems will disappear once we apply a strictly dynamic approach in accordance with the irreversible directionality of **ex**istence.

---

53 There is no reason to expand the analysis; with X3 and Z3 as our transition, the system will unacceptably yield a break after X3 and no break before Z3 in Group 1, while there are no breaks after X3 and before Z3 in these transitions in Groups 2 and 3.

Appendix

A summary of the developmental series

		X	Z	X2	Z2	X4	Z4		
Group 1			X1	Z1	X3	Z3			
Level > level		a3	a4	a5	a6	α1	α2	α3	α4	III>VI
Level > rising		æ2	æ3	æ4	æ5	α	æ1	ɛ	e	I>II
Level > falling		ɛ2	ɛ3	ɛ4	ɛ5	e	ɛ1	æ	α	VIII>II
Group 2				X1	Z1	X3	Z3			
		X	Z	X2	Z2					
Level > rising		a3	a4	a5	a6	α	æ	ɛ	e	I>II
Level > falling		u3	u4	u5	u6	ọ	o	ǫ	ɔ	III>IV
		X	Z	X2	Z2					
			X1	Z1	X3	Z3				
Rising > level		α	æ	ɛ	e1	e2	e3	e4	e5	II>III
Falling > level		ɛ	æ	α	a1	a2	a3	a4	a5	IV>VII
Group 1										
Falling > level		e	ɛ	æ1	α	æ2	æ3	æ4	æ5	IV>VIIb
Rising > level		α	æ	ɛ1	e	ɛ2	ɛ3	ɛ4	ɛ5	II>VIII
Group 3										
Rising > falling		α1	æ1	ɛ1	e	ɛ2	æ2	α2	a	II>V
Falling > rising		e1	ɛ1	æ1	α	æ2	ɛ2	e2	ẹ	V>IX
			X	Z	X2	Z2	X4	Z4		
				X1	Z1	X3	Z3			

5.5. The empirical relevance of the notion of *small* sign-functional *differences*: grammaticalization theory and in general synchronic linguistics operate with the notion. The sign-functional nature of the historical theory is adumbrated: the basic meaning of change is change of meaning. Parmenides was both right and wrong. The *attitudinal system* is the prime mover in the existence of the everyday language

In chapter 5.3. and 5.4. two definitions that the synchronic linguist's 'synchronic cut' requires were established, and they were applied statically to the transitions of various developmental series. It was demonstrated as an indirect criticism of the episodic nature of mainstream historical linguistics (a) that a period of four elements was needed in order to provide an adequate description of the autonomous entity that the synchronic cut is supposed to 'tie off' – from the stream of history, and (b) that the period is irreducible and its (phonetic) substance is more than mere raw material for change (6.1.). The concept of the developmental series implies that the development of a language cannot be grasped on the basis of events that are episodes in a particular language's existence. On reading the history of the Indo-European dialects one is given the impression that language history is the episodic treatment of the types of change that were mentioned in 5.2.; only recently – and this mainly due to grammaticalization studies – has this parochial field of vision been broadened to other non-Indo-European areas, but it does not conceal the fact that such studies are episodic, too, wanting periodic coherence[54].

Samuels (1972) provides us with a general theory of language development under which grammaticalization – in his diction the cyclic swing from synthesis to analysis – is subsumed as a particular type of change; therefore Samuels sees (English) grammaticalization processes within the context of sound change and implies a structural approach based on the very notion of small meaningful differences, this approach being the basis of the theory to be advanced here. Hopper and Traugott, it is true, do accept a structural approach – albeit implicitly, too, – when they work on the basis of this simple three-state variant-model (1993:10; 36):

5.5.1i   A > B > B
              A

5.5.1ii   Pre-Latin   Latin   French
          *?
          *kanta bhumos >   cantabimus
                              cantare habemus >   chanterons
                                                       allons chanter > ?

---

[54] Both Hopper and Traugott (1993) and Heine *et al.* (1992) have been built up on single large-scale changes in various languages. They do not answer intrasystemic questions such as *Why more?* and *Why sooner?* (Samuels 1972:ch.6), with the corollary type of question: why have /u:/ in English been continued from the beginning of historic times (c. 7/800 A.D.) to ab. 1400? – and short *e* as in *bed, let* (OE *lettan*) only moved within the space of Cardinal 2 and Cardinal 3?

a model that does not correspond to their generalization of the co-existence principle, where we find 'small differences', too. The theoretical problem is that Hopper and Traugott *a priori* abandoned the possibility of creating a historical theory within which the explicit period-chronology concept in 5.5.1ii had a place. Furthermore, the models in 5.5.1[55] do not agree; the abstract model implies, what I have been arguing for, the continuation of the 'past' into the present, a part of the historical process that the implemented model leaves out. Much as I agree with Hopper and Traugott's view (p. 10) of the interplay of variants competing for supremacy, I must underline the fact that their concept of *some nuance of meaning* separating two variants is a descriptive statement of historical inevitability and that a historical theory should take its point of departure here, *i.e.* in the dynamic and systematic interplay between the small meaning-differences of variants staged against the horizon of synonymy and polysemy. Apparently, Hopper and Traugott do not see that synonymy must be one of the theoretical pillars of a historical theory; and once this has been recognized, the theoretical importance of polysemy will follow[56], and not least the necessity of a critical approach to the dualistic basis of synonymy and its place in synchronic theorizing. Samuels interprets his small differences of meaning in relation to the phonological system and therefore makes do with a two-state-model *à la* the one in 5.5.1.

With the introduction of synonymy or, generally speaking, variation, questions like: *what is the origin of a 'meaning-nuance'* and *how does it rise?* impose themselves with necessity. The type of answer that I have been advocating takes its point of departure from the structural concept of contextualization, with a form acquiring part of its meaning in its synonymy field. Here I shall briefly illustrate my position from another angle. Consider the modern English so-called passive as in: *Caesar was killed by Brutus.* What is passive about this construction? Which of the five lexemes and/or the morphemes carries – or carry – the content of passivity? Traugott (1972:82; 144) erroneously claims that *be* + a past participle is a marker of passivity. None of the involved elements are *per se* unambiguous carriers of passivity:

> Caesar was smart // Caesar was walking // The shop was closed // Brutus saw Caesar // Caesar saw Brutus (*position*)
> Brutus killed Caesar // Caesar has killed many enemies
> Caesar was standing by the window, when ... // Caesar had been standing by the window for two hours, when ... // Caesar came in by the window // all by myself // she's got two children by her former husband
> (...) as though he had been doused by cold water (John Briley, *Cry Freedom*, Penguin (1987), p. 10); *compare*: Brutus killed Caesar :: *Cold water had doused him.
> -ed: *neither as a tense nor as a participle morpheme is its content 'passivity'*:
> Brutus has walked // Brutus walked // Caesar was killed //Caesar has killed many enemies // Caesar killed many enemies
> Caesar has been killed // Caesar has been smart // Caesar has been standing by the window for two hours // Caesar had been admiring Calpurnia for a long time

---

55 The implemented model illustrates my view of the syncretism involved in the A>B model, a syncretism that the abstract model above does not explicitize.
56 Their treatment of English *to be going to* is a proof of the obvious role of polysemy in their theory.

5.5. The empirical relevance of small sign-functional differences

That it is the over-all[57] context that ascribes passivity to the sentence *Caesar was killed by Brutus* may be a possibility that the synchronic point of view may be able to argue for; but it is obvious that the sentence's passive content is also due to the synonymy field of which it is a member: *Brutus killed Caesar*. And with this synonymy-field we are back where we started in chapter 4: in respect of what are these two clause-variants synonyms? Is it possible to set up or even point to a cognitive content (= /.../) that synonymously relates the two sentences, which the two sentences manifest, or which exists independently of its manifestations in the English language? If we apply the Hopper-Traugott concept, then we shall end up with this paradoxical analysis that two grammatical categories merely express a 'meaning-nuance'.

With Bache and Davidsen-Nielsen's *roughly* and Hopper and Traugott's *nuance of meaning* we are sent back to Bloomfield's notion of there being no such thing as a small difference, and I shall now begin to construct the basis of my sign-functional theory of language, resuming the discussion from the *Prologue*: my argument is that small differences do exist for the simple reason that they are functional in the existence of the everyday language. I shall first recapitulate the theoretical primitives:

1) the theory of language is subsumed under the general theory of existence
2) the everyday language is a necessary constituent of **ex**istence
3) the everyday language is sign-functional
4) any language phenomenon is a sign[58]
5) language development is sign-functional
6) sound change occurs against the horizon of sign contents
7) content change occurs against the horizon of sign expressions
8) **ex**istence is the present's development into the present
9) **ex**istence creates synonymy and polysemy
10) synonymy and polysemy occur against the horizon of structural interaction
11) history and structure are correlative concepts

---

57 Bache and Davidsen-Nielsen (1997:205) is right in assuming that *voice is a category of the entire clause*; but adding dualistically that it is also *two realizations of the predicator (e.g. made vs. was made)*, they need to point to the predicator-entity of which *was made* and *made* are realizations, their example being *No public explanation or apology was made by the Hazelton police*, a sentence *roughly synonymous* with its active counterpart. What is the theoretical content of the synchronic term *roughly*? But how is this clause-category defined formally, if not contextually! Furthermore, more differentiation is needed in the light of *The shop closed // He closed the shop* as against the ambiguity of *The shop was closed*; cf. also: *a room to let = a room to be let*; and Ælfric's *gif sum dysig man þas boc ræt oþþe rædan hyrþ* with *rædan* an infinitive and *read* in 'If a foolish person reads that book or hears it read' a past participle.

58 This is a corollary that Hjelmslev (1928 and 1972a:36;116) envisaged in these early works. Hjelmslev does not explain why his sign theory abandons this consequence, operating instead with the concept of a non-sign (1963:46; 1976:42), which term, incidentally is not a part of Glossematic theory. Note, however, that the English 1963-translation of the Danish version has left out nearly a page (p. 42) of the latter (1976), where it would have been relevant to discuss the repercussions of the notion of non-signs on the explanation of language development. Hjelmslev merely states that at some stage linguistic analysis proceeds from the levels of signs to a level of non-signs. Martinét's 'double articulation' (1970:8; 13; 7-41; cf. 27-31) elaborates this decisive step in synchronic analysis; the step is an economical – and ?*didactique* (p.33) – precondition for the communicative viability of the everyday language in that a small finite number of entities is sufficient to express *l'immense variété des besoins communicatifs de l'humanité* (p. 13; cf. p. 33). – Ingrid Dal has some pertinent critical remarks on Hjelmslev's sign-theory (see Willems 1995).

How does a small difference of meaning arise? An intersection obtains in its origin an endogenously motivated property from its predecessor (external endogeneity), and this successor-turned-into-predecessor will in turn be influenced katagenously by its successor: the possible alternative motivates the positive alternative in a vertical synthesis: as Hopper and Traugott's abstract model above indicates, a predecessor becomes a part of the present, but this model does not see the variants' structural interplay. A successor will be called the *originary successor* of the predecessor that motivates it endogenously; the motivating predecessor will be called an *original predecessor*.

The sign-functional theory of change also provides a solution to the Parmenidean paradox (cf. Barnes 1979:184[59]); Parmenides was right in pointing out the absurdity of this type of change [æ] > [a], with one sound developing into another sound; [æ] cannot change into or become [a], for the simple reason that if [æ] is (produced), [a] is not, and vice versa; the palimpsest nature of the spoken chain makes Parmenides right. However, [æ] does one thing: it develops into itself, being continued under the empirical aegis of development *qua* structural change, not under that of stability; so when [a] is produced, [æ] is also: the structural existence of the historical phenomenon makes Parmenides wrong:

5.5.2   æ > a   (Cf. 5.5.1i)
           ↘
            æ

The production conditions for [æ] and [a] are different. Less articulatory *qua* pulmonary energy is needed to produce the originary successor through internal endogeneity than its original predecessor. Thus the content of /CaC/ will be different from that of /CæC/; the content that will be associated with the individual sounds ('variants') in *parole* enters into an *attitudinal system*. This system consists of purely structurally motivated contents, based on a fixed correlation between content and expression. The system is rooted in the connotation, where the speaker's attitude is conveyed by the sounds *qua* signs of the spoken utterance, and it is a purely relational system, where the substance of the particular 'nodes' carries with it no astructurally acquired content.

This system of 'attitudes' is thoroughly structural: the existence of its elements hinges on the smallest distinguishing differential potential that can organize the elements into a system of sign-functional contrasts. What is typical of the attitudinal system's elements is that they bypass the synchronic step from signs to nonsigns – Martinét's second articulation. An attitudinal element has a **double existence**: an attitudinal element has a discourse existence – in

---

[59] Based as it is on formal logical analysis, it is hardly surprising that Barnes's argument leads him to conclude on a rather critical note (pp. 194, 230). In my view Parmenides's extreme position is due to the simple fact that he commits the Phenomenal Error, explaining an empirical fact logically (non-empirically) and thereafter identifying logic and reality. Christiansen (1998:37) is very close to a solution, saying that the language (*i.e.* the theory) of change is not to be a 'picture' of change: hence if mind and language are inseparable, Parmenides conflates – confuses – language and reality; in Hermann Paul's word: Parmenides had not liberated himself from the tyranny of language. – Parmenides also tried to explain what he denied reality (non-being) with reality.

5.5. The empirical relevance of small sign-functional differences

the spoken chain, and an existence at the level of *langue* in the language continuum[60]. There is a necessary relation between the attitudinal element and *parole* by way of the word. The connection between phonetics and *parole* is severed by the synchronic dualism, but has been lying more or less latent since the beginnings of phonetics (see Sievers (1850:279) in Labov (1994:221;261); cf. also references above to Horn and Luick[61]), to be re-discovered and worked into a theory by Samuels (1972); following Samuels I shall develop the theory's sign-functional basis, adding an overtly structural layer to it.

The synchronic level of phonology is ambiguous in that it conceals two systems, a system of nonsigns (phonemes; or distinctive features), and the attitudinal system, which it cannot handle theoretically, or which it tries to subsume under itself. Now we have a solution to the synchronic dilemmatic, breach-conception of change: the attitudinal system is a prime mover in the **ex**istence of the everyday language; its function is synthetic: it conveys the 'attitude' of the speaker towards the constitutive factors of the concrete speech situation. The universal fact that it cannot be controlled or explained by the phonemic system testifies to its independence of this system of nonsigns.

The intralinguistic function of the attitudinal elements is to ensure the temporal coherence of the speaker's language and an equilibrium in the group's languages. No synchronic theory has yet been set up that preserves the coherence of a given system of nonsigns through time, let alone the equilibrium of a given group of speakers' language.

One of the implications of a supraindividual, nay, a national language, is that all other languages, dialects (regional or social), are regarded as translations of it into particular codes. The attitudinal system is not such a translation of a system of nonsigns into an emotive ('affective') code. As Lévi-Strauss (1945:43) puts it: *la diversité des attitudes possibles dans le domaine des relations inter-individuelles est pratiquement illimitée*. And no system of nonsigns (phonemes) can convey all the 'meaning-nuances' that a speaker needs to express. To fulfil this dynamic function – as indicated above – independent attitudinal systems are required. The attitudinal system that will be developed now is – to follow Samuels's diction – stylistic[62], consisting of the dynamic interaction of forceful style (= FS), relaxed style (= RS) and normal style (= NS), each style ('speaker-attitude') being characterized by signs that cannot be regarded as predictable variants of a given system of nonsigns. Having a direct relation with the spoken chain[63], each element is a sign of (larger) signs, this in contra-

---

60 Lévi-Strauss (1945:38-39) pointed out a significant difference between phonology and anthropology: the phonemes belong to the level of phonological theory, *i.e.* synchronic linguistics, in that phonemes are inferred from words isolated from the concrete spoken chain; phonemes do not have a double existence: *Il n'y a pas de relations nécessaires à l'étage du vocabulaire* (p. 39). In contrast, the 'smallest' elements of sociology, the system of kinship terms, belong both to the level of sociological theory and the level of discourse.
61 Significantly, Labov emphasizes the mechanical aspect of the principle, saying that it follows from the *Eigennatur* of long vowels that speakers *tend to overshoot the target*, without considering possible functional conditions for this. Without repeating Labov's synchronic Bull's Eye stance, my attitudinal system will fuse both the mechanical and functional aspects of this principle (Sievers's) into a dynamic theory of history.
62 Gimson (1994:267-269) defines *stylistic variation* differently.
63 In contrast to the system of phonemes, which does not transcend the level of words, where minimal pairs are the domain of the commutation test.

distinction to the synchronic double articulation, where a sign *qua* a word is a sign of non-signs. But then we must ask: how can the combination of, say, four phonemes /k/ + /a:/ + /n/ + /t/ produce a sign as well as its content: [ka:nt]rC?

Those three types of contents are relationally motivated and are mechanically correlated with an articulatory 'gesture' of the abovementioned type: in relation to [CæC] the form [CaC] requires less subglottal energy in order to be produced, and as an originary successor [CaC] the form will acquire the attitudinal content of 'Relaxed Style'; as regards its original predecessor, we do not know its attitudinal content, and as such the inadequacy of a two-state model is again demonstrated. The sign-functional model in 5.5.3 will illustrate this as well as the impossibility of precise retrodiction of the past on the basis of the present[64].

5.5.3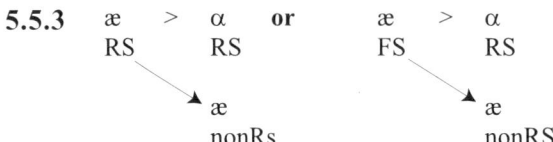

The period will dissolve this two-state alternative, which has a third possibility if the predecessor has the content of 'normal style' (cf. below). Let's suppose that the developmental series is characterized by nonlevel inertial processes as in 5.5.4

5.5.4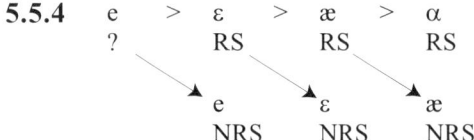

then we can infer on the basis of the original predecessor of [æ] that [æ] has a RS content. The structural change in the 5.5.4 model is **change of meaning**, *in casu*, ærRS > ærNRS.

When an originary successor develops into a new present, the new intersection creates coherence simultaneously with its separation from its original predecessor and its developmental series: through internal endogeneity [α] breaks away from its original predecessor; through its content conditioned by external endogeneity – the fact that [æ] is a relaxed style form implies the same attitudinal content of [α] – [α] breaks into (the series of) [æ], and through its katagenous motivation of /æ/, the continuation of the predecessor now *qua* the positive alternative of the present, [α] both **does not break away from** and **breaks away from its predecessor**.

The structural principle of contextualisation has asserted itself at the level of endogeneity and katageneity, and so has the concept of a historical phenomenon's *Eigennatur*. A structur-

---

64 The ambiguity of data is Janus-faced: the present is ambiguous in respect of retrodiction – and prediction (cf. below) – but the past is also ambiguous in the sense that it involves a choice that the historical theory resolves.

al phenomenon is more than its origin or (the sum of) its synchronic constituent parts: there is no automatic ('analytic') correlation between a given amount of subglottal energy and the content of a sound. In conclusion I shall set up the third type of developmental series:

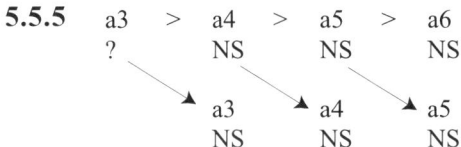

5.5.5

The content of inertial development with no change is neither relaxed-unemphatic nor forceful-emphatic styles, the speaker's attitude being the logical alternative of these two styles, namely normal style. This means that the possible and positive alternatives merge from the point of view phonetic substance in an irresoluble syncretism; the speaker will have other means at his disposal to *synonymously* express the other style extremes, *e.g.* redundancy, word order or lexical collocations (cf. Sten 1969:86).

To conclude: **existence** involves a **choice**, the choice between what becomes the **positive alternative** and the **possible alternative**; synonymy is involved in the structural process of katageneity: just as we say that modern English *great* acquires part of its meaning from the other adjectives in its synonymy field, so existence constitutes a **synonymy field** based on the above **attitudinal system**. – The place of level inertial development (as in 5.5.5) within this conceptual scheme will be developed in chapter 6.1., where this discussion will be resumed.

5.6.

The final definitions and the concluding analysis of the transitions of the two-state model, seen against the horizon of Archimedean omniscience and (only) the material aspect of the transitions. The concepts *a break before* and *a break after* are not the respective contradictories of *no break before* and *no break after*; the break-concept is mediated by *a potential break*. History is not episodic

Definitions 5.3.6 and 5.3.11 formalize the two concepts *no break around* and *a break around* – around a historical phenomenon. The latter definition did not indicate the whereabouts of the break in relation to the phenomenon. The following exposition leads up to such a type of definition and as such constitutes the last part of the critical analysis of the static synchronic state concept.

Def. 5.6.1 A fullblooded (FB) intersection is an intersection around which no break occurs (= Def. 5.3.6).

Def. 5.6.2 A nonfullblooded (NFB) intersection is an intersection around which a break occurs (= Def. 5.3.11).

With these definitions we can set up the following inferential equivalences

5.6.3 1) *fullblooded* is equivalent to *with no break around*
2) *no break around* is equivalent to *no break after* and *no break before*
3) *no break before* implies either *a break after* or *no break after*
4) *a break after* implies *no break before*
5) *no break after* implies either *a break before* or *no break before*
6) *a break before* implies *no break after*
7) *nonfullblooded* is equivalent to *with a break around*
8) *a break around* is equivalent to *a break after and no break before* <u>or</u> *a break before and no break after*[65]
9) *a fullblooded intersection* does not imply *a NFB predecessor* <u>and</u> *a NFB successor*.
10) *a nonfullblooded intersection* does not imply *a NFB predecessor* <u>and</u> *a NFB successor*.

As sample transitions the analysis will use those of III>VI (of Group 1), **a6 > α1**, and of II>III (of Group 2), **e1 > e2**. In order to make the two-state-model's predecessor and successor equiempirical the two transitions' respective predecessors and successors must be added. Finally, this four-state scenario will be increased to a period of six elements so as to underline the weaknesses of the static point of view. – A developmental series from Group 3 will not add anything to the analysis.

III>VI   a4 > a5 > **a6** > **α1** > α2 > α3   level > level
II>III   æ > ɛ > **e1** > **e2** > e3 > e4   rising > level
(II>V   æ1 > ɛ1 > **e** > **ɛ2** > æ2 > α2   rising > falling)

The question is: how does the static application of the definitions establish breaks among the relevant intersections, breaks (in boldface) that inhere in the calculus (point view)? The breaks established in the preceding discussion were due to the fact that we knew what the developmental series looked like in their calculus existence. The transitions appear from these '*parole*' combinations. My argument will be couched in the diction of the everyday language with no definitions; I shall point out the weaknesses immediately they appear, and despite the possibility of conducting the analysis at one fell swoop – from our omniscient static stance, I shall introduce some degree of empirically relevant directionality:

III>VI   a4 > a5 > **a6** > **α1** > α2 > α3   to be be interpreted in accordance with Def.s 5.6.1 and 5.6.2 as:
         ?FB FB NFB NFB FB

With a6 as the successor, no break occurs around **a5**, a FB intersection; and we say redundantly that no break occurs after it and no break occurs before it: <= a5 =>. Assuming that its predecessor is also a FB intersection, this transition type appears: a4=> <=a5.

With α1 as the successor, a break occurs around **a6**, a NFB intersection; and we say equivocally that there may be a break either before or after it: ≈a6≈. If we only had a two-state transition and assumed that our knowledge of a5 as FB and of a6 as NFB had been estab-

---

65 Equivalences 7) and 8) will be modified below.

5.6. The final definitions and the concluding analysis of transisitions

lished independently of the transition's predecessor and successor, then we were bound to say that the break occurs BEFORE ≈a6 – in contravention of the alternative of Equivalence 8) above: a5=>≈a6.

With $\alpha 2$ as the successor, a break occurs arround the NFB intersection $\alpha 1$; hence there may be a break either before or after it: ≈α1≈. The strictly two-state model would entail a break before α1, and with the break before a6, the transition a6 > α1 would yield a curious transition: ≈a6 ≈α1. However, as an autonomous transition the breaks would be before- and after-breaks: a6≈ ≈ α1.

With $\alpha 3$ as the successor, there is no break around the FB intersection $\alpha 2$. Since the α1>α2 transition consists of a NFB and a FB intersection, the system would imply a break after α1: α1≈<=α2.

As can be seen from this analysis, the two-state approach creates redundancies and inconsistencies. It is to be inferred from this that the break-concept is complex and requires definitions that consider larger stretches of the developmental series. And the fact that an intersection has been characterized as nonfullblooded is not sufficient to place the break in question, and we must redefine Def. 5.3.11 slightly: a nonfullblooded intersection is characterized by a **potential break around** it. In other words, the contradictory of **no break around** is not **a break around**. Finally, the definitions also indicate the ambiguous nature of a6 and α1 in respect of their developmental series: the potential breaks around them indicate that they are less closely related to their respective developmental series than their respective predecessor and successor; secondly, the analysis does not make them belong to the same (sub)series. The independence of the six inertial series's *Eigen*-existence in the calculus has as such been modified, and this change arises in *parole* – when calculus elements are combined into larger stretches. – The other type of break, which is found in Groups 2 and 3, corroborates the above findings:

II>III   æ > ε > e1 > e2 > e3 > e4 to be interpreted in accordance with Def.s 5.6.1 and 5.6.2 as:
        ?FB FB NFB FB   FB
(II>V) æ1 > ε1 > e > ε2 > æ2 > α2)

With **e1** as the successor, no break occurs around ε a FB intersection; and we say redundantly that no break occurs after it and no break occurs before it: <= ε =>. Assuming that its predecessor is also a FB intersection, this transition type appears: æ=> <=ε.

With **e2** as the successor, a break occurs around **e1**, a NFB intersection; and we say equivocally that there may be a break either before or after it: ≈e1≈. If we only had a two-state transition and assumed that our knowledge of ε as FB and of e1 as NFB had been established independently of the transition's predecessor and successor, then we were bound to say that the break occurs BEFORE ≈e1 – in contravention of the alternative of Equivalence 8) above: ε=>≈e1.

With **e3** as the successor, no break occurs around the FB intersection **e2**, <=e2=>. The strictly two-state model would entail a break before **e1**, and with the break before e1, the transition e1 > e2 would yield a curious transition: ≈e1<=e2. However, as an autonomous transition the break would be an after-break: e1≈<=e2.

With e4 as the successor, there is no break around the FB intersection **e3** and the transition will be characterized by: e2=><=e3.

The concept of the potential break asserts itself; and while the first analysis continued to establish a relatively sharp break between the calculus's two subseries, no such break can be established in the other series: in fact we do not know which subseries **e1** as well as **e** of II>V belongs to, knowledge that would unequivocally follow from the calculus. The function of **e1** remains equivocal.

The two analyses corroborate what has been argued for previously: coherence and separation are two 'gradual' concepts, and the two types of break permit us to conclude that the gap in the a6>α1 transition is more abrupt than the one in e1>e2 or for that matter in ε>e1. I shall therefore keep the break-concept to denote the gap between a transition's two nonfullblooded intersections; and since we cannot uphold a **break before a6** and a **break after α1**, the logical geography of our original contradictories, 'a break' and 'no break', will be summed up as follows: the concept of *a break* is mediated by that of *a potential break*; the concept of *no break* may be mediated by that of *a potential break*: 1) *a potential break* is the cause of itself, may be the cause of *no break*, and is the cause of *a break*; 2) *no break* is the cause of itself, and may be caused by *a potential break*; 3) *a break* is not caused by itself. – (The *no break* concept is mediated in relation to a6, where the *Eigennatur* of the successor changes the potential break before a6 into a coherence with a5, and the same goes for the potential break after α1, which is changed into a coherence with α2 because of the predecessor of α1.)

The static 'omniscient' two-state model results in an unprincipled description of the past: from a presentic stance, the a6>α1 transition is to be characterized as abrupt: the break after a6 was mediated by the potential break around it and established in relation to its predecessor function. Conversely, the break before α1 was not established in relation to its predecessor function, but in relation to *its* predecessor. The fact that a nonfullblooded intersection is succeeded by a fullblooded one does not entail a break before it; cf. ≈e1≈<=e2.

The outcome of the analysis so far is (1) that all knowledge of a given developmental series cannot be inferred from this series's existence in a calculus; (2) that the nature of the two-state transition cannot be generalized (cf. below); (3) that directionality is an analytical precondition; and (4) that an individual element's nature is dependent on its structural relationships; and perhaps most importantly (5) that the linguist's 'synchronic cut' may interfere significantly with data. The theoretical implication of the last item is that historical analysis should not be *episodic*. Applied to what may be called the Grand Change approach in historical linguistics and the approach that we see in grammaticalization studies, both of which are episodic, Item (5) is a warning against the identification of cross-language processes, where two processes in two different languages are subsumed under the purportedly same explanatory principles; the over-all chronological, non-episodic existence of such two changes determines their respective 'natures' and therefore requires different explanatory principles.

That the transition-concept is not uniform will appear from this summary of transitions, where '≈' now represents a potential break:

5.6. The final definitions and the concluding analysis of transisitions

5.6.4

A	>	B	A		B
a5	>	a6	a5=>		≈a6
a6	>	α1	a6≈		≈α1
α1	>	α2	α1≈		<=α2
ε	>	e1	ε=>		≈e1
e1	>	e2	e1≈		<=e2
e2	>	e3	e2=>		<=e3

And in contrast to the logical possibilitites set up on the basis of contradiction in 5.4.4 above, 5.6.4 reduces to the following four possibilities:

5.6.5

A	B		
No break after	No break before	e2=>	<=e3
A potential break after	No break before	α1≈	<=α2
No break after	A potential break before	a5=>	≈a6
A potential break after	A potential break before	a6≈	≈α1

The definition of the nonfullblooded intersection will be changed into 5.6.6 and definition 5.3.11 does not define *a break around*, but *a potential break around*:

5.6.6 A nonfullblooded intersection is an intersection around which a potential break occurs.

This redefinition agrees with the existential priority of continuity *qua* coherence: discontinuities are to be seen against the horizon of continuity and coherence. And the inherent absolutism of the static approach of synchrony has been replaced by the dynamic relativity that characterizes **ex**istence:

> the concept of *the potential break* implies either itself or (1) no break before and a break after (cf. a6) or (2) no break after and a break before (cf. α1)

> the concepts of *no break before* and *no break after* appear from either the concept of *no break around* or that of *the potential break*.

In conclusion: we cannot always be certain of the nature or the properties of a given phenomenon, if we only know it analytically as an autonomous, *sui generis* entity, be it an element or a transition: *the episodic view of history does not entail historical accuracy*. Coherence and historical continuity constitute the empirical horizon against which separation and discontinuity occur: and far from it being the analytical mind that calls the synthetic mind into its service, – when the latter is needed, it is the analytical mind that is made to subserve the requirements of discursive reasoning, not in an ancillary function, but as an integrated part of our historical sense.

CHAPTER 6

The immanent theory of history is implemented as a sign-functional theory of sound change: language development is the processes that constitute developmental series through time. It is semantic change through katageneity and endogeneity, and formal variation (replacement) through endogeneity. Concluding arguments

---

> *(...) historical linguistics remains the conceptually primary domain from which synchronic facts must be abstracted.* Anttila 1979:48.

6.1. Isolative shifts, both A>B and continuation A>A, are explained by the sign-functional theory of history. The theory explains short- and long-term development that consists of various types of ertial and inertial processes. The attitudinal system *en ergeiai* defines the coherence relations of the developmental series

IN CHAPTER 5 the material part of the attitudinal system was developed including an outline of the dynamics of content development. The theory assumes that this dynamic system has a mental correlate so that the adumbrated mechanical correlation between content and expression also involves a degree of predictability because the sign's purported arbitrariness has been suspended on language's own premises (cf. Martinét 1970:21 and end of 6.1.). The full sign-functional system will now be implemented and its 'energy' demonstrates how theoretical discontinuity and continuity are comprehended against the horizon of continuity. This continuity corresponds to the fact that we do not experience breaks or discontinuity in the existence of our everyday languages, only differences that make us comprehend **ex**istence: how the past is created by the present. It is these differences that the theory of history explains as different types of (dis)continuity. Since we do not *see* (sense) change, let alone causes, the concept of change is an innate category that can only come from our knowledge of our everyday language, our inborn, but developing knowledge of how to exist historically (see 6.2.D below). It is in and through our language that we understand reality as a non-atomistic process.

While the standard system of phonology consisting, in theory at any rate, of a relatively small finite number of calculus-like elements cannot cope with the facts of change adequately, the attitudinal system resolves the contradictions that the phonological system of synchrony incurs when applied to the day-to-day **ex**istence of the everyday language. It follows

6.1. Isolative shifts are sign-functional processes

from the nature of the synchronic system of phonemes that – through the manifestations of the phonemes – it is a communicatively economical instrument, but the attitudinal system has a larger communicative-synthetic function, seeing that it bridges the gaps[1] that are involved in the present's constant development into and creation of the present: our historical sense tells us both that the present that is becomes different from the present that was, and that there is no empirical gap in the process of **ex**istence, despite our historical knowledge of the difference between the present and its past. The historical theory illustrates this cognitive-existential paradox that where a difference is, there is no synchronic gap.

The attitudinal system connotes a content that the limited number of a set of phonemes cannot convey: the standard system copes with change only as recombination of its preexisting entities and/or as new realizations of allowable potentialities, while the attitudinal system ensures a 'gradual' process of development without alerting us to the existence of language. This 'gradualness' explains why – in Hjelmslev's words – *it is in the nature of language to be overlooked* (1963:5), it ensures that despite the fact that language carries with it the indelible mark of individuality and subjectivity, these unsocial differences exist against the horizon of the dynamic coherences that make up man's transcendence into reality. If the everyday language were not overlooked in man's normal speech-situation, the speaker/hearer attention would focus on the means of communication, and not on its 'meaning'; man's **ex**istence would lose its dynamics and energy. Normally, we are not particularly alert to the elements of the L-continuum, but we are certainly aware of the connotations of *parole*. It is they that making the speech-situation rational make interpersonal communication empirically relevant in the sense that the speech-situation's interlocutors are not just any speaker and any hearer, but some speaker and some hearer: the speech-situation is proprial, not any speech-situation.

Thus it is absurd to suppose that a small number of systemic elements, about whose number and nature no agreement can be found among phonologists, should constitute the fundamentum of the **ex**istential function of the everyday language, the hallmark of mankind, in that a phoneme cannot become – in Hjelmslev's diction – *the distinctive mark of the personality* (1963:3). In general terms what pre-exists in a calculus as 'already developed' is unable to exhaust the meaning of 'combined' phenomena. This will appear from this chapter's investigations.

The attitudinal system is expounded by the structural networks that its five basic contents, FS, RS, NS as well as NonForceful Style (NFS) and NonRelaxed Style (NRS) enter into with their respective material correlates; being dynamic, the system is not defined in virtue of the possible static conception of these 5 elements, – this would involve a generalizable definition with a fixed, nonstructural correlation between expression and content, – but in accordance with their **ex**istential function as **created/continued NS, FS, RS**, and **created NRS, NFS**.

The synchronic linguist has a data-constitution problem; since the historian's data is no longer, it must be constructed. And it is over this constructed data that the historian theorizes.

---

1 The attitudinal system assumes the function of bridging the distinctions (*Grenzlinien*) that the synchronic system makes (Paul 1968:33) by finding the *Zwischenstufen* between the tradition's categories (see items (2) and (3) immediately below). The system also expounds Luick's notions of *Werdegang* and *Verbindungslinien*.

Only on this condition can we assert the equilogical nature of the relates in the transcendence problem: the notion of historical accuracy is contained in such theoretical pronouncements as *This is what happened // That is not what happened* – falsification of an existence-postulate, while the historical theory can be rejected both for internal reasons and for being empirically irrelevant; thus discussions about the primacy of data (empiricism) or theory (rationalism) should not be conducted along trench warfare lines. The synchronic linguist tends to unwittingly conflate the two constructional processes: his theory enunciates empirically relevant and accurate *statements* – in a generalizing movement **away from** (says the historian), **into** (says the synchronic linguist) data – *of* what is the case *immediately* and what is 'there' for him to see analytically. As existence is always materially formed, the historical theory proceeds from 'language-formed substance' to 'language-formed substance' (its constructed data), not from a linguistically constructed essence to its concrete and unstable appearances[2]. The generalizing movement creates shortened knowledge – with the historian and synchronic linguist in agreement for different reasons: the so-called shortened version expounds what can be scientifically known about a phenomenon, namely its essence; the historian criticizes this for providing nonexhaustive knowledge about our phenomenon.

Historical theory is not a causal statement of episodic A>B-changes, and the so-called actuation- or inception-problem as well as the notion of long-term change will also disappear with the introduction of the distinction between the theory and aetiology of historical development in connection with the abandonment of historical eschatologies: A>B-change as a *riddle* (Weinreich *et al.* 1968:186-187) *insoluble* according to Milroy (1992:76), follows partly from the biassed question: *Why does language change?*, partly from the view of change as being disruptive and language development as being discontinuous, and therefore something in need of being explained causally through the *mechanism of change* (Weinreich *et al.* 1968:149). Paul (1968:32) saw nothing surprising in the fact of change, *Die eigentliche Ursache für die Veränderung*, being *nichts anders als die gewöhnliche Sprechtätigkeit*, a position that can be recognized in Samuels (1972) and Coseriu (1974) by way of the sociologist's succinctly causal-descriptive statements: *A chaque fois qu'une expression est employée, elle devient moins surprenante pour celui qui l'entend et plus aisée à reproduire pour celui qui l émet (...)* and *La valeur expressive des mots s'atténue par l'emploi, la force en diminue* (Meillet 1966:20[3]). The point to be emphasized as regards the Paul- and Meillet-quotes is that they consider the language elements – be they sounds or words of the L-continuum – in relation to their existence in *parole*; like the anthropologist they recognize the dual existence of their (theoretical) elements, but they do not proceed to the creation of a theory of, with reference to Meillet, *devaluation*[4].

---

2   Cf. Descartes's *cera nuda*, without its *externæ formæ* or *vestes*. Descartes idealizes his *cera* as an abstract entity that carries certain properties. This is an example of how, in the historian's eyes, an empirically irrelevant dichotomy between rationalism and empiricism arises.

3   Meillet is here describing the devaluation process that Samuels incorporated into his theory. The concept of devaluation will play an important role in how the A>B- and A>A-types of change will be united. Paul (1968:27) anticipated Meillet's views on devaluation.

4   Paul (1968:33) made the coherence aspect his primary objective: *Auf alle Gebieten des Sprachlebens ist eine allmähliche abgestufte Entwickelung möglich. Diese sanfte Abstufung zeigt sich einerseits in den Modifikationen, welche die Individualsprachen erfahren, anderseits in dem Verhalten der Individualsprachen zu einander. Dies im einzelnen zu zeigen ist die Aufgabe meines ganzes Werkes*. Paul sees his

6.1. Isolative shifts are sign-functional processes

The nonepisodic period so relativizes the two-state transition that the reduction of history to being a function of autonomous states remains a metaphysical postulate and a structural inconsistency (5.6.4).

In deference to the synchronic tradition the data of the historical theory has been constructed in accordance with the calculus point of view: a number of well-defined developmental series constitute the system's entities and these series may be combined into larger series (sequences) (cf. 5.4. Appendix). In chapter 5 we saw how this manifestation process reveals that all meaning is not contained pre-existingly in the systemic entities in that the structural interaction that arises in the combined series creates meaning that did not inhere in the individual series' calculus-existence. In the present chapter I shall substantiate those findings through a more empirically relevant and dynamic description against the horizon of **exi**stence. In either case the structural tenet that a (relative) totality is, perhaps not 'more' than, but certainly **different from** the sum of its parts asserts itself as a historico-dynamic principle. I shall now describe inertial and ertial development sign-functionally.

The below developmental series is a materially accurate description of the existence of the two OE sounds *a:* in *ham* 'home' and *u* in *-butan* 'but' from 700 to 1500+, in contrast to the tradition's syncratic or telescoped two state model: A>B, supplemented by an inaccurate statement of the type that OE *a:* did not change in Old English (about 400 years of uninterrupted development), a statement that is incompatible with the fact of constant change:

6.1.1   OE 700 > 800 > 900 > 1000 > 1100 > 1200 > 1300 > 1400 > 1500 >
        a   a   a   a   a/ɔ   ɔ   ɔ   ɔ   o
        u   u   u   u   u   u/ʊ   ʊ   ʊ   ə

The level inertial developments with ertial change in 6.1.1 will here receive theoretical treatment, because this data is as empirically relevant (accurate) as the trajectories that the tradition constructs for the individual sounds of a 'major' sound change such as the 'Great Vowel Shift'. In view of our tradition's formulas A>A and A>B, representing continuations and simple isolative shifts, it is legitimate to question the empirical relevance – historical accuracy – of the way the data of the Great Vowel Shift is constructed, where synchronic states, systems and cuts are ignored as well as the A>B notation. The distinctions appear through arbitrary application of eschatological and teleological assumptions with the related concept of on-going or long-term change. The way I have constructed the system of developmental series follows the tradition's acceptance of nonlevel inertial development, but my data does envisage these three hypotheses:

1) Both the level developmental series and the nonlevel series *qua* inertial change are empirically accurate, and both may develop ertial change.
2) Only the level developmental series are empirically accurate and the nonlevel series are syncratic series which must be resolved into a number of level developmental series with

---

'*Prinzipien*' as a corrective to the abstract, static and categorical stance of synchrony; the attitudinal system in particular and the construction of a historical theory (metalanguage) in general are similar necessary correctives.

ertial change. The I>II series immediately below will therefore be resolved into a level series with **ertial** changes:
> a3 > a4 > a5 > **a6** > α > α > α >æ > æ > æ > ɛ > ɛ > ɛ > ɛ > e > e > e > ɪ > ɪ > ɪ > ɪ >

3) The nonlevel inertial type of developmental series is a theoretical statement of the epitome of gradual development, *infinitesimal change* or perhaps *cumulative change*. This stage will represent some kind of subliminal process connecting the two elements of ertial change so that the A>B type is a resoluble syncretism: the fronting of ME *a:* towards a Cardinal 4 vowel α may be represented in the below series, where the underlined section represents nonlevel inertial development, occurring subliminally over time:
> a3 > a4 > a5 > a6 > <u>a > a > a > a > α > α > α > α > α</u> > α1 > α2 > α3 > α4 >

Despite its attractiveness the third possibility remains a mentalistic hypothesis. The important point is, however, that my theory can cope with all three possibilities, and is as such not only a theory of simple isolative shifts, but also a theory of the individual trajectories in our tradition's chain shifts.

The series set up in chapter 5.2. will be regrouped into two large groups with predecessor series characterized as (1) a level series and as (2) a nonlevel series.

The first subgroup consists of the below level series, each of which develops into a nonlevel series, *i.e.* from stability to change, due to the underlined successors; each of them creates some kind of break with its predecessor and is the start of a *non-inertial* developmental series. The last of the III>VI series will represent the rest.

6.1.2	level	+	rising/falling/level
I>II	a3 > a4 > a5 > a6	>	<u>α</u> > æ > ɛ > e > ɪ
or	æ2 > æ3 > æ4 > æ5	>	<u>α</u> > æ1 > ɛ > e4
III>IV	u3 > u4 > u5 > u6	>	<u>ʊ</u> > o̞ > o > o̞ > ɔ
VIII>II	ɛ2 > ɛ3 > ɛ4 > ɛ5	>	<u>e</u> > ɛ1 > æ > α > a
III>VI	α1 > α2 > α3 > α4	>	<u>a3</u> > a4 > a5 > a6
or	ɛ1 > ɛ2 > ɛ3 > ɛ4	>	<u>e1</u> > e2 > e3 > e4
or	e1 > e2 > e3 > e4	>	<u>ɛ1</u> > ɛ2 > ɛ3 > ɛ4
or	**a3 > a4 > a5 > a6**	>	<u>**α1**</u> **> α2 > α3 > α4**
	FB FB NFB		Cf. Def.s 5.3.6 and 5.3.11

Following the directionality of existence, we shall say: (1) *developing into a5, a4 becomes a FB intersection*, (2) *becoming the originary successor of a4 a5 makes a4 a FB original predecessor*. Consequently the originary successor of a given transition like α1 here will be the cutting here-and-now intersection, defined by *the involved possible and positive alternatives*. The notion of *a transition* is also redundant.

Our sample series will develop attitudinal contents, and the structure of the series will be the participative one I have been arguing for in the preceding sections:

$$A > B \quad \text{or} \quad A > A \quad \text{possible alternative} > \text{possible alternative}$$
$$\searrow \downarrow \qquad\qquad \searrow \downarrow \qquad\qquad\qquad\qquad \searrow \downarrow$$
$$A' \qquad\qquad\quad A' \qquad\qquad\qquad\qquad\quad \text{positive alternative}$$

6.1. Isolative shifts are sign-functional processes

6.1.3 a3 > a4 > a5 > a6 > α1 > α2 > α3 > α4
         NS   NS   NS   FS
         NS   NS   NS   NFS
         a3   a4   a5   a6

The structural process of **continuation** is language existence and will be described dynamically as follows: *the possible alternative of the originary successor generates its content endogenously from what has become its original predecessor, which, developing into the present, is influenced by its successor katagenously*. In the level inertial series neither **endogenous variation** nor **katagenous change** (to be determined below) occurs between a predecessor and its successor: a4rNS > a5rNS (possible alternatives); a4rNS > a4rNS (possible alternative > positive alternative). Continuation in the inertial series, A>A, is defined in 6.1.4[5]:

6.1.4 **Continued normal style** is the production of a present successor constituted by a merger in which the possible alternative is endogenously[6] similar to the positive alternative, while the possible alternative *qua* an originary successor differs from its positive alternative *qua* its original predecessor in virtue of different degrees of temporal (endogenous) devaluation. The possible alternative does not devalue its positive alternative distributionally. Continued normal style constitutes a nonambiguous irresoluble merger.

Comments: the term **devaluation** will be split into **temporal devaluation** and **distributional devaluation**; the term **nonambiguous** will be defined below, while **an irresoluble merger** is a here-and-now intersection whose possible alternative and positive alternative cannot be resolved into a synonymy field. The intersections until a6 in our sample series do not increase (ambiguate) or decrease (make precise) the contents of their respective positive alternatives, the result being the gradual *temporal devaluation* of the originary successors. The positive alternative, on the other hand, does not contribute to the precision of the possible alternative[7]. The former is devalued by its successor, while the successor is, as it were, born devalued (endogenously).

The cutting element is a potential historical phenomenon; it is a synchronic structural merger; it creates a subseries of its predecessors, of which it is no immediate part *qua* a historical phenomenon. It completes its meaning, its historical potential, as it develops into the present. – Paul (1968:28) described this process in psychological terms, not in the language of an immanent theory of history.

With α1 the conditioning factors (internal endogeneity) of the developmental series have changed; being the successor of a6, α1 makes a6 different from its predecessors through katagenous motivation, and internal endogeneity makes the attitudinal content of α1 different from the NS-content of a6; it develops a FS content. To specify: the FS-content is not correlated with this particular *low front* sound regarded as an autonomous entity: the content

---

5 This and the following type-like definitions are too synchronic to establish sufficient distinctions.
6 = 'in virtue of articulatory substance, over-all acoustic prominence and content'.
7 This may be put in distributional terms: the successor neither restricts nor expands the scope of its predecessor; the successor and predecessor enter into an overlapping distribution.

is generated dynamically through the structural difference-relations that α1 contracts with its predecessor. So instead of using static, synchronic diction: *low, front, peripheral Cardinal 4*, the historian avails himself of a structural-dynamic vocabulary: α1 has *become a raised and fronted sound in relation to its origin*; the process creates a structural difference correlated with an attitudinal content that cannot be the same as the predecessor's: being necessarily produced with greater subglottal force, the sound will assume a forceful-style content. – This is the first illustrative intimation of what I have called *empirical necessity* (see below).

Developing into the present, a6 undergoes katagenous change: a6rNS > a6rNFS; its distributional scope has been increased (= **distributional devaluation**), and created forceful style will be defined in 6.1.5:

6.1.5 **Created forceful style** is the production of a present successor constituted by a merger in which the possible alternative is not endogenously similar to the positive alternative, while the possible alternative *qua* an originary successor is attitudinally precise and devalues its positive alternative distributionally. Created forceful style constitutes a resoluble merger. = **Created relaxed style**.

The resoluble merger constitutes a tensional field through the structural interplay between the possible and positive alternatives: α1rFS :: a6rNFS; the successor α1 with its precise content ambiguates its predecessor by increasing its content katagenously: NFS = RS + NS; the predecessor as the possible alternative may now appear in relaxed-style contexts and normal-style contexts.

As a fullfledged historical phenomenon a6 differs from its predecessor a5 and creates a 'gap' between itself and a5 that is different from the gap between a5 and a4. The no-break before a6 established statically and mediately in chapter 5 will be explained dynamically through the coherences that a6 creates with its predecessor: the katagenous change of a6rNS into a6rNFS makes the latter both different from – in terms of RS – and similar to – in terms of NS – itself as a possible alternative.

The present's development into the present is a sign-functional, endogenous and katagenous process; and following Saussure, the sign-functional processes consist in displacements of sign-relates. So far we can infer two such sign-functional processes: (1) **endogenous** development is **continuation** with or without displacement of sign-functional relates and may be accompanied by **devaluation**. Endogenous displacement of sign-relates is a **replacement** process that involves **variation** or **change**. The former is **material variation** (the expression varies); change is **semantic change** (the content changes). (2) **Katagenous** development, only displacement of the content relate, is **continuation** with or without semantic change, involving no variation, but **devaluation**.

The further development of 6.1.3 will now be told; it is a process that by way of a NFS state develops into a level NS series with devaluation-type added[8]:

---

8 Distributional devalution (+D) occurs when α1 develops into α2 (NFS) and <u>narrowing</u> (-D) occurs when α2 develops into α3 (= NS). The development of α1rFS into α2rNFS begins the temporal devaluation of the expression [α] (+T).

6.1. Isolative shifts are sign-functional processes

6.1.6

	a3 >	a4 >	a5 >	a6 >	α1 >	α2 >	α3 >	α4 >
		FB	FB	NFB	NFB	FB	FB	
		NS	NS	NS	FS	NFS	NS	NS
		NS	NS	NS	NFS	NFS	NS	NS
		a3	a4	a5	a6	α1	α2	α3
Devaluation		-D+T	-D+T	-D+T	-D-T	+D+T	<u>-D+T</u>	-D+T

The cutting intersection α2 creates an **ertial** series, making α1 a NFB intersection, while further development makes α2 and α3 FB intersections.

The trajectory of this series describes the **circular** development, concealed by the syncratic A>B and A>A, of a developmental series in which a level inertial series develops into another level inertial series[9]. The whole process is sign-functional and relational, no particular sound/content being mechanically correlated with a *pre-given* content/expression: the 'mysterious' relationship between content and expression has now been dissolved into a logico-structural process whose basis is material and in which a sign function is created in relation to the sign function's synchronico-historical context. Our tradition's notion of circular shift cannot be upheld in its material version (cf. the diagram in Samuels 1972:42): apart from their possible articulatory similarity with ME /ai/, /au/, modern English /ai/ and /au/ differ in almost all other respects: contextual systemic relations, including symmetry-relations; diffusion in lexis, functional load, and not least in respect of manifestations: the Middle English sounds were monophthongized into higher front and back vowels, respectively, while the modern sounds exhibit diphthongal and monophthongal forms that do not allow us to predict the same clear-cut developments as we saw in late middle and early modern English; to underline one significant difference: *for many speakers, the first element of /ai/ and /au/ may in fact be identical* (Gimson 1994:127; 126-128 and 121-123)[10]. If such a purportedly circular shift is to be regarded as *homeostatic regulation*, the view needs to explain why in the first place the Middle English diphthongal system should change at all, how the Middle English diphthongization of /i:/ and /u:/ entered into, not a conspiracy, but some kind of homeostatic deal, with ME /ai/ and /au/[11]?

The next definition is **created nonforceful style**; in the series above in 6.1.6 the development of α1 into α2 creates an attitudinal sign with a NFS content.

---

9 The histories of Southumbrian and Northumbrian OE *a:* as well as the first step in the development of ME *a:* as a part of the Great Vowel Shift have been explained theoretically. The theory can be falsified for internal reasons and the narrative rejected if it is not empirically accurate: *That is not what happened*.

10 Note also the merger – into a low back sound, /a:/, – of the sounds in *shire, shower, Shah; tyre, tower, tar; byre, buyer, bower, bar* (Gimson, p. 129).

11 That the processes should be subject to universals of the type: no language has no diphthongs; all languages have diphthongal and monophthongal systems, does not explain a long-term correlation – if there is one – between our two sets of sounds. And Samuels's non-systemic conclusion (1972:43) must still be criticized for operating with a purely subjective concept such as 'instability', *i.e.* there is a period of instability in the history of English between the diphthongization of ME /i:/ and /u:/ and /ai/ and /au/ and the sounds' respective historic products, /ai/ and /au/ and /e:/ and /o:/ (between C6 and C7) in the 18th century. Furthermore, proponents of such long-term shifts do not put the processes in relation to the systems concept, *i.e.* the process 'the step from variant to change': is the generalization of the diphthongal variants in the 15th century [ɪi] and [ʊu] into /ɪi/ and /ʊu/ – or for that matter [əɪ], [əʊ] into /əɪ/ and /əʊ/ – 'causally' related with the monophthongization of ME /ai/ and /au/ during the same period?

6.1.7 Created nonforceful style is the production of a present successor constituted by a merger in which the possible alternative is endogenously similar to the positive alternative, while the possible alternative *qua* an originary successor is attitudinally ambiguous and devalues its positive alternative distributionally. The possible alternative begins to become devalued endogenously. Created nonforceful style constitutes an irresoluble merger. = **Created nonrelaxed style**.

The rise of a NFS sign is a necessary state in the process of creating a NS sign. The structural conditions for the rise of α2rNFS are different from those that created its predecessor, and such historical differences are matched by different structural phenomena. The same goes for the rise of the normal style sign: viewed from the point of view of internal endogeneity α1, α2 and α3 are articulatorily similar, but the empirical context in which they are produced so differ as to produce different signs.

Apparently, we have the sixth developmental possibility (F) in 5.2.4; the same sound – α1, α2 and α3 – is produced by structurally different endogenous motivation[12], but the important thing is that this identity (or seeming stability) occurs against the existential horizon of constant 'change'[13]. The structural context – external endogeneity – changes the semantic effect of the internally endogenous forces that produce α2 – and α3. To elaborate: since α1 assumes a FS content in relation to a6, α2 cannot continue this content and becomes a NFS sign[14], thereby being both distributionally and temporally devalued in relation to its predecessor. a6 will be continued katagenously as a NFS positive alternative, thereby creating **a tensional field**: FS (precision) :: NFS (ambiguity); since α2 becomes a NFS sign, there is no empirical or logical reason why α1 should not change katagenously into α1rNFS, and **an ambiguous field** is established at this state: while a6 is 'naturally' RS in virtue of its NFS content in relation to its successor, a6 cannot become a FS sign in relation to α2; 'synchronically' it is a natural RS sign, but its semantic potential (RS + NS) has been engulfed by α2rNFS, and the sound may not be continued[15]; an ertial process has been created.

---

12 Note the similarity with Labov's near-mergers, where the listener perceives two articulatorily different sounds as the same.
13 This conclusion throws some light on *reanalysis* as a historical process in grammaticalization, which *does not involve any immediate or intrinsic modification of its* [a form's] *surface manifestation*. To mention few examples: compound constructions becoming derivational ones such as *freedom* from ?(Pre)OE *freo-dom* 'realm of freedom', *man-lic* 'body of a man, likeness of a man' > *manly*. An *of*-genitive [[*back*] *of the barn*] is reanalyzed as a prepositional phrase [*back of* [*the barn*]]. Does reanalysis really expound this F-scenario? The answer is No. The functional change from 'constructional head' to 'complex preposition' is ertial change, in which change of 'form' (endogenous variation) accompanies change of content: *they were hiding in the back of the car // standing at the back of the barn* are certainly different from *they were hiding in back of the car // ... back of the car // standing (in) back of the barn* 'behind'. This is also what the modifying adjectives in the above dualism-addicted quote says: how is *immediate* and *intrinsic* as well as *modification* to be understood? The preposition *in back of* or *back of* is American English. – Cf. *OED* sub *back* subst.VI. 23 and *sub back* (the aphetic form of OE *on bæc* 'into he rear') adv. IV.14.? – Cf. also note 1 in chapter 6.2.
14 Its existence conditions cannot produce a sign with a RS or a FS content. – NFS is a default content.
15 Conditions for the (dis)continuation of a predecessor's predecessor will not be given, beyond what will be said in 6.2.10/11 and was said in O.4. and in principles 1.4.5 and 1.4.6 in terms of general existential conditions.

6.1. Isolative shifts are sign-functional processes 401

The present successor α3 acquires neither a FS content nor a RS content; and being different from the external endogeneity of α2rNFS, α3 can only become a NS sign, which will be defined as follows:

6.1.8 **Created normal style** is the production of a present successor constituted by a merger in which the possible alternative is endogenously similar to the positive alternative, while the possible alternative *qua* an originary successor differs from its positive alternative *qua* its original predecessor in virtue of different degrees of temporal (endogenous) devaluation. The possible alternative is attitudinally nonambiguous and does not devalue its positive alternative distributionally. Created normal style constitutes an irresoluble merger.

The definition does not differ from that of continued normal style; but when seen not as a synchronic phenomenon, the content of created NS is different from continued NS in virtue of the **narrowing** of its distributional scope. This narrowing occurs both endogenously: α2rNFS > α3rNS, and katagenously α2rNFS > α2rNS, whereas the endogenous and katagenous processes continuing a normal style sign are these: a5rNS > a6rNS and a5rNS > a5rNS, respectively. The tensional field α3rNS :: α2rNS becomes nonambiguous. The continued NS sign is, the created NS sign is not, temporally devalued.

The setting for the final analysis with definitions of the break-concept has now been laid out. The developmental series in 6.1.6 created the following transitions *qua* different 'breaks', whose differences will be specified now:

6.1.9 a3 > a4 > a5 > a6 > α1 > α2 > α3 > α4 > α5 > α6
         FB = FB ≠ NFB ≠ NFB ≠ FB ≠ FB ≠ FB = FB

Our focus will be the fullfledged historical intersection **a5** in a4 > **a5** > a6:

6.1.10
1) **a5** coheres with a4 (P) through [-First] & [-First] and coheres with a6 (S) through [-Last] & [-Last][16]. With a4 having the same specifications we say that there is no break between a4 and a5.
2) **a6** coheres with a5 (P) through [-First] & [-First] and coheres with α1 (S) through [+Last] & [+Last][17]. – *All successors so far create inertial development.*
3) **α1** coheres with a6 (P) through [+First] & [+First] and coheres with α2 (S) through [+Last] & [-Last]. – *The successor creates potential ertial development.*
4) **α2** coheres with **α1** (P) through [+First] & [-First] and coheres with α3 (S) through [+Last] & [-Last]. – *The successor creates ertial development*[18].

---

16 These specifications come from the metalanguage: a5 does **not break away from** a4 in virtue of its content = [-First] and **breaks into** a4 in virtue of its expression = [-First]; and a5 is **not made to break away from** a6 in virtue of its content = [-Last] and is **made to break into** a6 in virtue of its expression = [-Last].
17 a6 is made to break away from its successor in virtue of its content and not made to break into it in virtue of its expression.
18 I shall not characterize all the various transitions beyond this, but see Appendix 1.

5) α3 coheres with α2 (P) through [+First] & [-First] and coheres with α4 (S) through [-Last] & [-Last].
6) α4 coheres with α3 (P) through [-First] & [-First] and coheres with α5 (S) through [-Last] & [-Last]. – α5 (and α6) will be specified similarly.

With α5 as the original (FB) predecessor of α6 the whole developmental series constitutes a circular shift, seeing that the transition of the two fullblooded elements α4>α5 is formally and semantically the same as the transition of a4>a5[19], the transition of the two fullblooded elements of α3>α4 differs structurally from α4>α5. From the synchronico-phonetic point of view, a circular shift would have been established earlier, namely as soon as a transition of two internally endogenous elements occur: a4>a5 = α1>α2, while the concepts of non/fullbloodedness would postpone the completion by one more state: α2>α3.

In chapter 5 we set out from a simple calculus with a number of formally identical series to be combined into larger stretches; the dynamic analysis of such a combination has demonstrated that what an Archimedes may look on from the vantage point of the present and turn into a static picture of formally identical two-state transitions, appears on closer inspection with the help of a historical theory to represent a motley and varied, but dynamically coherent and rational picture of a historical period *qua* a number of transitions whose internal differences constantly change the nature of the series through the different breaks defined in 6.1.10 by the notion of coherence relation.

The systemic combination rule that can be inferred from these processes is that an originary NS-successor may develop into either a RS/FS successor or a NS successor; NFS/NRS successors are forbidden, in fact they are **impossible** (to be demonstrated below).

The second group of combinatory possibilities, level series followed by nonlevel series, will demonstrate that **directionality** is not inferable from the series existing in the calculus: A + B is empirically different from B + A, hence [AB] and [BA] are irreducible to /A/, /B/.

The first element of the nonlevel series becomes the originary successor of the last element of the level predecessor series. These originary successors make their original predecessors NFB elements with attitudinal contents that are either RS or FS:

6.1.11 A   level   + rising
  I>IIii   a3 > a4 > a5 > a6 > α >   æ > ɛ > e > ɪ   Nonlevel FS series
  I>IIi   æ2 > æ3 > æ4 > æ5 > α >   æ1 > ɛ > e4   RS>FS (see below)
       B   level   + falling
  III>IV   u3 > u4 > u5 > u6 > ʊ >   ǫ > o > ǫ > ɔ   Nonlevel RS series
  VIII>II ɛ2 > ɛ3 > ɛ4 > ɛ5 > e >   ɛ1 > æ > α > a   FS>RS (see below)

6.1.12   level   + falling
                 u3 >   u4 >   u5 >   u6 > ʊ >   ǫ >   o >   ǫ >   ɔ
  III>IV   FB   FB   **NFB** FB   FB   FB   FB
                 NS   NS   NS   RS   RS   RS   RS   RS
                 NS   NS   NS   NRS NRS   NRS   NRS   NRS
                 u3   u4   u5   u6   ʊ   ǫ   o   ǫ
  Devaluation   -D+T   -D+T   -D+T   -D-T -D-T   -D-T   -D-T   -D-T

6.1. Isolative shifts are sign-functional processes 403

The I>IIii level-rising series, in which a FS successor series is created, is structurally similar to the one in 6.1.12, in which a RS successor series is created.

The continued existence of the potential break around u6/a6, created by the successors υ/α is the reason why the NFB intersections u6/a6 are drawn into the successor series, contributing to the constitution of υ/α as fullblooded intersections in combination with o̜/æ, their successors. The other type looks like this:

6.1.13

	level			+ falling				
VIII>II	ε2 >	ε3 >	ε4 >	ε5 >	e >	ε1 >	æ >	α >	a
		FB	FB	NFB	NFB	FB	FB	FB	
		NS	NS	NS	FS	RS	RS	RS	RS
		NS	NS	NS	NFS	NRS	NRS	NRS	NRS
		ε2	ε3	ε4	ε5	e	ε1	æ	α
Devaluation		-D+T	-D+T	-D+T	-D -T	-D-T	-D-T	-D-T	-D-T

As in the first case the two series, VIII>II and I>IIi, are structurally similar and sign-functionally different. However, the decisive point is that the two pairs of rising or falling successor-series are identical as preestablished calculus entities, but become different when combined, with VIII>II different from III>IV and I>IIi different from I>IIii.

The 6.1.12 (I>IIii = III>IV)-series are less sign-functionally complex than the series in 6.1.13 (VIII>II = I>IIi), a complexity that will be determined by more or fewer different transitions defined by coherence relations in 6.1.14, here with **u5** as the focus of a subseries, u4 > **u5** > u6, of the straightforward level-falling series in 6.1.12:

6.1.14
1) **u5** coheres with u4 (P) through [-First] & [-First] and coheres with u6 (S) through [-Last] & [-Last]. With u4 having the same specifications we say that there is no break between u4 and u5.
2) **u6** coheres with u5 (P) through [-First] & [-First] and coheres with υ (S) through [+Last] & [+Last].
3) **υ** coheres with u6 (P) through [+First] & [+First] and coheres with o̜ (S) through [-Last] & [+Last].-*The successor creates non-inertial and potential ertial development.*
4) **o̜** coheres with υ (P) through [-First] & [+First] and coheres with o (S) through [-Last] & [+Last]. – o and o̜ develop the same type of coherence relations.

With **ε4** as the focus of a subseries, ε3 > **ε4** > ε5, of the more complex level-falling series in 6.1.13, the following coherence relations will be set up:

---

19 If we must split a hair, they may vary as to degrees of temporal devaluation. I shall illustrate the circle as follows: (a3>) a4-FB>a5-FB :: a6 = (α3>) α4-FB>α5-FB :: α6.

6.1.15

1) $\varepsilon 4$ coheres with $\varepsilon 3$ (P) through [-First] & [-First] and coheres with $\varepsilon 5$ (S) through [-Last] & [-Last]. With $\varepsilon 3$ having the same specifications we say that there is no break between $\varepsilon 3$ and $\varepsilon 4$.
2) $\varepsilon 5$ coheres with $\varepsilon 4$ (P) through [-First] & [-First] and coheres with e (S) through [+Last] & [+Last].
3) e coheres with $\varepsilon 5$ (P) through [+First] & [+First] and coheres with $\varepsilon 1$ (S) through [+Last] & [+Last].
4) $\varepsilon 1$ coheres with e (P) through [+First] & [+First] and coheres with æ (S) through [-Last] & [+Last].
5) æ coheres with $\varepsilon 1$ (P) through [-First] & [+First] and coheres with α (S) through [-Last] & [+Last].
6) α coheres with æ (P) through [-First] & [+First] and coheres with a (S) through [-Last] & [+Last].

These coherence relations result in the following differences between the series, differences that do not come from their systemic existence in the calculus:

6.1.16

(6.1.9)   a3 > a4 > a5 > **a6** > **α1** > α2 > α3 > α4 > α5 > α6
           FB = FB ≠ **NFB** ≠ **NFB** ≠ FB ≠ FB ≠ FB = FB
                        NS   **FS**   **NFS** **NS**   **NS**

from 6.1.13   æ2> æ3 > æ4 > **æ5** > **α** > æ1 > ɛ > e4 > ẹ
              FB = FB ≠ **NFB** ≠ **NFB** ≠ FB ≠ FB = FB
                           NS   **RS**   **FS** **FS**   **FS**

from 6.1.12   u3 > u4 > u5 > **u6** > υ > ǫ > o > ǫ > ɔ
              FB = FB ≠ **NFB** ≠ **FB** ≠ FB = FB = FB
                           NS   **RS**   **RS** **RS**   **RS**

Preliminary conclusions: firstly, the ambiguous successors NFS/NRS are always short-lived; they are not continued because their upholding endogenous conditions cannot be continued (empirical necessity); ambiguity and the notion of transitory existence go hand in hand here. Secondly, the FS and NS contents of the ambiguous intersections – α2rNFS – provide a gradual transition to the NS successor through the narrowing of the successor's distributional scope: FS+NS > NS. Thirdly, we now have a criterion for distinguishing between Continued and Created Normal Style intersections: the former process (6.1.4) generates temporal and non-distributional devaluation, while the latter process (6.1.8) generates distributional narrowing. Lastly, and perhaps not unexpectedly, a successor with the two precise RS/FS contents may be short-lived, too.

The synchronic object has also been given a point-in-time description: a short-lived phenomenon with a special property: either precise or ambiguous.

The next subgroup (cf. 5.4. Appendix) consists of initial series characterized by sign-functional nonlevel series developing into three types of series: two types of level series, (A) and (B), and one rising/falling series (C). The significant consequence of the analysis will again

6.1. Isolative shifts are sign-functional processes

be *the irreducibility of directionality*: [AB] is so different from [BA] that the two *parole* sequences do not reduce to *possible combinations* of the parts as calculus entities: /A/, /B/.

A. SIGN-FUNCTIONAL INTERPRETATION OF RISING/FALLING NONLEVEL DEVELOPMENT INTO LEVEL DEVELOPMENT

6.1.17

II>III	a >	α >	æ >	ε >	e1 >	e2 >	e3 >	e4 >	e5	= IV>VII
	FB	FB	FB	**NFB**	**FB**	FB	FB			
	FS	FS	FS	**FS**	**NFS**	**NS**	NS	NS		
	NFS	NFS	NFS	NFS	NFS	NS	NS	NS		
	a	α	æ	ε	e1	e2	e3	e4		

This type of development defined by **continued forceful style** signs (= continued relaxed style), the **rise of nonforceful style** signs (= the rise of nonrelaxed style) and the **rise of normal style** does not require much comment beyond the remark that in the transition, e1 > e2, the internal-endogenous conditions for both elements are the same, say, unrounded, front-peripheral, Cardinal-2 sounds, and these seemingly identical realization conditions become subject to structural influence in *parole*.

6.1.18 **Continued forceful style** is the production of a present successor constituted by a merger in which the possible alternative is not endogenously similar to the positive alternative, while the possible alternative *qua* an originary successor is attitudinally precise and devalues its positive alternative distributionally. Continued forceful style constitutes a resoluble merger. = **Continued relaxed style**.

Since definition 6.1.18 is not different from the definition of 'created forceful style' (see 6.1.5), a differentiating factor is required, and this factor inheres in the respective structural coherence relations of the two processes: created FS creates a break with its predecessor that is characterized by [+First] & [+First], while continued FS creates a break with its predecessor, developing these predecessor coherence relations: [-First] & [+First].

The II>III (= IV>VII) series creates the following coherence relations. I shall begin with ε as the focus of ... > æ > ε > e1 > ...

6.1.19
1) ε coheres with æ (P) through [-First] & [+First] and with e1 (S) through [-Last] & [+Last]. Its predecessor æ develops the same specification.
2) e1 coheres with ε (P) through [-First] & [+First] and with e2 (S) through [+Last] & [-Last].
3) e2 coheres with e1 (P) through [+First] & [-First] and with e3 (S) through [+Last] & [-Last].
4) e3 coheres with e2 (P) through [+First] & [-First] and with e4 (S) through [-Last] & [-Last].

5)   **e4** coheres with e3 (P) through [-First] & [-First] and with e5 (S) through [-Last] & [-Last]. – This specification is also developed by e5.

The transitions of the series do not exhibit ertial change, as the material nature of e3 makes its predecessor e2 a fullblooded intersection so that e1 (NFB) becomes both the last element of what precedes it and the first element in what it develops into.

B.   SIGN-FUNCTIONAL INTERPRETATION OF RISING/FALLING NONLEVEL DEVELOPMENT INTO LEVEL DEVELOPMENT

6.1.20

IV>VIIb	i >	e >	ε >	æ1 >	α >	æ2>	æ3 >	æ4 >	æ5
					RS	RS	FS		
II>VIII	a >	α >	æ >	ε1 >	e >	ε2 >	ε3 >	ε4 >	ε5
		FB	FB	FB	**NFB**	**NFB**	FB	FB	(FB)
		FS	FS	FS	**FS**	**RS**	**NRS**	**NS**	NS
		NFS	NFS	NFS	NFS	NRS	NRS	NS	NS
	a	α	æ	ε1	e	ε2	ε3	ε4	

While the last intersection of the predecessor series in II>III, e1, became formally ambiguous because of the structural interplay in the whole series, neither αrRS nor erFS in the above two series acquires this formal ambiguity; as transitional elements they belong to the respective predecessor series. This type of development requires definitions of **continued forceful style** signs (= continued relaxed style), the **rise of relaxed/forceful style** signs, the **rise of nonforceful/nonrelaxed style** signs and the **rise of normal style signs** and is as such more complex than the previous type of development. The coherence-relations of the series corroborate this difference.

I shall begin with ε1 as the focus in the subseries ... > æ > **ε 1** > e >...

6.1.21
1)   **ε 1** coheres with æ (P) through [-First] & [+First] and with e (S) through [-Last] & [+Last]. Its predecessor æ develops the same specification.
2)   e coheres with ε1 (P) through [-First] & [+First] and with ε2 (S) through [+Last] & [+Last].
3)   **ε 2** coheres with e (P) through [+First] & [+First] and with ε3 (S) through [+Last] & [-Last].
4)   **ε 3** coheres with ε2 (P) through [+First] & [-First] and with ε4 (S) through [+Last] & [-Last].
5)   **ε 4** coheres with ε3 (P) through [+First] & [-First] and with ε5 (S) through [-Last] & [-Last].

6.1. *Isolative shifts are sign-functional processes*   407

6) ε5 coheres with ε4 (P) through [-First] & [-First] and with ε6 (S) through [-Last] & [-Last]. Under the same developmental conditions ε6 will acquire the same specification.

C. SIGN-FUNCTIONAL INTERPRETATION OF NONLEVEL, RISING/FALLING
    DEVELOPMENT INTO NONLEVEL, FALLING/RISING DEVELOPMENT

This group consists of two series of which either is the mirror development of the other. In both series the only nonfullblooded element becomes the material pivot of the whole series: this element assumes the function of both the end and the start of each subseries. The two series are II>V and V>IX

6.1.22

II>V	a	>	α	>	æ1	>	ε1	>	e	>	ε2	>	æ2	>	α2	>	a2
							FS		FS		RS		RS				
V>IX	ẹ	>	e	>	ε1	>	æ1	>	α	>	æ2	>	ε2	>	e1	>	ẹ1
	FB		FB		FB		FB		**NFB**		**FB**		FB		FB		
	RS		RS		RS		**RS**		**FS**		FS		FS		FS		FS
	NRS		NRS		NRS		NRS		NFS		NFS		NFS		NFS		NFS
	ẹ		e		ε1		æ1		α		æ2		ε2		e1		

Comments: The two types (V>IX, II>V) may be the closest we get to *reversible directionality*. However, the formal identity (through internal endogeneity) is resolved by the full *Eigennatur* of each intersection, namely as signs in their own right: æ1rRS and æ2rFS, and by the coherence relations. As such the tradition's problems with reversible change and irreversibility of merger turn out to be irrelevant once we look at change as a sign-functional process occurring under the structural conditions of **ex**istence.

The coherence relations will be specified as follows with **æ1** as our focus in the subseries ... > ε1 > **æ1** > α > ...

6.1.23

1) æ1 coheres with ε1 (P) through [-First] & [+First] and with α (S) through [-Last] & [+Last]. The same specification applies to its predecessor ε1.
2) α coheres with æ1 (P) through [-First] & [+First] and with æ2 (S) through [+Last] & [+Last].
3) æ2 coheres with α (P) through [+First] & [+First] and with ε2 (S) through [-Last] & [+Last].
4) ε2 coheres with æ2 (P) through [-First] & [+First] and with e1 (S) through [-Last] & [+Last]. Its successor, e1, acquires the same specification.

The three sets of transitional difference are summed up as follows:

6.1.24

II>VIII	a	>	α	>	æ	>	ε1	>	e	>	ε2	>	ε3	>	ε4	>	ε5	>	ε6
	FB	=	FB	=	FB	≠	NFB	≠	NFB	≠	FB	≠	FB	≠	FB				
					FS		FS		RS		NRS		NS		NS				

II>III	a	>	α	>	æ	>	ε	>	e1	>	e2	>	e3	>	e4	>	e5	>	e6
	FB	=	FB	=	FB	≠	NFB	≠	FB	≠	FB	≠	FB	=	FB				
					FS		FS		NFS		NS		NS						

V>IX	ẹ1	>	e1	>	ε1	>	æ1	>	α	>	æ2	>	ε2	>	e2	>	ẹ2
	FB	=	FB	=	FB	≠	NFB	≠	FB	≠	FB	=	FB				
					RS		RS		FS		FS						

With this we have completed the narratives of the logically possible series that the calculus (of series) permits us to set up. It has been demonstrated that a theory of sign-functional sequentiality *qua* existence through time is possible, which makes us return to the opening questions in chapter 1:

1) How and why does the purported scientific status of the systems concept determined as a finite set of preestablished entities characterized by **simultaneous** – point-in-time – **existence** in an **abstract space** entail the unscientific status of history with its phenomena existing through time?
2) How can this synchronic point of view make history scientific – through its synchronization of history into a two-state process?

Brief answers:

ad 1) by the adoption of noncogent ideologies: stability (state) and the systems concept are autonomous with the corollary that change and development (historical existence) are dependent on them.
ad 2) by the rejection of the structural interplay between a successor and what it makes its predecessor, thereby making the energy of structural-historical existence static and inert.

In order to substantiate my answers it has been necessary to create a historical language that is devoid of ideologies that have dominated scientific or rational thinking since the dawn of our Western Civilization. As far as I know no other science than linguistics has been so adamantly faithful to an ancient nomenclature, a nomenclature that carries with it the metaphysics/ontologies of Plato and Aristotle, perhaps reinforced by Christianity. In the literature we find no agreement on how to define this dated, but still current language: what is a sentence, a subject, a noun, a pronoun; what is a language system, what is a language's grammar etc. How are *a sentence, a noun, a pronoun; a system, a grammar* etc. defined? Such synchronic terms cannot but pre-shape our way of comprehending reality in a way that leaves no room for the inborn compatibility of the dynamics of a discursively synthesizing mind and historical existence.

Neither our sense of history nor grasp of reality requires an abstract, unchangeable system of preexisting static elements 1) as a condition on cognition and intelligibility and 2) in order to create or recognize empirical coherence through time.

6.2. The process of change. Predictability: Empirical impossibility and empirical necessity. The Functionalist parameter reinterpreted. Examples of impossible developments. Arbitrariness. Conformity. Isomorphism

The argument in chapters 5 and 6.1. has demonstrated that the logical geography of the line-point imagery of synchrony cannot be transferred to historical phenomena, and as such Luick's call for a chronological approach has been implemented theoretically by means of the line(-metaphor)'s irreducibility; a synchronic point with no extension leaves no 'room' for the structural interactivity that makes up **ex**istence. The theory has also placed the cause-concept at the level of data, where it belongs as a non-idealized realization condition for a given historical event. Collingwood put it in straightforward words: when the historian *knows what happened, he already knows why it happened.* (...); his diction is too idealizing, identifying the cause with *the inside of the event itself* (1973:214), where the language historian says that the cause is the phenomenon's realization conditions *qua* intersecting lines.

The series has become a scientifically legitimate analysandum not to be grasped analytically – in and through one single act of the mind's eye, but synthetically and discursively. The conclusion is therefore that we have two equilogical positions which boil down to a difference of degree: the line – *parole* – is to be understood with reference to its point-like elements existing in the L-Continuum. However, since the point is to be understood and exists by the lines which constitute it – it must be understood on the analogy of the way we understand the line, and as such it has empirical extension: there is no empirically relevant entity as a point with no extension existing in an abstract space. Therefore the *langue-parole* dualism is no dichotomy.

A corollary of the argument is that existing in *parole* is a precondition for the **ex**istence of what we call structural and historical phenomena. Knowledge of such phenomena implies no static goal, as we saw in chapter 6.1., only the event itself (Collingwood). The eschatology that is implied in the traditional notion of 'the completion' of a process (a change) has therefore been criticized: the historical process[20] *qua* two or more variants in competition

---

20 Grammaticalization is a process that implies completion; but no 'completed' grammatical phenomenon has ever been seen to escape from becoming 'uncompleted' (cf. 6.1. note 13). – The 'step from variant to change' as well as the interpretation of a change in terms of either the predecessor system or the successor system also involves an eschatology that is incompatible with the universal of constant change. On the other hand, the fact that modern English *–es* in *she goes* etc., has not been generalized to all types of verbs tells against the notion of developmental completion; the extremely incomplete replacement of the Old English adverbial ending *–e* by *–ly* (< OE *–lice*) is another case in point, cf. Modern English *she hardly works* and *she works hard* (OE *hearde, heardlice*): *þe wolf (...) gon to siken harde and stronge* 'the wolf (...) sighed hard (heavily) and deeply', from *The Vox and the Wolf* in Bruce Dickins & R.M. Wilson (eds.) *Early Middle English Texts* (London: Bowes & Bowes. Fifth Impresion. 1961) Text XII, lines 194-195. In the Anglo-Saxon Chronicle anno 741A we read: *Her (...) Cuþræd (...) heardlice gewon wiþ Æþelbald cyning* 'In this year (...) Cuthræd fought vigorously against king Æthelbald'. My point here is that when we compare Samuel Johnson's 1828-entry for *hardly* (and *hard*) and the corresponding entries in a modern dictionary, we see how the word *hardly* (as well as *hard*) has been narrowed semantically; in other words, the grammaticalization of this ending could not be brought to completion because of overriding semantic forces: *hardly* is no longer an adverb of manner; *hard* is an adverb of manner. Cf. *Longman* 1995 (*sub hard* and *hardly* (´*Hardly* is not the adverb of *hard*´)). Note that *Adv.* 1960 has 'severely, unkindly; with difficulty' define *hardly*. – Another often cited process is the purported grammaticalization of *to be going to*

does not fulfil its purpose or complete its meaning in or with the so-called result that it – the single long- or short-term process – ends in. Historical existence is constant and has nothing 'eschaton' about it[21] and Uniformitarianism does not provide us with an empirically relevant stand point from which to interpret the processes that led forward to the present: to labour the issue with an example that cannot but bring out the absurdity of a motivating telos: is the grammar of so-called modern, present-day English that is described in any one particular synchronic grammar book the result or goal in which the grammatical system of Old English comes to its natural end[22]? As Hopper and Traugott put it: historical linguistics lacks its own metalanguage, and the main stumbling block to a dynamic understanding of the everyday language is absence of an explanatory apparatus (theory/language) which bypasses the eschatologies inherent in the various synchronic point-in-time terminologies. The present essay has established a number of dynamic terms that discard the synchronic *langue-parole* nomenclature, terms that reinterpret the mainstay of the so-called Neogrammarian Paradigm, sound change, be it conditioned or isolative. The advanced *existential* theory is **functional** in the sense that its definitions apply to signs and **mechanical** in the sense that the *Eigennatur* of the signs is not re-duced to a 'higher' level of essences.

The investigations so far have in an indirect manner substantiated Saussure's implicitly complex view of change as relate-displacement. The process involves **continuation** – of NS signs, of FS signs and RS signs, **creation** – of FS signs, RS signs, NFS signs, NRS signs and NS signs. The process of 'change' consists of **endogenous variation** (replacement), **endogenous change**, **endogenous continuation** and **katagenous continuation** and **katagenous change**. The coherences involved in these processes were defined above and described with four dynamic concepts, '(not) to break away from' and '(not) to break into'. These sign-functional processes are summed up by four types of relate-displacement[23]:

---

expressing 'immediate future' in modern English. Unquestionably some semantic devaluation has hit this verb form; but is the process complete? When was it completed? It still has some 'modal' content, due to the synonymy field that it exists in. Synchronically, the change is complete when the form either establishes a *new* grammatical category or enters into a *continued* category. Bache and Davidsen-Nielsen (1997:285) classify *to be going to* as a *semi-auxiliary* within a continuum from *clear auxiliaries* to *clear lexical verbs*. Is the process complete when the form attains to the same status as *may, be, have*, etc.? Togeby (1965:519) very pertinently calls attention to the fact that in French *vais* + an infinitive establishes a grammatically overt connection between the present and the future: *il va mourir* as against *il mourra*, whose content is not seen from the stand point of the present. The same must be said about English *he is/was going to die* … . Another question is: what degree of material attrition is a precondition for grammaticalization? *What does that mean? – What's that mean?* and in John Le Carré *The Honourable Schoolboy* (Hodder and Stoughton. 1977) we read on pp. 190, 191: "'Hell's that mean?' he demanded … ", "'Devil's happened to the extractors?' Enderby demanded crossly." The next step in such an attrition process could be: "'And Tiu goes to the Mainland regularly?" (p. 186), with the type "Anybody see him?" as the final stage, where the grammaticalization of a lexical verb reaches its ?logical conclusion: *zero* (cf. Heine *et al.* 1991:13). – "'*Facilities!*' Wilbrahan echoed. Temptations more like.' (p. 186): is *more like* a resuscitated *dated* form (*Adv.* 1992 *sub like* adv.) or is it a dated form as against *likely* with the sense 'probably'? Or is it a shortening due to ease of articulation?

21 a fact that Hopper and Traugott (1993:17) also envisage; but why, then, this talk about *change that is incompletely achieved at any given stage of a language? – incompletely achieved* being a concept that makes no sense within a consistent completion-theory.
22 The modification 'to its temporary' salvages nothing.
23 Cf. chapter 5.4. Appendix. **Group I**: 1) III>VI; 2) II>VIII = IV>VIIb; 3) I>IIi = VIII>II. – **Group 2**: III>IV = 1>IIii; 2) II>III = IV>VII- **Group 3**: V>IX = II>V

6.2. *Empirical impossibility and necessity*

6.2.1   **Sign-Functional Displacement of Relates**

	C1 > C1 E1 > E1	C1 > C2 E1 > E1	C1 > C1 E1 > E2	C1 > C2 E1 > E2
GROUP 1				
II>VIII	X	X	X	X
III>VI	X	X	Ø	X
I>IIi	X	Ø	X	X
GROUP 2				
III>IV	X	Ø	X	X
II>III	X	X	X	Ø
GROUP 3				
II>V	Ø	Ø	X	X

The first column with (seemingly) no relate-displacement is subsumed under the concept of temporal devaluation and the types of (endogenous) change may be specified as follows with four logical possibilities matched by four empirically relevant types:

6.2.2i   1) +expression variation & +content change   a6rNS > α1rFS
            2) -expression variation & +content change   α1rFS > α2rNFS, α2rNFS > α3rNS
            3) +expression variation & -content change   ærRS > αrRS
            4) -expression variation & -content change   a5rNS > a6rNS, a4rNS > a5rNS

These four endogenous developments characterize four types of coherence in which the sign-functional relates take on the role of bridging or separating two transitional elements:

6.2.2ii   1) expression separation & content separation
             2) expression coherence & content separation
             3) expression separation & content coherence
             4) expression coherence & content coherence

As regards the extremes, 1) and 4) are not absolute, for the theoretical reason that part of the past is continued into the present and because of the nature of **ex**istence. The coherence relations will formalize this; I shall make do with one example, III>VI, where the bold-faced transitions mark the relations of the two extremes

6.2.2iii

a3 >	a4 >	a5 >	**a6** >	**α1** >	α2 >	α3 >	α4
FB	FB	NFB	NFB	FB	FB	FB	
-First	-First	-First	+First	+First	+First	-First	
-First	-First	-First	+First	-First	-First	-First	
-Last	-Last	+Last	+Last	+Last	-Last	-Last	
-Last	-Last	+Last	- Last	-Last	-Last	-Last	

The system permits a certain degree of predictability on the basis of the synchronic element: a NRS sign comes from a RS sign and will develop into a NS sign; but neither a NS sign nor FS/RS signs permit us to predict anything about the respective predecessors or successors;

but the historical theory enables us to 'predict' on the basis of the parameter: empirical possibility and **empirical impossibility**. My argument will consist of two steps: the first is an exposition based on the theory's dynamic concepts of endogeneity and katageneity as well as its five attitudinal signs. The second part of the argument hinges on the coherence relations between transitional elements.

A. As a SYNCHRONIC FACT the attitudinal sign contains a historical potential that makes its development unpredictable within a two-state model. This 'unpredictability' is not to be identified with chaos, mysticism or with order-infringement – historical development becoming chaotic, mysterious or destructive – because our historical sense *qua* a historical theory brings order to **ex**istence, *i.e.* on condition that we make our existential or cognitive starting point synthetical, thereby transcending both the synchronic autonomous datum and the synchronic two-state conception: development into the present makes no sense without the (previous) present, and in absolute contrast to Saussure's view of the strait-jacket function of the Law of Tradition, it is the latter – the (previous) present – that makes **ex**istence free in virtue of choice – a choice within the theory in question:

1) NS > NS or RS/FS
2) RS > RS or FS or NRS

as against the necessity in

3) NRS/NFS > NS

and

4) NRS < RS; NFS < FS

This historical potential expounds the choice factor that is grounded on synonymy, and it is this choice-synonymy that makes the attitudinal system cohere with the spoken chain (*parole*): the speaker could have said *[bæd]rRS* instead of *[bɛd]rFS*. Let's take a look at series III>VI in order to demonstrate the internal structure of the attitudinal system in terms of necessity and where choice is involved in terms of a rational choice.

$$a4rNS > a5rNS > a6rNS > \alpha1rFS > \alpha2rNFS > \alpha3rNS > \alpha4rNS >$$
$$a4rNS \quad a5rNS \quad a6rNFS \quad \alpha1rNFS \quad \alpha2rNS > \alpha3rNS >$$

The endogenous motivation of $\alpha1$ entails a FS content; hence if the positive alternative, a6, is to be spoken, its endogenous motivation cannot entail a FS content, and due to the FS content of $\alpha1$ a6 changes its content into NFS. The material form of $\alpha1$ is continued into $\alpha2$, and in spite of the fact that the internally endogenous motivating factors are the same for both sounds, the structural influence of its original predecessor – the externally endogenous motivating factors of $\alpha2$ – does not entail a FS content; and since its internal endogeneity entails neither a NS content nor a RS content, there is only one possibility and that is a NFS content. The katagenous development of $\alpha1rFS$ cannot but produce a positive alternative with a NFS content: it is impossible that as a possible alternative the expression α acquires a NFS content while as a positive alternative it develops either a RS content or a NS content. The further development of $\alpha2rNFS$ is also necessary.

6.2. Empirical impossibility and necessity

It is impossible for α2rNFS to be continued: why is this so? Why does α2rNFS not develop into *α3rRS or *α3rFS?

The first answer is 'simple': the endogenous conditions for the production of α3rNFS which produced α2rNFS, cannot be repeated. I shall amplify this decisive point: the endogenous conditions for α1 make this originary successor FS in relation to its original predecessor, a6. α1 develops into α2 under the same internally endogenous conditions as α1 was produced under, but the structural nature (the historical context) of this internal endogenous motivation differs, hence α2 becomes a NFS sign. When α2rNFS develops into α3, the internal endogenous motivation is the same, but again its structural nature is different from that of α2 and α1: α2 develops from α1rFS, while α3 develops from α2rNFS. There being no alternative within the system, α3 **must** develop a NS content; internal endogeneity does not mark it off as either a RS sign or a FS sign (or a NFS sign). Followingly, there is no reason why the endogenous motivation that continues α3rNS should not produce a NS sign, α4rNS.

The NFS/NRS signs are necessarily short-lived; so may RS/FS signs be, while NS signs cannot be short-lived in accordance with the theory's definitions[24]: a short-lived phenomenon is **monotemporal** and as such a true synchronic phenomenon, existing at a point in time. A NS sign is not monotemporal, but **pluritemporal**, while RS- and FS-signs are **both** monotemporal **and** pluritemporal[25].

What makes an attitudinal sign short-lived? The structural nature of its successor's conditioning factors α>æ1, and this in combination with its predecessor: the fact that e2rNFS comes from e1rFS makes it develop into e3rNS. Consider the following series:

...>	æ4 >	æ5 >	α >	æ1 >	ε1 >	e1 >	e2 >	e3 >	e4 > ...
	NS	NS	**RS**	FS	FS	FS	**NFS**	NS	NS
	NS	NS	**NRS**	**NFS**	NFS	NFS	**NFS**	NS	NS

The short-lived intersection does not become autonomous in the sense that its conditioning factors cannot be repeated due to the historico-structural nature of the developmental series in question. The two logical extremes, the ambiguous NFS/NRS fields and the precise FS/RS fields, appear to be entities whose historical potential is and may be precarious, respectively. The difference between the two fields is that α**rRS** is better integrated into the series than **e2rNFS** is: the endogenous motivation of α**rRS** makes it amenable to cohering with its successor through its katagenous change into α**rNFS** with its RS and NS potential being compatible with the endogenous nature of æ1 as a FS sign. Similarly, the katagenous change of æ5rNS into æ5rNRS equips it with a sign-functional potential, NS and FS, that is

---

[24] One could argue that this series is possible: a4 > a5 > a6 > α1 > α2 > α3 as above, and then α3rNS develops into a FS sign: ... α3 > æ > .... But this is not allowed by the system's material definitions in chapter 5.6.1, 5.6.2, 5.6.6; 5.6.3(9).

[25] As anticipated above, item 3) of the group of complex properties that an intersection acquires is replaced by these properties, which are less static than (i) *nonambiguous* (the NS sign) = *pluritemporal*, (ii) *ambiguous and not precise* (the NFS/NFS signs) = *monotemporal*, and (iii) *precise* (the RS/FS signs) = *pluritemporal and monotemporal*.

compatible with αrRS: α develops a 'natural'[26] RS content in relation to its predecessor and successor.

In the other case the semantic potential of **e2rNFS** (RS + NS) militates against its endogenous motivation in relation to its predecessor e1rFS and successor e3rNS: the fact that the expression [e2] has the semantic potential of functioning as a sign with a RS style contradicts the contents that the expression has as an original predecessor [e1]rFS and as an originary successor [e3]rNS. The ambiguous sign **e2rNFS** is not so semantically well-integrated into its over-all historical context.

The fact that the ambiguous sign cannot be continued will be interpreted as follows: being a semantically devalued field and on its way to being temporally devalued, it expounds a pressure point or instability which can be 'remedied' by means of a simple process of distributional narrowing. It is in fact a *transitional* phenomenon and may be regarded as the implementation of the *stage* concept. But my point here is that the tradition's notions of *instability, pressure point* or *transitional stage* have been defined theoretically, and the remedy concept is not a subjective term such as an indeterminate need-concept or a drive for greater clarity, seeing that the semantic process is grounded on the theory's empirical facts. In deference to the linguistic tradition's terminology, distributional narrowing from e2rNFS to e3rNS resuscitates the sound [e] as a sign.

What is empirically impossible has been explained sign-functionally and it follows that the NS sign's contradictory cannot be NonNS; let the logically possible content NonNS be resolved by both RS and FS, then the endogenous conditions under which a RS sign is created are incompatible with the conditions under which a FS sign is created; and the resolution of NonNS into NFS and/or NRS leads to another absurdity in that NFS/NRS is defined by NS + RS/FS. This reflects on the synchronic parameter **marked :: unmarked**, where the latter corresponds to some normalcy-concept and 'unmarked' refers to the 'unusual', either of two – or more extremes. A marked form cannot be a 'non-normal'. The historical theory has shown that logical contradiction does not define empirical contrasts; other attitudinal systems (speaker-related systems) such as Formal Speech vs. Informal Speech, Standard vs. Nonstandard, 'Socially Acceptable' vs. 'Socially Unacceptable, Polite vs. Impolite, etc., may turn out to be contrary contrasts along these lines

	FS :: NFS	Formal :: Informal
=	FS :: RS+NS	Formal :: X+Z
=	FS :: RS and FS :: NS	Formal :: X and Formal :: Z

Thus Samuels's diagram (1972:137) must be reinterpreted along these lines: Stockwell and Minkova criticized – correctly in my opinion – Samuels for leaving unexplained how two contrasting style variants FS and RS could develop into a normal system. Samuels may have presupposed that the moment when a variant attains to changehood, being adopted into the system, it acquires the content of 'normalcy' (?unmarkedness). This is tantamount to com-

---

26 The adjective *natural* is not to be understood in the light of the inconsistent parameter of Natural Phonology, where the phonologist is bound to operate with natural and unnatural (more or less natural/(un)likely) processes (see *e.g.* Berg 1995:1977; Kenstowicz 1993:103, where – in both works – the natural is a question of statistical frequency).

6.2. Empirical impossibility and necessity

mitting the Phenomenal Error. What is normal can hardly be separated from *parole*, having as much variant-existence as Samuels's other two style variants. Samuels has not explained or illustrated how a phonetic variant whose origin is due to, *e.g.*, relaxed style factors (endogeneity) all of a sudden can be produced under endogenous conditions that produce a forceful style variant. Selection of *relaxed variants* for *expressive needs* or selection of *forceful variants* for *economy* seems to confuse the two sides of the involved signs, their respective expressions and contents. Thus the Functionalists' simple 'Tensional Field': *Zwecktätigkeit* and *Ausdruckstätigkeit*, could be reinterpreted into:

> Normal style :: *Zwecktätigkeit* (expediency) and *Ausdruckstätigkeit* (expressiveness)
> *Zwecktätigkeit* exists in a tensional field with both *Ausdruckstätigkeit* and Normal Style
> *Ausdruckstätigkeit* exists in a tensional field with both *Zwecktätigkeit* and Normal Style

Historical existence does not 'occur' against the general horizons of *ein Streben nach grösserer Deutlichkeit* implemented by a choice between two contradictory modes of speech, or of Martinét's equally simple dictum: *On parle pour être compris* (1970:28). Both views commit the Phenomenal Error, defining their object, the content and expression planes of the everyday language, through one part, the content plane.

B. THE COHERENCE RELATIONS of the individual intersections in relation to their predecessors and successors were specified in chapter 6.1. (cf. 6.1.10/14/15/19 and note 16) by the four directional features [+First] 'pointing forward'($\Rightarrow$), [-First] 'pointing backward' ($\Leftarrow$) in relation to a predecessor, and [+Last] 'pointing backward', [-Last] 'pointing forward' in relation to a successor.

The *Eigennatur* of each intersection together with its place ('context') in its developmental series makes certain developments impossible: the development of an element into an immanent autonomous entity in absolute separation from its predecessor and subsequently from its successor is theoretically impossible. Such an entity would imply sign-functional self-denotation, so that its denotative power would not transcend itself. An absolute sign-functional break between two elements (states) is also impossible (cf. 6.2.5) and the autonomous transition of the two-state model, [A>B] is also impossible (cf. 6.2.4). The theory permits us to set up three coherence-relations; the bracketed specifications below characterize the two transitional elements A and B in the transition A>B, so F points back to predecessors and L forward to successors:

1) *A partial coherence* is defined by [-F, -F], [+F, +F], [-L, -L], or [+L, +L]: $\Leftarrow$A$\Leftarrow$B
2) *An absolute separation* is defined by [+L, -L] or [-F, +F]: $\Leftarrow$A B$\Rightarrow$
3) *A mutual coherence* is defined by [-L, +L] or [+F, -F]: A$\Rightarrow$$\Leftarrow$B

This system (see Appendix 1) yields a continuum of 10 types of coherence relations of which some are impossible, marked by an asterisk[27]

---

27 The following calculus permits more types, but the present continuum is based on partial coherences in combination with either of the two others, plus type 9.

6.2.3   1)   *4 mutual coherences
2)   *3 mutual coherences   2a)   *3 mutual coherences
      1 partial coherence   1 absolute separation

3)   2 mutual coherences (Groups 1 & 2)   3a)   *2 mutual coherences
      2 partial coherences   2 absolute separations

4)   1 mutual coherence (Groups 1 & 2 & 3) 4a) *1 mutual coherence
      3 partial coherences   3 absolute separations

5)   Ø mutual coherence (Groups 1 & 2 & 3)   ...
      4 partial coherences   ...
                                          ...a) *2 mutual coherences
6)   1 absolute separation (Groups 1 & 2)   1 partial coherence
      3 partial coherences   1 absolute separation[28]

7)   2 absolute separations (Groups 1 & 3)   ...
      2 partial coherences   ...

8)   3 absolute separations (Groups 1 & 2)
      1 partial coherence   ...

9)   1 absolute separation (Group 2)   9a)   *2 absolute separations
      1 mutual coherence   1 mutual coherence
      2 partial coherences   1 partial coherence

              10)   *4 absolute separations.

The first conclusion to be drawn is that no empirically relevant development can occur without the creation of at least one partial coherence. Otherwise the focus of the remainder of this chapter is on some impossible processes, which leads to a redefinition of the synchronic concept of sign-functional arbitrariness.

The first impossible state transition is the one that the two-state model of synchrony implies, the autonomy of transitions; at the level of elements it says that the predecessor has no predecessor and the successor no succcessor:

6.2.4   Four mutual coherences
             A   ⇒⇐   B   Example: *ærNS > ɛrFS > ɛrFS > ærRS
Type 1)   [+F   -F]
            [+F   -F]
            [-L   +L]
            [-L   +L]

---

28 Purely reversible developments such as 'e > ɛ > æ > æ > ɛ > e' do not follow from the calculus of (sub)seriess; but their impossibility is also predicted by the theory.

6.2. *Empirical impossibility and necessity*   417

The predecessor has these sign-functional properties:
> A breaks away from its predecessor X in virtue of content
> A does not break into X in virtue of expression
> A is not made to break away from its successor B in virtue of content
> A is made to break into B in virtue of expression

The successor has these sign-functional properties:
> B does not break away from its predecessor A in virtue of content
> B breaks into A in virtue of expression
> B is made to break away from its successor Z in virtue of content
> B is not made to break into Z in virtue of expression

These properties yield the following type of developmental series, implemented above.

X	A	B	Z
æ	ɛ	ɛ // e	æ
NS	FS	NFS//FS	RS

The sound (sign) ɛ creates a gap before it, and its successor B must be a FS intersection, e, which however, cannot make ɛ break into it in virtue of expression; similarly with ɛ as a successor, this sound is bound to break away from ɛ in virtue of content. Finally, the successor of B must differ from B in virtue of content and in virtue of expression (this will be elaborated below).

The next impossible transition type is the one that is made by the synchronic cut: absolute separation between two states or two elements; the predecessor has no successor, the successor no predecessor.

6.2.5 Four absolute separations   *ærNS > ærNS > ɛrFS > ɛrFS
           A   ⇐⇒   B
Type 10)   [-F   +F]
           [-F   +F]
           [+L   -L]
           [+L   -L]

The predecessor has these sign-functional properties:
> A does not break away from its predecessor X in virtue of content
> A breaks into X in virtue of expression
> A is made to break away from its successor B in virtue of content
> A is not made to break into B in virtue of expression.

The successor has these sign-functional properties:
> B breaks away from its predecessor A in virtue of content
> B does not break into A in virtue of expression
> B is not made to break away from its successor Z in virtue of content
> B is made to break into Z in virtue of expression.

These processes yield this type of developmental series:

X	A	B	Z
æ	æ	ε // α	ε // α
NS	NS	FS//RS	FS//RS

Here the successor Z of the transition represents a sign-functional impossibility: if the element is to comply with the structural content requirements, the expression must be changed to either a higher sound than ε or a lower, more restracted sound than α. No successor satisfies such conditions.

It may seem surprising that type 2) is impossible, in view of the partial coherence. But let the specification be distributed in this way:

6.2.6 Three mutual coherences and one partial coherence*

	A	B
Type 2)	[+F	- F]
	[+F	+F]*
	[-L	+L]
	[-L	+L]

The predecessor has these sign-functional properties:
A breaks away from its predecessor X in virtue of content
A does not break into X in virtue of expression
A is not made to break away from its successor B in virtue of content
A is made to break into B in virtue of expression.

The successor has these sign-functional properties:
B does not break away from its predecessor A in virtue of content
B does not break into A in virtue of expression
B is made to break away from its successor Z in virtue of content
B is not made to break into Z in virtue of expression

X	A	B	Z
æ	ε	e // ε	ε
NS	FS	FS//NFS	RS

The dilemma arises from the predecessor; εrFS cannot develop into a successor, creating two mutual coherence relations; the successor cannot be a sign that has the same expression and the same content as its predecessor.

The last scenario to be analyzed in order to demonstrate the empirically relevant concept of empirical, sign-functional impossibility is an…a) scenario of this type:

6.2.8

	A	B	
Type … a)	[+F	-F]	a mutual coherence
	[-F	-F]	a partial coherence
	[+L	-L]	an absolute separation
	[-L	+L]	a mutual coherence

The transitional elements will have these sign-functional properties:

Predecessor:
> *A breaks away from its predecessor X in virtue of content*
> *A breaks into X in virtue of expression*
> *A is made to break away from its successor B in virtue of content*
> *A is made to break into B in virtue of expression*

Successor:
> *B does not break away from its predecessor A in virtue of content*
> *B breaks into A in virtue of expression*
> *B is not made to break away from its successor Z in virtue of content*
> *B is not made to break into Z in virtue of expression*

And the specifications yield this type of developmental series:

X	A	B	Z
æ	æ	æ	æ //ɛ
FS	NFS	NS	NS//FS

The predecessor ærNFS may both continue ærFS and develop into ærNS, but then the successor B is bound to break away from its predecessor contentwise, which it does not do in acccordance with the specifications: B cannot continue its predecessor expressionwise while at the same time avoiding breaking away from A expressionwise.

The order of the coherence-specifications is also relevant; type 9) permits some variation with possible and impossible sign combinations:

6.2.9 1 absolute separation, 1 mutual coherence and 2 partial coherences

While these specifications are empirically relevant

A	>	B		A	>	B
[-F		-F]		[-F		+F]
[+F		+F]		[+F		-F]
[-L		+L]		[+L		+L]
[+L		-L]		[-L		-L]

… > erRS > <u>ɛrRS > ærRS</u> > ærNRS > …   … > erRS > <u>ɛrRS > ɛrNRS</u> > ɛrNS > …

These are disallowed:

A	>	B		A	>	B
[-F		+F]		[+F		+F]
[-F		-F]		[+F		+F]
[-L		-L]		[-L		+L]
[-L		+L]		[+L		-L]

The former implies a development in which A coheres with its predecessor in virtue of both content and expression, while B coheres with A in virtue of expression, but not content: The predecessor εrNS (A) may continue εrNS (X), but cannot develop into an e-sound with the content RS, FS or NRS or NFS.

In the other disallowed series the predecessor may be εrFS, differing from its predecessor εrRS in virtue of both content and expression; and this sign can develop into a successor which differs from it in virtue of both content and expression: ... > εrRS > εrFS > εrRS > ... . But A cannot then be made not to break away from B contentwise.

It was said above under 6.2.4 that the autonomous transition is impossible; there is a counter-argument against this to the effect that the system allows the successor (B) of the transition to become a NFS-sign, if the system permits a series with more than two nonfullblooded intersections in succession: ... FB > FB > NFB > NFB > NFB > ... > FB .... The following comparison between a series of the III>VI-type with two NFB transitional intersections and a series with more NFB intersections will demonstrate the sign-functional impossibility of the latter type of developmental series, once we expand the field of vision to more than a two-state model:

6.2.10i   FB   NFB NFB   FB   FB   FB   (The allowed type)

...> æ1 > æ2 > æ3 > æ4 > ε1 >   ε2 >   ε3 > ε4 > ε5 > ...
      NS   NS   NS   NS   FS   NFS   NS   NS   NS
      NS   NS   NS   NFS   NFS   NS   NS   NS
      æ1   æ2   æ3   æ4   ε1   ε2   ε3   ε4
     T-1   T1   T2   T3   T4   T5   T6   T7

6.2.11i   FB   NFB NFB   NFB   NFB FB   FB   (The disallowed type)

...> æ1 > æ2 > æ3 > æ4 > ε1 >   ε2 >   æ5 > æ6 > æ7 > æ8 > ...
      NS   NS   NS   NS   FS   NFS   RS   NRS NS   NS
      NS   NS   NS   NFS   NFS   NRS NRS NS   NS
      æ1   æ2   æ3   æ4   ε1   ε2   æ5   æ6   æ7

I shall compare the sign possibilities from T1 onwards so that a synonymy field will be established by the possible alternative (1st row) and the positive alternative (2nd row) of each point in time as well as the continuation of the preceding positive alternative (3rd row).

6.2.10ii

	T1	T2	T3	T4	T5	T6
	æ3rNS	æ4rNS	ε1rFS	ε2rNFS	ε3rNS	ε4rNS
	æ2rNS	æ3rNS	æ4NFS	ε1rNFS	ε2rNS	ε3rNS
(T-1)	æ1rNS	æ2rNS	æ3rNS	æ4rNFS	ε1rNFS	ε2rNS

6.2. Empirical impossibility and necessity

At T1 and T2 no synonymy or polysemy fields are created, just a NS-sign at various stages of temporal devaluation.

At T3 the positive alternatives constitute a **compatible** polysemy field, and no synonymy field is established – within the attitudinal system.

At T4 the positive alternatives constitute a synonymy field

At T5 the positive alternatives constitute a compatible polysemy field, and no synonymy field is established.

At T6 no synonymy or polysemy fields are established.

6.2.11ii

	T1	T2	T3	T4	**T5**	T6	T7	T8
	æ3rNS	æ4rNS	ε1rFS	ε2rNFS	**æ5rRS**	æ6rNRS	æ7rNS	æ8rNS
	æ2rNS	æ3rNS	æ4rNFS	ε1rNFS	**ε2rNRS**	æ5rNRS	æ6rNS	æ7rNs
(T-1)	æ1rNS	æ2rNS	æ3rNS	æ4rNFS	**ε1rNFS**	ε2rNRS	æ5rNRS	æ6rNS

At T1 and T2 no synonymy or polysemy fields are created, just a NS-sign at various stages of temporal devaluation.

At T3 the positive alternatives constitute a compatible polysemy field, and no synonymy field is established.

At T4 the positive alternatives constitute a synonymy field.

At T5 the positive alternatives constitute a **noncompatible** polysemy field, and no synonymy field is established (see below).

At T6 a synonymy field is established

At T7 a compatible polysemy field is established.

At T8 no synonymy or polysemy fields are established.

The crucial difference between the two series is created at T5. In the empirically relevant series the polysemy field is created by the compatible merger of the created sign εrNS (nontensional field) at T5 and the continuation of the sign ε1rNFS (RS +NS). In the other series the conflict consists in the two contents NRS and NFS; one and the same expression ε cannot develop two contradictory contents RS from NFS (= RS + NS) and FS from NRS (= FS + NS). This is structurally impossible. The incompatible polysemy field is defined as follows: if a polysemy field is created by a resoluble merger of a created precise sign, say, æ5rRS (a tensional field with ε2rNRS), at Tx and a continuation of an ambiguous sign ε1rNFS (RS +NS), this merger is polysemously *incompatible*, and the developmental series in which this process occurs is empirically irrelevant.

As was stated in the opening paragraph of chapter 6, the mechanical correlation between content and expression found in the attitudinal system as well as the empirical relevance of necessity and (im)possibility of that system suspends the arbitrariness of the sign. The attitudinal sign is neither arbitrary nor necessary; it contains the freedom of choice within the confines of what is empirically possible. It is this freedom that does not make the expression plane and the content plane of the system **conformal** in a Glossematic static sense. If we find conformity between the two planes of a sign-functional system, the system belongs to the class of nonlanguages (Hjelmslev 1963:112). There is a one-for-one relation between the expression and content of the attitudinal sign; this one-for-one relationship, however, cannot be analyzed into a synchronic class of expressions and a synchronic class of contents so that selection of a FS-content always selects a particular expression, and the selection of this expression [e] will always select a FS-content. The sign-relates are dynamically conformable and the particular sign arises through the thorough **contextualisation** that is the existence of historical phenomena[29], not out of the combination-rules of an 'already developed' system's entities. **Isomorphism**, the parallelism between the expression and the content – another Glossematic pillar and criticized by Martinét (1970:27-28) – also finds its empirical relevance in the attitudinal system, seeing that at this level *un sens déterminé* cannot be read into the solidarity between the two attitudinal relates. It is impossible to prove that it is the content relate that takes the expression relate into its service, thereby subordinating the latter to its needs and purposes, and not the necessity inhering in the constant variability of the expression which takes the limited number of attitudinal contents into its service. I base this isomorphism on the synonymy and polysemy that permeate the attitudinal system: [i, e, ɛ, æ, ɑ] r FS, and, *e.g.*, [e] r RS, FS, NS, NRS, NFS. And in contradistinction to the truism that Martinét adopts[30], it was demonstrated in chapter 3.2. how it is the structural interplay (here: in the developmental series) among expressions (the matter of language) that makes [e], [æ] etc. FS signs and [e] *e.g.*, a FS or RS sign.

6.3. Summaries of the sign-functional dynamics of the theory with consequences. Polysemy, synonymy, transcendence and consequences. Definition of traditional historical terms. The synchronic linguist confuses seeing and judging. The definition of constant change in terms of devaluation: continuity and gradualness. The logic of *use* as the cause of constant change

A.   A SIGN-FUNCTIONAL SUMMARY OF THE THEORY'S ELEMENTS WITH CONSEQUENCES

Historical linguistics has experienced a swing from Positivism to 'hermeneutical' approaches[31], but the pendulum can hardly be said to have stopped at the latter (extreme). Positivism is still the main course of the linguistic menu – material forms detached from their sign-functional contexts are the purportedly unanalyzed whole of *analytical-deductive* stu-

---

29 (...) *any sign is defined relatively, not absolutely, and only by its place in the context. (...). In absolute isolation no sign has any meaning; any sign meaning arises in context* (Hjelmslev 1963:44-45;107). Note how the synchronic structuralist has made a dynamic verb, *to arise*, static.

30 Martinét claims that it is the 'meaning' that takes *parole* into its service; this may be reasonable from the point of synonymy, but as regards polysemy, it is equally reasonable to argue that *parole* takes into its service one of the meanings of a given form.

31 If defined as 'a discursive exposition – interpretation – of what is' (Ottosson 1996:3-4; 49), the present essay *qua* a sign-functional interpretation of sound change may fall under this broad definition. But that has not been a deliberate aim on my part, I would then be committing the Phenomenal Error. Therefore my theory is not a deliberate compromise between, *e.g.* Anttila's two claims (p. 400) that linguistics is a *pure hermeneutical discipline*, hence *non-empirical*, and that linguistics also ends with providing *theoretical knowledge* (p. 399): historical linguistics is an immanent discipline with an empirical theory of its own, into which synchrony has been integrated.

6.3. Summaries of the sign-functional dynamics of the theory

dy, and Stuart Mill's 19th century definition of the historical method[32] still covers how historical linguistics is being conducted.

The terms *a (non)tensional field* emphasize the structural and intersectional nature of the historical phenomenon, transcending the object-hooked terminology of linguistics[33]. Sound change is sign-functional, otherwise there is nothing to be understood; neither the form alone nor the content without an expression can, in the words of Appleby et al. (1995:246), *shape consciousness*. The exclusive focus on form perpetuates the *tyranny of positivism* (*loc.cit.*), while the search for only the intention or thought behind the sensuous event leads to idealism[34]; the 'hermeneutical' approach as advocated by Anttila disregards the fact that in order to grasp the intention of an action, there must be a fullfledged sign to grasp. Attitudinal signs are signs that constitute a dynamic system which, in turn, is no calculus of pre-existing elements. Historical development is not (re)combination of the system's elements ('already developed'), but existence is a meaning-creation process that contributes to the constant **form**-ation of man-made reality out of **form**-ed reality. The everyday language is one of the forming factors of our reality, and contributes to the creation of a present different from the previous present, and it is that differentiating process – reinforced by spatial differentiation – which our linguistic-cognitive tradition has interpreted katadynamically, thereby creating an ideal supraindividual entity and transitions or breaks in the empirical continuum. The advanced theory has demonstrated without relying on an empirically spurious systemic entity (1) how such discontinuities occur against continuity, and (2) that they belong to the level of theory, being cognitive contributions of the human mind.

As a structural field the present successor is the originary successor of an original predecessor. Its **possible alternative** is a sign and the expression of another sign, the **positive alternative** of the present successor, say, **(α1rFS)E r C(a6rNFS)**. The possible alternative 'points to' its positive alternative. The possible alternative develops through endogenous processes – variation and/or change – and may undergo katagenous change. The communicat-

---

32. which he defines as 'the inverse deductive method' (1970:594) with the historical method being, broadly speaking the search for causes, which may then be formulated into 'laws' whose generality exceeds the single language or single language state. The search for causes is the strait-jacket that prevents a given methodology from transcending the two-state stage.
33. Anttila (1989:404) continues not only an eschatological tradition, but regards the object of history as a *spectaculum*, a stance with overt naturalistic overtones and criticized by Collingwood (1973:214); Anttila, however, does accept a processual entity such as *a pattern in the making*, although this process must result in *a stable result* qua *a pattern once established*.
34. Clausen (1963:43) very pertinently comments on Collingwood's view of the object of history: a leap that transcends the form to an independent content (or mind) is mysterious (cf. below). Otherwise, I think that Clausen is not quite fair to Collingwood; Collingwood does distinguish between the *outside* and *the inside of an event* (1973:213); but the historian is not concerned with either the outside or the inside of the event to the exclusion of the other. On the contrary, the historian's field of study is *actions*, an action being *the unity of the outside and inside of an event*. But where Collingwood sees the task of the historian as to *think himself into* [a given] *action, to discern the thought of its agent*, a stance that Anttila (1989:404) copies in: *(...) one has to reach the intention behind the action*, the language historian's *Verstehen* involves the creation of a historical theory. And here I must agree with Clausen: Collingwood overlooks the empirical interdependence of the outside and the inside of an event in that he concludes that *at bottom* [the historian] *is concerned with thoughts alone*, whose *outward expression* is mere raw-material for the narrative in question (1973:217).

ive potential of the possible alternative's content expounds the speaker's attitude **precisely**, **nonambiguously** or **ambiguously** with each attitude having these intralinguistic properties, respectively: mono- and pluritemporal and [-D, -T], pluritemporal and [-D, +T]; monotemporal and [+D, +T].

When a possible alternative becomes a fullfledged historical phenomenon, it becomes subject to devaluation by its successor and has made its predecessor subject to devaluation; these processes expound part of the meaning of *constant change* (see section E below): no matter how the possible alternative of a successor develops into the present, it will undergo temporal and/or distributional devaluation in the vertical synthesis of katageneity.

An originary successor coheres with its predecessor and as such points to its origin; an original predecessor does not always cohere with its successor, but points to the fact that it comes from something. These sign-functional coherence relations have been defined by the primitives [+/-First] and [+/-Last] in combination with the four 'break'-concepts. The coherence relations formalize the successor's 'pointing' functions in the sense that the positive alternative, C(a6rNFS), becomes an expression denoting its past *qua* its content: $(C(a6rNFS))E\ r\ C(a6r?)$[35], while the possible alternative *qua* an expression creates a complex sign-functional field with a number of contents, its connotatum, its denotatum, and with its endogenous replacement, which becomes a sign endogenously and katagenously motivated.

The sign function behind the (non)tensional field of the historical phenomenon is irreducible to the simple abstract ErC-structure: **neither sign-functional relate is not a sign**. If our sample sign $(\alpha1rFS)E\ r\ C(a6rNFS)$ is reduced to /$\alpha1rFS$/ and /a6rNFS/ it is easy to see how shortened the analysis is: why is /$\alpha1rFS$/ a forceful-style sign? The synchronic linguist can only answer circularly, because its positive alternative is /a6rNFS/. The answer leaves out the fact that for the two sign functions to be constituted as a tensional field, they must observe what is empirically possible and what is empirically impossible, because the formed substance [$\alpha1rFS$] 'comes from' another formed substance', not katadynamically from /ErC/[36]. The calculus in question does not contain that knowledge or meaning (cf. 6.3.1 below).

The historical phenomenon **ex**ists in process of becoming within the totality of the L-Continuum's developmental series – *in casu*, vocalic series. As it is sign-functional, we create it and grasp it by means of our everyday language. This is a non-dualistic and explicitly anthropocentric way of stating the process that the dualism turns into a static *manifestatio*-process: *(...) l'usage qui est fait de la langue contribue à la transformer*[37] (Meillet 1966:20;

---

35 The possessive pronoun is used in the above-suggested senses: we know historical phenomena have a specific origin, but we do not know it.

36 This kind of process is not the exclusive property of phenomenology, the latter's object being phenomena *qua* 'something that appears before somebody as something' (Ottosson, p.3; Hansen 1984).

37 Coseriu (1974:65; cf. p. 64) voices a similar view in his 'fourth' language: *Man kann sagen, dass in einem Dialog, der in ein und derselben "historischen Sprache" geführt wird, immer vier verschiedene "Sprachen" eingeschlossen sind: a) das Wissen des Sprechers; b) das Wissen des Hörers; c) das gemeinsame dieser beiden Wissen; d) die neue Sprache, die sich aus dem Dialog ergibt.* (My emphasis). The difference between my view and Coseriu's is that the primary function of the everyday language, its *Mitteilungszweck* manifested in the dialogue (cf. Martinét below), is a position that is bound to make *die Mitteilung* an object and preexisting, while I subordinate the communicative function to the existential function of the everyday language, which makes the *Mitteilung* complete its meaning in its being spoken.

6.3. Summaries of the sign-functional dynamics of the theory

and above ch. 3.4.(f), ch. 3.2. note 10), and which Paul (1966:32) sums up by: *Wenn durch die Sprechtätigkeit der [Sprach]Usus verschoben wird, (...) so beruht das natürlich darauf, dass der Usus die Sprechtätigkeit nicht vollkommen beherrscht (...)*; and by the question: *Wie wird diese [die Sprechtätigkeit] durch jenen [den Usus] bestimmt und wie wirkt sie umgekehrt auf ihn zurück?* (Paul, pp. 33; 29; and above chapter 5.1.). Paul's use of *Usus* reveals how difficult it is to escape static Realistic overtones.

The sign-functional processes that constitute the speech situation will be summed up as follows. The **processual** structure of a present successor, α1rFS, is specified in 6.3.1i, while the abstract form in 6.3.1ii exemplifies the shortened process of moving from data to 'form'.

6.3.1i   (a5 r NS) <u>a6 r NS</u> (a5 r NS)   >   (a6 r NS) <u>α1 r FS</u> (a6 r NFS)
6.3.1ii (E r^{end}P C) <u>E r C</u> (E r^{kat} C)   >   (E r^{end}P C) <u>E r C</u> (E r^{kat} C)

Spoken by Primus (= P), the present successor, α1rFS *replaces* a6rNS through internal endogeneity and develops its content through both internal and external endogeneity in the katagenous synthesis with what has been continued of the past, a6 r NFS. The moment Primus utters a sentence, of which Primus becomes the connotative content (= con), his language (*das Wissen des Sprechers*)[38] develops into the present making the involved elements of the L-Continuum (= L) the respective predecessors of the new present successors. This process is sign-functionally represented in 6.3.2:

6.3.2   (E r^{den}L C) E r^{con} CP   Primus's unspoken language (L-Continuum) at T1; a connotation.
(The attitudinal system connects the denotative sign function of the L-Continuum and the utterance (*parole*)): Primus speaks at T2:

{(E r^{den}L C) E r^{con} C} C rP **E** [(E r^{end}P C) <u>E r C</u> (E r^{kat} C)]
The utterance **E** [...] also becomes the hearer's property; it becomes a relate of a new connotation, the hearer's (*das Wissen des Hörers*).

E [(E r^{end}P C) <u>E r C</u> (E r^{kat} C)*] r C(= Secundus)

In this sign-functional process the three signs (Er^{den}LC), (Er^{end}C) and (Er^{kat}C) are dynamically similar, a6rNS, which leaves us with this alternative: *synchronically*, the katagenous sign (Er^{kat}C), a6rNFS of T2, manifests the L-Continuum's denotative sign *qua* a *langue/Usus* element (Er^{den}LC), /a6rNS/, and is itself a *parole* phenomenon *qua* the positive alternative of this originary successor, while (Er^{end}C) represents the possible alternative, produced by the speaker, [α1rFS], an allophonic variant of the *langue* element /a6rNS/. This interpretation supposes that the *langue* sign will continue unchanged through time and we are back to our original starting point, the slant-bracket notation of the dualism: /X/ [FS], [NFS] implemented by /a6/ [α1rFS], [a6rNFS], which leaves unexplained where the contents come from:

---

[38] I must complete the above Coseriu-quote: *Doch ist die Sprache (das sprachliche Wissen) des Sprechers nie vollkommen identisch mit der des Hörers, während das gesprochene Wort – nach Montaigne – immer "zur Hälfte vom Sprecher und zur Hälfte vom Hörer" ist (und sein muss)* (p. 65).

as an abstract form /big/ does not explain where Jakobson's two variants, [bi:g] 'marked' – [big] 'unmarked', get their meaning from.

Dynamically, the communicative situation is interpreted in accordance with the theory of history: the denotative sign function ($Er^{den}LC$) exists in the speaker's L-Continuum as the temporary successor of a developmental series. When it is made to develop into the present, it enters into a horizontal relationship of endogeneity ($Er^{end}PC$), itself being continued in the katagenously motivated vertical synthesis ($Er^{kat}C$)] of the (non)tensional field in question. This dynamic model anchors the elements of the spoken chain and their predecessors anthropocentrically and the originary successor of the present expounds the fundamental properties of the everyday language, that of constant change, and those of polysemy and synonymy through the speaker-hearer connotations. When the utterance becomes the hearer's, it is understood in accordance with the attitudinal system of his L-Continuum, here represented by katagenous relation marked by an *. Anticipating the argument in section B, we shall say that **the content of the speaker's utterance is also the hearer's meaning**. This dynamic model does not require the idealizing notion of something supraindividual, only the notion of temporal and simultaneous differences in the transindividual utterance being anchored in the speaker and hearer.

B.   THE SCOPE OF THE SIGN FUNCTION: SYNONYMY AND POLYSEMY VERSUS METAPHOR. THE SPEECH SITUATION AND ANALOGIES

The structure of the connotation is similar to that of the Glossematic metaphor in 6.3.3 (Hjelmslev 1961), but the metaphor is dynamically similar to the attitudinal sign.

6.3.3   (ErC)ErC   (/father/ r Pater familias)E r Deus

This structure, in which the denotative sign connotes another content, is too simple, because no sign-relate is not a sign. The content Deus must contract a sign function with an expression, *in casu*, God in modern English, and 6.3.3 is to be expanded into the structure in 6.3.4.

6.3.4   (ErC)ErC(ErC)   (/father/ r Pater familias)E r C(/God/ r Deus)

a sign function that agrees with the supposition of the fundamentality of synonymy in its structural interplay with polysemy. The complex structure in 6.3.4 exhibits the same type of dynamic interplay that constitutes a possible alternative and its positive alternative as an originary successor's (non)tensional field:

6.3.5   (a6 r NS)α1 r FS(a6 r NFS)   (ErC)ErC(ErC).

On the analogy of this and with reference to Thomas Aquinas (*Summa* 1a. 1, 9-10; pp. 142-146), where we find this seeming paradox: 'the metaphorical sense of a metaphor is its literal sense', we say that 'the synchronic content of a synchronic sign is its historical content'. In my participative diction: literal sense :: metaphorical sense(literal sense), and historical content :: synchronic content(historical content).

The *father-God* metaphor would not exist as a metaphor without the 'literal' sign God

6.3. Summaries of the sign-functional dynamics of the theory

with the content Deus. When we try to understand a metaphor, we must not stop at the metaphor itself, but move on to the sign of the literal content, which exists in a synonymous field. Understanding the synchronic content of a present intersection involves the sign of the historical content: when we analyse [a6] or [α1] as synchronic entities (motivated by internal endogeneity), we shall also be considering their respective predecessors in order to understand the FS content of [α1] and [a6]'s NS content as well as their degree of temporal devaluation.

And with reference to the dualism (cf. Meillet and Paul above) the systemic content of the spoken utterance is its existential content (chapter 4.3.): the hearer does not stop at the synchronic linguist's preestablished system (as Wittgenstein did (chapter 5.1.)) or componential analysis (cf. Bache and Davidsen-Nielsen 1997:191), but moves on to the utterance's sign-functional constitution of the present (the sign that the denotatum becomes the content of (4.3.2 onwards)).

The two below examples may sum up the dual type of problem (inconsistency) that the synchronic grammarian lands himself in: as regards ambiguity, the synchronic system only provides alternative 'solutions': if, instead, we say that the (purported) ambiguity of a sign (polysemy) is its precision (synonymy), then in *He left his wife to deal with the creditors* (from Bache and Davidsen-Nielsen 1997 = Schibsbye 1970:30-31) no speaker, let alone a hearer, will stop at what the system may yield, but consider the structural relationips that suchlike utterances enter into, with *e.g. He left it to his wife to deal with the creditors, He left the room so that his wife could deal with the creditors alone*; or: *He left his wife in order that he could deal with the creditors*. Were these synonymous constructions not possible (cf. 1.5.2 above), the original utterance would not be possible. This precision does not inhere as a systemic potentiality, but in the actuality (in Aristotle's sense *Met.* 1049) of the synonymy field in question. Knowledge of what is actual precedes knowledge of the potential; a potentiality is only a potentiality because it can be realized!

While the synchronic linguist is much concerned with that type of precision, he seems to ignore the ambiguity that his theory creates: King Ælfred translates *Boethius* (Mitchell and Robinson 1986:211): *he Iohannes þone papan het ofslean = he ordered John the pope to be slain* or *he ordered X to slay John the pope* or *he ordered (the) slaying (of) John the pope*. In a word: the synchronic theory must choose between various 'solutions' on the basis of what its concrete material offers, thereby leaving the stage to synonymy or polysemy: is the 'infinitive' a verb or a noun or neither? Is it an active/passive or a nominal form or neither[39]?

The **transcendence** problem may be voiced sign-functionally as follows: the scientific content of a scientific sign (= a theory/a language) is the content of its object sign (= the everyday language); the scientific analysis of the everyday language must not stop at the the-

---

39 With reference to Lévi-Strauss Gellner (1969:154-155) discusses how meaning may originate with elements in the general speech situation, not from a language system shared by two (groups of) people, so that *the burden of comprehension* may be said to hinge on *the communication itself*, in my diction: '… on the general context of the speech-situation and the spoken utterance itself' – *communication* being too vague a word – seeing that the spoken utterance in its existential function is able to embraciate and exbraciate elements of the entire speech-situation. – Lévi-Strauss had reported a situation in which two groups of Brazilian indians managed to form an *effective cooperating group*, in spite of the fact that their respective languages were not mutually intelligible (not 'the same').

ory itself, the *Selbstverstehen* of the nominalist's theory. In other words, we shall not be able to grasp the object-content of the scientific language unless we know the object language. This means that strict nominalism, with data purely contingent and with truth and falsity only obeying a scientific principle such as the Empirical Principle of Glossematics, must transcend its own *Selbstverstehen*, and *suppose* data as necessary, not contingent, and this transcendence is a simple corollary of the sign function: no matter whether the nominalistic theory is 'empty': (ErC)E r ..., with all meaning coming from data, or 'full': (ErC)C r ..., with all meaning coming from the theory, the sign-functional interdependence makes the missing relates C and E, respectively, necessary.

Platonic idealism meets with the same participative fate: the meaning of the transcendent world of Ideas is the meaning of the sensuous world of constant change.

Hence, the historical content of a historical narrative is the empirical content of the historical phenomena in question, because when we read a historical narrative of the French Revolution we do not stop at the narrative itself, but move on to its denotatum, the reality whose creation the French language contributed to in the last decades of the 18th century. The gap between the language of the narrative and the language of the event is bridged by the historical theory of language that retrieves the original meaning.

On the analogy of the speech situation in 6.3.2 the historical narrative, like a poem or a grammar book, is also read by other people who now become hearers, and I shall draw some consequences from this analogy, observing a theoretical parallelism between the different phenomena. At the same time the ultimate consequence of the advanced theory of the existential function of language will be corroboration of Glossematics's sweeping view of language.

1) It follows from the existential function of the everyday language that the historian's writing of his narrative, the poet's writing of his poem, and the synchronic grammarian's description (explanation) of the grammar of a language, as well as the literary critic's critical evaluation of an author's novel contribute to the present's development into the present. These human activities are sign-functional; that is why they can be understood immediately and performed without anybody having recourse to (pre)theoretical manipulation that makes them amenable to a given approach.

The last but one sentence in section A above is a consequence of the universality of synonymy and polysemy; therefore three main lines intersect the speaker's utterance contributing to its meaning: the content of the speaker's L-Continuum, the denotatum of the utterance, and the hearer's meaning (how he 'translates' the utterance).

The historian's narrative is intersected by three similar lines, with the content of the narrative becoming the reader's meaning.

The grammatical description is intersected by the writer's L-Continuum (including his particular theory), by the denotatum, and by the reader's content-contribution.

The literary critic's essay is intersected by the critic's L-Continuum (including his particular theory), by the novel or poem *qua* the essay's denotatum, and by the reader's content contribution.

This obliges me to qualify what was said in chapter 4.3. Whether a poet's poem or a painter's painting is intersected by a denotatum-line may be a moot point that I shall not pursue here. Relying on Winterson (1996:104), who says that all art is *more than its subject matter*, I also believe that a work of art signifies something beyond itself and points to more than

6.3. Summaries of the sign-functional dynamics of the theory

its origin. Keats's *Grecian Urn* poem certainly refers to something extralinguistic. What I am about to say here is that if the work by a writer of fiction is also a theory-based statement of reality, which both Wordsworth (1968:242;257), Winterson (1996:11) and Keats's truth-beauty identification say, then the nominalism of the synchronic grammarian's theory obliges the theoretician to explain why his theory (grammar book) is not fiction. That the scientific theory enunciates no existence postulates does not automatically make it scientific or place it on a par with the strictly formal domain of logic, geometry and mathematics.

2) Since each of our sign-functional phenomena above is historical, the here-and-now meaning of a poem, a novel, a critical essay, a grammar book, a historical narrative is also their respective historical meanings, which is their fourth intersecting line. Neither the individual poem, the individual narrative, the individual essay, the invidual grammar book can be grasped outside the horizontal syntheses in which they exist; each contributes to the **ex**istence of their respective developmental series.

Could Keats, Alasdair Gray, Frank Kermode have written *Ode on a Grecian Urn*, *A History Maker* (London 1995), *The Sense of an Ending* (Oxford 1967), respectively, and could a synchronic grammarian have written his grammar book without predecessors, and do the respective successors as well as predecessors contribute to the meaning of those works? The first question will be left dangling, pending further investigations into the nature of historical existence (cf. chapters 2.5. and 2.6.).

In accordance with the theory of history the second answer must be in the affirmative, and I shall use the synchronic grammar book (Ga, Gb, etc.) and its author (C1, C2, etc.) as my illustrative material; I am not implying that the processes at this level occur in accordance with the advanced attitudinal system, only the structure in 6.3.6 is similar to that system's structure. – Glossematics envisages a parallel history between the continuation of non-scientific semiotics and the continuation of scientific semiotics.

6.3.6 GarC(onnotatum)1 > GbrC2 >   GcrC3 >   GdrC4 >   GerC5 >

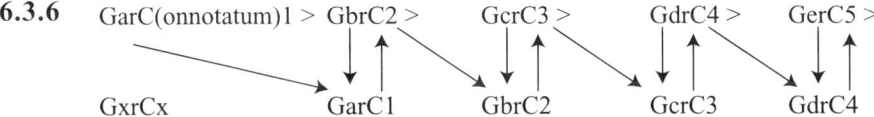

GxrCx   GarC1   GbrC2   GcrC3   GdrC4

The meaning of a grammar book is therefore the structural product of its own existence in a tradition *qua* developmental series, where it is motivated endogenously and katagenously, and just as Gb is to be understood through its effect on its Gc successor, so it is understood through its predecessor's effect on it. It has not completed its historical meaning potential until a successor, carrying it into the new present, makes it a predecessor; the past's explanatory function (historicism and biographical methodology) and the present's similar function (uniformitarianism) have been united.

This conclusion is of course a mechanical extension of my view of history as existence in developmental series, but it agrees with the synchronic *tout-se-tient* truism, and since it rejects the possibility of man-made phenomena being islands unto themselves, it has the cognitive-pedagogical consequence that the study and learning of a particular grammar book requires the study of other grammar books, the former's immediate historical context.

Neither the literary, grammmatical, nor historical traditions may regard their objects as autonomous entities existing in a time- and placeless vacuum. I cannot put it better than Winterson, who even manages to include a moral: *By making islands of separation out of the unbreakable chain of human creativity, we are able to set up false comparisons, false expectations* ... (p. 11), a moral that I prefer to couch in the historian's ideology-free diction: a realistic grasp of reality presupposes the ability to see how the present (past) is continued in the present.

Like the *true artist*, the linguist and the historian should be *connected* (Winterson, p.12). This connection is a living connectedness with the past. The artist studies the past, *not as a copyist*[40], but in order to 'connect' with the reader, *because it is connection that we seek* (Winterson, p. 13). The connection is sign-functional: the writer of fiction makes art speak although *the language of art, all art, is not our mother-tongue* (Winterson 1996:4). Therefore the artist is a translator (Winterson, p. 146) who has learnt the languages that are reality, in order to translate them into her own language (art), a language which the reader or critic must then translate into his or her language. Existence, as a scientific process, a literary/fictional process, a cognitive process, a communicative process, a teaching-learning process is the creation of connection. And since structural and historical connections are inevitable meaning-creators and shapers, structuralism and history will never become dated cognitive principles.

That was one way of stating the Glossematic generalization, whose narrow version that all non-semiotics are components of semiotics has been further generalized and adopted here in its non-technical consequence: *linguistic theory* 'illuminates' all objects (Hjelmslev 1963:127) on three conditions, (1) that the ultimate metalanguage of the theory of language is the everyday language, (2) that the everyday language is co-extensive with our grasp of reality, including our sense of history, and (3) that all objects to be illuminated are (made) sign-functional.

C.   REDEFINITION OF SOME BASIC TERMS

The fact of change and development is not a positivistic datum, a *spectaculum*, to be seen, heard or felt, but something to be understood. Change is a cognitive category rooted in our mental make up, applied by our *historical sense* when we **ex**ist and applied as an epistemic concept when we want to grasp reality; it is a fundamental property of the existential function of the everyday language, and is the language itself (cf. Aristotle *loc.cit.*). This necessitates some redefining of traditional concepts used in synchrony and historical linguistics, and I shall start with the *use*-concept (cf. section F below): the everyday language changes because it is used; use necessarily entails devaluation, some wear-and-tear which in turn entails a need for remedial processes: change entails change. This intralinguistic pleonasm is matched by an extralinguistic one: language change is entailed by extralinguistic change. The everyday language is worn down by use, eventually reaching a stage when it must be re-

---

40 Alain Robbe-Grillet voiced similar sentiments about the role of the past, the present's connection with the past in his 1955-essay *A quoi servent les théories* (1963:9-10; 11). Blind copying of the past and the pastiche are denials of existence, because *Loin de respecter des formes immuables, chaque nouveau livre tend à constituer ses lois de fonctionnement en même temps qu'à produire leur destruction* (pp. 12-13; a view repeated on p. 146 in the form of *une évolution constante du genre romanesque*).

6.3. Redefinition of some basic terms

dressed in order to survive, the logic being that if the everyday language does not reach this stage, there is no reason why it should change – in the other sense of the word. The wear-and-tear is not virtual, the grammatical rule is in fact broken: is the rise of the verbal *-es*-ending in modern English remedial change? Is the zero-ending in the first person singular remedial (< OE *-e/-u*)? The eschatology implied by the notion of (remedial) change is in perfect accordance with the idealism of our historical heritage from Ancient Greece (cf. chapter 3.2. note 13) and is summed up in Descartes's religious myth and his wax-example, both having become a conceptual *Nachlass* in modern linguistics. In his *Third Meditation* (45.9-14) Descartes writes[41]:

> *Itaque sola restat idea Dei, in qua considerandum est an aliquid sit quod a me ipso non potuerit proficisci. Dei nomine intelligo substantiam quandam <u>infinitam</u>, <u>independentem</u>, <u>summe intelligentem</u>, <u>summe potentem</u>, & a qua tum ego ipse, tum aliud omne, si quid aliud extat, quodcumque extat, est creatum.*

The French translation (p. 45) runs as follows:

> Partant il ne reste que la seule idée de Dieu, dans laquelle il faut considérer s'il y a quelque chose qui n'ait pu venir de moi-même. Par le nom de Dieu j'entends une substance infinie, <u>éternelle</u>, <u>immuable</u>, indépendante, toute connaissante, toute puissante, et par laquelle moi-même, et toutes les autres choses qui sont (s'il est vrai qu'il y en ait qui existent) ont été créées et produites.

Both additions to the Latin text are justified in a footnote (cf. also 40.16-18; see Hansen 1983:67-68): *Descartes présente toujours l'immutabilité de Dieu comme une évidence naturelle; ... tout changement suppose un manque d'être et un mouvement pour acquérir ce qui manque.*

The coupling of change and imperfection is evident: imperfect man must change in order to attain to imperfection, or to a lesser degree of imperfection than previously; deficiency is the ultimate cause of change in conjunction with the unattainable perfect Telos. Man being deficient and his use of things therefore imperfect are – with Ancient Greece's negative-imperfect view of sensuous reality – the prime shapers of Western Man's conception of his own existence. Modern linguistics has continued this value-loaded view of history.

Since no language has ever been 'heard' to break down, since no language – state or stage – has ever been heard to be less efficient than another language, since language – any state or stage – has always been seen to preserve its individual and collective orientation and continue itself equiempirically and equifunctionally through time, there is only one conclusion: the everyday language is **not** worn down and consequently **not** changed in order to 'survive' **because** it is used, but the everyday language is only worn down and in need of change if it is not used: from which it follows that use (= change) ensures the survival, existence of any

---

41 I have underlined the properties in both the French and the Latin texts. Kant (1963:44) makes immutability a 'character' of necessity. Descartes might be in trouble if he attributed necessity to imperfect man's development towards perfection as well as to God. Since we cannot but exist, empirical necessity is a property of existence.

everyday language. The notion of 'remedial change' is either redundant or belongs to the level of theory.

Theoretically linguistics must scrap that part of the historical and synchronic terminology that perpetuates religious or dated metaphysical myths, and my argument leads up to the following, if not redefinitions, then determinations of some metaterms:

1) *Change* denotes content change: C1 > C2.
2) *Variation* denotes expression replacement: E1 > E2.
   1) and 2) resolve Parmenides's dilemma, if divested of ideological connotations.
3) *To be historical* or more precisely *to become historical* is an acquired property that a present orginary successor acquires when it develops into the present.
4) *To be synchronic* denotes a part of the historical phenomenon or present successor. The property is defined by internal endogeneity and is a part of an intersectional whole. Seen in isolation, it is a *flatus vocis*, but is synonymous with *to be static*.
5) *To have a history* cannot be detached from the concept of the developmental series; predicated of a carrier-entity, X, it is no more meaningful than to say of X that 'X has a synchrony'. OE *æ* as in *dæg* 'a day' has a history no more than PreOE *a* or ME *a* has. *To have a history* is a macro-concept, if it means 'to have or fall under a historical theory' and a micro-concept when it denotes endogenously and katagenously motivated processes; in this sense it is synonymous with *to exist* or *to be in process of becoming*.
6) *To have an origin* may be predicated of both the possible and the positive alternatives and is as such not sufficient to make the possible alternative historical. It is a synonym of 'to **ex**ist' or 'to develop into the present' and therefore implies *common origin*. In the Jones-Hypothesis its full cognitive force is apparent as an interpretational category, and the L-Continuum constitutes the intrasystemic or idiolectal origin of the spoken chain. In a language group it is that which guarantees mutual intelligibility and as such it merges with existence *qua* an intelligible process. – The concept is made static through the reification of its implied process: Meillet turned an object which has been *parlée antérieurement* (1966:16) into an origin and the same applies to Glossematics's mother language *vis-à-vis* its daughter languages.
7) *Cause* has been mentioned above, where it was replaced by 'to be effective in'. – Mechanical causes participate in the effect and do not differ from other motivating factors (intersecting lines). Due to the existential function of the everyday language it is redundant; cause is a value-loaded concept which is invoked when something happens unexpectedly in respect of a pre-given goal.
8) *To become* is perhaps the greatest stumbling block to a proper historical understanding of reality. It is so polysemous as to become antinomical: at the surface it is a dynamic word, suggesting process and change. At the same time it suspends this development, enunciating a Cartesian dualism: 'A becoming B' implies that something is stable (essence), while 'it' changes exchanging accidental properties, so that our fundamental cognitive category becomes: to see and find identity in (the) differences (which are A and B). Its immanently historical sense is illustrated as follows: OE *cyning* **is** the successor of PreOE **kuning-*, and it **became** a successor in a process different from the process that made **kuning-* the predecessor of *cyning*. *To become a successor* implies no identity in differences between two endogenously motivated possible alternatives; *to become*

a predecessor implies a cognitive process in which a possible alternative becomes (develops into) a positive alternative and through its katagenously acquired content contributes to the creation of the present; the concept implies two entities or one structural entity (a (non)tensional field), which makes no *analytical* sense, if *analyzed into* its purportedly component parts, say, a6rNS and a6rNFS. In a successor, concept and phenomenon merge[42] in the sense that *to be a successor* is no sign (predicate) by which we assign a property to a phenomenon *qua* a preexisting carrier entity.

In sum, the fundamental existential distinction, which our tradition has subjected to value-loaded interpretations thereby suspending time, may be stated by this nonsymmetrical relationship: **all successors create predecessors and all predecessors have a successor**, into which we must not read any static-katadynamic sense: a successor is not **really** an entity with the property of being a successor. No successor necessarily has or implies (the creation of) a successor.

D   CHANGE AND CAUSE APPEAR FROM A JUDGEMENT, A MENTALLY DISCURSIVE PROCESS

In section C I tried to determine some terms that have achieved metatheoretical, if not axiomatic status in the linguistic literature; the concepts of *system*, *state* and *totality* have been criticized all through the essay for being metaphysical concepts, while *change* and *stability* have been treated as equiempirical. *Structure* has been redefined as a dynamic concept and *synchrony* and *simultaneity* have found a place within the theory of history. Many such concepts remain defined by the way they are used in the literature so that we do not really know when they refer to the level of theory or that of data. In this concluding argument I shall illustrate why I have attached so much importance to the distinction between **theory** and **data**.

Cause and *change* enter into a particularly close cognitive nexus and both have nearly been struck from the theory's nomenclature. In the developmental series of isolative shifts they are redundant. But they deserve further investigation because of their general contribution to the dominant view of how the human mind works in the interplay between the faculties of seeing and judging.

1) Cause

A cause is part and parcel of the individual events; there is no reason why one part of an event should be promoted to the status of Explanatory Principle. That something is a cause appears from a judgement, and as such it belongs to the level of theory and cannot be a part of an empirical *Kausalnexus*. To make a judgement is an analytical mental process that assigns a property to an entity (Kant 1762:44). Can we find just one property of a cause without which a cause cannot be? Hardly. To be a cause, on the contrary, is a property of an en-

---

42 Cf. the merger of the property 'to be synchronic' with 'linguistic theory' in modern linguistics; the object of linguistic theorizing is *synchronic*, hence the theory of it is synchronic. Or truth with existence: truth becomes a cognitive-epistemic concept at the logical level where falsehood emerges, otherwise 'what is truth is existence' and 'what exists is truth'. Thus Coseriu's displacement of the synchronic-diachronic parameter to the level of theory does not catch the whole truth.

tity, judging by the way it is used in the linguistic literature: to be causal is a property of Pre-OE *i*- and *j*-sounds: **kuning-* > OE *cyning*; **satjan-* > OE *settan*. But this property is, to repeat, no more causal than other factors in the rise of OE *cyning*, and the concept remains at the theoretical level and polysemous: PreOE *-j-* is also the cause of gemination: *sæt-j-an > *sættjan, making the new geminated consonants **-tt-* the cause of its loss. But the concept of cause is superfluous. With Collingwood (1973:214-215) I shall say that we comprehend historical events by finding the *thought qua* the content *expressed in* them *qua* sign-functional phenomena.

My point is that the rise of the concept of cause appears from a complex mental operation that presupposes **time**. In the Old English example above, we establish a temporal connection between two facts: what caught our attention at T2 and what was the case at T1 before the second fact, and this comparison between two virtually reified entities leads to a part of the former entity being singled out which we judge to be the cause; now 'causal' has become a property of a 'thing'. However, it is not the analytic mind alone that grasps the fact that our **-i/j* sounds are causal at T1; that knowledge presupposes the discursive-synthesizing faculty of our mind: our historical sense or sense of time. Secondly, if cause comes from a property-assigning judgement, it is clear that historical explanation carries with it the synchronic two-state model, with the result that we only acquire a shortened understanding of the event in question, because it exists in a developemental series.

2) Change

Descartes set great store by the distinction between judgement and immediate sensuous perception, almost anticipating Hermann Paul: if we say that we *see* the wax (Descartes: *videmus*) – that we *see* a change (the linguist) -, then we have not liberated ourselves from (the *Herrschaft* of) the words, but are deceived by the way we speak (*usus loquendi*). We do not see the same wax at T1 and T2, we **judge** the two different lumps to be the same (*dicimus nos ceram ipsammet adesse judicare*) – we judge something to be a change or that a change has occurred. As above, knowledge of change requires a judgement: *sola judicandi facultate (...) comprehendo* (Descartes *Med*. II.32). Descartes could not judge that the wax is the same through time and change, unless his mind was able to span – and to grasp – the periodic existence of the wax, and thereby to establish a necessary connection between what appeared to him as the two forms of his lump of wax. Similarly, the variationist point of view in linguistics, exemplified by /A/ [a], [b] (cf. 5.5.1), requires the passing of time. The existence of /A/ presupposes the passing of time which sees the rise of two variants, say, modern English [telephone] and [phone] in addition to being the result of a judgement; hence /A/ does not enter into empirical *Kausalnexus*, and if /A/ is synchronic, it suspends the time between [a] and [b].

Having the synonymous language signs *corpus* and *cera nuda* denote an entity without properties (*vestes*), Descartes was indeed deceived by his language; what the two signs are made to denote, a so-called carrier of properties (Kant agrees), is no empirical entity. It is an entity of the mind's eye, hence a linguistic entity; and the everyday language's *Erkenntnisswert* is not in doubt. More importantly, a judgement – *Cecil is a smith* – is a way of using the everyday language (*usus loquendi*), and since to form a judgement about human existence is to comprehend human existence through language, the *Erkenntniswert* of the everyday language is undeniable, and thus the faculty of judging and speaking presupposes the temporal structure of the human mind or our **historical sense**. This entails, however, that to form a judgement is not necessarily an analytical mental process. Judgement and comprehension are not synonymous when it comes to historical phenomena. When we comprehend the historical phenomenon in its developmental series, we do it through our sense of history expounded by a given theory of history; our sense of history makes **ex**istence rational, ordered, non-chaotic, because it enables us to create the present (and therefore the past) in accordance with rational *qua* sign-functional principles. The everyday language forms reality from formed reality, therefore reality is eminently suitable to being grasped by the everyday language: object and means are 'amenable'.

The analytical conception of **cause** establishes coherence between two temporally separated phenomena; this is the 'synchronic' origin of the two-state-model: *that -i- in OE cyning is the cause of -y- presupposes the incorporation of knowledge of* *kuning- *as the immediate predecessor into the judgement*. The analytico-synchronic dualistic conception of reality makes spatially separated and simultaneously existing phenomena cohere through the linguistically motivated class concept (Descartes *cera nuda*), established by an atemporal judgement to the effect that [a] and [b] are properties *qua* members of a class. On the other hand, **change** as an existential process creates coherence with no empirical breaks, and change as an epistemic-cognitive concept establishes various degrees of coherence in the empirical continuum of differentiation. Consequently, man's cognitive task – in general terms – of creating coherence, where none seems to be, tallies perfectly with the nature of human existence on condition that the two-state limits of the analytic conception of judgement are abandoned: 'to be a change' is not a property that comes from an S-is-P judgement. At best a change is a property of a transsynchronic phenomenon, but judging **kuning > OE cyning* to have the property of change implies far more cognitive processes than the judgement itself, and is virtually meaningless when we come to this complex phenomenon: *PreOE of 525 *kuning- > PreOE of 575 *kuning-*.

Change and for that matter cause come from the everyday language, not from a judgement. Neither human existence nor our grasp of it transcends into a linguistic haven of unchanging and permanent entities. Neither seeing nor judging provides knowledge about historical phenomena. The judgement turns them into something which they are not, first into reified objects, then into ideal entities away from their proper historical domain.

E. CONSTANT CHANGE: CONTINUITY, GRADUALNESS AND DEVALUATION

Historical discontinuity, separation and break processes occur against the horizon of attachment and continuation. The theoretical emphasis on the partial-coherence type of attachment follows a Saussure-like law-of-tradition concept, observing the facts that no dialect's, no idiolect's, or no language state's **ex**istence has been seen to occur against the horizon of a break, separation or discontinuity. The other two types of coherence, mutual coherence and absolute separation, have no independent existence and either does not exclude the other. The break between the denotata of such synchronic terms as *Old English* and *Early Middle English* is traditionally established with reference to a matrix of features as in Nielsen (1998:210;196-211); but as Nielsen correctly suggests, the assignment of a text to one of the two periods denoted by the abovementioned terms is not always *an entirely clear-cut one* (p. 211). The question is: why should there be empirically relevant 'clear-cut' lines of demarcation betweeen two texts/periods? The notion of 'a clear-cut' distinction with that of the synchronic cut and Nielsen's matrix methodology belong to the class of synchronic metaterms, empirically relevant when seen against the horizon of continuation – with differences and breaks being the necessary result of how we comprehend a language's **ex**istence. The abruptness entailed by synchronic metaterms is *per se* not empirically relevant – to the historian. Gradualness is understood by means of the matrix of the features that the advanced theory of history has defined, a theory which aims to subsume historical development under an empirically relevant, and ideology-free, *in casu* sign-functional view of language existence, **ex**istence.

Gradual development involves neither absolute separation nor absolute coherence between two elements of a developmental series, but varying degrees of attachment and separation. These degrees of coherence are caused by the elements' *structural differentness* motivated by their constant development into the present, and as such they will take us back to the material definitions in chapter 5.3., where I set up various nonchronological *internal* relationships; these definitions will now be reformulated in the light of the concept of devaluation, as this concept dissolves the similarities that the material definitions created between the elements of the nonlevel series. The concept also implies a developmental series of more than the three elements which the material definitions require.

Historical existence is not only a split process (synonymy) occurring against the horizon of polysemy, it is also processes of devaluation and nondevaluation – (temporal) **devaluation**[43] characterizing the structural processes of the nonambiguous (and ambiguous) nontensional fields of level development, **nondevaluation** the structural processes of the tensional field of nonlevel development. As was indicated above in chapter 6.1., devaluation is a commonplace in the linguistic literature, and with a final glance at Saussure's *Cours* (1972:112; 113), where we find this metaphysical statement: *le temps altère toutes choses* and *la continuité implique nécessairement l'alteration*, I shall formulate the principle as follows: the successor in the nonambiguous tensional field of a level series becomes more devalued than its predecessor tensional field: all other fields are not temporally devalued (the ambiguous nontensional field is also distributionally devalued). – The definitions are based on endogenous motivation and will therefore combine the synchronic nature (internal endogeneity) and the historical nature (external endogeneity) of the historical phenomenon in question:

6.3.7 INERTIAL DEVELOPMENT
An internally contrasting non-chronological relationship = df
Cf. 5.3.5: ... > ε > æ1 > a > ...   **The material definition**

1a)   A given element is relationally <u>similar</u> to two other <u>mutually different</u> elements in relation to a specific property – motivated through internal endogeneity – in respect of which it is also <u>different</u> from those elements.

The devaluation definition

1b)   A given element is relationally <u>similar</u> to two other <u>not mutually different</u> elements

---

43 Temporal devaluation comes from the resolution of an A>A syncretism: OE 700 *a:* > 1100 *a:*.

6.3. Last definitions

in relation to a specific property (-D) – motivated through external endogeneity – in respect of which it is <u>not different</u> from those elements[44].

An internally noncontrasting nonchronological relationship = df
Cf. 5.3.4: ... > a4 > a5 > a6 > ... **The material definition**
2a)   A given element is relationally <u>similar</u> to two other <u>not mutually different</u> elements in relation to a specific property – motivated through internal endogeneity – in respect of which it is <u>not different</u> from those elements.
The devaluation definition
2b)   A given element is relationally <u>similar</u> to two other <u>mutually different</u> elements in relation to a specific property (+D)[45] – motivated through external endogeneity – in respect of which it is also <u>different</u> from those elements.

NONINERTIAL DEVELOPMENT
An internally contrasting nonchronological relationship = df
Cf. 5.3.7 (Successor): ... > a5 > a6 > $\alpha 1$ > ...; and 5.3.9 (Predecessor): ... > **a6** > $\alpha 1$ > $\alpha 2$ ... .
The material definition
3a)   A given element ($\alpha 1$; **a6**) is relationally <u>similar</u> to two other <u>not mutually different</u> elements in relation to a specific property – motivated through internal endogeneity – in respect of which it is also <u>different</u> from both those elements.
The devaluation definition
3b)   A given element is relationally <u>similar</u> to two other <u>mutually different</u> elements in relation to a specific property ($\alpha 1$ = -D; a6 = +D) – motivated through external endogeneity – in respect of which it is also <u>different</u> from those elements.

The next definitions dissolve the ambiguity that the material *break-* and *no-break*-definitions (Def. 5.3.6 and Def. 5.3.11) created seeing that they established a break and no break around an element that entered into *symmetrical noncontrasting* similarity- and difference-relationships. Thus the below sets of definitions will distinguish between 5.3.4 and 5.3.8/5.3.10

NONINERTIAL DEVELOPMENT
An internally noncontrasting nonchronological relationship = df
Cf. 5.3.8 ...> ϵ > <u>æ1</u> > <u>a</u> > <u>æ2</u> > æ3 > ... and 5.3.10... > æ4 > <u>æ5</u> > <u>α</u> > <u>æ1</u> > ϵ
The material definition
4a)   A given element (æ1/æ2 and æ5/æ1) is relationally <u>not similar</u> to both of two other <u>mutually different</u> elements in relation to a specific property – motivated through internal endogeneity – in respect of which it is <u>not different</u> from both those elements.

---

44 The three intersections are all FS signs and suffer no temporal devaluation. The content of the predecessor of the series, εrFS, presupposes the influence of its predecessor.
45 The predecessor a4 is less devalued than either of its successors and a6 more devalued than its predecessors.

The devaluation definitions

4b)   A given element is relationally <u>similar</u> to two other <u>not mutually different</u> elements in relation to a specific property (-D) – motivated through external endogeneity – in respect of which it is <u>not different</u> from both those elements. – (= 5.3.8).

4c)   A given element is relationally <u>similar</u> to two other <u>not mutually</u> different elements in relation to a specific property (+D) – motivated through external endogeneity – in respect of which it is also <u>different</u> from both those elements. – (= 5.3.10).

In 5.3.8 none of the underlined elements is devalued; the speaker will know this because of his knowledge of the predecessor of æ1. That the predecessor æ5 in 5.3.10 is more devalued than its successors is knowledge coming from the speaker's historical sense – knowledge of the predecessor of æ5.

F.   THE LOGIC OF THE INSTRUMENTALIST-VIEW OF CHANGE

In our linguistic tradition **use** is the ultimate cause of and metacondition for change – intra- as well extralinguistic: the everyday language must change **because** it is a tool in constant use. The view has become an axiom, only supported by the ideological postulate that time is a kind of destroyer. Man's existence is basically a negative one, occurring contingently, accidentally and chaotically on its inevitable journey to destruction, and in order to survive man must see history against the speculative horizon of a better, ordered reality – necessarily transcendent. The axiom appears in two versions 6.3.7 and 6.3.8/9.

Grammaticalization theory formulates the instrumentalist view as follows: (…) *grammaticalization offers an important parameter for understanding linguistic behavior* [and is itself] *motivated by extralinguistic factors, above all by cognition* (Heine *et al.* 1991:27). Let us look at a consequence of this view within the framework of the present study:

6.3.8   Grammaticalization enables man to grasp linguistic behaviour
<u>Linguistic behaviour is expounded by the change of OE *a:* into ME *ɔ:*</u>
… Grammaticalization enables man to grasp the change of OE *a:* into *ɔ:*
<u>Cognition and other extralinguistic factors motivate grammaticalization</u>
… Cognition and other extralinguistic factors motivate the change of OE *a:* into *ɔ:*

The argument brings out the problems that were mentioned in note 2 in *The Prologue*. As regards the first conclusion we may ask: (1) how does grammaticalization as a *theory* explain the sound change of OE *ham* into ME *hom* 'home'? (2) How does grammaticalization as an *empirical process* explain another empirical process? Grammaticalization does not apply to sound processes, whereas a dualistic theory behind *the step from variant to change* could replace grammaticalization theory. Furthermore, grammaticalization theory does not seem to consider continuations: how is the Northumbrian continuation of OE *ham* to be motivated by cognition?

The last conclusion may be subsumed under the present theory: **ex**istence is a cognitive process controlled by our historical sense, but then the logic of grammaticalization arrives at a sound conclusion by way of an unsound one.

6.3. Change: use, impossibility, arbitrariness

The logic of the traditional formulation of the use-argument in 6.3.9 will also reveal some weaknesses. In 6.3.9 it appears as a mediated axiom (cf. section D above):

6.3.9 If a language is used, then it is worn down   S is M
<u>If it is worn down, then it is changed</u>   <u>M is P</u>
If a language is used, then it is changed   S is P

The argument is logically flawless, but it rests on the above negative assumption that change is therapeutic or homeostatic regulation. In contrast it is just as empirically correct to assume that if a language is not used, then it will suffer attrition and devaluation (nonS is M), from which it follows that if the everyday language does not suffer attrition and devaluation, then it (is because it) is used, nonM is S (nonS is M is logically synonymous with nonM is S), and since this is the conclusion that we are forced to infer from empirical experience, the argument in 6.3.10 is the empirically more relevant of the two, because it denies the empirical possibility of the wearing-down of any everyday language; the middle term could be replaced by *is always in order as it is*

6.3.10 If a language is not worn down, then it is used   Non-M is P
<u>If a language is changed, then it is not worn down</u>   <u>S is Non-M</u>
If a language is changed, then it is used   S is P

If use is the sufficient reason for change in 6.3.9, then change is the sufficient reason for use in 6.3.10: for language to be used (or to exist), it must change.

The use argument has now been given a logical and empirically relevant foundation within the advanced theory and without our Western Civilization's negative view of history and of sensuous reality: *use*, *change* and *existence* are co-extensive terms not in need of being redefined by a transcendent theory in order to become empirically relevant terms in such a theory.

G. IMPOSSIBILITY AND ARBITRARINESS

Language impossibility has been grounded on sign-functional properties which disallow the yoking together of historico-structurally incompatible expression- and content-relates. As the *simultaneous* combination of the two subglottal and glottal features of aspiration and voice is impossible[46], so an endogenously motivated expression is not accompanied by any type of content. Two principles must be emphasized here: in isolation no sound is correlated with one and only one type of content: an expression is not born with a special content, a given content does not inhere in a particular sound. Any sound can be correlated with the type of content that follows *from the structurally possible forces that control its* **ex***istence.*

---

46 There is a psychological correlate here: it is impossible for a person to be angry with and scold another person while smiling. So one must disagree with Trevisa's claim that if two persons speaking different languages meet, then either may just as well curse at the other instead of saying a kind 'Good morning': *For jangle þe on nevere so vaste, þet oþer ys nevere þe wyser, þey a schrewe hym in stude of good morowe.* (Burrow and Turville-Petre 1992:214-215)

The articulatory continuum of so-called asymmetrical phonological space (Labov 1994:256) will therefore consist of two sounds both mutually contrasting and contrasting with all other sounds: C5 as the 'lowest' expression of an attitudinal sign cannot develop a forceful style content, only relaxed- or normal style contents. C1 as the 'highest' expression of an attitudinal sign cannot develop a relaxed style content, only forceful or normal style contents. This conclusion agrees with Labov's view of C1 [i] as a kind of developmental terminus (p. 257; 258), while *the ambivalence of [a]* (C5), will not be corroborated by the attitudinal system; paradoxically, the present theory gives Labov's view of C5 a sign-functional foundation: it functions as the original predecessor of a forceful attitudinal sign in both the front and back regions: arNS > α1rFS or arNS > ɔrFS; conversely, it develops as a relaxed style originary successor from both front and back original predecessors: α1rRS > arRS or ɔrRS > arRS. *Thus the attitudinal system is empirically grounded on the Eigennatur of the sounds of the everyday language*, the existing (= changing) sounds being signs so that there is a 'natural' correlation between expression and content.

The relation contracted by the sign function's expression relate and content relate is **not arbitrary**. If it seems arbitrary synchronically, it is because that point of view prevents us from comprehending the structurally complex nature of the everyday language; as Saussure so correctly observed, the synchronic conception of *la langue* – being merely *viable* in itself – must be placed within the existential conception of the everyday language, so that *la langue* can be placed in its historical context and become *vivante*.

Since a *sui-generis* entity (as in Penzl 1972:3) can be established only arbitrarily or as a 'shortened' empirical phenomenon, the synchronic state must be transferred to the theoretical level, at which level we also find the synchrony-diachrony dichotomy (Coseriu 1974:9), a dichotomy that this essay's theory of history bridges (cf. Coseriu, p. 10h)). Thus it is the synchronic linguist's task to establish the (degree of) empirical relevance that the concept of the immanent state may have, and at the same fell swoop to remove the synchronic system 'behind' the everyday language from the arbitrary clutches of the many existing synchronic theories.

With the arbitrariness principle gone, arbitrariness as a factor that makes for both immutability and mutability is gone, too. Secondly, the principle of contingency must also be disappeared, seeing that the purported arbitrariness of the sign function cannot make the transposition of the modern French sign *un cheval* to Old French empirically relevant. Being an element that contributes to the nature of modern French as well as being determined by and determining the other elements of this modern everyday language, it cannot be torn out of its *geschichtlische Wirklichkeit* without either losing its *Eigennatur* – then it is no longer modern French *un cheval* that is being transposed – or without taking all the other elements with it on its time journey – and then the transposition is absurd. By the same token Saussure's *soeur – sister* thought-experiment is made absurd and Whitney's language learning infant baby has only the choices that her developing historical sense allows her.

Structuralism itself both prevents such transpositions and limits the infant baby's purported legion of choices; history and structure are two sides of the same coin; hence the structuralist's hallmark, *le refus de l'histoire*, is a contradiction in terms and

who says structure, says history

APPENDIX I

An overview of the transitions of the various developmental series

The following overview is a summary of the different types of transition that the developmental series produce (cf. chapter 5.4. Appendix). The typographical lay-out is meant to facilitate the reading of the overview, which should be read in conjunction with 6.1.2, 6.1.11, 6.1.17, 6.1.20, 6.1.22 and the summaries of the developmental series in 6.1.16 and 6.1.24. The underlined terms in the left-hand column on the next page indicate the theory's continuum of coherences and breaks.

The breaks and coherences of series III>VI are described in 6.1.6, 6.1.9, 6.1.10.
The breaks and coherences of series <u>III>IV</u> are described in 6.1.12, 6.1.14.
The breaks and coherences of series **VIII>II** are described in 6.1.13, 6.1.15.
The breaks and coherences of series *II>III* are described in 6.1.17, 6.1.19.
The breaks and coherences of series ***II>VIII*** are described in 6.1.20, 6.1.21.
The breaks and coherences of series V>IX are described in 6.1.22, 6.1.23.

III>VI	a3 > <u>a4 > a5</u> > a6 > α1 > α2 > α3 > <u>α4 > α5</u>	five breaks 6.1.9
II>VIII	a > α > <u>æ > ε1</u> > e > ε2 > ε3 > ε4 > <u>ε5 > ε6</u>	five breaks 6.1.24

With the a4>a5/æ>ε1 transitions as our starting points, characterized by inertial development with no breaks around the elements, the breaks of the succeeding transitions are defined by 1) potential ertial coherence, 2) ertial coherence, 3) nonertial coherence, 4) near-potential inertial coherence, and 5) potential inertial coherence, and the developmental series become inertial with the transitions <u>α4 > α5/ε5>ε6</u>.

With five types of break, the different transitions cannot all be described by means of the four types that sign-relate replacement makes possible.

VIII>II	ε2 > <u>ε3 > ε4</u> > ε5 > e > ε1 > <u>æ > α</u> > a	four breaks 6.1.16 = 6.1.13
II>III	a > α > <u>æ > ε</u> > e1 > e2 > e3 > <u>e4 > e5</u> > e6	four breaks 6.1.24

With the ε3>ε4/æ>ε inertial transitions as our starting point, the breaks of the succeeding transitions are 1) potential ertial coherence, 2) ertial coherence (VIII>II), nonertial coherence (II>III), 3) near-potential inertial coherence, and 4) potential inertial coherence, and the developmental series become inertial with the transitions <u>æ>α/e4>e5</u>.

With the u4>u5/æ1>α inertial transitions as our starting point, the breaks of the succeeding transitions are 1) potential ertial coherence, 2) near-potential inertial coherence, and 3) potential inertial coherence, and the developmental series become inertial with the transitions <u>ǫ>o/e4>e5</u>.

III>IV	u3 > <u>u4 > u5</u> > u6 > ʋ > <u>ǫ > o</u> > ǫ > ɔ	three breaks 6.1.16
V>IX	e > ę > ε1 > <u>æ1 > α</u> > æ2 > ε2 > <u>ę1 > e1</u> > ę	three breaks 6.1.24.

	C1 > C1 E1 > E1	C1 > C2 E1 > E1	C1 > C1 E1 > E2	C1 > C2 E1 > E2
<u>Inertial development</u>				
Endogeneity	III>VI. – <u>III>IV</u> **VIII>II**		<u>*II>III*</u>, *II>VIII* V>IX	
Katageneity	III>VI. – <u>III>IV</u> **VIII>II**	<u>*II>III*</u> *II>VIII* V>IX		
<u>Potential ertial development</u>				
Endogeneity	III>VI. – <u>III>IV</u> **VIII>II**		<u>*II>III*</u>, *II>VIII* V>IX	
Katageneity	III>VI. – <u>III>IV</u> **VIII>II**	<u>*II>III*</u>, *II>VIII* V>IX		
<u>Ertial development</u>				
Endogeneity				III>VI, *II>VIII* **VIII>II**
Katageneity		III>VI, **VIII>II** *II>VIII*		
<u>Nonertial development</u>				
Endogeneity		III>VI, <u>*II>III*</u> *II>VIII*		
Katageneity		III>VI <u>*II>III*</u>, <u>III>IV</u> *II>VIII*		
<u>Near-potential inertial development</u>				
Endogeneity		III>VI, <u>*II>III*</u> *II>VIII*	V>IX, <u>III>IV</u> **VIII>II**	
Katageneity		III>VI, <u>*II>III*</u>, <u>III>IV</u> *II>VIII*, **VIII>II**, V>IX		
<u>Potential inertial development</u>				
Endogeneity	III>VI, <u>*II>III*</u> *II>VIII*		**VIII>II**, <u>III>IV</u> V>IX	
Katageneity	III>VI, <u>*II>III*</u> *II>VIII*	**VIII>II** – <u>III>IV</u> V>IX		
<u>Inertial development</u>				
Endogeneity	III>VI, <u>*II>III*</u> *II>VIII*		**VIII>II**, <u>III>IV</u> V>IX	
Katageneity	III>VI, <u>*II>III*</u> *II>VIII*	**VIII>II**, <u>III>IV</u> V>IX		

APPENDIX 2

Explanation of abbreviations, symbols used, and their correlative concepts

OE	Old English.
PreOE	Prehistoric Old English.
ME	Middle English; EME = Early Middle English; LME = Late Middle English.
PE	Present-Day English.
mod	modern as in ModEng (modern English), modGerm (modern German), etc.
A>B	A two-state or two-element transition, *e.g.* a sound change.
A>A	A two-state or two-element transition, *e.g.* a sound continuation.
A :: B	A synchronic function, systemic or nonsystemic (subphonemic): A and B exist simultaneously in different places or in an abstract space.
A>B>C >	A developmental series (with B as its focus).
A>B :: C	A transition seen from the point of view of its successor.
=	Is (are) similar to; sign of logical identity.
≠	Is (are) different from.
⇒	A chronological series is interpreted as a nonchronological series by means of difference-relations and similarity-relations (in respect of a given phonetic property):

$a4 > a5 > a6$ ⇒ *Predecessor*: $a4 = a5 = a6$: $a4$ is similar to $a5$ and similar to $a6$
⇒ *Successor*: $a6 = a5 = a4$: $a6$ is similar to $a5$ and similar to $a4$.

$æ1 > a1 > a2$ ⇒ *Predecessor*: $æ1 = a1 = a2$ and $æ1 ≠ a1 ≠ æ2$: $æ1$ is <u>both</u> similar to $a1$ and similar to $a2$ <u>and</u> different from $a1$ and different from $a2$.
⇒ *Successor*: $a2 = a1 ≠ æ1$: $a2$ is similar to $a1$ and different from $æ1$.

A⇒	No break after A.
⇐A	No break before A.
A⇒ ⇐ B	No breaks in the transition A>B.
A ⇒ ≈ B	No break after A and a potential break before B in the transition A>B.
A ≈ ≈ B	Potential breaks after A and before B in the transition A>B.
≈ B ≈	Potential breaks around B.

A : B(A)	A participative relationship, to be interpreted ationally as 'A * B'. In the relationship contracted by A and B, A participates in (the existence or constitution of) B. – A develops into B under the aegis of (its own) change. – A determines B and is determined by B. – The individual creates a totality (or a supraindividual phenomenon) and is (subsequently) determined by that totality (or phenomenon): $I \Rightarrow T(I)$.
A * B	The aition(al function): A (is the case), but not only if B (is the case). A becomes a successor, but not only if B has become its predecessor // A comes from a predecessor, but not only if this predecessor is B. A is a predecessor, but not only if B is its successor // A develops into a successor, but not only if this successor is B. A word is a noun in modern English, but not only if it observes the number rule: $\emptyset :: $ -es, as in *kiss :: kisses*. The word *ox* is a grammatically correct and acceptable noun in modern English, but not only if it observes the synchronic number rule $\emptyset : $ -es. What is a precondition for something is also a constituent part of that something.
A]	A break is created after A.
[A	A break is created before A.
[A]	A break is created around A.
EL	Extralinguistic reality. – The denotata of a denoting everyday language.
L	An everyday language or a language state; often used interchangeably with S.
S	A given (synchronic) system or state. – A successor in relation to a predecessor, P. S1, S2, etc. The numbers indicate predecessors and successors. S2 is a predecessor in relation to S3 and a successor in relation to S1.
P	Place. A predecessor in relation to a successor, S. Primus, the essay's speaking interlocutor.
T	Time or point in time; $A(T1) > B(T2)$: A develops into B; $T1+1 = T2$: to underline (the) immediate development (of A into B). Totality in relation to parts. Temporal devaluation in connection with +/- as in -T = no temporal devaluation.
D	Distributional devaluation in connection with +/- as in -D = no distributional devaluation or distributional narrowing.
E	The expression relate of complex sign functions, *e.g.* CrE(ErC), and of the ideal sign function ErC. The material (phonetic) form of an element in the spoken chain.
C	The content relate of complex sign functions, *e.g.* CrE(ErC), and of the ideal sign function ErC.
r	The interdependent relation of sign functions. Superscripts *den* or *con* specify a sign-functional relationship as *a denotative relationship* and *a connotative relationship*, respectively.

R	A relation between two elements, *e.g.* aRb.
F *and* L	First *and* Last: primitives of the theory, specifying the coherences in the developmental series in chapter 6; in chapter 1.5. they specify (the directionality of the) various types of assimilation-based coherence in the spoken chain (exogenous motivation). The use of F and L indicates the affinity between the sequential nature of the spoken chain and that of the developmental series. – They are short-hand symbols for the dynamic terms, *(not) to break into, (not) to break away from*: +F: an element or intersection **breaks away from** its predecessor in virtue of its content. *And*: an element or intersection **does not break into** its predecessor in virtue of its expression. -F: an element or intersection **does not break away from** its predecessor in virtue of its content. *And*: an element or intersection **breaks into** its predecessor in virtue of its expression. +L: an element or intersection is **made to break away from** its successor in virtue of its content. *And*: an element or intersection is **not made to break into** its successor in virtue of its expression. -L: an element or intersection is **not made to break away from** its successor in virtue of its content. *And*: an element or intersection is **made to break into** its successor in virtue of its expression. The first +/-F and the first +/- L refer to the content coherence of an element with its predecessor and successor, respectively. The second +/-F and the second +/-L refer to the expression coherence of an element with its predecessor and successor, respectively: *e.g.* [-F, +F, +L, -L].
/X/	Slants: the systemic entity X. – Entities of *la langue*. The epitome of essentialism. Does /X/ denote a class as one or a class as many?
[x], [x]	Brackets: (at least two) concrete entities subsystemic in relation to /X/. /A/ [a], [b]: the variants a and b of the phoneme or systemic entity A. What is put between the slants create coherence between the relevant [a] and [b] of sensuous reality. [a] and [b] are differentiated either temporally or spatially.
RS	A relaxed-style sign or the **precise** content of such an attitudinal sign.
FS	A forceful-style sign or the **precise** content of such an attitudinal sign.
NS	A normal-style sign or the **nonambiguous** content of such an attitudinal sign.
NRS	A nonrelaxed-style sign or the **ambiguous** content of such an attitudinal sign. – Defined by FS+NS.
NFS	A nonforceful-style sign or the **ambiguous** content of such an attitudinal sign. – Defined by RS+NS.
FB	A fullblooded intersection or element in a developmental series.
NFB	A nonfullblooded intersection or element in a developmental series

la langue In the synchronic dualism's various versions *la langue* is the systems relate of *la parole*.

In the L-continuum of the essay's theory *langue* and *parole* are relative terms representing language elements (stretches) of varying degrees of length to be produced in the spoken chain as *the positive alternatives* of the here-and-now intersection's *possible alternative*.

la parole *Parole* is the everyday language's development into the present, thereby forming the physical substance of the historical phenomenon or creating a transindividual phenomenon in relation to its cooperating individuals (the makers of such a (historico-cultural) phenomenon).

What is put between the synchronic tradition's brackets, [....]. In the theory of history it is that which has no preexistence, being the possible alternative of a present here-and-now intersection. It enters into a replacement relationship with its predecessor.

existence as in **ex**istence when used in its original, dynamic sense: Latin *ex* and *sistere* (<*stare, sto*) 'to stand out of, appear from; to arise, become; to be(come) visible'. – Note that the static and logical sense 'to follow from' develops out of the word's historical sense (Cicero *De Fato* 18.18-20: *ut sine causa fiat aliquid, ex quo existet, ut de nihilo quippiam fiat* 'that something occurs without a cause, from which (fact) it follows that something arises out of nothing').

The existential function of language refers to the constant reality-creating function of the everyday language; it is a constituent and effective part of our historical existence.

Such man-made existence is sign-functional in virtue of the participative role of language in man's historical existence.

development as in *the present*'s *development into the present* is the superconcept of the theory, expounding historical existence. Development is a differentiating process that creates the present turning the present into its past. The symbols FIRST and LAST are derived from it and so are both the concepts of A BREAK BEFORE and A BREAK AFTER and the notion of 'DISTANCE *qua* ATTACHMENT and SEPARATION. *Existence* and *development* are synonyms, although *development* is also used synonymously with *to come from :: to develop from* at the level of theory.

distance The distance-concept is expounded by the absolute gap in the conceptual pairs that make up the **trancendence** complex: 1) there is an absolute gap between a cause and its effects, between a reason (a ground) and its consequences; 2) a theory and its data (denotata); 3) language and what it denotes (nonlanguage); and 4) we cognize at a distance of the cognized. In synchronic theorizing the synchronic language (state/system), being supraindividual exists at a distance of – in a state of separation or separability from – its speakers. The same applies to the social fact. This means that we can identify three separable elements which the analytic mind is able to analyze separately: (1) a theory, a language, a system, a state, a sociofact; (2) data, denotata, speakers/hearers, language users, and (3) the relation that establishes some kind of coherence between (1) and (2). – Cf. *a class as many* and *as one*.

In the theory this distance-concept will be reconciled with the concept of continuity (= historical continuity) in the sense that there are no gaps in man's historical existence. Participation is the bridge-making concept. – The nominalistic theory exists independently of what it is a theory about.

a class as many

expounds a totality concept that is linguistically motivated: a sign denotes a plurality of entities. In *the class of shoes* there is no totality *per se*, existing independently of the single shoes in question. – A class as many epitomizes the notion of extensionality *qua* the relation of denotation constituted by a sign (L) and its denotata (EL). There is a gap (distance) between linguistic reality and nonlinguistic reality. The concept epitomizes nominalism.

a class as one

If it expounds a totality in its own right, *qua* independently existing, so that a given totality sign is claimed to denote, then it belongs to a dated metaphysics (Realism): *the English language* does not denote (such an independently existing entity); in synchronic theorizing *a subject* or *an object* or *a finite verb* – *a noun phrase* or *a verb phrase* – or *a phoneme* does not denote entities other than particular phenomena in a particular language. If they denote preestablished entities in relation to such particular phenomena, the synchronic theory in question has an empirical reckoning. – In the present essay a class as one is a phenomenon (a merger) that we comprehend synthetically as a structure and which is resoluble or irresoluble. All man-made, historical phenomena are mergers whose *Eigennatur* is comprehended synthetically as structures. Hence the conceptual pairs mentioned under *distance*, items 1) to 5), are related *participatively*: the 'cause' participates in its effect. – The historical process of existence is and appears as a class as one which the historian tries to resolve into as many processual 'transitions' as possible: >A>B>: > A> X > Z >B. – To the historian A>B is always a syncretism, as an existential process it is not: when the present develops into the present, there is no gap or discontinuity between the present that the present makes its past.

The elements of the L-continuum are also classes as one, each of which is resolved in *parole* into a possible altenative and a positive alternative.

transcendence

1) In the *transcendence-problem* the word refers to the transition from/to theory to/from data and sums up the question: how do theory and data relate? 2) In the Glossematic sense the word refers to the use of undefined theoretical terms, hence to the possibility that the everyday language has *Erkenntnisswert*. 3) In the *transcendent use* of, say, the logical conditional, it refers to the use of

logical functions beyond the proper domain of symbolic logic. – To transcend away from/into reality describes two modes of **ex**istence with transcendence-into the empirical horizon against which we exist. The two processes are aitionally related: transcendence-into :: transcendence-from(transcendence-into).

the aition   The aitional relationship has been inferred from (1) the everyday language's predicate clause *John is a smith*, (2) formal logic's categorical statement *All S is P*, and (3) the truth-functional definition of the logical conditional, *If S, then P* ($S \supset P$), seeing that there are more smiths in the world than the person John, that there are more mammals, more mortal entities than a lion (the class of lions) and Socrates (the class of human beings), respectively. It formalizes the notion of multiconditional motivation: A, but not only if B, as well as the participative relationships, as *e.g.* in: effect : cause(effect), which says that the cause is a constituent part of the effect. – The aition looks at the conditional of symbolic logic from the point of view of the consequent as in *q, but not only if p* ($= q * p$) – instead of 'If p, then q'. The significance of the fact that the logical conditional can be deduced from the aition will not be dealt with in this essay. – The truth-functional definition of the aition and the logical consequent as one of its theorems:

q	*	p	$\supset$	p	$\supset$	q
T	T	T	T	T	T	T
T	T	F	T	F	T	T
F	F	T	T	T	F	F
F	F	F	T	F	T	F

The essay's 'aitional' point is that the truth-functional definition of the conditional makes the latter both too powerful and too arbitrary for its transcendent use to be relevant in empirical studies; I shall illustrate this with some examples that are relevant to the essay's argument in chapter 6.3.F:

$p \supset q$		if it rains	then the streets are wet	
1) T T T	*The fact that*	it rains	& the streets are wet	*can be the case (is true)*
2) F T T	*The fact that*	it does not rain	& the streets are wet	*can be the case (is true)*
3) T F F	*The fact that*	it rains	& the streets are not wet	*cannot be the case (is false)*
4) F T F	*The fact that*	it does not rain	& the streets are not wet	*can be the case (is true)*

Empirically speaking, rows two and three are the moot points: the possibility of row two is no more cogent (natural) than the possibility of row three: the streets can be wet for other reasons than rain; the streets can be dry, because they have been protected from the rain! As a definition of the cause-effect parameter, the logical conditional fares no better; let's proceed from the standard conception: 'No effect without a cause':

	if effect X	then cause Y	
The fact that	X is present	& Y is present	*can be the case (is true)*
The fact that	X is absent	& Y is present	*can be the case (is true)*
The fact that	X is present	& Y is absent	*cannot be the case (is false)*
The fact that	X is absent	& Y is absent	*can be the case (is true)*

Appendix II 449

Here the second row idealizes the cause seeing that it can appear 'alone', at a distance of its potential effect. If linguistic causes are transstat-ic, with the 'same' cause explaining processes in, say, English, French and Swahili, or in Old English and modern English, then causes become supraindividual and the theory in question has an empirical reckoning.

The alternative interpretation is equally unacceptable, where row two makes an effect present without its cause:

	if cause Y		then effect X	
The fact that	Y is present	&	X is present	*can be the case (is true)*
The fact that	Y is absent	&	X is present	*can be the case (is true)*
The fact that	Y is present	&	X is absent	*cannot be the case (is false)*
The fact that	Y is absent	&	X is absent	*can be the case (is true)*

And if we implement these two causal schemata with the two opposed theories of language change, the Functional one and Mechanical one – each explaining the interplay of loss of (grammatical) function and weakening (the loss) of grammatical flexives in the synthesis-analysis process of, say, Old English > Middle English -, then the logical weaknesses appear once again:

If Lack of function[1] (*Funktionslosigkeit*		then Weakening *Abschwächung*)
lack of function	*entails*	phonetic weakening
no lack of function	*entails*	phonetic weakening
lack of function	*entails*	no phonetic weakening
no lack of function	*entails*	no phonetic weakening
If Weakening		then Lack of function
phonetic weakening	*entails*	lack of function
no phonetic weakening	*entails*	lack of function
phonetic weakening	*entails*	no lack of function
no phonetic weakening	*entails*	no lack of function

---

1 Short for the Functional view that in a prepositional phrase in which the noun complement also contains a case suffix the preposition is sufficient to mark the syntactico-grammatical relations in question, and the supposedly redundant case ending may therefore weaken and drop: the Parker-manuscript of *AS*: *On þissum geare* (Year 889), *on þys geare* (Year 894), *þy ilcan geare* (Years 892; 777 (the Laud-ms)); *nigon nihtum ær middum sumere* 'nine nights before midsummer' (Year 898); in ME *Sir Orfeo* l.15 (Burrow and Turville-Petre, p. 112) *aventours þat fel bi dayes* 'adventures that happened once (upon a time)'; *The Peterborough Chronicle* 1137 (*op.cit.*, p. 75) *bathe be nihtes and be dæies* 'both by night and by day'. Or: in *oþ-þæt hie hine ofslægenne hæfdon* postverbal placement of the object is communicatively more efficient than the OE inflectionally motivated position of it; but note the sign-functional development: the expression: (OE) > *until they had killed him* and the content: (OE) 'until they had him in a state of slain-hood' > 'until they had killed him'.

In the former case phonetic weakening may occur irrespectively of the grammatical function of the flexive in question, while in the second lack of grammatical function may be caused by other factors than weakening. In a word, the conditional implementation of the cause-effect parameter implies the concept of multiconditional motivation and presupposes the meaning of the aition: a given effect/cause, but not only if one particular cause/effect.

BIBLIOGRAPHY

Den Store Danske Udtaleordbog
    1989   Munksgaards ordbøger. Af Lars Brink, Jørn Lund, Steffen Heger, J. Normann Jørgensen. København: Munksgaard.

Adv   *The Advanced Learner's Dictionary of Current English.*
    1960   Twelfth impression. (1948). London: Oxford University Press.
    1992   Seventh impression. (Third edition 1974). (*Oxford Advanced Learner's Dictionary*). Oxford: Oxford University Press.

AS Chronicle
    1965   *Earle's Two Saxon Chronicles Parallel.* A revised text. Edited by Charles Plummer on the basis of an edition by John Earle. Vol. 1. Oxford: At the Clarendon Press.

Longman   *Longman Dictionary of Contemporary English.*
    1978   (First edition). London: Longman.
    1995   Third edition. London: Longman Dictionaries.

Johnson   *A Dictionary of the English Language*. By Samuel Johnson. London: Joseph Ogle
    1828   Robinson, 42, Poultry.

OED   *The Oxford English Dictionary*. Reprinted 1961 (1933); With *A Supplement to 'The Oxford English Dictionary'*. 4 vols. 1972-. Oxford: at the Clarendon Press.

Allwood, Jens S.– Andersson, Lars-Gunnar – Dahl, Östen
    1972   *Logik för lingvister*. Lund: Studentlitteratur.

Andersen, Henning
    1973   'Abductive and Deductive Change.' *Language* 49:765-793.

Anderson, John
    1992   'Exceptionality and non-specification in the history of English phonology.' In Rissanen *et al.* 1992:103-116.

Anttila, Raimo
    1979   'Generative Grammar and Language Change: Irreconcilable Concepts.' In *Amsterdam Studies in the Theory and History of Linguistic Science*. Series IV. Current Issues in Linguistic Theory. Volume 11. (Ed. Bela Brogyanyi). Pp.35-51.
    1989   *Historical and Comparative Linguistics.* Second revised edition. Amsterdam/Philadephia: John Benjamins Publishing Company.
    1992   'The Return to Philology in Linguistics.' In *Thirty Years of Linguistic Evolution.* Pp. 313-332. Edited by Martin Pütz. Philadelphia/Amsterdam:John Benjamins Publishing Company.

Appleby, Joyce – Hunt, Lynn – Jacob, Margaret
    1995   *Telling the Truth about History*. New York: W.W. Norton & Company

Aquinas, Thomas
    1964   *Summa Theologiæ*. Vol. 1 (1a. 1); vol. 2 (1a. 2-11); vol. 3 (1a. 12-13). Latin text, English translation. By Thomas Gilby O.P. London: Eyre & Spottiswoode.

Arens, Hans
    1969   *Sprachwissenschaft. Der Gang ihrer Entwicklung von der Antike bis zur Gegenwart*. Zweite, durchgesehene und stark erweiterte Auflage. München: Verlag Karl Alber Freiburg.
Aristotle
    1967   *The Categories. On interpretation. Prior Analytics*. Translated by Harold P. Cooke; Hugh Tredennick. The Loeb Classical Library 1. London: William Heinemann.
    1968   *The Metaphysics*. Translated by Hugh Tredennick. The Loeb Classical Library 17-18. London: William Heinemann.
Arlotto, Anthony
    1972   *Introduction to Historical Linguistics*. Boston: Houghton Mifflin Company
Bache, Carl – Davidsen-Nielsen, Niels
    1997   *Mastering English Grammar*. Berlin/New York: Mouton de Gruyter.
Back, M.
    1979   'Sonorität und Lautwandel.' In *Amsterdam Studies in the Theory and History of Linguistic Science*. Series IV. Current Issues in Linguistic Theory. Volume 11. (Ed. Bela Brogyanyi). Pp. 53-69.
Barnes, Jonathan
    1979   *The Presocratic Philosophers. Thales to Zeno*. Vol. 1. London: Routledge & Kegan Paul.
Barthes, Roland
    1970   *Mythologies*. Paris: Éditions du Seuil.
Baskin, Wade
    1974   *Course in General Linguistics*. Translated from the French. Glasgow: William Collins Sons and Co Ltd.
Beck, L.J.
    1967   *The Metaphysics of Descartes. A study of the Meditations*. Oxford: At the Clarendon Press.
Beekes, Robert S. P.
    1995   *Comparative Indo-European Linguistics. An Introduction*. Amsterdam/Philadelphia: John Benjamins Publishing Company.
Berg, Thomas
    1995   'Word-Final Voicing in the History of English.' *English Studies* 76.2:185-201.
Bergson, Henri
    1946   *La Pensée et le Mouvant*. Oeuvres complètes, vol. 7. Genève: Éditions Albert Skira.
Berlin, Isaiah
    1976   *Vico and Herder. Two Studies in the History of Ideas*. London: The Hogarth Press.
Bernsen, Niels Ole
    1978   *Knowledge. A Treatise of our Cognitive Situation*. Odense: Odense University Press.
Bessermann, Lawrence
    1998   'Chaucer's Multi-Verbs: An Historical Introduction and Illustrative Material.' *Nowele* 34:99-153.
Blake, N.F.
    1996   *A History of the English Language*. London: Longman.

Bibliography 453

Bloch, Marc
    1964   *Apologie pour l'histoire: ou métier d'historien.* (1949). Paris: Librairie Armand Collin.
    1973   *Forsvar for historien eller Historikerens håndværk.* København: Gyldendal.
Bloomfield, Leonard
    1926   'A Set of Postulates for the Science of Language.' *Language* 2: 153-164.
    1967   *Language.* (1933). London: George Allen & Unwin Ltd.
    1971   'Linguistic Aspects of Science.' *Foundations of the Unity of Science. Toward an International Encyclopedia of Unified Science.* Vol. 1, Nos. 1-10. Pp. 215-277. (1938) Edited by Otto Neurath, Rudolph Carnap, Charles Morris.
Bohr, Niels
    1971   'Analysis and Synthesis in Science.' *Foundations of the Unity of Science. Toward an International Encyclopedia of Unified Science.* Vol. 1, Nos. 1-10. Pp. 215-277. (1938) Edited by Otto Neurath, Rudolph Carnap, Charles Morris.
Bolinger, Dwight
    1973   'Truth is a linguistic question.' *Language* 49:539-550.
Boudon, Raymond
    1968   *A quoi sert la notion de "structure"?* Paris: Éditions Gallimard.
Brugmann, Karl – Osthoff, Hermann
    1878   'Vorwort.' *Morphologische Untersuchungen auf dem Gebiete der indogermanischen Sprachen.* Erster Teil. Pp. iii-xx. Nachdruck der Ausgabe 1878-90. Leipzig: Verlag von S. Hirzel. Leutershausen:Strauss & Cramer.
Brunner, Karl
    1960   *Die englische Sprache.* Erster Band. Tübingen: Max Niemeyer Verlag.
    1965   *Altenglische Grammatik.* Nach der angelsächsischen Grammatik von Eduard Sievers. Dritte, neubearbeitete Auflage. Tübingen: Max Niemeyer Verlag.
Brøndal, Viggo
    1940   'Præpositionernes Theori.' *Festskrift* udgivet af Københavns Universitet November 1940. København: Bianco Lunos Bogtrykkeri A/S.
    1943   'Linguistique structurale.' In *Essais de linguistique générale.* Pp. 90-97. Copenhague: Munksgaard
Brüder-Grimm
    1980   *Kinder- und Hausmärchen.* Band 2. (1857). Stuttgart: Reclam.
Bukdahl, Jørgen K.
    1973   'Sprog og erkendelse. En historisk indledning.' *Sprog og Virkelighed.* Udg. Malte Jacobsen. Pp. 9-28. København: Gyldendal.
    1979/80   'Strukturalismen.' *Filosofien efter Hegel.* Udg. Jørgen K. Bukdahl *et al.* Pp. 260-306. København: Gyldendal.
Bulhof, Ilse N.
    1976   'Structure and Change in Wilhelm Dilthey's Philosophy of History.' *History and Theory.* XV.1:21-32.
Burrow, J.A. – Turville-Petre, Thorlac
    1992   *A Book of Middle English.* Oxford: Blackwell.

Bübner, Rüdiger
   1976   'Is Transcendental Hermeneutics Possible?' In *Essays on Explanation and Understanding*. Edited by Juha Manninen and Raimo Tuomela. Pp. 59-77. Dordrecht/Boston: D. Reidel Publishing Company.

Bynon, Theodora
   1977   *Historical Linguistics*. Cambridge: Cambridge University Press.

Børch, Marianne Novrup
   1993   *Chaucer's Poetics. Seeing and Asking*. Vol. 1. Bagsværd: (Eget forlag).

Campbell, A.
   1964   *Old English Grammar*. Reprint. Oxford: At the Clarendon Press.

Cassirer, Ernst
   1942   'The Influence of Language upon the Development of Scientific Thought.' *The Journal of Philosophy*. 39.12:309-327.
   1945   'Structuralism in Modern Linguistics.' *Word* 1:99-120.
   1953   *Substance and Function. And Albert Einstein's Theory of Relativity*. (1923). Chicago: Dover Publications, Inc.

Cheshire, Jenny
   1982   *Variation in an English Dialect*. Cambridge: Cambridge University Press.

Chomsky, Noam
   1965   *Aspects of the Theory of Syntax*. Massachusetts: The M.I.T. Press.
   1966   *Cartesian Linguistics. A chapter in the history of rationalist thought*. New York: Harper and Row.
   1986   *Knowledge of Language: its nature, origin and use*. New York: Praeger.

Chomsky, Noam – Halle, Morris
   1968   *The Sound Pattern of English*. New York: Harper & Row, Publishers.

Christensen, Johny
   1962   *An Essay on the Unity of Stoic Philosophy*. København: Munksgaard.

Christiansen, Lars
   1997   *Metafysikkens Historie*. 2. oplag. København: Museum Tusculanums Forlag. Københavns Universitet.

Church, Alonzo
   1956   *Introduction to Mathematical Logic*. Volume 1. Princeton: Princeton University Press.

Cicero, Marcus Tullius
   De Fato Herausgegeben von Karl Bayer. 2. Auflage. 1976. Darmstadt: Wissenschaftliches Buchgesellschaft.

Clausen, H.P.
   1963   *Hvad er historie?* København: Berlingske Forlag.

Climo, T.A. – Howells, P.G.A.
   1976   'Possible Worlds in Historical Explanation.' *History and Theory* XV.1:1-20.

Collier, L.W.
   1987   'The Chronology of *i*-Umlaut and Breaking in Pre-Old English.' *NOWELE* 9:33-45.

Collinge, N. E.
   1985   *The Laws of Indo-European*. CILT. Series IV. Vol 35. Amsterdam: John Benjamins Publishing Company.

Collingwood, R.G.
- 1964 *The Idea of Nature*. Reprint of the second impression. Oxford: At the Clarendon Press.
- 1973 *The Idea of History*. Reprint. Oxford: Oxford University Press.

Comrie, Bernard
- 1989 *Language Universals and Linguistic Typology*. Second Edition. Oxford: Basil Blackwell.

Coseriu, Eugenio
- 1970 *Sprache. Strukturen und Funktionen*. Tübinger Beiträge zur Linguistik 2. Tübingen.
- 1974 *Synchronie, Diachronie und Geschichte. Das Problem des Sprachwandels*. Übersetzt von Helga Sohre. München: Wilhelm Fink Verlag.
- 1975 *Sprachtheorie und allgemeine Sprachwissenschaft*. 5 Studien. München: Wilhelm Fink Verlag.

Culler, Jonathan
- 1982 'The Linguistic Basis of Structuralism.' In David Robey 1982:20-36.

Davidsen-Nielsen, Niels
- 1994 *An Outline of English Pronunciation*. Translated and revised by Fritz Larsen and Hans Frede Nielsen. Odense: Odense University Press.

Delong, Howard
- 1971 *A Profile of Mathematical Logic*. Reading, Massachusetts/London: Addison-Wesley Publishing Company.

Descartes
- 1966 *Descartes. Regler og Metafysiske Meditationer*. Oversat af Povl Dalsgård-Hansen. København: Berlingske Forlag.
- 1968 *Descartes. Discourse on Method and Other Writings*. Translated by F.E. Sutcliffe. Harmondsworth: Penguin Books Ltd.

Descartes *Med.*
- 1970 *Meditationes de prima philosophia*. Paris: Librairies Philosophique J. Vrin.

Diderichsen, Paul
- 1966 *Helhed og Struktur. Udvalgte sprogvidenskabelige afhandlinger*. G.E.C.Gads Forlag: København.

Dobson, E.J.
- 1968 *English Pronunciation. 1500-1700*. 2 vols. Oxford: At the Clarendon Press.

Dray, William
- 1957 *Laws and Explanation in History*. Third Impression. Oxford: Oxford University Press.

Ducrot, Oswald
- 1968 *Le structuralisme en linguistique*. Paris: Éditions du Seuil.

Durkheim, Émile
- 1927 *Les Règles de la méthode sociologique*. (1896). Huitième édition. Paris: Librairie Félix Alcan.
- 1974 *Sociologie et philosophie*. Paris: Presses Universitaires de France.

Eaton, Roger — Fischer, Olga — Koopman, Willem — van der Leek, Frederike
- 1985 *Papers from the 4th International Conference on English Historical Linguistics*. Amsterdam/Philadelphia: John Benjamins Publishing Company.

Eco, Umberto
- 1982 'Social Life as a Sign System.' In David Robey 1982:57-72.

Erslev, Kr.
   1975   *Historisk Teknik.* Anden udgave. Niende oplag. København: Gyldendal.
Fahrendorff, Lars
   1995   'Ranke og den historiske tænkning.' *Kritik* 130:47-51.
Favrholdt, David
   1965   *An Interpretation and Critique of Wittgenstein's Tractatus.* Copenhagen: Munksgaard.
Francis, W. Nelson
   1958   *The Structure of American English. With a Chapter on American English Dialects* by Raven I. McDavid, Jr. New York: The Ronald press Company.
Fischer-Jørgensen, Eli
   1975   *Trends in Phonological Theory. A Historical Introduction.* Copenhagen: Akademisk Forlag.
Frisch, Hartvig
   1962   *Europas Kulturhistorie.* Bind 4. Bearbejdet af Svend Erik Stybe. København: Politikens Forlag.
Gadamer, Hans-Georg
   1972   *Kleine Schriften* III: *Idee und Sprache.* Tübingen: J.C.B. Mohr (Paul Siebeck).
   1975   *Wahrheit und Methode.* 4. Auflage. Tübingen: J.C.B. Mohr (Paul Siebeck).
Gardiner, Patrick
   1969   *Theories of History.* Ninth printing. New York/London: The Free Press/Collier-Macmillan.
Geeraerts, Dirk
   1992   'The Return of Hermeneutics to Lexical Semantics.' In *Thirty Years of Linguistic Evolution.* Pp. 257. Edited by Martin Pütz. Philadelphia/Amsterdam: John Bejamins Publishing Company.
Gellner, Ernest
   1969   *Thought and Change.* Second impression. London: Weidenfeld and Nicolson.
Gellrich, Jess M.
   1988   *The Idea of the Book in the Middle Ages. Language Theory, Mythology, and Fiction.* Ithaca: Cornell University Press.
Gilliam, Harriet
   1976   'Dialectics of Realism and Idealism.' *History and Theory* XV:231-256.
Gilbers, Dickey – de Hoop, Helen
   1998   'Conflicting Constraints: An introduction to Optimality Theory.' *Lingua* 104:1-12.
Gimson, A.C.
   1964   *An Introduction to the Pronunciation of English.* Reprint. London: Edward Arnold.
   1994   *Gimson's Pronunciation of English.* Fifth Edition. Revised by Alan Cruttenden. London: Edward Arnold.
Givón, T.
   1973   'The Time-axis phenomenon.' *Language* 49.4
   1990   *Syntax. A Functional-typological Introduction.* Vol. II. Amsterdam/Philadelphia: John Benjamins Publishing Company.
   1995   *Functionalism and Grammar.* Amsterdam/Philadelphia: John Benjamins Publishing Company.

Goldmann, Lucien
- 1966 *Sciences humaines et philosophie.* Suivi de *Structuralisme génétique et création littéraire.* Paris: Éditions Gonthier.

De Graaf, Tjeerd
- 1982 'Vowel Duration and Vowel Quality.' *Proceedings of the XIIIth International Congress of Linguistics, Tokyo.*

Greenbaum, Sidney
- 1994 *An Introduction to English Grammar.* Fifth impression. London: Longman.

Greenberg, Joseph H.
- 1979 'Rethinking Linguistics Diachronically.' *Language* 55:275-290.

Gregersen, Frans
- 1991 *Sociolingvstikkens (u)mulighed.* Bind 1+2. København: Tiderne Skifter.

Guthrie, W.K.C.
- 1967 *A History of Greek Philosophy.* Volume 1. Cambridge: Cambridge University Press.

Haack, Susan
- 1974 *Deviant Logic. Some Philosophical Issues.* Cambridge: Cambridge University Press.
- 1980 *Philosophy of Logic.* Cambridge: Cambridge University Press.

Hamilton, Terrell H.
- 1967 *Process and Pattern in Evolution.* Current Concepts in Biology Series. London: The Macmillan Company.

Hammarström, Göran
- 1982 'Diachrony in synchrony.' In Maher 1982:51-64.

Hammerich, L.L.
- 1935 *Indledning til tysk Grammatik.* København: G.E.C. Gads Forlag.

Hansen, Erik
- 1982 'On the Purported Relationship between Historical Linguistics and Synchronic Linguistics.' *Sprachwissenschaft* 7.1:58-74.
- 1983 *Om historie og sproghistorie. Skitse til en udviklingslære for historiske fænomener. Forandring 1.* Odense: Odense Universitetsforlag.
- 1983a 'Change and Glossematics. On Consistency and Exhaustiveness.' *Sprachwissenschaft* 8.4:456-476.
- 1985 'K.E. Løgstrup's sprogsyn. Religiøs tydning: supplement til sprogfilosofi og sprogteori?' *Danske Studier.* Pp. 70-102.
- 1986 'Logical Methods and History. On Generalization and Contradiction.' *NOWELE* 7: 99-110.
- 1987 'On the Logic of the First and Second Germanic Consonant Shifts. Five Postulates.' *Althochdeutsch.* Band 1, herausgegeben von Rolf Bergmann, Heinrich Tiefenbach, Lothar Voetz. Heidelberg: Carl Winter Universitätsverlag.
- 1997 'On the logic of implicational universals.' *Language History and Linguistic Modelling.* Edited by Raymond Hickey and Stanislaw Puppel. Pp. 1593-1612. Trends in Linguistics. Studies and Monographs 101. Berlin: Mouton de Gruyter.

Hansen, Erik – Nielsen, Hans F.
- 1986 *Irregularities in Modern English.* Odense: Odense University Press.

Hardt, Manfred
 1981 'Die Zeit als Faktor der Sinnkonstruktion. Ein Entwurf.' Ed. Annemarie Lange-Seidl *Zeichenkonstitution*. Akten des 2. Semiotischen Kolloquiums Regensburg. Pp. 280-288. Berlin: Walter de Gruyter.

Harris, John
 1994 *English Sound Structure*. Oxford: Blackwell

Harris, Randy A.
 1993 *The Linguistics Wars*. Oxford: Oxford University Press.

Harris, Roy
 1987 *Reading Saussure*. London: Duckworth.

Harris, Zellig S.
 1968 *Structural Linguistics*. (1951). Eighth Impression. Chicago: The University of Chicago Press.

Hass, Jørgen
 1979/80 'Friedrich Nietzsche'. In *Filosofien efter Hegel*. Udg. Jørgen K. Bukdahl *et al*. Pp. 95-154. København: Gyldendal.

Hauge, Hans
 1985 'Grundtvig, Løgstrup og af-kanoniseringen af litteraturen.' *Præsteforeningens Blad* 37:661-666.

Hawkins, John A.
 1979 'Implicational Universals as Predictors of Word Order Change.' *Language* 55:618-648.
 1983 *Word Order Universals*. New York: Academic Press.

Heine, Bernd – Claudi, Ulrike – Hünnemeyer, Friederike
 1991 *Grammaticalization: A Conceptual Framework*. Chicago: University of Chicago Press.

Hempel, Carl G.
 1942 'The Function of General laws in History.' In Gardiner 1969:344-356.

Hjelmslev, Louis
 1928 *Principes de grammaire générale*, Det Kgl. Danske Videnskabernes Selskab. Historisk-filologiske Meddelelser. XVI,1. København: And. Fred. Høst & Søn.
 1941 'A Causerie on Linguistic Theory.' *Essais linguistiques*. II. Travaux du cercle linguistique du Copenhague. XIV. 1973. Copenhague: Nordisk Sprog- og Kulturforlag.
 1961 'Some Reflexions on Practice and Theory in Structural Semantics. *Language and Society*. Pp. 55-63. Copenhagen: Det Berlingske Bogtrykkeri.
 1963 *Prolegomena to a Theory of Language*. Translated by Francis J. Whitfield. København: The University of Wisconsin Press.
 1970 *Language. An Introduction*. Translated from the Danish by Francis J. Whitfield. Second printing. Madison: The University of Wisconsin Press.
 1971 *Essais linguistiques*. Paris: Les Éditions de Minuit.
 1972a *Sprogsystem og Sprogforandring*. Travaux du cercle linguistique du Copenhague: Nordisk Sprog- og Kulturforlag.
 1972b *La catégorie des cas. Étude de grammaire générale*. Zweite verbesserte und mit Korrekturen des Autors versehene Auflage der Ausgabe Kopenhagen 1935-1937. München: Wilhelm Fink.
 1973 *Essais linguistiques*. II. Travaux du cercle linguistique du Copenhague. Vol. XIV. Copenhague: Nordisk Sprog- og Kulturforlag.

1973b   *Sproget. En Introduktion.* 2. forøget udgave. København: Berlingske Forlag.
1975   *Résumé of a Theory of Language.* Edited and translated with an introduction by Francis J. Whitfield. København: The University of Wisconsin Press.
1976   *Omkring Sprogteoriens Grundlæggelse.* (1943). København: Akademisk Forlag.

Hobbes, Thomas
1971   *Leviathan.* Edited by C.B. Macpherson. London: Penguin Books Ltd.

Hock, Hans Henrich
1991   *Principles of Historical Linguistics.* Second revised and updated edition. Berlin: Mouton de Gruyter.

Hockett, Charles F.
1958   *A Course in Modern Linguistics.* New York: The Macmillan Company.

Hoenigswald, Henry M.
1966   *Language Change and Linguistic Reconstruction.* Fourth impression. First Phoenix edition 1965. Chicago: The University of Chicago Press.

Hogg, Richard M. (ed.)
1992   *The Cambridge History of the English Language.* Vol. 1. *The Beginnings to 1066.* Cambridge: Cambridge University Press.
1992   'General Editor's Preface.' *The Cambridge History of the English Language.* Vol. 1. *The Beginnings to 1066.* Pp. xv-xvii. Cambridge: Cambridge University Press.

Hopper, Paul J.– Traugott, Elizabeth Closs
1993   *Grammaticalization.* Cambridge: Cambridge University Press.

Horn, Wilhelm
1923   'Sprachkörper und Sprachfunktion.' Zweite Auflage. *Palaestra* 135. Leipzig: Mayer & Müller.
1936   'Experimentalphonetik und Sprachgeschichte.' *Proceedings of the 2nd International Congress of Phonetic Sciences. London, 1935.* Pp. 12-18.

Horn, Wilhelm – Lehnert, Martin
1954   *Laut und Leben. Englische Lautgeschichte der neueren Zeit (1400-1950).* Berlin: Deutscher Verlag der Wissenschaften.

Hughes, Arthur – Trudgill, Peter
1988   *English Accents and Dialects.* Second Edition. Reprint. London: Edward Arnold.

Høffding, Harald
1910   *Den menneskelige Tanke. Dens Former og dens Opgaver.* København: Gyldendalske Boghandel.

Itkonen, Esa
1978   *Grammatical Theory and Metascience.* Amsterdam Studies in the Theory and History of Linguistic Science. IV. Amsterdam: John Benjamins B.V.

Jacobsen, Erling
1976   *De psykiske Grundprocesser.* København: Berlingske Forlag.

Jacobsen, Malte (ed.)
1973   *Sprog og virkelighed.* København: Gyldendal.

Jakobson, Roman
1962   *Selected Writings.* Vol. 1. *Phonological Studies.* Second, expanded edition. 'S-Gravenhage: Mouton & Co.

1968   'Linguistics and Poetics.' In Thomas A. Sebeok *Style in Language*. Pp. 350-377. Second paperback printing. Cambridge: The M.I.T. Press.
1968b   'Prinzipien der historischen Phonologie.' *Travaux du cercle linguistique de Prague.* Vol. 4/5:247-267. Prague 1931. Kraus Reprint. Nendeln/Liechtenstein: Kraus Thomson Organization Limited.
1972   *Child Language, Aphasia and Phonological Universals*. The Hague: Mouton.

Jakobson, Roman – Halle, Morris
1971   *Fundamentals of Language.* Second, revised edition. The Hague: Mouton.

Jankowsky, Kurt R.
1972   *The Neogrammarians*. The Hague: Mouton.

Jespersen *MEG*
        Otto Jespersen *A Modern English Grammar. On Historical Principles*. Vols. 1-7. Reprint. 1961/1965. *Sounds and Spellings*. Vol 1. London: George Allen & Unwin Ltd.

Jespersen, Otto
1941   *Efficiency in Linguistic Change*. In *Selected Writings of Otto Jespersen*. Pp. 346-426. (No date of publication). London: George Allen & Unwin Ltd.
1967   *How to Teach a Foreign Language.* 13th impression. London: George Allen & Unwin Ltd.
1969   *Language. Its Nature, Development & Origin*. Fourteenth Impression. London: Unwin University Books.
1970   *Linguistica. Selected Papers.* ('Zur Lautgesetzfrage. A.' (1886); 'Nachtrag. B.' (1904); 'Letzte Worte. C.' (1933). Reprint). Pp. 160-228. New York: McGrath Publishing Company.
1995   *A Linguist's Life.* An English translation of Otto Jespersen's autobiography. Edited by Arne Juul, Hans. F. Nielsen, Jørgen Erik Nielsen. Odense: Odense University Press.

Jones, Charles
1989   *A History of English Philology*. London: Longman.
1993   *Historical Linguistics: Problems and Perspectives*. London: Longman.

Jones, Daniel
1962   *An Outline of English Phonetics*. Ninth edition. Cambridge: Cambridge University Press.
1963   *Everyman's English Pronouncing Dictionary.* London: J.M. Dent & Sons Ltd.
1976   *The Phoneme. Its Nature and Use*. Third edition. Cambridge: Cambridge University Press.
1997   *English Pronouncing Dictionary*. 15th edition. Edited by Peter Roach & James Hartman. Cambridge: Cambridge University Press.

Jones, William
1786   'The Third Anniversary Discourse. delivered 2 February, 1786.' *Asiatick Researches* (Calcutta 1789) 1:415-431.

Jones, W.T.
1969   *A History of Western Philosophy. The Medieval Mind*. Second edition. New York: Harcourt Brace Jovanovich, Inc.
1969b   *A History of Western Philosophy. Hobbes to Hume*. Second edition. New York: Harcourt Brace Jovanovich, Inc.

1970   *A History of Western Philosophy. The Classical Mind.* Second edition. New York: Harcourt Brace Jovanovich, Inc.

1975   *A History of Western Philosophy. The Twentieth Century to Wittgenstein and Sartre.* Second edition. New York: Harcourt Brace Jovanovich, Inc.

Joseph, Brian D – Wallace, Rex E.
1994   'Proto-Indo-European Voiced Aspirates in Italic: A test for the 'Glottalic Theory.' *Historische Sprachforschung* 107:244-261.

Jørgensen, Jørgen
1931   *A Treatise of Formal Logic. Its evolution and main branches with its relations to mathematics and philosophy.* Vols. 1-3. London: Oxford University Press.

Kanngiesser, Siegfried
1972   *Aspekte der synchronen und diakronen Linguistik.* Tübingen: Max Niemeyer Verlag.

Kant, Immanuel
1762   'The False and Subtlety of the Four Syllogistic Figures.' In *Kant* (1963) by G. Rabel. Pp. 44-45.

1764   'Investigation into the Evidence of the Principles of Natural Theology and Morals.' In *Kant* (1963) by G. Rabel. Pp. 64-67.

1923   *Kant's gesammelte Schriften: Erste Abteilung: Werke.* Band VIII. Abhandlungen nach 1781. Hereausgegeben von der Königlich Preussischen Akademie der Wissenschaften. Berlin und Leipzig: Walter de Gruyter & Co.

1923a   'Idee zu einer allgemeinen Geschichte in weltbürgerlicher Absicht.' In Kant 1923: 15-32.

1923b   'Über den Gebrauch teleologischer Principien in der Philosophie.' In Kant 1923: 157-184.

1956   *Kritik der reinen Vernunft.* Herausgegeben von Raymund Schmidt. Hamburg: Felix Meiner.

1963   *Kant.* (Translated and edited) By Gabrielle Rabel. Oxford: at the Clarendon Press.

1969   *Prolegomena zu einer jeden künftigen Metaphysik, die als Wissenschaft wird auftreten können.* Heraugegeben von Karl Vorländer. Hamburg: Verlag von Felix Meiner. (English translation: *Prolegomena to any Future Metaphysics.* By Lewis White Beck. 1950. Eigth Printing. Indianapolis/New York: The Bobbs-Merrill Company, Inc.)

Keesing, Roger
1992   'Linguistics and anthropology.' In *Thirty Years of Linguistic Evolution.* Pp. 593-609. Edited by Martin Pütz Philadelphia/Amsterdam: John Benjamins Publishing Company.

Keller, Rudi
1985   'Towards a Theory of Linguistic Change.' *Linguistic Dynamics.* Research in Text Theory. Edited by Thomas T. Ballmer. Berlin: Walter de Gruyter.

Kellner, Hans
1987   'Narrativity in History: Post-Structuralism.' *History and Theory.* Beiheft 26.4:1-29.

Kenstowicz, Michael – Kisseberth, Charles
1979   *Generative Phionology. Description and Theory.* New York: Academic Press.

Kenstowicz, Michael
1994   *Phonology in Generative Grammar.* Oxford: Blackwell.

Kierkegaard, Søren
- 1963 'Gjentagelsen.' In *Samlede Værker.* Bind 5. Udgivet af A.B. Drachmann, J.L. Heiberg, H.O. Lange (1901-1906/1920-1936). 2. udgave gennemset og ajourført af Peter P. Rhode. Købenavn: Gyldendal.

Kiparsky, Paul
- 1995 'The Phonological Basis of Sound Change.' (Ed.) John A. Goldsmith *The Handbook of Phonological Theory.* Pp. 640-670. Oxford: Blackwell.

Kitto, H.D.F.
- 1965 *Poiesis. Structure and Thought.* Berkeley: University of California Press.

Koerner, E.F.K
- 1976 'Towards a Historiography of Linguistics.' *History of Linguistic Thought and Contemporary Linguistics.* Edited by Herman Parret. Pp. 685-718. Berlin: Walter de Gruyter.

Kosso, Peter
- 1993 'Historical Evidence and Epistemic Justification.' *History and Theory* 32.1:1-13.

Krampen, Martin
- 1981 'Struktur und Geschichte.' Ed. Annemarie Lange-Seidl *Zeichenkonstitution.* Akten des 2. Semiotischen Kolloquiums. Regensburg. Pp. 19-25. Berlin: Walter de Gruyter.

Kucharczik, Kerstin
- 1996 'Zur Organismusmetaphorik in der Sprachwissenschaft des 19. Jahrhunderts.' *Sprachwissenschaft* 23.1:85-111.

Kuhn, Thomas S.
- 1971 *The Structure of Scientific Revolutions.* Second edition. Enlarged. Third impression. Chicago: The University of Chicago Press.

Kuiper, Koenraed – Allan, W. Scott
- 1996 *An Introduction to English Language. Sound, Word and Sentence.* London: Macmillan Press Ltd.

Labov, William
- 1978 *Sociolinguistic Patterns.* Oxford: Basil Blackwell.
- 1992 'Evidence for regular sound change in English dialect geography.' In Rissanen 1992:42-71.
- 1994 *Principles of Linguistic Change.* Vol. 1: *Internal Factors.* Oxford: Blackwell.

Ladefoged, Peter
- 1975 *A Course in Phonetics.* New York: Harcourt, Brace, Jovanovich, Inc.

Lass, Roger
- 1976 *English Phonology and Phonological Theory. Synchronic and Diachronic Studies.* Cambridge: Cambridge University Press.
- 1980 *On Explaining Language Change.* Cambridge: Cambridge University Press.
- 1987 *The Shape of English Structure and History.* London: J.M. Dent & Sons Ltd.
- 1988 'Vowel Shifts, great and otherwise: Remarks on Stockwell and Minokova.' (Eds.) D. Kastovsky & G. Bauer *Luick Revisited: papers read at the Luick-Symposium at Schloss Liechtenstein, 15.-18.9.1985.* Pp. 395-410. Tübingen: Gunter Narr.
- 1990 'How to Do Things with Junk: Exaptation in Language Evolution.' *Journal of Linguistics* 26:79-102.

    1997   *Historical Linguistics and Language Change.* Cambridge: Cambridge University Press.
Lass, R. – Anderson, John M.
    1975   *Old English Phonology.* Cambridge: Cambridge University Press.
Leach, Edmund
    1982   'Structuralism in Social Anthropology.' In David Robey 1982:37-56.
Lear, Jonathan
    1988   *Aristotle: the desire to understand.* Cambridge: Cambridge University Press.
Lehmann, Winfred P.
    1966   *Proto-Indo-European Phonology.* Fifth printing. Austin: The University of Texas Press.
    1992   *Historical Linguistics.* 3rd edition. London: Routledge.
Lehmann, W.P. – Malkiel, Y. (eds.)
    1968   *Directions for Historical Linguistics: a Symposium.* Austin: University of Texas Press.
Lenneberg, Eric H.
    1967   *Biological Foundations of Language.* New York: John Wiley & sons.
Lepschy, Giulio C.
    1972   *A Survey of Structural Linguistics.* London: Faber and Faber.
Leskien, A.
    1876   *Die Declination im Slavisch-Litauischen und Germanischen.* Leipzig: Bei S. Hirzel.
Lévi-Strauss, Claude
    1945   'l'analyse structurale en linguistique et en anthropologie.' *Word* 1:33-53.
    1962   *La pensée sauvage.* Paris: Plon.
Lewis, C.I.
    1912   'Implications and the Algebra of Logic.' *Mind* 21:522-531. New series.
Lira, Héctor Ortiz
    1976   'Some phonetic correlates of the rapid colloquial style of pronunciation in RP.' *Journal of the International Phonetic Association* 6.1:12-22.
Lockwood, W.B.
    1977   *An Introduction to Modern Faroese.* 3. Printing. Tórshavn: Føroya Skúlabókgrunnur.
Lommel, Hermann
    1967   *Grundfragen der allgemeinen Sprachwissenschaft.* 2. Auflage. Mit einem Nachwort von Peter v. Polenz. (1931). Berlin: Walter de Gruyter & Co.
Luick, Karl
    1903   'Studien zur englischen Lautgeschichte.' *Wiener Beiträge zur englischen Philologie.* Band 17:1-218
    1912   'Über die neuenglische Vokalverschiebung.' *Englische Studien* 45:432-437.
    1922   'Sprachkörper und Sprachfunktion.' *Englische Studien* 56:185-203.
    1923   'Experimentalphonetik und Sprachwissenschaft.' *Romanische Monatsschrift* XI:257-270.
    1927   'Emphatische Betonung als Quelle neuer Wortformen.' *Englishce Studien* 62:17-24
    1930   'Über einige Zukunftsaufgaben der englischen Sprachwissenschaft.' *Germanisch-Romanische Monatsschrift* XVIII:365-376.
    1930a   Besprechung: *Gunther Ipsen und Fritz Karg, Schallanalytische Versuche.* Heidelberg, Winter, 1928. *Beiblatt zur Anglia* Bd. XLI, Nr. XII:353-357.

1930/31 'Über den Einfluss der Intonation auf die Vokalqualität.' *Englische Studien* 65:337-342.
1964 *Historische Grammatik der englischen Sprache.* 1.i – 1.ii. (1914-1940). Reprint. Oxford: Basil Blackwell.

Lukasiewicz, Jan
1967 *Aristotle's syllogistic. From the standpoint of modern formal logic.* Second edition. Enlarged. Oxford: At the Clarendon Press.

Lyons, John
1968 *Introduction to Theoretical Linguistics.* Cambridge: Cambridge University Press.
1977 *Semantics.* Vol. 1. Cambridge: Cambridge University Press.
1982 'Structuralism and Linguistics.' In Robey 1982:5-19.

Lübcke, Poul
1981 *Tidsbegrebet. Et metafysisk essay.* København: G.E.C. Gad.

Løgstrup, K.E.
1976 *Norm og spontaneitet.* Andet oplag. København: Gyldendal.
1976b *Vidde og prægnans. Metafysik 1. Sprogfilosofiske betragtninger.* København: Gyldendal.
1978 *Skabelse og tilintetgørelse. Religionsfilosofiske betragtninger. Metafysik IV.* København: Gyldendal.
1984 *Ophav og omgivelse. Betragtninger over historie og natur. Metafysik III.* København: Gyldendal.

Macnamara, John
1992 'The take over of psychology by biology or the devaluation of reference in psychology.' In *Thirthy Years of Linguistic Evolution.* Pp. 545-570. Edited by Martin Pütz. Philadelphia/Amsterdam: John Benjamins Publishing Company.

Maher, Peter J. – Bomhard, Allan R. – Koerner, E.F. Konrad
1982 *Papers from the 3rd International Conference on Historical Linguistics.* Amsterdam: John Benjamins B.V.

Makkreel, Rudolf A.
1990 'Traditional Historicism, Contemporary Interpretations of Historicity, and the History of Philosophy.' *New Literary History* 21.4:977-991.

Martin, Raymond
1993 'Objectivity and Meaning in Historical Studies: Towards a Post-Analytic View.' *History and Theory* 32.1:25-50.

Martin, Rex
1976 'Explanation and Understanding in History.' Eds. Juha Manninen and Raimo Tuomela *Essays on Explanation and Understanding.* Pp. 305-334. Dordrecht-Boston: D. Reidel Publishing Company.

Martinét, André
1962 *A Functional View of Language.* Oxford: At the Clarendon Press.
1964 *Économie des changements phonetiques. Traité de phonologie diachronique.* (1955). Deuxième édition. Bern. A. Francke AG Verlag.
1970 *La linguistique synchronique.* Paris: Presses Universitaires de France.

McMahon, M.S.

1994   *Understanding language change.* Cambridge and New York: Cambridge University Press.

Meillet, A.
1922   *Les dialectes indo-européennes.* Paris: Librairie Ancienne Édouard Champion, Éditeur.
1951/52 *Linguistique historique et linguistique générale.* Tome II. Noveau tirage. Paris: Librairie C. Klincksieck.
1966   *Introduction à l'étude comparative des langues indo-européennes.* Second printing. Alabama: University of Alabama Press.
1975   *Linguistique historique et linguistique générale.* (1921). Paris: Éditions Champion.
1977   'Der gegenwärtige Stand der Forschungen auf dem Gebiet der allgemeinen Sprachwissenschaft.' *Sprachwissenschaft des 19. Jahrhunderts.* Pp. 314-333. Herausgegeben von Hans Helmuth Christmann. Darmstadt: Wissenschaftliche Buchgesellschaft.

Mepham, John
1982   'The Structuralist Sciences and Philosophy.' In David Robey 1982:104-137.

Mikkelsen, Kr.
1975   *Dansk Ordföjningslære. Med sproghistorisk Tillæg.* København: Hans Reitzels Forlag.

Mill, John Stuart
1970   *A System of Logic.* New impression. London: Longman.

Milroy, James
1992   'A Social Model for the Interpretation of Language Change.' In Rissanen *et al.* 1992:72-91.
1993   'On the Social Origins of Language Change.' In Jones 1993:215-236.

Milroy, John – Harris, John
1980   'When is a merger not a merger? The MEAT/MATE problem in present-day English vernacular.' *English World-Wide* 1:2:199-210.

Mitchell, Bruce
1964   'Syntax and word-order in the *Peterborough Chronicle* 1122-1154.' *Neuphilologische Mitteilungen* 65:113-144.
1992   'How to Study Old English Syntax?' In Rissanen *et al.* 1992:92-100.

Mitchell, Bruce – Robinson, Fred C.
1986   *A Guide to Old English. Revised with Prose and Verse. Texts and Glosary.* 4th edition. Oxford: Blackwell.

Muller, Henri F.
1945   'Phénomènes sociaux et linguistiques.' *Word* 1:121-131.

Nelson, Leonard
1970   *Progress and Regress in Philosophy.* Vol. 1. Edited posthumously by Julius Kraft. Translated by Humphrey Palmer. Oxford: Basil Blackwell.

Nielsen, Hans Frede
1998   *The Continental Backgrounds of English and its Insular Development until 1154.* Odense: Odense University Press.

Nietzsche, Friedrich
1966   'Vom Nutzen und Nachteil der Historie für das Leben.' *Werke.* Werke in drei Bänden. Erster Band. Pp.209-285. Darmstadt: Wisssenschaftliche Buchgesellschaft.

1966b   'Aus dem Nachlass der Achtzigerjahre.' Dritter Band. Pp. 415-925
1996   'Friedrich Nietzsche. Historiens Nytte.' *Moderne Tænkere.* På dansk ved Helge Hultberg. København: Gyldendal.

Oakes, Guy
1977   'The Verstehen Thesis and the Foundations of Max Weber's Methodology.' *History and Theory* 16.1:11-29.

Ogden, C.K. – Richards, I.A.
1969   *The Meaning of Meaning.* Tenth edition. Seventh impression. London: Routledge & Kegan Paul Ltd.

Olsen, Jens Henrik
1979   *Historiens fortællere. Grundrids af europæisk historietænknings udvikling.* København: G.E.C. Gad.

Osthoff, Hermann – Brugmann, Karl
1878   'Vorwort.' *Morphologische Untersuchungen auf dem Gebiete der indogermanischen Sprachen.* I. Teile 1, 2, 3. Pp. iii-xx. (1878-90. 1-5). Hildesheim/New York: Georg Olms Verlag.

Ottossen, Ulf
1996   *Kausalitet och Semantik. Ett bidrag til belysningen av förhållandet mellan lingvistisk språkteori och hermeneutisk fenomenologi.* Göteborg: Acta Universitatis Gothoburgensis.

Palmer, Leonard R.
1972   *Descriptive and Comparative Linguistics. A Critical Introduction.* London: Faber and Faber.

Parmenter, C.E. – Trevino, S.N. – Bevans, C.A.
1933   'The Influence of a Change in Pitch on the Articulation of Vowels.' *Language* 9:72-81.

Passmore, John
1987   'Narratives and Events.' *History and Theory* 26.4:68-74.

Paul, Hermann
1968   *Prinzipien der Sprachgeschichte.* (1880). Achte unveränderte Auflage. Tübingen: Max Niemeyer Verlag.

Pedersen, Holger
1978   *Videnskaben om Sproget. Historisk Sprogvidenskab i det 19. Århundrede.* Fotografisk optryk af *Sprogvidenskaben i det nittende. Århundrede. Metoder og Resultater.* Århus: Arkona.

Peeters, Bert
1992   'Le principe de l'économie linguistique et ses manifestations en phonologie diachronique.' *Cahiers Ferdinand de Saussure* 48:59-72.

Penzl, Herbert
1972   'Methods of comparative Germanic linguistics.' In Frans von Coetsem & Herbert L. Kufner *Towards a Grammar of Proto-Germanic.* Pp. 1-42. Tübingen: Max Niemeyer Verlag.

Peterson, Gordon E.
1961   'Parameters of Vowel Quality.' *Journal of Speech and Hearing* 4.1:10-29.

Piaget, Jean
1970   *Le structuralisme.* In '*Que sais-je?*' No. 1311. Paris: Presses Universitaires de France.

Pilch, Herbert
    1970   *Altenglische Grammatik.* München: Max Hueber Verlag.

Pinborg, Jan
    1976   'Some Problems of Semantic Representations in Medieval Logic.' Ed. Herman Parret *History of Linguistic Thought and Contemporary Linguistics.* Pp. 254-278. Berlin: Walter de Gruyter.

Pingaud, Bernard
    1966   'Introduction.' (Otherwise untitled). In *l'Arc.* 30:1-4.

Plato
    1970   *The Dialogues of Plato.* Translated by Benjamin Jowett. 4. vols. (*Phaedo* in vol. 1; *Phaedrus* in vol 2; *The Republic* in vol 4). Edited by R.M. Hare & D.A. Russell. London: Sphere Books Limited.

Popper, Karl
    1961   *The Poverty of Historicism.* (1957). Reprint. London: Routledge & Kegan Paul.
    1974   *The Logic of Scientific Discovery.* Seventh impression. (1959/1934). London: Hutchinson.

Preisler, Bent
    1992   *A Handbook of English Grammar on Functional Principles.* Aarhus: Aarhus University Press.

Prieto, Luis J.
    1981   'l'idéologie structuraliste et les origins du structuralisme.' Ed. Annemarie Lange-Seidl *Zeichenkonstitution.* Akten des 2. Semiotischen Kolloquiums Regensburg. Pp. 26-30 Berlin: Walter de Gruyter.

Prokosch, E.
    1939   *A Comparative Germanic Grammar.* Philadelphia:University of Pennsylvania.

Quine, Willard van Orman
    1971   *Set Theory and its Logic.* Revised edition. Cambridge: The Belknap Press of Harvard University Press.
    1972   *Methods of Logic.* Third edition. New York: Holt, Rinehart and Winston, Inc.

Quirk, Randolph
    1957   'Relative Clauses in Educated Spoken English.' *English Studies* 38:97-109.
    1970   'Aspect and Variant Inflection in English Verbs.' *Language* 46:300-311.

Quirk, Randolph – Greenbaum, Sidney
    1989   *A University Grammar of English.*Twenty second impression. London: Longman

Rasmussen, E. Tranekjær
    1956   *Bevidsthedsliv og erkendelse.* København: Munksgaard.

Rasmussen, Michael
    1992   *Hjelmslevs sprogteori.* Odense: Odense Universitets Forlag.

van Reenen, Pieter – Coetze, Anna
    1996   'Afrikaans, a Daughter of Dutch.' *Proceedings from the Second Rasmus Rask Colloquium, Odense University, November 1994.* Edited by Hans F. Nielsen and Lene Schøsler. Pp. 71-102. (*NOWELE* Supplement Vol. 17). Odense: Odense University Press.

Reill, Peter Hanns
   1986   'Narration and Structure in Late eighteenth-Century Historical Thought.' *History and Theory* XXV.3:286-298.

Renfrew, Colin
   1989   *Archaeology and Language. The Puzzle of Indo-European Origins.* (1987). London: Penguin Books Ltd.

Rescher, Nicholas
   1973   *Conceptual Idealism.* Oxford: Basil Blackwell.

Rescher, Nicholas – Urquhart, Alasdair
   1971   *Temporal Logic.* Wien: Springer-Verlag.

Riedel, Manfred
   1976   'Causal and Historical Explanation.' Eds. Juha Manninen and Raimo Tuomela *Essays on Explanation and Understanding.* Pp. 3-26. Dordrecht-Boston: D. Reidel Publishing Company.

Rissanen, Matti – Ihalainen, Ossi – Nevalainen, Terttu – Taavitsainen, Irma
   1992   *History of Englishes. New Methods and Interpretations in Historical Linguistics.* Berlin: Mouton de Gruyter

Robbe-Grillet, Alain
   1963   *Pour un nouveau roman.* Paris: Gallimard. Les Editions de Minuit.

Robey, David (ed.)
   1982   *Structuralism: an Introduction.* Oxford: At the Clarendon Press.

Robinson, Orrin W.
   1993   *Old English and its Closest Relatives.* London: Routledge.

Rozov, Nikolai S.
   1997   'An Apologia for Theoretical History.' *History and Theory* 36.3:336-352.

Ryle, Gilbert
   1970   *The Concept of Mind.* (1949). Harmondsworth: Penguin Books Ltd.

Sacksteder, William
   1981   'Some ways of doing language philosophy: Nominalism, Hobbes, and the Linguistic Turn.' *The Review of metaphysics* XXXIV.No. 3:459-485.

Samuels, M.L.
   1972   *Linguistic Evolution. With special reference to English.* Cambridge: Cambridge University Press.

Sankoff, David
   1979   'Categorical Context and Variable Rule.' In *Papers from the Scandinavian Symposium on Syntactic Variation.* Stockholm, May 18-19, 1979. Edited by Sven Jacobson. Pp. 7-22. Stockholm: Almqvist & Wiksell International.

Sapir, Edward
   1925   'Sound Patterns in Language.' *Language* 1:37-51.
   1949   *Language.* (1921). New York: Harcourt, Brace & World, Inc.

Sartre, Jean-Paul
   1966   'Jean-Paul Sartre répond.' In *l'Arc.* 30:87-96.

Saumjan, S.K.
   1971   *Principles of Structural Linguistics.* Translated from the Russian. The Hague: Mouton

de Saussure, Ferdinand

1879 *Mémoire sur le système primitif des voyelles dans les langues indo-européennes.* Leipsick: en ente chez B.B. Teubner.
1972 *Cours de linguistique générale.* Édition critique préparée par Tullio de Mauro. Paris: Payot.

Schibsbye, Knud
1970 *A Modern English Grammar. With an Appendix on Semantically Related Prepositions.* Second edition. London: Oxford University Press.
1977 *Origin and Development of the English Language.* Vol. III. (3 vols. 1972-1977). København: Nordisk Sprog- og Kulturforlag.

Schiwy, Günther
1973 *Der französische Strukturalismus.* Hamburg: Rowohlt.

Schlegel, Friedrich
1808 *Ueber die Sprache und Weisheit der Indier.* Heidelberg.

Schuchardt, Hugo
1885 'Über die Lautgesetze: Gegen die Junggrammatiker.' In Vennemann and Wilbur 1972.

Scriven, M.
1959 'Explanation and Prediction in Evolutionary Theory.' *Science* 130:477-482.

Sears, Laurence
1942 'The Meanings of History.' *The Journal of Philosophy* XXXIX. No. 15:393-401.

Siertsma, B
1965 *A Study of Glossematics.* Second edition. The Hague: Martinus Nijhoff.

Steffens, Henrich
1968 *Indledning til philosophiske Forelæsninger.* (1803). København: Gyldendals Forlagstrykkeri.

Sten, H.
1969 *Manuel de phonétique francaise.* 5e édition. Copenhague: Munksgaard.

Stiles, Patrick V.
1988 'Questions of phonologization and relative chronology, with illustrations from Old English.' In (Eds.) D. Kastovsky & G. Bauer *Luick Revisited: papers read at the Luick-Symposium at Schloss Liechtenstein, 15.-18.9.1985.* Pp. 335-354. Tübingen: Gunter Narr.

Stockwell, Robert P. – Macauley, Ronald S.K.
1972 *Linguistic Change and Generative Theory.* Bloomington: Indiana University Press.

Stockwell, Robert P – Minkova, Donka
1988 'The English Vowel Shift: problems of coherence of explanation.' (Eds.) D. Kastovsky & G. Bauer *Luick Revisited: papers read at the Luick-Symposium at Schloss Liechtenstein, 15.-18.9.1985.* Pp. 355-393. Tübingen: Gunter Narr.

Strang, Barbara M.H.
1970 *A History of English.* London: Methuen & Co Ltd.

Streitberg, Wilhelm
1963 *Urgermanische Grammatik.* Dritte, unveränderte Auflage. Heidelberg: Carl Winter Universitätsverlag.

Sweet, Henry
   1968/71   *A New English Grammar. Logical and Historical*. Parts 1 & 2. (1891). Oxford: the Clarendon Press.
Swiggers, Pierre
   1992   'Linguistic Theory and Epistemology of Linguistics.' In *Thirty Years of Linguistic Evolution.* Pp. 573-589. Edited by Martin Pütz. Philadelphia/Amsterdam: John Bejamins Publishing Company.
Sørensen, Holger Steen
   1958   *Word-Classes in Modern English. With special reference to Proper Names. With an introductory theory of grammar, meaning and reference.* Copenhagen: G.E.C. Gad Publisher.
Syrett, Martin
   1994   'The Unaccented Vowels of Proto-Norse'. *Nowele.* Supplement vol. 11. Odense: Odense University Press.
Thèses 1929
   1968   *Travaux du cercle linguistique de Prague.* Vol. 1:5-29. Prague 1929. Kraus reprint. Nendeln/Liechtenstein: Kraus-Thomson Organization Limited.
Thrane, Torben
   1980   'Typology and Genetics of Language.' *Travaux du cercle linguistique de Copenhague.* Vol. XX. Edited by Torben Thrane, Vibeke Winge, Lachlan Mackenzie, Una Canger, Niels Ege. Copenhagen: The Linguistic Circle of Copenhagen/Reitzel's Boghandel.
Togeby, Knud
   1965a   *Structure immanente de la langue française.* Paris: Librairie Larousse.
   1965   *Fransk grammatik.* København: Gyldendals Forlag.
   1974   *Précis historique de grammaire française.* København: Akademisk Forlag.
Topolski, Jerzy
   1976   *Methodology and History.* Translated from the Polish by Olgierd Wojtasiewicz. Synthese Library. Vol. 88. Boston: D. Reidel Publishing Company.
   1987   'Historical Narrative: Towards a Coherent Structure.' *History and Theory* 26.4:68-74. Beiheft 26.
Toynbee, Arnold J.
   1963   *A Study of History.* Vol. 1. (1934). Eighth impression. London: Oxford Unviersity Press.
Traugott, Elizabeth Closs
   1972   *A History of English Syntax. A Transformational Approach to the History of English Sentence Structure.* New York: Holt, Rinehart and Winston, Inc.
   1992   'Syntax.' In Hogg 1992:168-289.
Trager, George 1. & Henry Lee Smith, Jr
   1975   'An Outline of English Structure.' Eighth Printing. *Studies in Linguistics: Occasional Papers.* 3. New York: Johnson Reprint Corporation.
Trask, R.L.
   1996   *Historical Linguistics.* London: Arnold.
Trudgill, Peter
   1974   *The social differentiation of English in Norwich.* Cambridge: at the University Press.

Twaddell, W. Freeman
   1967   'On Defining the Phoneme.' *Readings in Linguistics* 1. (Fourth Edition). Second Impression. Edited by Martin Joos. Chicago: The University of Chicago Press. Pp. 55-80.

Ulldall, H.J.
   1957   *Outline of Glossematics. A Study in the Methodology of the Humanities with special Reference to Linguistics*. Part 1: *General Theory.* Copenhagen: Nordisk Sprog- og Kulturforlag.

Vachek, Josef
   1972   'The Linguistic Theory of the Prague School.' Ed. V. Fried *The Prague School of Linguistics and Language Teaching.* Pp. 11-28. London: Oxford University Press.

Vann, Richard, T.
   1987   'Louis Mink's Linguistic Turn.' *History and Theory* 26.1:68-74.

Vennemann, Th. – Wilbur, T.H.
   1972   *Schuchardt, the Neogrammarians, and the transformational theory of phonological change*. Frankfurt: Athenäum.

Verner, Karl
   1877   'Eine ausnahme der ersten lautverschiebung.' *Zeitschrift für vergleichenden Sprachforschung*. XXIII. Neue Folge Band III:97-130.

Vestergaard, Torben
   1993   *Engelsk Grammatik*. 2. oplag. Det Schønbergske Forlag.

Wang, William S-Y – Lien, Chinfa
   1993   'Bidirectional diffusion in sound change.' In *Historical Linguistics. Problems and Perspectives*. Edited by Charles Jones. Pp. 345-400. London: Longman.

Wardhaugh, Ronald
   1994   *Investigating Language. Central Problems in Linguistics*. Oxford: Blackwell

Weber, Max
   1978   *Economy and Society. An Outline of Interpretive Sociology*. A Translation of Max Weber, *Wirtschaft und Gesellschaft. Grundriss der verstehenden Soziologie*, based on the fourth German edition, Johannes Winckelmann (ed.). Edited by Guenther Roth and Claus Wittich. (1968). Berkeley: University of California Press.

Weinreich, Uriel – Labov, William – Herzog, Marvin I.
   1968   'Empirical Foundations for a Theory of Language Change.' Eds. W.P. Lehmann and Yakov Malkiel *Directions for Historical Linguistics.* Pp. 95-188. Austin: University of Texas Press.

Whitehead, Alfred North
   1967   *Science and the Modern World*. New York: The Free Press.

Whitney, William Dwight
   1970   *The Life and Growth of Language*. (1875). Hildesheim/New York: Georg Olms Verlag.

Wilden, Anthony
   1980   *System and Structure. Essays in Communication and Exchange*. Second edition. London: Tavistock Publications.

Willems, Klaas
   1995   'Der Erkenntniswert der Sprache.' *Sprachwissenschaft* 20(2):168-186.

Williams, Raymond
    1976   *Keywords. A Vocabulary of Culture and Society*. London: Fontana/Croom Helm.
Winters, Margaret E.
    1992   'Diachrony within Synchrony: The Challenge of Cognitive Grammar.' In *Thirty Years of Linguistic Evolution.* Pp. 503-512. Edited by Martin Pütz. Philadelphia/Amsterdam: John Bejamins Publishing Company.
Winterson, Jeanette
    1996   *Art Objects. Essays on Ecstacy and Effrontery*. London: Vintage.
Wisdom, J.O.
    1976   'General Explanation in History.' *History and Theory* 15:257-266.
Wittgenstein, Ludwig
    1963   *Philosophische Untersuchungen*. Second edition. Reprinted 1963. Oxford: Basil Blackwell.
    1972   *Preliminary studies for the "Philosophical Investigastions". Generally known as the Blue and Brown Books.* Oxford: Basil Blackwell.
Wiwell, H.-G.
    1901   *Synspunkter for dansk Sproglære*. København: Gyldendal.
Wolfe, Patricia M.
    1972   *Linguistic Change and the Great Vowel Shift.* Los Angeles. University of California Press.
Wordsworth, William
    1968   *Lyrical Ballads. Wordsworth and Coleridge.* Edited by R.L. Brett and A.R. Jones. Reprint. London: Methuen & Co.
Zeller, Eduard
    1963   *Die Philosophie der Griechen in ihrer geschichtlichen Entwicklung.* Erster Teil. Erste Abteilung. 6. Auflage. (Leipzig 1919). Darmstadt: Wissenschaftliche Buchgesellschaft.

INDEX OF DEFINITIONS AND TECHNICAL TERMS

The terms of the index refer primarily to the definitions of terms of the essay's historical theory as well as their less formal determinations and summary statements.

Single numerals 1., 2. etc. refer to chapters; 1.4., 2.3. etc. refer to chapter sections; 1.4.2, 2.3.1 etc. refer to summary statements, definitions, figures and diagrams in the relevant chapter sections.

abduction 0.3.1, 0.3.1.1
Abstraktion 0.2.2, 1.4.7, 4.2.
aition, the 1.2., 1.2.17, 1.2.25-1.2.26, 2.5., Appendix 2.
alternatives, possible and positive 5.1., 5.1.7-5.1.13,
ambiguity 0.3.1, 4.2. (polysemy), 4.2.3-4.2.9,
amenability = raw material
analogy, false 1.1.2, analogical process Def. 13 (in 1.5.).
analysis 1.4., 4.1., 4.1.1, 4.4., change 2.3.2
Arbeitsteilung 1.3., 1.4., 5.2.note 18
Arbeitsvereinigung 1.3., 1.4.
arbitrariness 3.4., 3.4.1, 3.4.2-3.4.4, 3.4.5, 6.2., 6.2.B, 6.3.G
assimilation 1.5.9-1.5.16, progressive Def. 21 (in 1.5.), regressive Defs. 20, 22 (in 1.5.), 1.6.
asymmetry 0.3.2, 0.3.3
attitudinal system, the 0.5., 5.1., 5.5., 5.5.3-5.5.5, 6.1., definition of its elements 6.1.4, 6.1.5, 6.1.7, 6.1.8, 6.1.18
become 6.3.C
Bill-Peters Effect, the 0.5.
biuniformal = cause
break 2.5., 5.1. (*Introduction*), 5.2., 5.3.1, no break around Def. 5.3.6, 5.6.3, a break around Def. 5.3.11, 5.6.3, breaks (logical) 5.4.4, 5.6., 5.6.3, 5.6.5, potential break 5.6., Def. 5.6.6f.; *(not) break away from, (not) break into* Appendix 2, *sub* First and Last. Break = degrees of coherence, *sub* transition
but-not-only-if = aition, the

cause 1.2., 1.4., 2.1., 2.1.3-2.1.6 biuniformal 1.4.5/1.4.6, 1.4.8, 1.5., 2.1., 2.1.2, 2.3.,2.4., 2.4.1, 2.6.4, 5.1., 6.3.D
change, negative 0.1.2, 1.1.4-1.1.17, conditioned 0.4.3, 2.1.3-2.1.6, 2.2.2, isolative 0.4.4, 1.1.3-1.1.9, 1.1.11, 2.2.4-2.2.5, 2.5. – 1.1.17, 1.4.6, 1.5.16, 3.1., 3.3.6-3.3.9, 3.4. (Saussure), 3.4.2-3.4.6, 3.4.9-3.4.10, 4.3., 4.3.10, 5.2., ertial 5.2., 5.2.3, 5.2.4 inertial c. 5.2., 5.2.3, 5.2.4, 5.5., 5.5.2 (Parmenides); see *sub* endogeneity, exogeneity, katageneity; ertial 6.1.2, 6.1.6, 6.1.10, 6.2., 6.2.2, 6.3., 6.3.1, 6.3.C, 6.3.D, (constant) 6.3.E. Appendix 1
chronological series 5.1.1iv, 5.3., (non)chronological series 5.3.4-5-3-10, 5.4., 5.4.1—5.4.2
class, as one, as many 4.1. Appendix 2
coherence, participative Def. 1.5.7, cohesion 1.5.10-1.5.13, Defs. 20, 21 (in 1.5.), complex cohesion 1.5.14 (Def. 22 (in 1.5.)), gap cohesion Def. 23 (in 1.5.); = change 1.5.16, 5.3.2, coherence relations (calculus of) 6.1.16, 6.1.24, (*sub* narrative), 6.2., 6.2.B, 6.2.3, 6.2.4, 6.2.6, 6.2.8, 6.2.9. *Sub* transition
collective 3.3., 4.2.17
common origin = origin
commutation 0.5., 2.6.
competence/performance 0.1.1 note 8
conditioned change 0.4.3A
conformity 6.2.B (end of chapter)
connotation 1.3., 1.5., 2.2.1, 6.3., 6.3.1, 6.3.2, *sub* metaphor, polysemy
consequences of the theory 6.3.
contextualisation 6.2.B (end of chapter) and note

contingency 2.2., 2.3., 2.5., 2.5.4, Requirements A and B 2.5., 2.6., 2.6.4, 3.1.
continuation (A>A) 5.2., 5.2.3, 6.1.
continued forceful style Def. 6.1.18
continued normal style Def. 6.1.4
continued relaxed style (= Def. 6.1.18 *mutatis mutandis*)
continuity 2.5., 6.1.1, 6.1.(1-3)
contrast = distinctiveness
created forceful style Def. 6.1.5
created nonforceful style Def. 6.1.7
created nonrelaxed style (= Def. 6.1.7 *mutatis mutandis*)
created normal style Def. 6.1.8
created relaxed style (= Def. 6.1.5 *mutatis mutandis*)
Critico-Philological Method 1.1., 1.4.5, 1.4.6
degrees of coherence Appendix 1
denotation 1.1.20
dependency 1.2., 2.4., (logical conditional) Appendix 2
devaluation 0.4.3 note 160, 6.1., 6.1.2, 6.1.3; 6.1.6, 6.1.12, 6.1.13 (temporal d.), 6.1.6 (narrowing), +/-temporal devaluation 6.3.E, 6.3.6 (definitions), distributional devaluation 6.1.6, 6.1.12, 6.1.13, 6.3.A, (definitions) 6.3.7
developmental series, 0.4.2, 0.4.3, 0.4.4, 1.5., Defs. 1.5.1, 14, 2.3., 5.2., 5.2.4, 5.4. (*Appendix*). 6.1., 6.1.1, 6.1.2, 6.1.3, 6.1.6, 6.1.11, 6.1.12, 6.1.13, 6.1.17, 6.1.20, 6.1.22
diachrony, diachronic linguistics (the terms are not used)
differences, small 0.5., 5.5.
directionality *sub* irreversibility
displacement of sign relates 6.1., 6.2.1; *sub* change; Appendix 1.
distance 1.2.
distinctiveness 0.5.
dualism, 1.1., 1.1.3-1.1.9, 1.1.11-1.1.17, 1.1.19, 1.3., 1.6., 2.2., 2.2.2, ontological d. 1.1., 1.3., 1.4., 2.1.,4.2., 4.2.10, *sub* ideology
endogeneity Defs. 6, 7, 8 (1.5.), 2.1.2, Def. 31b (in 2.1.), 5.1., 5.2., 6.1., 6.1.3, 6.2.
Erkenntniswert, of the everyday language 0.4.3 and note 137, 1.2., 4.2., 4.4.
ertial *sub* change
essentialism 1.1.10, 1.1.18, 1.2.2, 1.3.2.-1.3.3

existence (existential function) 0.4.1, 1.4., 1.4., 1.5., 1.5.9-1.5.16, 1.6., 2.4., 3.1., 3.2., 4.3., 5. (Introduction), 5.2., 5.3., 5.3.1, 6.1.
exogeneity Defs. 5, 9, 10 (in 1.5.), 2.1.2, Def. 31b (in 2.1.)
extensionality 4.1. (a class as many)
external nonsymmetry 5.3.4, 5.3.5, 5.3.7
external symmetry 5.3.4, 5.3.5
fait social, un 3.3. (Durkheim), 3.3.1-3.3.6
First (a theoretical primitive) Appendix 2
focus, the historian's 5.3.
fullblooded 5.6.1, 5.6.2, 5.6.3
function *sub* weakening
gap *sub* break and coherence
geneticism 1.4., 1.5., genetic process Def. 11 (1.5.), 2.1., 2.3., 2.4.
gradualness 6.3.E.
'Great Vowel Shift' 0.3.2, 0.3.3, 0.4.1 note 104.
here-and-now intersection Def. 33 (in 2.1.)
hermeneutics, 0.0., 0.4.3 note 140, 6.3 and note 1
historical sense, man's 0.4.2, explanation 2.3.1, phenomenon 1.5.2, 2.1.,2.5.2, 3.4., 3.4.2-3.4.4, 4.3., 4.3.2, 4.4.(narrative), accuracy 4.3., 4.3.1, directionality 5.1., 5.1.1iv, history 1.6.4, generic 2.3., 3.1. (Sartre), 3.4.12-3.4.13 (Saussure), signfunctional 4.3.2-4.3.6, 4.3.9, 4.3.10
homonymy (= polysemy) 0.3.1 and note 58.
horizontal synthesis Defs. 1, 2, 3, 4 (1.5.), 2.1.
ideology 1.1. (Plato), 1.6.1.
idiolectal networks 5.1., 5.1.1, 5.1.2-5.1.6, 5.1.7-5.1.13.
impossibility (empirical impossibility, impossible developments) 6.2., 6.2.A, 6.2.3, 6.2.4-6.2.11, 6.3.G
inertial and noninertial developments 5,2., 5.2.3, Def. 5.2, 5.2.4, 6.1.20, *sub* change; Appendix 1
internal contrast 5.3.4, 5.3.5, 5.3.7, 6.3.6 (Defs. of inertial and noninertial developments)
internal law 0.2.1, 0.3.
internal noncontrast 5.3.4, 5.3.5, 5.3.7, 6.3.6 (Defs. of inertial and noninertial development)
intersection 1.5.2, a here-and-now i. Def. 16 (in 1.5.), Def. 31a (in 2.1.), Def. 33 (2.1.), fullblooded Def. 5.6.1, 5.6.3; (in the developmental series of the implemented theory) 6.1.; nonfullblooded Def. 5.6.2, 5.6.3, 6.1.13, 6.1.14, 6.2.9, 6.3.4, 6.3.13, 6.3.14, 6.3.23
irreversibility and reversibility 5.1., 5.1.1i-5.1.1iv, 5.1.2, 5.2.; (in the developmental series) 6.1.

Index of definitions and technical terms 475

isolative change 0.4.3B, 6.1., 6.1.1
isolative continuation 6.1., 6.1.1
isomorphism 6.2.B (end of chapter)
Jones Hypothesis 0.2.1
katageneity Def. 32 (in 2.1.), 2.1.2, Def. 31 (in 2.1.), 5.1., 6.1.3, 6.1.5
katalysis 1.6.
Kausalzusammenhang 0.2.2, *sub* cause.
language learning 0.3.1, 4.4.
Last (a theoretical primitive) Appendix 2
L-Continuum, the 1.6., 1.6.2, 1.6.4, 6.3.A
'lines', intersecting lines 1.5.2, 1.5.5,
linguistic motivation Def. 30 (1.6.)
linguisticism 1.1., 1.3., 1.3.4, 1.6.4
logic 1.1.,1.1.18, 1.2.2, 1.2.3-1.2.24, 1.4.2, 2.5., 2.6., 4.2., 4.3.8, (logical conditional) Appendix 2
loan Def. 12 (in 1.5.)
manifestatio ontology 0.3.1, 6.3.A
markedness 0.5., 5.1., 5.1.13
material definitions 5.4., 5.5., 5.6., 6.3.7
meaning 2.3., completion 1.5., Def. 31b (in 2.1.)
Metahistorical Precondition 1.4.3, metaconditions 1, 2, 3, 4, 5, 6 (in 1.5.)
metagenous motivation Def. 29 (in 1.6.).
metaphor 6.3.B
Method, the Critico-Philological 1.1., 1.4., 1.4.5
Moment-when, the 0.1.1
monotemporal sign 6.2.A and note 6
motivation *sub* endogeneity, exogeneity, katageneity, paragenous, metagenous, linguistic.
narratives 1.3.5, 6.1., 6.1.10, (of the developmental series) 6.1.14, 6.1.15, 6.1.19, 6.1.21, 6.1.23, 6.3.B
necessity, empirical 0.2.1 note 36, 1.1.10; *sub* impossibility
near-merger 0.5.
network, synchronic 5.1.1., 5.1.2-5.1.6.
nominalism 0.2.2, 1.1.,1.2., 1.6.4, 4.1. (extensionality), 4.2., 4.3. (*vs.* the poet and the historian), 6.3.B, *sub* ideology,
nonfullblooded 5.6.2, 5.6.3
objects, Glossematic. 1.4.4
Optimal Theory 0.4.1 note 124
orientation, collective 5.1.
origin (common), 0.2.1, 0.2.1.1, 0.2.1.2, 0.2.2.1, 0.2.2.2, 1.4., 4.2.13 (originary), 5,1., 5.1.2-5.1.6, 6.3.C

palimpsest 1.5.
paragenous motivation Def. 28 (in 1.6.), Def. 32 (in 2.1.)
parole 0.2.2.2, 0.4.2, Def. 15 (1.5.), 2., 4.2., 4.2.16, 4.2.14, 4.2.17, 4.2.18, 5.1.
participation (participative) 0.2.1.1/2, 0.2.2.1, 0.2.2.2, 0.4.1, 0.5 note184, 1.1.1, 1.2., 1.2.6b, 1.5.7, 1.5.8, 1.5.9-1.5.16 participative cause 1.4.5, 1.4.6, 1.4.8, 2.4., 2.4.3, 2.6., 4.2.15, 4.3.1, 4.3.7, 4.3.10, 4.4., 4.4.2, 4.4.3
past/present 0.4.,1.5.3
period-ic (cf. developmental series) 0.4.4
peripherality 0.5.
Phenomenal Error 0.4.1, 1.1.19, 1.4., 1.4.11, 6.3. and note 1
phenomenon, historical 1.5.2, Def. 32 (2.1.), 2.1.2, 2.5.2, 4.3., 4.3.9-4.3.10
phonemic-phonetic 1.3., 1.6.
pluritemporal sign 6.2.A and note 6
polysemy 0.3.1 and note 58, 4.2., 4.2.3-4.2.9, 4.2.10-4.2.11, 4.4., 4.4.1, 6.3.B, 6.3.3-6.3.5
possibility empirical possibility *sub* impossibility and alternatives
potentiality 2.5., 2.6.
precision 4.2. (synonymy), 4.2.3, 4.2.3-4.2.9, 4.2.12
predecessor Def. 31a (2.1.), 2.5.3
present/past 0.4., 1.5.3, Def.31a (2.1.)
presupposition 1.2., 1.2.3-1.2.24, 2.6.3
process, historical 1.4.5, 1.4.6, 2.1.2, Def. 31a/b (2.1.), 2.3.3
proprialization Def. 18 (1.5.), Def. 31b (2.1.)
raw-material 1.2., 2.4.3, 4.4.
Realism (*sub* ideology) 0.2.1, 0.2.2, 1.1., 1.2., 1.6.4, 4.1.
reasoning, *sub* synthetic/synthesis, analytic/analysis, 6.3.D
reductionism 1.1.19, 1.2.2, 1.2.3-1.2.24, 1.3.2-1.3.3, 1.4.2
refus de l'histoire, le 3.1.
rehabilitation of substance 0.4.2, (material definitions) 6.3.6
relevance, empirical 1.3., 1.3.4, 1.4.1
reversibility *sub* irreversibility
rule, synchronic r. 3.3.6-3.3.9
selection (L-continuum) 1.6.4
semiotic hierarchy (Glossematic) 1.4.4

separation 1.2., 1.2.1, 6.2.2, 6.2.3, 6.2.5, 6.2.8, 6.2.9
serialism = developmental series
short-lived (synchronic) 6.2, 6.2.8 (in developmental series VIII>II and I>IIi), 6.3. (in developmental series II>III and II>VIII) *sub* monotemporal *and* pluritemporal.
sign-functional 1.1., 1.1.20, 1.2.24, 1.3.2-1.3.3, 1.4.11, 3.4., 4.1., 5.5., 5.5.3-5.5.5, (historical phenomenon), 6.3.
simultaneity 0.2.1.1/2, Def. 17 (1.5.), Defs. 31, 32, 33 (2.1.), 2.1.2, 2.1.7
sociology 3.3. (Durkheim)
sound law 1.2.25-1.2.26, 1.4., 1.4.7-1.4.9, 1.5.
speech situation 6.3.1-6.3.2 (as a sign function)
spoken chain, the = *parole*.
stability 1.1.3, 1.1.4-1.1.17, 2.2.4-2.2.5, 5.2., 5.2.3
stage *vs*. state 2.5. (the word *stage* denoting a property of the synchronic state is not used)
startling theorems *sub* 1.2.10
state (also = stability) 1.1.3-1.1.9, 1.1.11-1.1.17, 1.2.18-1.2.24, 1.6.3, two-state model 1.5.9, 2.2.3, 2.5., 3.2.2, 3.4.12-3.4.13
structuralism 3., 3.1 (overview and Sartre), 3.2. (linguistic discussion), 3.2.1 (definitions), 3.3. (Durkheim; Meillet), 3.4. (Sausure), 4.1. (7 tenets), 4.2.4, (rehabilitation of) 6.1., 6.3 (end of chapter)
styles 0.5., 2.1.6,
subsumption 1.4., *sub* extensionality
substance, rehabilitation 1.1.10, (definitions) 6.3.6
succession 1.5.9-1.5.16
successor Def. 31a (in 2.1.), 2.5.3
symmetry 0.3.2, 0.3.3, symmetrical relationship *sub* 5.3.4 and 5.3.5.
synchronic = short-lived
synchrony 0.0, 0.2.1, 6.3., 6.3.B, 6.3.C,6.3.6

synonymy 1.2.1, 4.2 (precision), 4.3., 4.4., 4.4.1, 5.5., 6.3.B, cf. 6.3.3-6.3.5
synthesis 1.4., 1.5.1, 4.4, change 2.3.2
system 1.3., static s. 1.5.5—1.5.6, 1.6., 1.6.2, 2.1., 2.2., .2.3, 2.2.4, 3.2., 5.1., 5.1.1 (its constitution)
tenseness 0.3.2
tensional field *sub* definitions of the signs of the attitudinal system; 6.1.
totality 4.1. (a class as many; a class as one), 4.1.1, 4.1.2, 4.1.3, 4.1.4, 4.2., 6.1
trace de l'homme 3.1., 3.2., 4.1.
transcendence 0.2.2, 1.2., 1.2.1, 1.4., 1.4.11, 1.6.2, 4.2., 4.4., (sign-functional), 6.3.B, 6.3.D
transindividual 0.2.2, 0.3.1, 0.3.2, 0.3.3,
transition 0.1.1, 3.1 (*dépassement*), 4.2., 5.2, 5.3., 5.3.12-5.3.14, 5.4., transitional definitions (static ones) 5.4.3, 5.4.5-5.4.15, 5.6., 5.6.4-5.6.5; dynamic ones 6.1., 6.1.19, (calculus of transitions) 6.2., 6.2.3. Appendix 1
two-state model 0.0, 0.4.3A/B, 1.5.6, 2.6., 2.1., 2.1.1, 2.2.2, 2.4., 2.6.1-2.6.4, 5.2., 5.2.1, 5.3.2, 5.4.3, 5.4.5-5.4.15, 5.5.1, 5.6.
uniformitarianism 0.5 note 184
universal (cf. polysemy, synonymy) 0.1.2
unmarked *sub* markedness
use 6.3F, 6.3.7 (cognitive), 6.3.8/9 (mechanical). Appendix 2
validation 2.5., 2.5.1
variability = variation
variant = variation
variation 0.1.2, 0.3.1, 0.4.3A/B, 0.4.4, 0.5., 1.3.1, 1.4.9, 1.6.5, 1.6.3, 2.1., 2.1.5-2.1.7, 2.2.1, 5.1., 5.5., 5.5.1, 6., 6.2., 6.2.2, 6.3.C
vertical synthesis Def. 19 (in 1.5.), Defs. 24-27 (1.6.), 2.1., Def. 31a (in 2.1.)
weakening and lack of function (its logic) Appendix 2

RESUMÉ PÅ DANSK

PROLOGEN

AFHANDLINGENS FORMÅL og tese opstilles – at demonstrere muligheden for at konstruere en immanent historieteori. Kritik af synkroniseringen af historien starter (0.1.) med en introduktion af den fremførte historieteoris grundlæggende struktur (0.2.). Teorien tager sit udgangspunkt i Hermann Pauls diskussion af den skjulte Realisme i sprogvidenskaben og dermed af transcendensproblemet, forholdet mellem teori og data. En kritisk gennemgang af tre synkront baserede fonologi-teorier, der viser, at generaliseringsværdien i synkroniseringen af historien ikke er klar (0.3.).

I 0.4. introduceres sprogets eksistentielt-kognitive funktion samt vor historiske sans som relevant begreb og faktor; substansen og dermed det transsynkrone rehabiliteres i lyset af en begyndende kritik af synkroniens dualisme-fundament således, at historieteoriens data bliver **perioden** forstået som en **udviklingsrække**. I 0.5. kritiseres synkroniens kontrast-, distinktivitet- og forskelsbegreber, og der konkluderes med en generalisering af tegnfunktionens begreb: alt sprogligt er tegn-funktionelt; alt tegn-funktionelt er afhængigt af hverdagssproget og mennesket og dermed historien.

KAPITEL I

I 1.1. GIVES EN OVERSIGT over dualismens klassiske baggrund; det arbitrære i den lingvistiske (samt filosofiske) traditions fremhævelse af stabilitet som forandringens overhoved påvises. Historieteorien hviler på et participativt grundlag og ikke et historie-fremmed grundlag, hvilket formuleres i to principper, Participations- og Den falske Analogis principper; hermed bliver historie betragtet som en tredje signifikationsmodus mellem nominalisme og Realisme. Diskussion af transcendensproblemet (1.2.) i en formallogisk kontekst fører til etablering af en ny logisk funktion, en Aition, og fortsættes i 1.3. med en kritik af synkronfonologiens reduktionisme, der ender i historie.

I 1.4. opstilles i lyset af traditionens synkroni-historie modsætning flere betingende principper for en immanent historieteori: 'Empirisk Relevans', 'Den metahistoriske Betingelse', 'Den kritisk-filologiske Metode', og 'Fænomenfejlen' samt Det 'biuniforme og participative Udviklingsbegreb'. Den såkaldt neogrammatiske lydlov omfortolkes aitionalt.

I 1.5. argumenteres for, at årsagsforklaring ikke er egentlig forklaring, men årsagen hører til i udviklingsrækken som datakonstituerende. Historiens genstand er ikke synkroniens 'punkt', objectum eller spectaculum, men et fænomen, der konstitueres som et krydsningspunkt af krydsende 'linjer'. Dette fører frem til konstruktionen af historieteoriens egne termer, der viser, at synkroniens tilstands- og samtidighedsbegreber har deres plads blandt historiens dynamiske termer. På basis af de opstillede definitioner påvises sammenhængen mellem 1) den 'blinde' lydforandring, 2) den betingede lydforandring, 3) analogidannelse og 4) lån. Definitionssystemet (1.5. og 1.6.) sættes op med henblik på det transsynkrone i udviklingsrækken, dvs. historieteoriens emne, hvorved lighedspunkter med talens forløb fremtræder. Via definitionen af regressiv og progressiv assimilation kan forandring delbestemmes som **sammenhængsskabende** faktor i talen. I 1.6. påviser definitionerne en sammenhæng mellem fire normalt adskilte ideologier, hvorefter lingvistikkens dualisme (System og Forløb) analyseres i en Glossematisk og ikke-glossematisk version (M.L. Samuels). Der gives et konkret eksempel på svagheder i synkroniens udefinerede tilstandsbegreb. Den fremførte kritik af dualismen medfører relativering af *langue-* og *parole-*begreberne i L-kontinuummet.

KAPITEL 2

I 2.1. OPSTILLES de sidste definitioner, hvis eksempler fører til en kritik af de fire aristoteliske årsager, der kædes sammen med synkroniens dualisme, symboliseret med fonologiens /A/ [a], [b] notation. Denne dualismeopfattelse, der har skabt synkroniens to-tilstandsmodel, kritiseres fortsat i 2.2. især i lyset af synkroniens kontingens. Der argumenteres for, at forandring og historie ikke er kontingent.

I 2.3. påvises det sammenhængsskabende i periodens transsynkronicitet ved en analyse af det nutidiges og det fortidiges begreb, der munder ud i en formalisering og dynamisk omfortolkning af Eugenio Coserius ide om 'den uophørlige skaben af nutid og fortid'. Herunder argumenteres fortsat for, at historien er et tegnfunktionelt domæne, der ikke skal opsplittes i tilstande, der derefter sammenknyttes i par som 'historiske overgange', for at blive en videnskabelig genstand. I 2.4. kritiseres det idealiserede årsagsbegrebs logiske geografi, der noget paradoksalt kommer til at implicere det mekaniske synspunkts flerfoldige årsagsbegreb (M.L. Samuels), og kritikken munder ud i afhandlingens participative og biuniformelle årsagsbegreb; som eksempel omfortolkes Aitkens Lov.

I 2.5. kritiseres synkroniens arbitrære selvvalidering samt dens udefinerede tilstandsbegreb, dens udefinerede 'synkrone snit' og forudsatte begreber om 'brud før' og 'brud efter' en tilstand. Den logiske analyse påviser det absurde i tilstandsbegrebet og synkroniens tavse forudsætninger, herunder især kontingensens rolle. Kontingensen og dens alternativ opstilles. I 2.6. fortsætter kritikken af to-tilstandsmodellens analyse af forandring A>B, og via etablering af begrebet empirisk nødvendighed og afvisning af kontingens kritiseres den formal-logiske fortolkning (Hjelmslevs) af to-tilstandsmodellens moder- og dattersprog. Periodebegrebets nødvendighed påvises, og forandringsbegrebet bestemmes yderligere som et **epistemisk-kognitivt** begreb. Det er forbavsende, at synkroniens kommutation, afgørende for dualismen, intet har at gøre i selve forandringsprocessen, dvs. i synkroniseringen af historien. Undersøgelsen fører frem til en omfortolkning af forholdet mellem det statiske og det foranderlige over for dualismens system (form) og forløb (substans): den talte substans etableres som et strukturelt fænomen.

KAPITEL 3

STRUKTURBEGREBET diskuteres fra forskellige standpunkter og begrundes som et historisk begreb. Sprogets antropocentriske basis implicerer nødvendigvis et menneskesyn. I 3.1. afvises strukturalismens menneskesyn som statisk og inaktivt, og en filosofisk kritik (Sartre) af strukturalismen kritiseres for at arbejde på grundlag af en traditionel, grundlæggende

Resumé på dansk 479

statisk og dualistisk, model af overgangsprocessen: start/stasis-process-slut/stasis, med et værdiladet menneskesyn, der primært lader mennesket være re-aktivt. I 3.2. konkluderes, at det strukturelle program i lingvistikken er udefineret, med de afgørende termer **tilstand**, **synkroni** og **struktur** *qua* en *tout-se-tient* helhed afhængige af ordenes hverdagssproglige betydning, mens **systemet** defineres tvetydigt som enten en sui-generis entitet eller et klassifikationsprincip. En undersøgelse af de fire termer resulterer i rehabilitering af syntesen A>B som et empirisk-sprogligt fænomen, og i menneskets virkelighedsopfattelse som forudsættende dets **historiske sans**. Det sociologiske *fait social* (E. Durkheim) undersøges i 3.3., og der drages sproglige paralleller hertil. Analysen påviser inkonsistenser og cirkularitet i Durkheims argumentation, samt at Durkheims *fait social* er et transindividuelt fænomen med en historisk oprindelse. Sociologiens *fait social* svarer snarere til lingvistikkens metatermer (kategorier, regler) end til hverdagssprogets fænomener. Sproget som *un fait social* kan ikke begrundes med henvisning til strukturalismens ophavsmand (3.4.). Saussures strukturalisme ender i historie: tegnets arbitraritet er både udefineret og forudsætter historie, som Whitney også så det. Det påvises, dels hvorledes Saussures analyse af tegnets foranderlighed og ikke-foranderlighed er logisk inkonsistent, dels at Saussures analyse ikke overskrider almindeligheder som 'tiden ændrer alt'. Historie og struktur er to sider af samme mønt.

KAPITEL 4

I 4.1. PÅVISES, hvorledes nogle af traditionens afgørende strukturelle sentenser dels er logisk uholdbare, dels ender i mennesket og historien. Det synkrone helhedsbegreb er en metafysisk størrelse, og kontekstbegrebet er cirkulært. Meningsbegrebet forudsætter begrebet om elementets egennatur. Typologien ender i abstraktion, mens historien forbliver empirisk. Det glossematiske tegnbegreb afvises og erstattes af det participative princip om, at ethvert tegnelement selv er et tegn. I 4.2. rehabiliteres de to *parole*-fænomener polysemi og synonymi, både som sprogkonstituerende universaler og som konstituerende for barnets sprogindlæring. De to universaler grundes i talerens og hørerens irreducible talesituation. Strukturelt er polysemi og synonymi ikke symmetriske, og dualismen grundes i polysemien. Logikerens kritik af sprogets tvetydighed (polysemi) og redundans (synonymi) afvises, og dets erkendelsesværdi hævdes i kraft af dets eksistentielle funktion som transcendensskabende. I 4.3. foretages en sammenligning mellem historikeren, nominalisten og 'digteren' (*poietés*), og undersøgelsen bekræfter synonymi og polysemi som universaler; såvel historikeren og sprogbrugeren som historikerens data og sproghistorikerens data sammenlignes og fortolkes tegnfunktionelt. Afsnittet munder ud i en bekræftelse af sprogets historiske natur med en aitionel fortolkning af synkroniens formet-uformet parameter, der ikke kan beskrive historisk eksistens. I 4.4. begrundes synonymi og polysemi som eksistentielle betingelser for barnets skabelse af sin virkelighedssans, og struktur/identitet i forskelle (synonymi) og polysemi/forskel i identiteten begrundes antropocentrisk, – i menneskets to modsatrettede træk: det socialt-totaliserende og det individuelt-adskillende; begge modsatrettede træk er eksistentielle bevægelser, der sker i lyset af den enkeltes konstante overskridelse af nu'et og dermed skabelse af nu'et. Denne eksistentielle transcendens er tegnfunktionel og forudsætter menneskets historiske sans og dermed hverdagssprogets erkendelsesværdi.

KAPITEL 5

I 5.1. TEMATISERES synkroniens forudsatte begreb om det 'synkrone snit' og dets korrelative begreber. Disse illustreres ved overgang fra et system til et andet. Systembegrebet er overflødigt og erstattes af vor historiske sans. Synkronien har et data-konstitueringsproblem. Begrebet fælles oprindelse defineres på basis af synkroniens data, og sprogets eksistens bestemmes yderligere i lyset af frihed/valg og nødvendighed. Fælles oprindelse er en nødvendig betingelse for begrebet om en sproggruppes gensidige forståelighed, for begrebet om det at tale samme sprog. I 5.2. opstilles teoriens data, mulige **udviklingsrækker**, med deres definitionssystem. Udgangspunktet er tilstandens og forandringens equilogiske og equiempiriske status; forandring fremtræder mere omfattende end tilstand. En synkron analyse af historikerens data viser den synkrone to-tilstandsmodels svagheder. I 5.3. vises, at to-tilstandsmodellens overgangsbegreb er polysemt. Definitionssystemet for 'ikke-brud omkring' og 'brud-omkring' samt for overgange opstilles på grundlag af ikke-kronologiske forhold. I 5.4. fortsætter definitionerne af de opstillede udviklingsrækker, og deres synkront-statiske anvendelse demonstrerer, at overgang er et strukturelt fænomen, der får sin betydning fra både overgangens forgænger og efterfølger. Synkroniens episodiske overgang erstattes af strukturens dynamiske periode. I 5.5. indsættes Bloomfields (og Labovs) 'små' forskelle i teoriens attitudinelle system, der skaber den forbindelse mellem L-kontinuummets elementer og *parole*, som dualismen ikke kan skabe mellem fonemet/systemet og lyden/forløbet. I 5.6. opsummeres den statiske overgangs tvetydighed i lyset af de opstillede definitioner for brud-begrebet, og det vises, hvorledes analysen og dens autonome datum ikke giver empirisk relevant viden om historiske fænomener.

KAPITEL 6

I 6.1. IMPLEMENTERES teorien, og det vises, hvorledes historieteorien med sit veldefinerede teoretiske sprog er i stand til at formalisere en periodes udviklingsrække af monoftonger fra f.eks. 7/800 i oldengelsk til moderne engelsk. Skellet mellem forandring og kontinuation er overflødigt, og den blinde lydlov A>B (OE *e:* > PE *i:*) og A>A (OE *e* > PE *e*) samt den irreducible periode rehabiliteres som videnskabelige genstande. Teorien opstiller muligheder for, hvorledes en udvikling som den nævnte type kan forløbe.

Det attitudinelle systems dynamik illustrerer, hvorledes to-tilstandsmodellen og dens forandringsbegreb giver en 'forkortet' erkendelse af data. Elementerne i det attitudinelle system defineres og den tegnfunktionelle fortolkning af perioden viser både lighed i forskellighederne og forskel i lighederne. Vor historiske sans behøver ikke et abstrakt, uforanderligt system af præeksisterende elementer til dels at skabe empirisk sammenhæng i tid, dels at være en forståelsesbetingelse.

I 6.2. gives en oversigt over forandringsbegrebets kompleksitet, og det empirisk relevante i de to begreber, **empirisk nødvendighed** og **empirisk (u)mulighed** *qua* dynamiske begreber, påvises på grundlag af udviklingsrækkens interne kohærenser: forholdet mellem udtryk og indhold er ikke vilkårligt i det attitudinelle system. Dette resulterer dels i en hævdelse af kontekstualiseringens nødvendige dynamik, dels i en omfortolkning af både synkroniens vilkårlighedsbegreb og af de to Glossematiske begreber udtryks- og indholdssidernes konformitet og isomorfi. I 6.3. søges ideen om den konstante forandring bestemt, og en række af teoriens konsekvenser drages: 1) tegnfunktionens generalitet og omfang tager sit udgangs-

Resumé på dansk 481

punkt i hverdagssproget og ikke i Glossematikkens snævrere semiotik-begreb; 2) polysemi, synonymi og metafor bestemmes tegnfunktionelt, og analogier til talesituationen og historikerens søgen efter sammenhænge påvises; 3) en sidste bestemmelse af visse grundlæggende termer; 4) forandring og årsag er diskursivt, ikke analytisk-betingede begreber og er reelt overflødige termer i teorien; denne konklusion perspektiveres ved henvisning til de to mentale processer sansning (perception) og dommen: årsag og forandring kommer fra sidstnævnte men under forudsætning af, at vor historiske sans giver sit væsentlige bidrag; 5) begrebet om den konstante forandring får sin tegnfunktionelle definition; 6) sprogbrug som kausalt overbegreb for forandring undersøges logisk og falder sammen med eksistensbegrebet; 7) den af eksistensen, ikke systemet formede substans, rehabiliteres i lyset af historiens nødvendighed og det vilkårliges afvisning. Empirisk nødvendighed, defineret af historieteorien, bliver den tredje mulighed mellem traditionens umulighed-muligheds parameter, mens det muliges begreb bliver gjort afhængigt af dets konkrete realisering i eksistensens udvikling fra og af formet til formet.

Afhandlingen bliver et bidrag til en enhedssprogvidenskab, hvor tilstand og forandring samt synkroni og historie ikke kan adskilles inden for den realiserede mulighed af en immanent historieteori, hvis tegnfunktionelle fænomener eksisterer både i en tilstand/på et tids-**punkt** og gennem et forløb/**over tid**.